Michael T. Vaughn

**Introduction to Mathematical
Physics**

1807–2007 Knowledge for Generations

Each generation has its unique needs and aspirations. When Charles Wiley first opened his small printing shop in lower Manhattan in 1807, it was a generation of boundless potential searching for an identity. And we were there, helping to define a new American literary tradition. Over half a century later, in the midst of the Second Industrial Revolution, it was a generation focused on building the future. Once again, we were there, supplying the critical scientific, technical, and engineering knowledge that helped frame the world. Throughout the 20th Century, and into the new millennium, nations began to reach out beyond their own borders and a new international community was born. Wiley was there, expanding its operations around the world to enable a global exchange of ideas, opinions, and know-how.

For 200 years, Wiley has been an integral part of each generation's journey, enabling the flow of information and understanding necessary to meet their needs and fulfill their aspirations. Today, bold new technologies are changing the way we live and learn. Wiley will be there, providing you the must-have knowledge you need to imagine new worlds, new possibilities, and new opportunities.

Generations come and go, but you can always count on Wiley to provide you the knowledge you need, when and where you need it!

William J. Pesce
President and Chief Executive Officer

Peter Booth Wiley
Chairman of the Board

Michael T. Vaughn

Introduction to Mathematical Physics

WILEY-VCH Verlag GmbH & Co. KGaA

The Author

Michael T. Vaughn
Physics Department - 111DA
Northeastern University Boston
Boston MA-02115
USA

1st Edition 2007

Library of Congress Card No.:
applied for

British Library Cataloguing-in-Publication Data
A catalogue record for this book is available from the British Library.

Bibliographic information published by the Deutsche Nationalbibliothek
Die Deutsche Nationalbibliothek lists this publication in the Deutsche Nationalbibliografie; detailed bibliographic data are available in the Internet at <http://dnb.d-nb.de>.

© 2007 WILEY-VCH Verlag GmbH & Co. KGaA, Weinheim

Typesetting Uwe Krieg, Berlin

Wiley Bicentennial Logo Richard J. Pacifico

Printed on acid-free paper

ISBN 978-3-527-40627-2

Contents

Introduction to Mathematical Physics. Michael T. Vaughn
Copyright © 2007 WILEY-VCH Verlag GmbH & Co. KGaA, Weinheim
ISBN: 978-3-527-40627-2

Preface

Mathematics is an essential ingredient in the education of a professional physicist, indeed in the education of any professional scientist or engineer in the 21st century. Yet when it comes to the specifics of what is needed, and when and how it should be taught, there is no broad consensus among educators. The crowded curricula of undergraduates, especially in North America where broad general education requirements are the rule, leave little room for formal mathematics beyond the standard introductory courses in calculus, linear algebra, and differential equations, with perhaps one advanced specialized course in a mathematics department, or a one-semester survey course in a physics department.

The situation in (post)-graduate education is perhaps more encouraging—there are many institutes of theoretical physics, in some cases joined with applied mathematics, where modern courses in mathematical physics are taught. Even in large university physics departments there is room to teach advanced mathematical physics courses, even if only as electives for students specializing in theoretical physics. But in small and medium physics departments, the teaching of mathematical physics often is restricted to a one-semester survey course that can do little more than cover the gaps in the mathematical preparation of its graduate students, leaving many important topics to be discussed, if at all, in the standard physics courses in classical and quantum mechanics, and electromagnetic theory, to the detriment of the physics content of those courses.

The purpose of the present book is to provide a comprehensive survey of the mathematics underlying theoretical physics at the level of graduate students entering research, with enough depth to allow a student to read introductions to the higher level mathematics relevant to specialized fields such as the statistical physics of lattice models, complex dynamical systems, or string theory. It is also intended to serve the research scientist or engineer who needs a quick refresher course in the subject of one or more chapters in the book.

We review the standard theories of ordinary differential equations, linear vector spaces, functions of a complex variable, partial differential equations and Green functions, and the special functions that arise from the solutions of the standard partial differential equations of physics. Beyond that, we introduce at an early stage modern topics in differential geometry arising from the study of differentiable manifolds, spaces whose points are characterized by smoothly varying coordinates, emphasizing the properties of these manifolds that are independent of a particular choice of coordinates. The geometrical concepts that follow lead to helpful insights into topics ranging from thermodynamics to classical dynamical systems to Einstein's classical theory of gravity (general relativity). The usefulness of these ideas is, in my opinion, as significant as the clarity added to Maxwell's equations by the use of vector notation in place of the original expressions in terms of individual components, for example.

Introduction to Mathematical Physics. Michael T. Vaughn
Copyright © 2007 WILEY-VCH Verlag GmbH & Co. KGaA, Weinheim
ISBN: 978-3-527-40627-2

Thus I believe that it is important to introduce students of science to geometrical methods as early as possible in their education.

The material in Chapters 1–8 can form the basis of a one-semester graduate course on mathematical methods, omitting some of the mathematical details in the discussion of Hilbert spaces in Chapters 6 and 7 if necessary. There are many examples interspersed with the main discussion, and exercises that the student should work out as part of the reading. There are additional problems at the end of each chapter; these are generally more challenging, but provide possible homework assignments for a course. The remaining two chapters introduce the theory of finite groups and Lie groups—topics that are important for the understanding of systems with symmetry, especially in the realm of condensed matter, atoms, nuclei, and subnuclear physics. But these topics can often be developed as needed in the study of particular systems, and are thus less essential in a first course. Nevertheless, they have been included in part because of my own research interests, and in part because group theory can be fun!

Each chapter begins with an overview that summarizes the topics discussed in the chapter—the student should read this through in order to get an idea of what is coming in the chapter, without being too concerned with the details that will be developed later. The examples and exercises are intended to be studied together with the material as it is presented. The problems at the end of the chapter are either more difficult, or require integration of more than one local idea. The diagram at the right provides a flow chart for the chapters of the book.

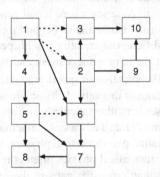

Flow chart for chapters of the book.

I would like to thank many people for their encouragement and advice during the long course of this work. Ron Aaron, George Alverson, Tom Kephart, and Henry Smith have read significant parts of the manuscript and contributed many helpful suggestions. Tony Devaney and Tom Taylor have used parts of the book in their courses and provided useful feedback. Peter Kahn reviewed an early version of the manuscript and made several important comments. Of course none of these people are responsible for any shortcomings of the book.

I have benefited from many interesting discussions over the years with colleagues and friends on mathematical topics. In addition to the people previously mentioned, I recall especially Ken Barnes, Haim Goldberg, Marie Machacek, Jeff Mandula, Bob Markiewicz, Pran Nath, Richard Slansky, K C Wali, P K Williams, Ian Jack, Tim Jones, Brian Wybourne, and my thesis adviser, David C Peaslee.

Michael T Vaughn

Boston, Massachusetts
October 2006

1 Infinite Sequences and Series

In experimental science and engineering, as well as in everyday life, we deal with integers, or at most rational numbers. Yet in theoretical analysis, we use real and complex numbers, as well as far more abstract mathematical constructs, fully expecting that this analysis will eventually provide useful models of natural phenomena. Hence we proceed through the construction of the real and complex numbers starting from the positive integers[1]. Understanding this construction will help the reader appreciate many basic ideas of analysis.

We start with the positive integers and zero, and introduce negative integers to allow subtraction of integers. Then we introduce *rational numbers* to permit division by integers. From arithmetic we proceed to analysis, which begins with the concept of *convergence* of infinite sequences of (rational) numbers, as defined here by the Cauchy criterion. Then we define *irrational numbers* as limits of convergent (Cauchy) sequences of rational numbers.

In order to solve algebraic equations in general, we must introduce *complex numbers* and the representation of complex numbers as points in the *complex plane*. The fundamental theorem of algebra states that every polynomial has at least one root in the complex plane, from which it follows that every polynomial of degree n has exactly n roots in the complex plane when these roots are suitably counted. We leave the proof of this theorem until we study functions of a complex variable at length in Chapter 4.

Once we understand convergence of infinite sequences, we can deal with infinite series of the form

$$\sum_{n=1}^{\infty} x_n$$

and the closely related infinite products of the form

$$\prod_{n=1}^{\infty} x_n$$

Infinite series are central to the study of solutions, both exact and approximate, to the differential equations that arise in every branch of physics. Many functions that arise in physics are defined only through infinite series, and it is important to understand the convergence properties of these series, both for theoretical analysis and for approximate evaluation of the functions.

---

[1] To paraphrase a remark attributed to Leopold Kronecker: "God created the positive integers; all the rest is human invention."

Introduction to Mathematical Physics. Michael T. Vaughn
Copyright © 2007 WILEY-VCH Verlag GmbH & Co. KGaA, Weinheim
ISBN: 978-3-527-40627-2

We review some of the standard tests (comparison test, ratio test, root test, integral test) for convergence of infinite series, and give some illustrative examples. We note that absolute convergence of an infinite series is necessary and sufficient to allow the terms of a series to be rearranged arbitrarily without changing the sum of the series.

Infinite sequences of functions have more subtle convergence properties. In addition to pointwise convergence of the sequence of values of the functions taken at a single point, there is a concept of *uniform convergence* on an interval of the real axis, or in a region of the complex plane. Uniform convergence guarantees that properties such as continuity and differentiability of the functions in the sequence are shared by the limit function. There is also a concept of *weak convergence*, defined in terms of the sequences of numbers generated by integrating each function of the sequence over a region with functions from a class of smooth functions (*test functions*). For example, the Dirac δ-function and its derivatives are defined in terms of weakly convergent sequences of well-behaved functions.

It is a short step from sequences of functions to consider infinite series of functions, especially *power series* of the form

$$\sum_{n=0}^{\infty} a_n z^n$$

in which the a_n are real or complex numbers and z is a complex variable. These series are central to the theory of functions of a complex variable. We show that a power series converges absolutely and uniformly inside a circle in the complex plane (the *circle of convergence*), with convergence *on* the circle of convergence an issue that must be decided separately for each particular series.

Even divergent series can be useful. We show some examples that illustrate the idea of a *semiconvergent*, or *asymptotic*, series. These can be used to determine the asymptotic behavior and approximate asymptotic values of a function, even though the series is actually divergent. We give a general description of the properties of such series, and explain Laplace's method for finding an asymptotic expansion of a function defined by an integral representation (Laplace integral) of the form

$$I(z) = \int_0^a f(t) e^{zh(t)} \, dt$$

Beyond the sequences and series generated by the mathematical functions that occur in solutions to differential equations of physics, there are sequences generated by dynamical systems themselves through the equations of motion of the system. These sequences can be viewed as *iterated maps* of the coordinate space of the system into itself; they arise in classical mechanics, for example, as successive intersections of a particle orbit with a fixed plane. They also arise naturally in population dynamics as a sequence of population counts at periodic intervals.

The asymptotic behavior of these sequences exhibits new phenomena beyond the simple convergence or divergence familiar from previous studies. In particular, there are sequences that converge, not to a single limit, but to a periodic limit cycle, or that diverge in such a way that the points in the sequence are dense in some region in a coordinate space.

An elementary prototype of such a sequence is the *logistic map* defined by

$$T_\lambda : x \to x_\lambda = \lambda x(1 - x)$$

This map generates a sequence of points $\{x_n\}$ with

$$x_{n+1} = \lambda x_n(1 - x_n)$$

$(0 < \lambda < 4)$ starting from a generic point x_0 in the interval $0 < x_0 < 1$. The behavior of this sequence as a function of the parameter λ as λ increases from 0 to 4 provides a simple illustration of the phenomena of *period doubling* and transition to *chaos* that have been an important focus of research in the past 30 years or so.

1.1   Real and Complex Numbers

1.1.1   Arithmetic

The construction of the real and complex number systems starting from the positive integers illustrates several of the structures studied extensively by mathematicians. The positive integers have the property that we can add, or we can multiply, two of them together and get a third. Each of these operations is *commutative*:

$$x \circ y = y \circ x \tag{1.1}$$

and *associative*:

$$x \circ (y \circ z) = (x \circ y) \circ z \tag{1.2}$$

(here $\circ$ denotes either addition or multiplication), but only for multiplication is there an *identity* element **e**, with the property that

$$\mathbf{e} \circ x = x = x \circ \mathbf{e} \tag{1.3}$$

Of course the identity element for addition is the number zero, but zero is not a positive integer. Properties (1.2) and (1.3) are enough to characterize the positive integers as a *semigroup* under multiplication, denoted by $\mathbf{Z}_*$ or, with the inclusion of zero, a semigroup under addition, denoted by $\mathbf{Z}_+$.

Neither addition nor multiplication has an inverse defined within the positive integers. In order to define an inverse for addition, it is necessary to include zero and the negative integers. *Zero* is defined as the identity for addition, so that

$$x + 0 = x = 0 + x \tag{1.4}$$

and the *negative* integer $-x$ is defined as the *inverse* of x under addition,

$$x + (-x) = 0 = (-x) + x \tag{1.5}$$

With the inclusion of the negative integers, the equation

$$p + x = q \tag{1.6}$$

has a unique integer solution x ($\equiv q - p$) for every pair of integers p, q. Properties (1.2)–(1.5) characterize the integers as a *group* $\mathbf{Z}$ under addition, with 0 as an identity element. The fact that addition is commutative makes $\mathbf{Z}$ a *commutative*, or *Abelian*, group. The combined operations of addition with zero as identity, and multiplication satisfying Eqs. (1.2) and (1.3) with 1 as identity, characterize $\mathbf{Z}$ as a *ring*, a *commutative* ring since multiplication is also commutative. To proceed further, we need an inverse for multiplication, which leads to the introduction of *fractions* of the form p/q (with integers p, q). One important property of fractions is that they can always be reduced to a form in which the integers p, q have no common factors[2]. Numbers of this form are *rational*. With both addition and multiplication having well-defined inverses (except for division by zero, which is undefined), and the *distributive* law

$$a * (x + y) = a * x + a * c = y \tag{1.7}$$

satisfied, the rational numbers form a *field*, denoted by $\mathbf{Q}$.

➔ **Exercise 1.1.** Let p be a prime number. Then $\sqrt{p}$ is not rational.   □

Note. Here and throughout the book we use the convention that when a proposition is simply stated, the problem is to prove it, or to give a counterexample that shows it is false.

1.1.2   Algebraic Equations

The rational numbers are adequate for the usual operations of arithmetic, but to solve algebraic (polynomial) equations, or to carry out the limiting operations of calculus, we need more. For example, the quadratic equation

$$x^2 - 2 = 0 \tag{1.8}$$

has no rational solution, yet it makes sense to enlarge the rational number system to include the roots of this equation. The real *algebraic* numbers are introduced as the real roots of polynomials of any degree with integer coefficients. The algebraic numbers also form a field.

➔ **Exercise 1.2.**   Show that the roots of a polynomial with rational coefficients can be expressed as roots of a polynomial with integer coefficients.   □

Complex numbers are introduced in order to solve algebraic equations that would otherwise have no real roots. For example, the equation

$$x^2 + 1 = 0 \tag{1.9}$$

has no real solutions; it is "solved" by introducing the imaginary unit $i \equiv \sqrt{-1}$ so that the roots are given by $x = \pm i$. Complex numbers are then introduced as ordered pairs $(x, y) \sim$

---

[2]The study of properties of the positive integers, and their factorization into products of *prime* numbers, belongs to a fascinating branch of pure mathematics known as *number theory*, in which the reducibility of fractions is one of the elementary results.

$x + iy$, of real numbers; x, y can be restricted to be rational (algebraic) to define the complex rational (algebraic) numbers.

Complex numbers can be represented as points (x, y) in a plane (the *complex plane*) in a natural way, and the *magnitude* of the complex number $x + iy$ is defined by

$$|x + iy| \equiv \sqrt{x^2 + y^2} \tag{1.10}$$

In view of the identity

$$e^{i\theta} = \cos\theta + i\sin\theta \tag{1.11}$$

we can also write

$$x + iy = re^{i\theta} \tag{1.12}$$

with $r = |x + iy|$ and $\tan\theta = y/x$. These relations have an obvious interpretation in terms of the polar coordinates of the point (x, y). We also define

$$\arg z \equiv \theta \tag{1.13}$$

for $z \neq 0$. The angle $\arg z$ is the *phase* of z. Evidently it can only be defined as mod 2π; adding any integer multiple of 2π to $\arg z$ does not change the complex number z, since

$$e^{2\pi i} = 1 \tag{1.14}$$

Equation (1.14) is one of the most remarkable equations of mathematics.

1.1.3 Infinite Sequences; Irrational Numbers

To complete the construction of the real and complex numbers, we need to look at some elementary properties of sequences, starting with the formal definitions:

Definition 1.1. A *sequence* of numbers (real or complex) is an ordered set of numbers in one-to-one correspondence with the positive integers; write $\{z_n\} \equiv \{z_1, z_2, \ldots\}$. ∎

Definition 1.2. The sequence $\{z_n\}$ is *bounded* if there is some positive number M such that $|z_n| < M$ for all positive integers n. ∎

Definition 1.3. The sequence $\{x_n\}$ of real numbers is *increasing (decreasing)* if $x_{n+1} > x_n$ ($x_{n+1} < x_n$) for every n. The sequence is *nondecreasing (nonincreasing)* if $x_{n+1} \geq x_n$ ($x_{n+1} \leq x_n$) for every n. A sequence belonging to one of these classes is *monotone (or monotonic)*. ∎

Remark. The preceding definition is restricted to real numbers because it is only for real numbers that we can define a "natural" ordering that is compatible with the standard measure of the distance between the numbers. □

Definition 1.4. The sequence $\{z_n\}$ is a *Cauchy sequence* if for every $\varepsilon > 0$ there is a positive integer N such that $|z_p - z_q| < \varepsilon$ whenever $p, q > N$. ∎

Definition 1.5. The sequence $\{z_n\}$ is *convergent* to the *limit* z (write $\{z_n\} \to z$) if for every $\varepsilon > 0$ there is a positive integer N such that $|z_n - z| < \varepsilon$ whenever $n > N$.   ∎

There is no guarantee that a Cauchy sequence of rational numbers converges to a rational, or even algebraic, limit. For example, the sequence $\{x_n\}$ defined by

$$x_n \equiv \left(1 + \frac{1}{n}\right)^n \tag{1.15}$$

converges to the limit $e = 2.71828\ldots$, the base of natural logarithms. It is true, though nontrivial to prove, that e is not an algebraic number. A real number that is not algebraic is *transcendental*. Another famous transcendental number is π, which is related to e through Eq. (1.14).

If we want to insure that every Cauchy sequence of rational numbers converges to a limit, we must include the *irrational numbers*, which can be *defined* as limits of Cauchy sequences of rational numbers. As examples of such sequences, imagine the infinite, nonterminating, nonperiodic decimal expansions of transcendental numbers such as e or π, or algebraic numbers such as $\sqrt{2}$. Countless computer cycles have been used in calculating the digits in these expansions.

The set of *real* numbers, denoted by **R**, can now be defined as the set containing rational numbers together with the limits of Cauchy sequences of rational numbers. The set of *complex* numbers, denoted by **C**, is then introduced as the set of all ordered pairs $(x, y) \sim x + iy$ of real numbers. Once we know that every Cauchy sequence of real (or rational) numbers converges to a real number, it is a simple exercise to show that every Cauchy sequence of complex numbers converges to a complex number.

Monotonic sequences are especially important, since they appear as partial sums of infinite series of positive terms. The key property is contained in the

Theorem 1.1. A monotonic sequence $\{x_n\}$ is convergent if and only if it is bounded.

Proof. If the sequence is unbounded, it will diverge to $\pm\infty$, which simply means that for any positive number M, no matter how large, there is an integer N such that $x_n > M$ (or $x_n < -M$ if the sequence is monotonic nonincreasing) for any $n \geq N$. This is true, since for any positive number M, there is at least one member x_N of the sequence with $x_N > M$ (or $x_N < -M$)—otherwise M would be a bound for the sequence—and hence $x_n > M$ (or $x_n < -M$) for any $n \geq N$ in view of the monotonic nature of the sequence.   ∎

If the monotonic nondecreasing sequence $\{x_n\}$ is bounded from above, then in order to have a limit, there must be a bound that is smaller than any other bound (such a bound is the *least upper bound* of the sequence). If the sequence has a limit X, then X is certainly the least upper bound of the sequence, while if a least upper bound $\overline{X}$ exists, then it must be the limit of the sequence. For if there is some $\varepsilon > 0$ such that $\overline{X} - x_n > \varepsilon$ for all n, then $\overline{X} - \varepsilon$ will be an upper bound to the sequence smaller than $\overline{X}$.

The existence of a least upper bound is intuitively plausible, but its existence cannot be proven from the concepts we have introduced so far. There are alternative axiomatic formulations of the real number system that guarantee the existence of the least upper bound; the convergence of any bounded monotonic nondecreasing sequence is then a consequence as just explained. The same argument applies to bounded monotonic nonincreasing sequences, which must then have a *greatest lower bound* to which the sequence converges.

1.1.4   Sets of Real and Complex Numbers

We also need some elementary definitions and results about sets of real and complex numbers that are generalized later to other structures.

Definition 1.6.   For real numbers, we can define an *open* interval:

$$(a, b) \equiv \{x \mid a < x < b\}$$

or a *closed* interval:

$$[a, b] \equiv \{x \mid a \leq x \leq b\}$$

as well as *semiopen* (or *semiclosed*) intervals:

$$(a, b] \equiv \{x \mid a < x \leq b\} \quad \text{and} \quad [a, b) \equiv \{x \mid a \leq x < b\}$$

A *neighborhood* of the real number x_0 is any open interval containing x_0. An ε-*neighborhood* of x_0 is the set of all points x such that

$$|x - x_0| < \varepsilon \tag{1.16}$$

This concept has an obvious extension to complex numbers: An (ε)-*neighborhood* of the complex number z_0, denoted by $N_\varepsilon(z_0)$, is the set of all points z such that

$$0 < |z - z_0| < \varepsilon \tag{1.17}$$

Note that for complex numbers, we exclude the point z_0 from the neighborhood $N_\varepsilon(z_0)$.   ∎

Definition 1.7.   The set S of real or complex numbers is *open* if for every x in S, there is a neighborhood of x lying entirely in S. S is *closed* if its complement is open. S is *bounded* if there is some positive M such that $x < M$ for every x in S (M is then a *bound* of S).   ∎

Definition 1.8. x is a *limit point* of the set S if every neighborhood of x contains at least one point of S.   ∎

While x itself need not be a member of the set S, this definition implies that every neighborhood of x in fact contains an infinite number of points of S. An alternative definition of a closed set can be given in terms of limit points, and one of the important results of analysis is that every bounded infinite set contains at least one limit point.

→ **Exercise 1.3.**   Show that the set S of real or complex numbers is closed if and only if every limit point of S is an element of S.   □

→ **Exercise 1.4.**   (Bolzano–Weierstrass theorem) Every bounded infinite set of real or complex numbers contains at least one limit point.   □

Definition 1.9.   The set S is *everywhere dense*, or simply *dense*, in a region $\mathcal{R}$ if there is at least one point of S in any neighborhood of every point in $\mathcal{R}$.   ∎

❑ **Example 1.1.**   The set of rational numbers is everywhere dense on the real axis.   ∎

1.2   Convergence of Infinite Series and Products

1.2.1   Convergence and Divergence; Absolute Convergence

If $\{z_k\}$ is a sequence of numbers (real or complex), the formal sum

$$S \equiv \sum_{k=1}^{\infty} z_k \qquad (1.18)$$

is an *infinite series*, whose *partial sums* are defined by

$$s_n \equiv \sum_{k=1}^{n} z_k \qquad (1.19)$$

The series $\sum z_k$ is *convergent* (to the *value s*) if the sequence $\{s_n\}$ of partial sums converges to s, otherwise *divergent*. The series is *absolutely convergent* if the series $\sum |z_k|$ is convergent; a series that is convergent but not absolutely convergent is *conditionally convergent*. Absolute convergence is an important property of a series, since it allows us to rearrange terms of the series without altering its value, while the sum of a conditionally convergent series can be changed by reordering it (this is proved later on).

→ **Exercise 1.5.**   If the series $\sum z_k$ is convergent, then the sequence $\{z_k\} \to 0$.   □

→ **Exercise 1.6.**   If the series $\sum z_k$ is absolutely convergent, then it is convergent.   □

To study absolute convergence, we need only consider a series $\sum x_k$ of positive real numbers ($\sum |z_k|$ is such a series). The sequence of partial sums of a series of positive real numbers is obviously nondecreasing. From the theorem on monotonic sequences in the previous section then follows

Theorem 1.2. The series $\sum x_k$ of positive real numbers is convergent if and only if the sequence of its partial sums is bounded.

❏ **Example 1.2.**   Consider the *geometric series*

$$S(x) \equiv \sum_{k=0}^{\infty} x^k \qquad (1.20)$$

for which the partial sums are given by

$$s_n = \sum_{k=0}^{n} x^k = \frac{1-x}{1-x^{n+1}} \qquad (1.21)$$

These partial sums are bounded if $0 \leq x < 1$, in which case

$$\{s_n\} \to \frac{1}{1-x} \qquad (1.22)$$

The series diverges for $x \geq 1$. The corresponding series

$$S(z) \equiv \sum_{k=0}^{\infty} z^k \tag{1.23}$$

for complex z is then absolutely convergent for $|z| < 1$, divergent for $|z| > 1$. The behavior on the *unit circle* $|z| = 1$ in the complex plane must be determined separately (the series actually diverges everywhere on the circle since the sequence $\{z^k\} \nrightarrow 0$; see Exercise 1.5). ∎

Remark.   We will see that the function $S(z)$ defined by the series (1.23) for $|z| < 1$ can be defined to be $1/(1-z)$ for complex $z \neq 1$, even outside the region of convergence of the series, using the properties of $S(z)$ as a function of the complex variable z. This is an example of a procedure known as *analytic continuation*, to be explained in Chapter 4. □

❑ **Example 1.3.**   The *Riemann ζ-function* is defined by

$$\zeta(s) \equiv \sum_{n=1}^{\infty} \frac{1}{n^s} \tag{1.24}$$

The series for $\zeta(s)$ with $s = \sigma + i\tau$ is absolutely convergent if and only if the series for $\zeta(\sigma)$ is convergent. Denote the partial sums of the latter series by

$$s_N(\sigma) = \sum_{n=1}^{N} \frac{1}{n^\sigma} \tag{1.25}$$

Then for $\sigma \leq 1$ and $N \geq 2^m$ (m integer), we have

$$s_N(\sigma) \geq s_N(1) \geq s_{2^m}(1) > s_{2^{m-1}}(1) + \frac{1}{2} > \cdots > \frac{m}{2} \tag{1.26}$$

Hence the sequence $\{s_N(\sigma)\}$ is unbounded and the series diverges. Note that for $s = 1$, Eq. (1.24) is the *harmonic series*, which is shown to diverge in elementary calculus courses. On the other hand, for $\sigma > 1$ and $N \leq 2^m$ with m integer, we have

$$s_N(\sigma) < s_{2^m}(\sigma) < s_{2^{m-1}}(\sigma) + \left(\frac{1}{2}\right)^{(m-1)(\sigma-1)} < \cdots$$

$$< \sum_{k=0}^{m-1} \left(\frac{1}{2}\right)^{k(\sigma-1)} < \frac{1}{1 - 2^{(1-\sigma)}} \tag{1.27}$$

Thus the sequence $\{s_N(\sigma)\}$ is bounded and hence converges, so that the series (1.24) for $\zeta(s)$ is absolutely convergent for $\sigma = \mathrm{Re}\, s > 1$. Again, we will see in Chapter 4 that $\zeta(s)$ can be defined for complex s beyond the range of convergence of the series (1.24) by analytic continuation. ∎

1.2.2   Tests for Convergence of an Infinite Series of Positive Terms

There are several standard tests for convergence of a series of positive terms:

Comparison test. Let $\sum x_k$ and $\sum y_k$ be two series of positive numbers, and suppose that for some integer $N > 0$ we have $y_k \leq x_k$ for all $k > N$. Then

(i) if $\sum x_k$ is convergent, $\sum y_k$ is also convergent, and

(ii) if $\sum y_k$ is divergent, $\sum x_k$ is also divergent.

This is fairly obvious, but to give a formal proof, let $\{s_n\}$ and $\{t_n\}$ denote the sequences of partial sums of $\sum x_k$ and $\sum y_k$, respectively. If $y_k \leq x_k$ for all $k > N$, then

$$t_n - t_N \leq s_n - s_N$$

for all $n > N$. Thus if $\{s_n\}$ is bounded, then $\{t_n\}$ is bounded, and if $\{t_n\}$ is unbounded, then $\{s_n\}$ is unbounded.

Remark.   The comparison test has been used implicitly in the discussion of the ζ-function to show the absolute convergence of the series 1.24 for $\sigma = \mathrm{Re}\, s > 1$.   □

Ratio test. Let $\sum x_k$ be a series of positive numbers, and let $r_k \equiv x_{k+1}/x_k$ be the ratios of successive terms. Then

(i) if only a finite number of $r_k > a$ for some a with $0 < a < 1$, then the series converges, and

(ii) if only a finite number of $r_k < 1$, then the series diverges.

In case (i), only a finite number of the r_k are larger than a, so there is some positive M such that $x_k < Ma^k$ for all k, and the series converges by comparison with the geometric series. In case (ii), the series diverges since the individual terms of the series do not tend to zero.

Remark.   The ratio test works if the largest limit point of the sequence $\{r_k\}$ is either greater than 1 or smaller than 1. If the largest limit point is exactly equal to 1, then the ratio test does not answer the question of convergence, as seen by the example of the ζ-function series (1.24).   □

Root test. Let $\sum x_k$ be a series of positive numbers, and let $\varrho_k \equiv \sqrt[k]{x_k}$. Then

(i) if only a finite number of $\varrho_k > a$ for some positive $a < 1$, then the series converges, and

(ii) if infinitely many $\varrho_k > 1$, the series diverges.

As with the ratio test, we can construct a comparison with the geometric series. In case (i), only a finite number of roots ϱ_k are bigger than a, so there is some positive M such that $x_k < Ma^k$ for all k, and the series converges by comparison with the geometric series. In case (ii), the series diverges since the individual terms of the series do not tend to zero.

Remark.   The root test, like the ratio test, works if the largest limit point of the sequence $\{\varrho_k\}$ is either greater than 1 or smaller than 1, but fails to decide convergence if the largest limit point is exactly equal to 1.   □

Integral test. Let $f(t)$ be a continuous, positive, and nonincreasing function for $t \geq 1$, and let $x_k \equiv f(k)$ $(k = 1, 2, \ldots)$. Then $\sum x_k$ converges if and only if the integral

$$I \equiv \int_1^\infty f(t)\, dt < \infty \tag{1.28}$$

also converges. To show this, note that

$$\int_k^{k+1} f(t)\,dt \le x_k \le \int_{k-1}^k f(t)\,dt \tag{1.29}$$

which is easy to see by drawing a graph. The partial sums s_n of the series then satisfy

$$\int_1^{n+1} f(t)\,dt \le s_n = \sum_{k=1}^n x_k \le x_1 + \int_1^n f(t)\,dt \tag{1.30}$$

and are bounded if and only if the integral (1.28) converges.

Remark. If the integral (1.28) converges, it provides a (very) rough estimate of the value of the infinite series, since

$$\int_{N+1}^\infty f(t)\,dt \le s - s_N = \sum_{k=N+1}^\infty x_k \le \int_N^\infty f(t)\,dt \tag{1.31}$$

1.2.3 Alternating Series and Rearrangements

In addition to a series of positive terms, we consider an *alternating series* of the form

$$S \equiv \sum_{k=0}^\infty (-1)^k x_k \tag{1.32}$$

with $x_k > 0$ for all k. Here there is a simple criterion (due to Leibnitz) for convergence: if the sequence $\{x_k\}$ is nonincreasing, then the series S converges if and only if $\{x_k\} \to 0$, and if S converges, its value lies between any two successive partial sums. This follows from the observation that for any n the partial sums s_n of the series (1.32) satisfy

$$s_{2n+1} < s_{2n+3} < \cdots < s_{2n+2} < s_{2n} \tag{1.33}$$

❑ **Example 1.4.** The alternating harmonic series

$$A \equiv 1 - \frac{1}{2} + \frac{1}{3} - \frac{1}{4} + \cdots = \sum_{k=0}^\infty \frac{(-1)^k}{k+1} \tag{1.34}$$

is convergent according to this criterion, even though it is not absolutely convergent (the series of absolute values is the harmonic series we have just seen to be divergent). In fact, evaluating the logarithmic series (Eq. (1.69) below) for $z = 1$ shows that $A = \ln 2$. ∎

Is there any significance of the ordering of terms in an infinite series? The short answer is that terms can be rearranged at will in an absolutely convergent series without changing the value of the sum, while changing the order of terms in a conditionally convergent series can change its value, or even make it diverge.

Definition 1.10. If $\{n_1, n_2, \ldots\}$ is a permutation of $\{1, 2, \ldots\}$, then the sequence $\{\zeta_k\}$ is a *rearrangement* of $\{z_k\}$ if

$$\zeta_k = z_{n_k} \tag{1.35}$$

for every k. Then also the series $\sum \zeta_k$ is a rearrangement of $\sum z_k$.   ∎

❑ **Example 1.5.** The alternating harmonic series (1.34) can be rearranged in the form

$$A' = \left(1 + \frac{1}{3} - \frac{1}{2}\right) + \left(\frac{1}{5} + \frac{1}{7} - \frac{1}{4}\right) + \cdots \tag{1.36}$$

which is still a convergent series, but its value is not the same as that of A (see below).   ∎

Theorem 1.3. If the series $\sum z_k$ is absolutely convergent, and $\sum \zeta_k$ is a rearrangement of $\sum z_k$, then $\sum \zeta_k$ is absolutely convergent.

Proof. Let $\{s_n\}$ and $\{\sigma_n\}$ denote the sequences of partial sums of $\sum z_k$ and $\sum \zeta_k$, respectively. If $\varepsilon > 0$, choose N such that $|s_n - s_m| < \varepsilon$ for all $n, m > N$, and let $Q \equiv \max\{n_1, \ldots, n_N\}$. Then $|\sigma_n - \sigma_m| < \varepsilon$ for all $n, m > Q$.   ∎

On the other hand, if a series in not absolutely convergent, then its value can be changed (almost at will) by rearrangement of its terms. For example, the alternating series in its original form (1.34) can be expressed as

$$A = \sum_{n=0}^{\infty} \left(\frac{1}{2n+1} - \frac{1}{2n+2}\right) = \sum_{n=0}^{\infty} \frac{1}{(2n+1)(2n+2)} \tag{1.37}$$

This is an absolutely convergent series of positive terms whose value is $\ln 2 = 0.693\ldots$, as already noted. On the other hand, the rearranged series (1.36) can be expressed as

$$A' = \sum_{n=0}^{\infty} \left(\frac{1}{4n+1} + \frac{1}{4n+3} - \frac{1}{2n+2}\right) = \sum_{n=0}^{\infty} \frac{8n+5}{2(n+1)(4n+1)(4n+3)} \tag{1.38}$$

which is another absolutely convergent series of positive terms. Including just the first term of this series shows that

$$A' > \frac{5}{6} > \ln 2 = A \tag{1.39}$$

In fact, any series that is not absolutely convergent can be rearranged into a divergent series.

Theorem 1.4. If the series $\sum x_k$ of real terms is conditionally convergent, then there is a divergent rearrangement of $\sum x_k$.

Proof. Let $\{\xi_1, \xi_2, \ldots\}$ be the sequence of positive terms in $\{x_k\}$, and $\{-\eta_1, -\eta_2, \ldots\}$ be the sequence of negative terms. Then at least one of the series $\sum \xi_k$, $\sum \eta_k$ is divergent (otherwise the series would be absolutely convergent). Suppose $\sum \xi_k$ is divergent. Then we can choose a sequence $n_1, n_2, \ldots$ such that

$$\sum_{k=n_m}^{n_{m+1}-1} \xi_k > 1 + \eta_m \tag{1.40}$$

$(m = 1, 2, \ldots)$, and the rearranged series

$$S' \equiv \sum_{k=n_1}^{n_2-1} \xi_k - \eta_1 + \sum_{k=n_2}^{n_3-1} \xi_k - \eta_2 + \cdots \tag{1.41}$$

is divergent. ∎

Remark. It follows as well that a conditionally convergent series $\sum z_k$ of complex terms must have a divergent rearrangement. For if $z_k = x_k + iy_k$, then either $\sum x_k$ or $\sum y_k$ is conditionally convergent, and hence has a divergent rearrangement. □

1.2.4 Infinite Products

Closely related to infinite series are *infinite products* of the form

$$\prod_{m=1}^{\infty} (1 + z_m) \tag{1.42}$$

($\{z_m\}$ is a sequence of complex numbers), with *partial products*

$$p_n \equiv \prod_{m=1}^{n} (1 + z_k) \tag{1.43}$$

The product $\prod (1 + z_m)$ is *convergent* (to the *value p*) if the sequence $\{p_n\}$ of partial products converges to $p \neq 0$, *convergent to zero* if a finite number of factors are 0, *divergent to zero* if $\{p_n\} \to 0$ with no vanishing p_n, and *divergent* if $\{p_n\}$ is divergent. The product is *absolutely convergent* if $\prod (1 + |z_m|)$ is convergent; a product that is convergent but not absolutely convergent is *conditionally convergent*.

The absolute convergence of a product is simply related to the absolute convergence of a related series: if $\{x_m\}$ is a sequence of positive real numbers, then the product $\prod (1 + x_m)$ is convergent if and only if the series $\sum x_m$ is convergent. This follows directly from the observation

$$\sum_{m=1}^{n} x_m < \prod_{m=1}^{n} (1 + x_m) < \exp\left(\sum_{m=1}^{n} x_m\right) \tag{1.44}$$

Also, the product $\prod (1 - x_m)$ is convergent if and only if the series $\sum x_m$ is convergent (show this).

❑ **Example 1.6.** Consider the infinite product

$$P \equiv \prod_{m=2}^{\infty} \left(\frac{m^3 - 1}{m^3 + 1}\right) < \prod_{m=2}^{\infty} \left(1 - \frac{1}{m^3}\right) \tag{1.45}$$

The product is (absolutely) convergent, since the series

$$\sum_{m=1}^{\infty} \frac{1}{m^3} = \zeta(3)$$

is convergent. Evaluation of the product is left as a problem. ∎

1.3   Sequences and Series of Functions

1.3.1   Pointwise Convergence and Uniform Convergence of Sequences of Functions

Questions of convergence of sequences and series of functions in some domain of variables can be answered at each point by the methods of the preceding section. However, the issues of continuity and differentiability of the limit function require more care, since the limiting procedures involved approaching a point in the domain need not be interchangeable with passing to the limit of the sequence or series (convergence of an infinite series of functions is defined in the usual way in terms of the convergence of the sequence of partial sums of the series). Thus we introduce

Definition 1.11.   The sequence $\{f_n(z)\}$ of functions of the variable z (real or complex) is (*pointwise*) *convergent* to the function $f(z)$ in the region $\mathcal{R}$:

$$\{f_n(z)\} \to f(z) \text{ in } \mathcal{S}$$

if the sequence $\{f_n(z_0)\} \to f(z_0)$ at every point z_0 in $\mathcal{R}$.

Definition 1.12. $\{f_n(z)\}$ is *uniformly convergent* to $f(z)$ in the closed, bounded $\mathcal{R}$:

$$\{f_n(z)\} \Rightarrow f(z) \text{ in } \mathcal{S}$$

if for every $\varepsilon > 0$ there is a positive integer N such that $|f_n(z) - f(z)| < \varepsilon$ for every $n > N$ and *every* point z in $\mathcal{R}$. ∎

Remark.   Note the use of different arrow symbols ($\to$ and $\Rightarrow$) to denote strong and uniform convergence, as well as the symbol ($\rightharpoonup$) introduced below to denote weak convergence. □

❑ **Example 1.7.**   Consider the sequence $\{x^n\}$. Evidently $\{x^n\} \to 0$ for $0 \le x < 1$. Also, the sequence $\{x^n\} \Rightarrow 0$ on any closed interval $0 \le x \le 1 - \delta$ ($0 < \delta < 1$), since for any such x, we have $|x^n| < \varepsilon$ for all $n > N$ if N is chosen so that $|1 - \delta|^N < \varepsilon$. However, we cannot say that the sequence is uniformly convergent on the open interval $0 < x < 1$, since if $0 < \varepsilon < 1$ and n is any positive integer, we can find some x in $(0, 1)$ such that $x^n > \varepsilon$. The point here is that to discuss uniform convergence, we need to consider a region that is closed and bounded, with no limit point at which the series is divergent. ∎

It is one of the standard theorems of advanced calculus that properties of continuity of the elements of a uniformly convergent sequence are shared by the limit of the sequence. Thus if $\{f_n(z)\} \Rightarrow f(z)$ in the region $\mathcal{R}$, and if each of the $f_n(z)$ is continuous in the closed bounded region $\mathcal{R}$, then the limit function $f(z)$ is also continuous in $\mathcal{R}$. Differentiability requires a separate check that the sequence of derivative functions $\{f_n'(z)\}$ is convergent, since it may not be. If the sequence of derivatives actually is uniformly convergent, then it converges to the derivative of the limit function $f(z)$.

❑ **Example 1.8.**   Consider the function $f(z)$ defined by the series

$$f(z) \equiv \sum_{n=1}^{\infty} \frac{1}{n^2} \sin n^2 \pi z \tag{1.46}$$

This series is absolutely and uniformly convergent on the entire real axis, since it is bounded by the convergent series

$$\zeta(2) = \sum_{n=1}^{\infty} \frac{1}{n^2} \tag{1.47}$$

However, the formal series

$$f'(z) \equiv \pi \sum_{n=1}^{\infty} \cos n^2 \pi z \tag{1.48}$$

converges nowhere, since the terms in the series do not tend to zero for large n. A similar example is the series

$$g(z) \equiv \sum_{n=1}^{\infty} a^n \sin 2^n \pi z \tag{1.49}$$

for which the convergence properties of the derivative can be worked out as an exercise. Functions of this type were introduced as illustrative examples by Weierstrass.   ∎

1.3.2   Weak Convergence; Generalized Functions

There is another type of convergent sequence, whose limit is not a function in the classical sense, but which defines a kind of generalized function widely used in physics. Suppose C is a class of well-behaved functions (*test functions*) on a region $\mathcal{R}$–typically functions that are continuous with continuous derivatives of suitably high order. Then the sequence of functions $\{f_n(z)\}$ (that need not themselves be in C) is *weakly convergent* (relative to C) if the sequence

$$\left\{ \int_{\mathcal{R}} f_n(z)\, g(z)\, d\tau \right\} \tag{1.50}$$

is convergent for every function $g(z)$ in the class C. The limit of a weakly convergent sequence is a *generalized function*, or *distribution*. It need not have a value at every point of $\mathcal{R}$. If

$$\left\{ \int_{\mathcal{R}} f_n(z)\, g(z)\, d\tau \right\} \to \int_{\mathcal{R}} f(z)\, g(z)\, d\tau \tag{1.51}$$

for every $g(z)$ in C, then $\{f_n(z)\} \rightharpoonup f(z)$ (the symbol $\rightharpoonup$ denotes weak convergence), but the weak convergence need not define the value of the limit $f(z)$ at discrete points.

❑ **Example 1.9.**   Consider the sequence $\{f_n(x)\}$ defined by

$$f_n(x) = \begin{cases} \dfrac{n}{2} & -\dfrac{1}{n} \leq x \leq \dfrac{1}{n} \\[2mm] 0, & \text{otherwise} \end{cases} \tag{1.52}$$

Then $\{f_n(x)\} \to 0$ for every $x \neq 0$, but

$$\int_{-\infty}^{\infty} f_n(x)\, dx = 1 \tag{1.53}$$

for $n = 1, 2, \ldots$, and, if $g(x)$ is continuous at $x = 0$,

$$\left\{ \int_{-\infty}^{\infty} f_n(x)\, g(x)\, dx \right\} \to g(0) \tag{1.54}$$

The weak limit of the sequence $\{f_n(x)\}$ thus has the properties attributed to the *Dirac δ-function* $\delta(x)$, defined here as a distribution on the class of functions continuous at $x = 0$. The derivative of the δ-function can be defined as a generalized function on the class of functions with continuous derivative at $x = 0$ using integration by parts to write

$$\int_{-\infty}^{\infty} \delta'(x)\, g(x)\, dx = -\int_{-\infty}^{\infty} \delta(x)\, g'(x)\, dx = -g'(0) \tag{1.55}$$

Similarly, the nth derivative of the δ-function is defined as a generalized function on the class of functions with the continuous nth derivative at $x = 0$ by

$$
\begin{aligned}
\int_{-\infty}^{\infty} \delta^{(n)}(x)\, g(x)\, dx &= -\int_{-\infty}^{\infty} \delta^{(n-1)}(x)\, g'(x)\, dx \\
&= \cdots = (-1)^n\, g^{(n)}(0)
\end{aligned}
\tag{1.56}
$$

using repeated integration by parts.   ∎

1.3.3   Infinite Series of Functions; Power Series

Convergence properties of infinite series

$$\sum_{k=0}^{\infty} f_k(z)$$

of functions are identified with those of the corresponding sequence

$$s_n(z) \equiv \sum_{k=0}^{n} f_k(z) \tag{1.57}$$

of partial sums. The series $\sum_k f_k(z)$ is (pointwise, uniformly, weakly) convergent on $\mathcal{R}$ if the sequence $\{s_n(z)\}$ is (pointwise, uniformly, weakly) convergent on $\mathcal{R}$, and absolutely convergent if the sum of absolute values,

$$\sum_k |f_k(z)|$$

is convergent.

An important class of infinite series of functions is the *power series*

$$S(z) \equiv \sum_{n=0}^{\infty} a_n z^n \tag{1.58}$$

in which $\{a_n\}$ is a sequence of complex numbers and z a complex variable. The basic convergence properties of power series are contained in

Theorem 1.5. Let $S(z) \equiv \sum_{n=0}^{\infty} a_n z^n$ be a power series, $\alpha_n \equiv \sqrt[n]{|a_n|}$, and let α be the largest limit point of the sequence $\{\alpha_n\}$. Then

(i) If $\alpha = 0$, then the series $S(z)$ is absolutely convergent for all z, and uniformly on any bounded region of the complex plane,

(ii) If α does not exist ($\alpha = \infty$), then $S(z)$ is divergent for any $z \neq 0$,

(iii) If $0 < \alpha < \infty$, then $S(z)$ is absolutely convergent for $|z| < r \equiv 1/\alpha$, uniformly within any circle $|z| \leq \rho < r$, and $S(z)$ is divergent for $|z| > r$.

Proof. Since $\sqrt[n]{|a_n z^n|} = \alpha_n |z|$, results (i)–(iii) follow directly from the root test. ∎

Thus the region of convergence of a power series is at least the interior of a circle in the complex plane, the *circle of convergence*, and r is the *radius of convergence*. Note that convergence tests other than the root test can be used to determine the radius of convergence of a given power series. The behavior of the series *on* the circle of convergence must be determined separately for each series; various possibilities are illustrated in the examples and problems.

Now suppose we have a function $f(z)$ defined by a power series

$$f(z) = \sum_{n=0}^{\infty} a_n z^n \tag{1.59}$$

with the radius of convergence $r > 0$. Then $f(z)$ is differentiable for $|z| < r$, and its derivative is given by the series

$$f'(z) = \sum_{n=0}^{\infty} (n+1) a_{n+1} z^n \tag{1.60}$$

which is absolutely convergent for $|z| < r$ (show this). Thus a power series can be differentiated term by term inside its circle of convergence. Furthermore, $f(z)$ is differentiable to any order for $|z| < r$, and the kth derivative is given by the series

$$f^{(k)}(z) = \sum_{n=0}^{\infty} \frac{(n+k)!}{n!} a_{n+k} z^n \tag{1.61}$$

since this series is also absolutely convergent for $|z| < r$. It follows that

$$a_k = f^{(k)}(0)/k! \tag{1.62}$$

Thus every power series with positive radius of convergence is a *Taylor series* defining a function with derivatives of any order. Such functions are *analytic functions*, which we study more deeply in Chapter 4.

❑ **Example 1.10.** Following are some standard power series; it is a useful exercise to verify the radius of convergence for each of these power series using the tests given here.   ∎

(i) The *binomial series* is

$$(1+z)^\alpha \equiv \sum_{n=0}^{\infty} \binom{\alpha}{n} z^n \tag{1.63}$$

where

$$\binom{\alpha}{n} \equiv \frac{\alpha(\alpha-1)\cdots(\alpha-n+1)}{n!} = \frac{\Gamma(\alpha+1)}{n!\,\Gamma(\alpha-n+1)} \tag{1.64}$$

is the *generalized binomial coefficient*. Here $\Gamma(z)$ is the Γ-function that generalizes the elementary factorial function; it is discussed at length in Appendix A. For $\alpha = m = 0, 1, 2, \ldots$, the series terminates after $m+1$ terms and thus converges for all z; otherwise, note that

$$\binom{\alpha}{n+1} \Big/ \binom{\alpha}{n} = \frac{\alpha-n}{n+1} \longrightarrow -1 \tag{1.65}$$

whence the series (1.63) has the radius of convergence $r = 1$.

(ii) The *exponential series*

$$e^z \equiv \sum_{n=0}^{\infty} \frac{z^n}{n!} \tag{1.66}$$

has infinite radius of convergence.

(iii) The *trigonometric functions* are given by the power series

$$\sin z \equiv \sum_{n=0}^{\infty} (-1)^n \frac{z^{2n+1}}{(2n+1)!} \tag{1.67}$$

$$\cos z \equiv \sum_{n=0}^{\infty} (-1)^n \frac{z^{2n}}{(2n)!} \tag{1.68}$$

with infinite radius of convergence.

(iv) The *logarithmic series*

$$\ln(1+z) \equiv \sum_{n=0}^{\infty} (-1)^n \frac{z^{n+1}}{n+1} \tag{1.69}$$

has the radius of convergence $r = 1$.

(v) The *arctangent series*

$$\tan^{-1} z \equiv \sum_{n=0}^{\infty} (-1)^n \frac{z^{2n+1}}{2n+1} \tag{1.70}$$

has the radius of convergence $r = 1$.

1.4 Asymptotic Series

1.4.1 The Exponential Integral

Consider the function $E_1(z)$ defined by

$$E_1(z) \equiv \int_z^\infty \frac{e^{-t}}{t}\, dt = e^{-z} \int_0^\infty \frac{e^{-u}}{u+z}\, du \equiv e^{-z} I(z) \tag{1.71}$$

$E_1(z)$ is the *exponential integral*, a tabulated function. An expansion of $E_1(z)$ about $z = 0$ is given by

$$E_1(z) = \int_1^\infty \frac{e^{-t}}{t}\, dt - \int_1^z \frac{e^{-t}}{t}\, dt = -\ln z + \int_1^z \frac{1-e^{-t}}{t}\, dt + \int_1^\infty \frac{e^{-t}}{t}\, dt \tag{1.72}$$

$$= -\ln z - \left[\int_0^1 \frac{1-e^{-t}}{t}\, dt - \int_1^\infty \frac{e^{-t}}{t}\, dt \right] - \sum_{n=1}^\infty \frac{(-1)^n}{n} \frac{z^n}{n!}$$

Here the term in the square brackets is the *Euler–Mascheroni* constant $\gamma = 0.5772\ldots$, and the power series has infinite radius of convergence.

Suppose now $|z|$ is large. Then the series (1.72) converges slowly, and a better estimate of the integral $I(z)$ can be obtained by introducing the expansion

$$\frac{1}{u+z} = \frac{1}{z} \sum_{n=0}^\infty (-1)^n \left(\frac{u}{z} \right)^n \tag{1.73}$$

into the integral (1.71). Then term-by-term integration leads to the series expansion

$$I(z) = \sum_{n=0}^\infty (-1)^n \frac{n!}{z^{n+1}} \tag{1.74}$$

Unfortunately, the formal power series (1.74) diverges for all finite z. This is due to the fact that the series expansion (1.73) of $1/(u+z)$ is not convergent over the entire range of integration $0 \le u < \infty$. However, the main contribution to the integral comes from the region of small u, where the expansion does converge, and, in fact, the successive terms of the series for $I(z)$ decrease in magnitude for $n+1 \le |z|$; only for $n+1 > |z|$ do they begin to diverge. This suggests that the series (1.74) might provide a useful approximation to the integral $I(z)$ if appropriately truncated.

To obtain an estimate of the error in truncating the series, note that repeated integration by parts in Eq. (1.71) gives

$$I(z) = \sum_{n=0}^N (-1)^n \frac{n!}{z^{n+1}} + (-1)^{N+1}(N+1)! \int_0^\infty \frac{e^{-u}}{(u+z)^{N+2}}\, du \tag{1.75}$$

$$\equiv S_N(z) + R_N(z)$$

If Re $z > 0$, then we can bound the remainder term $R_N(z)$ by

$$|R_N(z)| \leq \frac{(N+1)!}{|z|^{N+2}} \tag{1.76}$$

since $|u + z| \geq |z|$ for all $u \geq 0$ when Re $z \geq 0$. Hence the remainder term $R_N(z) \to 0$ as $z \to \infty$ in the right half of the complex plane, so that $I(z)$ can be approximated by $S_N(z)$ with a relative error that vanishes as $z \to \infty$ in the right half-plane. In fact, when Re $z < 0$ we have

$$|R_N(z)| \leq \frac{(N+1)!}{|\mathrm{Im}\, z|^{N+2}} \tag{1.77}$$

so that the series (1.74) is valid in any sector $-\delta \leq \arg z \leq \delta$ with $0 < \delta < \pi$. Note also that for fixed z, $|R_N(z)|$ has a minimum for $N + 1 \cong |z|$, so that we can obtain a "best" estimate of $I(z)$ by truncating the expansion after about $N + 1$ terms.

The series (1.74) is an *asymptotic* (or *semiconvergent*) series for the function $I(z)$ defined by Eq. (1.71). Asymptotic series are useful, often more useful than convergent series, in exhibiting the behavior of functions such as solutions to differential equations, for limiting values of their arguments. An asymptotic series can also provide a practical method for evaluating a function, even though it can never give the "exact" value of the function because it is divergent. The device of integration by parts, for which the illegal power series expansion of the integrand is a shortcut, is one method of generating an asymptotic series. Watson's lemma, introduced below, provides another.

1.4.2   Asymptotic Expansions; Asymptotic Series

Before looking at more examples, we introduce some standard terminology associated with asymptotic expansions.

Definition 1.13. $f(z)$ is *of order* $g(z)$ as $z \to z_0$, or $f(z) = O[g(z)]$ as $z \to z_0$, if there is some positive M such that $|f(z)| \leq M|g(z)|$ in some neighborhood of z_0. Also, $f(z) = o[g(z)]$ (read "$f(z)$ is *little o* $g(z)$") as $z \to z_0$ if

$$\lim_{z \to z_0} f(z)/g(z) = 0 \tag{1.78}$$

❑ **Example 1.11.**   We have

(i) $z^{n+1} = o(z^n)$ as $z \to 0$ for any n.

(ii) $e^{-z} = o(z^n)$ for any n as $z \to \infty$ in the right half of the complex plane.

(iii) $E_1(z) = O(e^{-z}/z)$, or $E_1(z) = o(e^{-z})$, as $z \to \infty$ in any sector $-\delta \leq \arg z \leq \delta$ with $0 < \delta < \pi$. Also, $E_1(z) = O(\ln z)$ as $z \to 0$.   ∎

Definition 1.14.   The sequence $\{f_n(z)\}$ is an *asymptotic sequence* for $z \to z_0$, if for each $n = 1, 2, \ldots$, we have $f_{n+1}(z) = o[f_n(z)]$ as $z \to z_0$.   ∎

❑ **Example 1.12.** We have

(i) $\{(z - z_0)^n\}$ is an asymptotic sequence for $z \to z_0$.

(ii) If $\{\lambda_n\}$ is a sequence of complex numbers such that Re $\lambda_{n+1} <$ Re λ_n for all n, then $\{z^{\lambda_n}\}$ is an asymptotic sequence for $z \to \infty$.

(iii) If $\{\lambda_n\}$ is any sequence of complex numbers, then $\{z^{\lambda_n} e^{-nz}\}$ is an asymptotic sequence for $z \to \infty$ in any sector $-\delta \le \arg z \le \delta$ with $0 < \delta < \frac{\pi}{2}$. ∎

Definition 1.15. If $\{f_n(z)\}$ is an asymptotic sequence for $z \to z_0$, then

$$f(z) \sim \sum_{n=1}^{N} a_n f_n(z) \tag{1.79}$$

is an *asymptotic expansion* (to N terms) of $f(z)$ as $z \to z_0$ if

$$f(z) - \sum_{n=1}^{N} a_n f_n(z) = o[f_N(z)] \tag{1.80}$$

as $z \to z_0$. The formal series

$$f(z) \sim \sum_{n=1}^{\infty} a_n f_n(z) \tag{1.81}$$

is an *asymptotic series* for $f(z)$ as $z \to z_0$ if

$$f(z) - \sum_{n=1}^{N} a_n f_n(z) = O[f_{N+1}(z)] \tag{1.82}$$

as $z \to z_0$ ($N = 1, 2, \ldots$). The series (1.82) may converge or diverge, but even if it converges, it need not actually converge to the function, since we say $f(z)$ is *asymptotically equal* to $g(z)$, or $f(z) \sim g(z)$, as $z \to z_0$ with respect to the asymptotic sequence $\{f_n(z)\}$ if

$$f(z) - g(z) = o[f_n(z)] \tag{1.83}$$

as $z \to z_0$ for $n = 1, 2, \ldots$. For example, we have

$$f(z) \sim f(z) + e^{-z} \tag{1.84}$$

with respect to the sequence $\{z^{-n}\}$ as $z \to \infty$ in any sector with Re $z > 0$. Thus a function need not be uniquely determined by its asymptotic series. ∎

Of special interest are asymptotic power series

$$f(z) \sim \sum_{n=0}^{\infty} \frac{a_n}{z^n} \tag{1.85}$$

for $z \to \infty$ (generally restricted to some sector in the complex plane). Such a series can be integrated term by term, so that if $F'(z) = f(z)$, then

$$F(z) \sim a_0 z + a_1 \ln z + c - \sum_{n=1}^{\infty} \frac{a_{n+1}}{nz^n} \qquad (1.86)$$

for $z \to \infty$. On the other hand, the derivative

$$f'(z) \sim -\sum_{n=1}^{\infty} \frac{na_n}{z^{n+1}} \qquad (1.87)$$

only if it is known that $f'(z)$ has an asymptotic power series expansion.

1.4.3 Laplace Integral; Watson's Lemma

Now consider the problem of finding an asymptotic expansion of the integral

$$J(x) = \int_0^a F(t)e^{-xt} \, dt \qquad (1.88)$$

for x large and positive (the variable is called x here to emphasize that it is real, although the series derived can often be extended into a sector of the complex plane). It should be clear that such an asymptotic expansion will depend on the behavior of $F(t)$ near $t = 0$, since that is where the exponential factor is the largest, especially in the limit of large positive x. The important result is contained in

Theorem 1.6. *(Watson's lemma)*. Suppose that the function $F(t)$ in Eq. (1.88) is integrable on $0 \le x \le a$, with an asymptotic expansion for $t \to 0+$ of the form

$$F(t) \sim t^b \sum_{n=0}^{\infty} c_n t^n \qquad (1.89)$$

with $b > -1$. Then an asymptotic expansion for $J(x)$ as $x \to \infty$ is

$$J(x) \sim \sum_{n=0}^{\infty} c_n \frac{\Gamma(n+b+1)}{x^{n+b+1}} \qquad (1.90)$$

Here

$$\Gamma(\xi + 1) \equiv \int_0^{\infty} t^{\xi} e^{-t} \, dt \qquad (1.91)$$

is the Γ-function, which will be discussed at length in Chapter 4. Note that for integer values of the argument, we have $\Gamma(n+1) = n!$.

Proof. To derive the series (1.90), let $0 < \varepsilon < a$, and consider the integral

$$J_\varepsilon(x) \equiv \int_0^\varepsilon F(t)e^{-xt} \, dt \sim \sum_{n=0}^{\infty} c_n \int_0^\varepsilon t^{n+b} \, e^{-xt} \, dt \qquad (1.92)$$

Note that

$$J(x) - J_\varepsilon(x) = \int_\varepsilon^a F(t)e^{-xt}\,dt = e^{-\varepsilon x}\int_0^{a-\varepsilon} F(\tau+\varepsilon)e^{-x\tau}\,d\tau \qquad (1.93)$$

is exponentially small compared to $J(x)$ for $x \to \infty$, since $F(t)$ is assumed to be integrable on $0 \le t \le a$. Hence $J(x)$ and $J_\varepsilon(x)$ are approximated by the same asymptotic power series.

The asymptotic character of the series (1.89) implies that for any N, we can choose ε small enough that the error term

$$\Delta_\varepsilon^N(x) \equiv \left| J_\varepsilon(x) - \sum_{n=0}^{N-1} c_n \int_0^\varepsilon t^{n+b}e^{-xt}\,dt \right| < C \int_0^\varepsilon t^{N+b}e^{-xt}\,dt \qquad (1.94)$$

for some constant C. But we also know that

$$\frac{\Gamma(N+b+1)}{x^{N+b+1}} - \int_0^\varepsilon t^{n+b}e^{-xt}\,dt = \int_\varepsilon^\infty t^{n+b}e^{-xt}\,dt$$
$$= e^{-\varepsilon x}\int_0^\infty (\tau+\varepsilon)^{n+b}e^{-x\tau}\,d\tau \qquad (1.95)$$

The right-hand side is exponentially small for $x \to \infty$; hence the error term is bounded by

$$\left|\Delta_\varepsilon^N(x)\right| < C\,\frac{\Gamma(N+b+1)}{x^{N+b+1}} \qquad (1.96)$$

Thus the series on the right-hand side of Eq. (1.90) is an asymptotic power series for $J_\varepsilon(x)$ and thus also for $J(x)$. ∎

We can use Watson's lemma to derive an asymptotic expansion for $z \to \infty$ of a function $I(z)$ defined by the integral representation (*Laplace integral*)

$$I(z) = \int_0^a f(t)e^{zh(t)}\,dt \qquad (1.97)$$

with $f(t)$ and $h(t)$ continuous real functions[3] on the interval $0 \le t \le a$. For z large and positive, we can expect that the most important contribution to the integral will be from the region in t near the maximum of $h(t)$, with contributions from outside this region being exponentially small in the limit $\mathrm{Re}\,z \to +\infty$. There is actually no loss of generality in assuming that the maximum of $h(t)$ occurs at $t = 0$.[4]

The integral (1.97) can be converted to the form (1.88) by introducing a new variable $u \equiv h(0) - h(t)$, and approximating $I(z)$ by

$$I(z) \sim I_\varepsilon(z) \equiv -e^{zh(0)}\int_0^\varepsilon \frac{f(t)}{h'(t)}e^{-zu}\,du \qquad (1.98)$$

---

[3]It is enough that $f(t)$ is integrable, but we will rarely be concerned about making the most general technical assumptions.

[4]Suppose $h(t)$ has a maximum at an interior point ($t = b$, say) of the interval of integration. Then we can split the integral (1.97) into two parts, the first an integral from 0 to b, the second an integral from b to a, and apply the present method to each part.

The asymptotic expansion of $I_\varepsilon(z)$ is then obtained from the expansion of $f(t)/h'(t)$ for $t \to 0+$, as just illustrated, provided that such an expansion in the form (1.89) exists. Note that the upper limit $\varepsilon \ (> 0)$ in this integral can be chosen at will. This method of generating asymptotic series is due to Laplace.

❑ **Example 1.13.** Consider the integral

$$I(z) = \int_0^\infty e^{-z \sinh t} \, dt \tag{1.99}$$

Changing the variable of integration to $u = \sinh t$ gives

$$I(z) = \int_0^\infty (1 + u^2)^{-1/2} \, e^{-zu} \, du \tag{1.100}$$

Expanding

$$(1 + u^2)^{-1/2} = \sum_{n=0}^\infty (-1)^n \frac{\Gamma(n + \frac{1}{2})}{n! \, \Gamma(\frac{1}{2})} \, u^{2n} \tag{1.101}$$

then gives the asymptotic series

$$I(z) \sim e^{-z} \sum_{n=0}^\infty (-1)^n \frac{\Gamma(n + \frac{1}{2})}{\Gamma(\frac{1}{2})} \frac{(2n)!}{n!} \frac{1}{z^{2n+1}} \tag{1.102}$$

for $z \to \infty$ with $|\arg z| \leq \frac{\pi}{2} - \delta$ and fixed $0 \leq \delta < \frac{\pi}{2}$.   ∎

Now suppose the function $h(t)$ in Eq. (1.97) has a maximum at $t = 0$, with the expansion

$$h(t) \sim h(0) - At^p + \cdots \tag{1.103}$$

for $t \cong 0$, with $A > 0$ and $p > 0$. Then we can introduce a new variable $u = t^p$ into Eq. (1.97), which gives

$$I(z) \sim \frac{1}{p} e^{zh(0)} \, f(0) \int_0^\infty u^{\frac{1}{p}} e^{-Auz} \frac{du}{u} = e^{zh(0)} \, f(0) \frac{\Gamma(\frac{1}{p})}{p \, (Az)^{\frac{1}{p}}} \tag{1.104}$$

Note that $\Gamma(\frac{1}{p}) = \Gamma(\frac{1}{2}) = \sqrt{\pi}$ in the important case $p = 2$ that corresponds to the usual quadratic behavior of a function near a maximum. In any case, the leading behavior of $I(z)$ for $z \to \infty$ (with Re $z > 0$) follows directly from the leading behavior of $h(t)$ near $t = 0$.

❑ **Example 1.14.** Consider the integral

$$K_0(z) = \int_0^\infty e^{-z \cosh t} \, dt \tag{1.105}$$

which is a known representation for the modified Bessel function $K_0(z)$. Since $\cosh t \cong 1 + t^2/2$ near $t = 0$, the leading term in the asymptotic expansion of $K_0(z)$ for $z \to \infty$ is given by

$$K_0(z) \sim e^{-z} \sqrt{\frac{\pi}{2z}} \tag{1.106}$$

Here the complete asymptotic expansion of $K_0(z)$ can be derived by changing the variable of integration to $u \equiv \cosh t$. This gives

$$K_0(z) = \int_1^\infty (u^2 - 1)^{-1/2} e^{-zu} \, du$$

$$= \sqrt{\tfrac{1}{2}} e^{-z} \int_0^\infty v^{-1/2} (1 + \tfrac{1}{2}v)^{-1/2} e^{-zv} \, dv \tag{1.107}$$

$(v = u - 1)$. Expanding $(1 + \tfrac{1}{2}v)^{-1/2}$ then provides the asymptotic series

$$K_0(z) \sim e^{-z} \sum_{n=0}^\infty (-1)^n \frac{[\Gamma(n + \tfrac{1}{2})]^2}{n!\,\Gamma(\tfrac{1}{2})} \left(\frac{1}{2z}\right)^{n + \frac{1}{2}} \tag{1.108}$$

again for $z \to \infty$ with $|\arg z| \le \frac{\pi}{2} - \delta$ and fixed $0 \le \delta < \frac{\pi}{2}$. ∎

The method introduced here must be further modified if either of the functions $f(t)$ or $h(t)$ in Eq. (1.97) does not have an asymptotic power series expansion for $t \to 0$. Consider, for example, the Γ-function introduced above in Eq. (1.91), which we can write in the form

$$\Gamma(\xi + 1) = \int_0^\infty e^{\xi \ln t} e^{-t} \, dt \tag{1.109}$$

The standard method to find an asymptotic expansion for $\xi \to +\infty$ does not work here, since $\ln t$ has no power series expansion for $t \to 0$. However, we can note that the argument $(-t + \xi \ln t)$ of the exponential has a maximum for $t = \xi$. Since the location of the maximum depends on ξ (it is a *moving maximum*), we change variables and let $t \equiv \xi u$, so that Eq. (1.109) becomes

$$\Gamma(\xi + 1) = \xi^{\xi+1} \int_0^\infty e^{\xi \ln u} e^{-\xi u} \, du \tag{1.110}$$

Now the argument in the exponent can be expanded about the maximum at $u = 1$ to give

$$\Gamma(\xi + 1) \cong \xi^{\xi+1} e^{-\xi} \int_0^\infty e^{-\frac{1}{2}\xi(u-1)^2} \, du \cong \sqrt{2\pi\xi} \left(\frac{\xi}{e}\right)^\xi \tag{1.111}$$

This is the first term in *Stirling's expansion* of the Γ-function; the remaining terms will be derived in Chapter 4.

A   Iterated Maps, Period Doubling, and Chaos

We have been concerned in this chapter with the properties of infinite sequences and series from the point of view of classical mathematics, in which the important question is whether or not the sequence or series converges, with asymptotic series recognized as useful for characterizing the limiting behavior of functions, and for approximate evaluation of the functions.

Sequences generated by dynamical systems can have a richer structure. For example, the successive intersections of a particle trajectory with a fixed plane through which the trajectory passes more or less periodically, or the population counts of various species in an ecosystem at definite time intervals, can be treated as sequences generated by a map T that takes each of the possible initial states of the system into its successor. The qualitative properties of such maps are interesting and varied.

As a simple prototype of such a map, consider the *logistic map*

$$T_\lambda : \ x \mapsto f_\lambda(x) \equiv \lambda x(1 - x) \tag{1.A1}$$

that maps the unit interval $0 < x < 1$ into itself for $0 < \lambda < 4$ (the maximum value of $x(1 - x)$ in the unit interval is $1/4$). Starting from a generic point x_0 in the unit interval, T_λ generates a sequence $\{x_n\}$ defined by

$$x_{n+1} = \lambda x_n (1 - x_n) \tag{1.A2}$$

If $\lambda < 1$, we have

$$x_{n+1} < \lambda x_n < \cdots < \lambda^{n+1} x_0 < \lambda^{n+1} \tag{1.A3}$$

and the sequence converges to 0. But the sequence does not converge to 0 if $\lambda > 1$, and the behavior of the sequence as λ increases is quite interesting.

Remark.   A generic map of the type (1.A1) that maps a coordinate space (or *manifold*, to be introduced in Chapter 3) into itself, defines a *discrete-time dynamical system* generated by iterations of the map. The bibliography at the end of the chapter has some suggestions for further reading.   ☐

To analyze the behavior of the sequence in general, note that the map (1.A1) has *fixed points* x_* (points for which $x_* = f_\lambda(x_*)$) at

$$x_* = 0, 1 - \tfrac{1}{\lambda} \tag{1.A4}$$

If the sequence (1.A2) starts at one of these points, it will remain there, but it is important to know how the sequence develops from an initial value of x near one of the fixed points. If an initial point close to the fixed point is driven toward the fixed point by successive iterations of the map, then the fixed point is *stable*; if it is driven away from the fixed point, then the fixed point is *unstable*. The sequence can only converge, if it converges at all, to one of its fixed points, and indeed only to a stable fixed point.

To determine the stability of the fixed points in Eq. (1.A4), note that

$$x_{n+1} = f_\lambda(x_n) \cong x_* + f'_\lambda(x_*)(x_n - x_*) \tag{1.A5}$$

for $x_n \cong x_*$, so that

$$\varrho_n \equiv \frac{x_{n+1} - x_*}{x_n - x_*} \cong f_\lambda'(x_*) \tag{1.A6}$$

Stability of the fixed point x_* requires $\{x_n\} \to x_*$ from a starting point sufficiently close to x_*. Hence it is necessary that $|\varrho_n| < 1$ for large n, which requires

$$-1 < f_\lambda'(x_*) < 1 \tag{1.A7}$$

This criterion for the stability of the fixed point is quite general. Note that if

$$f_\lambda'(x_*) = 0 \tag{1.A8}$$

the convergence of the sequence will be especially rapid. With $\varepsilon_n = x_n - x_*$, we have

$$\varepsilon_{n+1} \cong \tfrac{1}{2} f''(x_*) \varepsilon_n^2 \tag{1.A9}$$

and the convergence to the fixed point is exponential; the fixed point is *superstable*.

→ **Exercise 1.A1.**   Find the values of λ for which each of the fixed points in (1.A4) is superstable.   □

Remark.   The case $|f_\lambda'(x_*)| = 1$ requires special attention, since the ratio test fails. The fixed point may be stable in this case as well.   □

For the map defined by Eq. (1.A2), we have

$$f_\lambda'(x_*) = \lambda(1 - 2x_*) \tag{1.A10}$$

so the fixed point $x_* = 0$ is stable only for $\lambda \le 1$, while the fixed point $x_* = 1 - 1/\lambda$ is stable for $1 \le \lambda \le 3$. Hence for $1 \le \lambda \le 3$, the sequence $\{x_n\}$ converges,

$$\{x_n\} \to 1 - \tfrac{1}{\lambda} \equiv x_\lambda \tag{1.A11}$$

It requires proof that this is true for any initial value x_0 in the interval $(0, 1)$, but a numerical experiment starting from a few randomly chosen points may be convincing.

What happens for $\lambda > 3$? For λ slightly above 3 ($\lambda = 3.1$, say), a numerical study shows that the sequence begins to oscillate between two fixed numbers that vary continuously from $x_* = \tfrac{2}{3}$ as λ is increased above 3, and bracket the now unstable fixed point $x_* = x_\lambda$. To study this behavior analytically, consider the iterated sequence

$$x_{n+2} = \lambda^2 x_n (1 - x_n)[1 - \lambda x_n(1 - x_n)] = f(f(x_n)) \equiv f^{[2]}(x_n) \tag{1.A12}$$

This sequence still has the fixed points given by Eq. (1.A4), but two new fixed points

$$x_*^\pm \equiv \frac{1}{2\lambda}\{\lambda + 1 \pm \sqrt{(\lambda+1)(\lambda-3)}\,\} \tag{1.A13}$$

appear. These new fixed points are real for $\lambda > 3$, and the original sequence (1.A2) eventually appears to oscillate between them (with *period* 2) for $\lambda > 3$.

→ **Exercise 1.A2.**   Derive the result (1.A13) for the fixed points of the second iterate $f^{[2]}$ of the map (1.A1) as defined in Eq. (1.A12). Sketch on a graph the behavior of these fixed points, as well as the fixed points (1.A4) of the original map, as a function of λ for $3 \leq \lambda < 4$. Then derive the result (1.A16) for the value of λ at which these fixed points become unstable, leading to a bifurcation of the sequence into a limit cycle of period 4.   □

The derivative of the iterated map $f^{[2]}$ is given by

$$f^{[2]'}(x) = f'(f(x)) f'(x) \tag{1.A14}$$

which at the fixed points (1.A13) becomes

$$f^{[2]'}(x_*^\pm) = f'(x_*^+) f'(x_*^-) = 4 + 2\lambda - \lambda^2 \tag{1.A15}$$

Thus $f^{[2]'}(x_*^\pm) = 1$ at $\lambda = 3$, and decreases to -1 as λ increases from $\lambda = 3$ to

$$\lambda = 1 + \sqrt{6} \cong 3.4495\ldots \tag{1.A16}$$

when the sequence undergoes a second *bifurcation* into a stable cycle of length 4. Successive period doublings continue after shorter and shorter intervals of λ, until at

$$\lambda \cong 3.56994\ldots \equiv \lambda_c \tag{1.A17}$$

the sequence becomes *chaotic*. Iterations of the sequence starting from nearby points become widely separated, and the sequence does not approach any limiting cycle.

This is not quite the whole story, however. In the interval $\lambda_c < \lambda < 4$, there are islands of periodicity, in which the sequence converges to a cycle of period p for a range of λ, followed (as λ increases) by a series of period doublings to cycles of periods $2p$, $4p$, $8p$, $\ldots$ and eventual reversion to chaos. There is one island associated with period 3 and its doublings, which for the sequence (1.A2) begins at

$$\lambda = 1 + \sqrt{8} \cong 3.828\ldots \tag{1.A18}$$

and one or more islands with each integer as fundamental period together with the sequence of period doublings. In Fig. 1.1, the behavior of the iterates of the map is shown as a function of λ; the first three period doublings, as well as the interval with stable period 3, are clearly visible. For further details, see the book by Devaney cited in the bibliography at the end of the chapter.

The behavior of the iterates of the map (1.A2) as the parameter λ varies is not restricted to the logistic map, but is shared by a wide class of maps of the unit interval $I \equiv (0, 1)$ into itself. Let

$$T_\lambda : x \mapsto \lambda f(x) \tag{1.A19}$$

be a map $I \to I$ such that $f(x)$ is continuously differentiable and $f'(x)$ is nonincreasing on I, with $f(0) = 0 = f(1)$, $f'(0) > 0$, $f'(1) < 0$. These conditions mean that $f(x)$ is concave downward in the interval I, increasing monotonically from 0 to a single maximum in the interval, and then decreasing monotonically to 0 at the end of the interval.

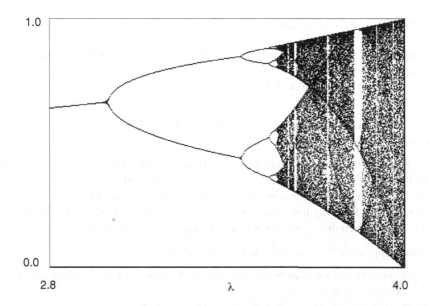

Figure 1.1: Iterates of the map (1.A2) for λ between 2.8 and 4.0. Shown are 100 iterates of the map after first iterating 200 times to let the dependence on the initial point die down.

If $f(x)$ satisfies these conditions, then T_λ shows the same qualitative behavior of period doubling, followed by a chaotic region with islands of periodicity, as a function of λ. Furthermore, if λ_n denotes the value of λ at which the nth period doubling occurs, then

$$\lim_{n \to \infty} \frac{\lambda_{n+1} - \lambda_n}{\lambda_{n+2} - \lambda_{n+1}} \equiv \delta = 4.6692 \ldots \tag{1.A20}$$

is a *universal* constant, discovered by Feigenbaum in the late 1970s, independent of the further details of the map.

A simple context in which the sequence (1.A2) arises is the model of a biological species whose population in generation $n + 1$ is related to the population in generation n by

$$p_{n+1} = r\, p_n - a\, p_n^2 \tag{1.A21}$$

Here $r > 0$ corresponds to the natural growth rate ($r > 1$ if the species is not to become extinct), and $a > 0$ corresponds to a natural limitation on the growth of the population (due to finite food supply, for example). Equation (1.A21) implies that the population is limited to

$$p < p_{\max} \equiv r/a \tag{1.A22}$$

and rescaling Eq. (1.A21) by defining $x_n \equiv p_n/p_{\max}$ leads precisely to Eq. (1.A2) with $\lambda = r$. While this model, as well as some related models given in the problems, is oversimplified, period doubling has actually been observed in biological systems. For examples, see Chapter 2 of the book by May cited in the bibliography.

Bibliography and Notes

The first three sections of this chapter are intended mainly as a review of topics that will be familiar to students who have taken a standard advanced calculus course, and no special references are given to textbooks at that level. A classic reference dealing with advanced methods of analysis is

> E. T. Whittaker and G. N. Watson, *A Course of Modern Analysis* (4th edition), Cambridge University Press (1958).

The first edition of this work was published in 1902 but it is valuable even today. In addition to its excellent and thorough treatment of the classical aspects of complex analysis and the theory of special functions, it contains many of the notorious Cambridge Tripos problems, which the modern reader may find even more challenging than the students of the time!

A basic reference on theory of convergence of sequences and series is

> Konrad Knopp, *Infinite Sequences and Series*, Dover (1956).

This compact monograph summarizes the useful tests for convergence of series, and gives a collection of elementary examples.

Two books that specialize in the study of asymptotic expansions are

> A. Erdélyi, *Asymptotic Expansions*, Dover (1955), and
> N. Bleistein and R. A. Handelsman, *Asymptotic Expansions of Integrals*, Dover (1986).

The first of these is a concise survey of methods of generating asymptotic expansions, both from integral representations and from differential equations. The second is a more comprehensive survey of the various methods used to generate asymptotic expansions of functions defined by integral representations. It also has many examples of the physical and mathematical contexts in which such integrals may occur. The book by Whittaker and Watson noted above, the book on functions of a complex variable by Carrier, Krook, and Pierson in Chapter 4 and the book on advanced differential equations by Bender and Orszag cited in Chapter 5 also deal with asymptotic expansions.

A readable introduction to the theory of bifurcation and chaos is

> R. L. Devaney, *An Introduction to Chaotic Dynamical Systems* (2nd edition), Westview Press (2003).

Starting at the level of a student who has studied ordinary differential equations, this book clearly explains the mathematical foundations of the phenomena that occur in the study of iterated maps. A comprehensive introduction to chaos in discrete and continuous dynamical systems is

> Kathleen T. Alligood, Tim D. Sauer, and James A. Yorke, *Chaos: An Introduction to Dynamical Systems*, Springer (1997).

Each chapter has a serious computer project at the end, as well as simpler exercises. A more advanced book is

> Edward Ott, *Chaos in Dynamical Systems*, Cambridge University Press (1993).

Two early collections of reprints and review articles on the relevance of these phenomena to physical and biological systems are

R. M. May (ed.), *Theoretical Ecology* (2nd edition), Blackwell Scientific Publishers, Oxford (1981), and

P. Cvitanoviç (ed.), *Universality in Chaos* (2nd edition), Adam Hilger Ltd., Bristol (1989).

The first of these is a collection of specially written articles by May and others on various aspects of theoretical ecology. Of special interest here is the observation of the phenomenon of period doubling in ecosystems. The second book contains a collection of reprints of classic articles by Lorenz, May, Hénon, Feigenbaum, and others, leading to the modern studies of chaos and related behavior of dynamical systems, with some useful introductory notes by Cvitanoviç.

The reader should be aware that the behavior of complex dynamical systems is an important area of ongoing research, so that it is important to look at the current literature to get an up-to-date view of the subject. Nevertheless, the concepts presented here and in later chapters are fundamental to the field. Further readings with greater emphasis on differential equations and partial differential equations are found at the end of Chapters 2 and 8.

Problems[5]

1. Show that

$$\lim_{n \to \infty} \left(1 + \frac{z}{n} \right)^n = \sum_{k=0}^{\infty} \frac{z^k}{k!} \; (= e^z)$$

2. Show that

$$\zeta(s) = \prod_{p} \left(1 - \frac{1}{p^s} \right)^{-1}$$

   where $\prod_p$ denotes a product over all *primes p*.

3. Investigate the convergence of the following series:

   (i) $$\sum_{n=1}^{\infty} \left\{ \frac{1}{n} - \ln \left(1 + \frac{1}{n} \right) \right\}$$

   (ii) $$\sum_{n=1}^{\infty} \left\{ 1 - n \ln \left(\frac{2n+1}{2n-1} \right) \right\}$$

   (iii) $$\sum_{n=2}^{\infty} \frac{1}{n^a (\ln n)^b}$$

   Explain how convergence depends on the complex numbers a, b in (iii).

---

[5] When a proposition is simply stated, the problem is to prove it, or to give a counterexample that shows it is false.

4. Investigate the convergence of the following series:

(i) $$\sum_{n=0}^{\infty} (n+1)^a\, z^n$$

(ii) $$\sum_{n=0}^{\infty} \frac{(n+n_1)!\,(n+n_2)!}{n!\,(n+n_3)!}\, z^n$$

(iii) $$\sum_{n=0}^{\infty} e^{-na}\, \cos(bn^2 z)$$

where a, b are real numbers, n_1, n_2, n_3 are positive integers, and z is a (variable) complex number. How do the convergence properties depend on a, b, z?

5. Find the sums of the following series:

(i) $$S = 1 + \frac{1}{4} - \frac{1}{16} - \frac{1}{64} + \frac{1}{256} + \cdots$$

(ii) $$S = \frac{1}{1\cdot 3} + \frac{1}{2\cdot 4} + \frac{1}{3\cdot 5} + \frac{1}{4\cdot 6} + \cdots$$

(iii) $$S = \frac{1}{0!} + \frac{2}{1!} + \frac{3}{2!} + \frac{4}{3!} + \cdots$$

(iv) $$f(z) = \sum_{n=0}^{\infty} (-1)^n \frac{(n+1)^2}{(2n+1)!}\, z^{2n+1}$$

6. Find a closed form expression for the sums of the series

$$S_p \equiv \sum_{n=1}^{\infty} \frac{1}{n(n+1)\cdots(n+p)}$$

$(p = 0, 1, 2, \ldots)$.

Remark. The result obtained here can be used to improve the rate of convergence of a series whose terms tend to zero like $1/n^{p+1}$ for large n; subtracting a suitably chosen multiple of S_p from the series will leave a series whose terms tend to zero at least as fast as $1/n^{p+2}$ for large n. As a further exercise, apply this method to accelerate the convergence of the series for the ζ-function $\zeta(p)$ with p integer.   □

7. The quantum states of a simple harmonic oscillator with frequency ν are the states $|n\rangle$ with energies

$$E_n = (n + \tfrac{1}{2})h\nu$$

$(n = 0, 1, 2, \ldots)$ where h is Planck's constant. For an ensemble of such oscillators in thermal equilibrium at temperature T, the probability of finding an oscillator in the state $|n\rangle$ is given by

$$P_n = A \exp\left(-\frac{E_n}{kT}\right)$$

where A is a constant to be determined, and the exponential factor is the standard *Boltzmann factor*.

(i) Evaluate the constant A by requiring

$$\sum_0^\infty P_n = 1$$

(ii) Find the average energy $\langle E(T) \rangle$ of a single oscillator of the ensemble.

Remark. These results are used in the study of blackbody radiation in Problem 4.9. □

8. Investigate the convergence of the following products:

(i) $$\prod_{m=1}^\infty \frac{m(m + a + b)}{(m + a)(m + b)}$$

(ii) $$\prod_{m=1}^\infty \left(1 - \frac{z^2}{m^2}\right)$$

where a, b are the real numbers and z is a (variable) complex number.

9. Evaluate the infinite product

$$\prod_{m=1}^\infty \left\{ \frac{1 + \exp(i\omega/2^m)}{2} \right\}$$

Hint. Note that $1 + e^{\frac{1}{2}i\omega} = (1 - e^{i\omega})/(1 - e^{\frac{1}{2}i\omega})$.

10. Evaluate the infinite products

(i) $$\prod_{n=1}^\infty \left\{ 1 - \frac{1}{(n + 1)^2} \right\}$$

(ii) $$\prod_{n=2}^\infty \left(\frac{n^3 - 1}{n^3 + 1} \right)$$

11. Show that

$$\prod_{m=0}^\infty \left(1 + z^{2^m}\right) = \frac{1}{1 - z}$$

12. The Euler–Mascheroni constant γ is defined by

$$\gamma \equiv \lim_{N \to \infty} \left\{ \sum_{n=1}^{N} \frac{1}{n} - \ln(N + 1) \right\}$$

(i) Show that γ is finite (i.e., the limit exists). *Hint.* Show that

$$\gamma = \lim_{N \to \infty} \sum_{n=1}^{N} \left\{ \frac{1}{n} - \ln\left(1 + \frac{1}{n}\right) \right\}$$

(ii) Show that

$$\sum_{n=1}^{N} \frac{1}{n} = \int_0^1 \frac{1 - (1 - t)^N}{t} \, dt$$

(iii) Show that

$$\int_0^1 \left(\frac{1 - e^{-t}}{t} \right) dt - \int_1^\infty \frac{e^{-t}}{t} \, dt = \gamma$$

13. The *error function* $\mathrm{erf}(z)$ is defined by

$$\mathrm{erf}(z) \equiv \frac{2}{\sqrt{\pi}} \int_0^z e^{-t^2} \, dt$$

(i) Find the power series expansion of $\mathrm{erf}(z)$ about $z = 0$.

(ii) Find an asymptotic expansion of $\mathrm{erf}(z)$ valid for z large and positive. For what range of $\arg z$ is this asymptotic series valid?

(iii) Find an asymptotic expansion of $\mathrm{erf}(z)$ valid for z large and negative. For what range of $\arg z$ is this asymptotic series valid?

14. Find an asymptotic power series expansion of

$$f(z) \equiv \int_0^\infty \frac{e^{-zt}}{1 + t^2} \, dt$$

valid for z large and positive. For what range of $\arg z$ is this expansion valid? Give an estimate of the error in truncating the series after N terms.

15. Find an asymptotic power series expansion of

$$f(z) \equiv \int_0^\infty \left(1 + \frac{t}{z} \right)^\alpha e^{-zt} \, dt$$

valid for z large and positive (here α is a complex constant). For what range of $\arg z$ is this expansion valid? Give an estimate of the error in truncating the series after N terms.

16. The modified Bessel function $K_\lambda(z)$ of order λ is defined by the integral representation

$$K_\lambda(z) = \int_0^\infty e^{-z \cosh t} \cosh \lambda t \, dt$$

(i) Find an asymptotic expansion of $K_\lambda(z)$ valid for $z \to \infty$ in the right half-plane with λ fixed.

(ii) Find an asymptotic expansion of $K_\lambda(z)$ for $\lambda \to \infty$ with z fixed in the right half of the complex plane (Re $z > 0$).

17. Find an asymptotic expansion of

$$f(z) \equiv \int_0^\infty e^{-zt - 1/t} \, dt$$

valid for z large and positive.

18. The reaction rate for the fusion of nuclei A and B in a hot gas (in the center of a star, for example) at temperature T can be expressed as

$$R(T) = \frac{C}{(kT)^{\frac{3}{2}}} \int_0^\infty S(E) \exp \left(-\frac{b}{\sqrt{E}} - \frac{E}{kT} \right) dE \qquad (*)$$

where $S(E)$ is often a smoothly varying function of the relative energy E. The exponential factor $\exp(-E/kT)$ in $(*)$ is the usual Boltzmann factor, while the factor $\exp(-b/\sqrt{E})$ is the probability of tunneling through the energy barrier created by the Coulomb repulsion between the two nuclei. The constant b is given by

$$b = \frac{Z_A Z_B e^2}{\hbar} \sqrt{\frac{2 m_A m_B}{m_A + m_B}}$$

where m_A, m_B are the masses, $Z_A e, Z_B e$ the charges of the two nuclei, and $\hbar$ is Planck's constant.

(i) Find the energy $E_* = E_*(T)$ at which the integrand in $(*)$ is a maximum, neglecting the energy dependence of $S(E)$.

(ii) Find the width $\Delta = \Delta(T)$ of the peak of the integrand near E_*, again neglecting the energy dependence of $S(E)$.

(iii) Find an approximate value for $R(T)$, assuming $\Delta \ll E_*$.

Remark. A detailed discussion of nuclear reaction rates in stars can be found in

C. E. Rolfs and W. S. Rodney, *Cauldrons in the Cosmos: Nuclear Astrophysics*, University of Chicago Press (1988).

among many other books on the physics of stars. □

19. Find the value(s) of λ for which the fixed points of the iterated logistic map (Eq. (1.A12)) are superstable.

20. Consider the sequence $\{x_n\}$ defined by

$$x_{n+1} = \alpha x_n (1 - x_n^2)$$

Find the fixed points of this sequence, and the ranges of α for which each fixed point is stable. Also find the values of α for which there is a superstable fixed point.

21. Consider the sequence $\{x_n\}$ defined by

$$x_{n+1} = x_n \, e^{\lambda(1-x_n)}$$

with $\lambda > 0$ real, and $x_0 > 0$.

(i) For what range of λ is $\{x_n\}$ bounded?

(ii) For what range of λ is $\{x_n\}$ convergent? Find the limit of the sequence, as a function of λ.

(iii) What can you say about the behavior of the sequence for $\lambda > \lambda_0$, where λ_0 is the largest value of λ for which the sequence converges?

(iv) Does the map

$$T_\lambda : x \mapsto x \, e^{\lambda(1-x)}$$

have any fixed point(s)? What can you say about the stability of the fixed point(s) for various values of λ?

22. Consider the sequence $\{x_n\}$ defined by

$$x_{n+1} = \frac{r \, x_n}{(1 + x_n)^b}$$

with $b, r > 0$ real, and $x_0 > 0$.

(i) For what range of b, r is $\{x_n\}$ bounded for all $x > 0$?

(ii) For what range of b, r is $\{x_n\}$ convergent? Find the limit of the sequence (as a function of b, r).

(iii) What can you say about the behavior of the sequence outside the region in the b–r plane for which the sequence converges?

(iv) Does the map

$$T_{r,b} : x \mapsto \frac{r \, x}{(1 + x)^b}$$

have any fixed point(s)? What can you say about the stability of the fixed point(s) for various values of b, r?

2 Finite-Dimensional Vector Spaces

Many physical systems are described by linear equations. The simple harmonic oscillator is the first example encountered by most students of physics. Electrical circuits with resistance, capacitance, and inductance are linear systems with many independent variables. Maxwell's equations for the electromagnetic fields, the heat equation and general wave equations are linear (partial) differential equations. A universal characteristic of systems described by linear equations is the superposition principle, which states that any linear superposition of solutions to the equations is also a solutions of the equations. The theory of linear vector spaces and linear operators provides a natural and elegant framework for the description of such systems.

In this chapter, we introduce the mathematical foundations of the theory of linear vector spaces, starting with a set of axioms that characterize such spaces. These axioms describe addition of vectors, and multiplication of vectors by scalars. Here scalars may be real or complex numbers; the distinction between real and complex vector spaces is important when solutions to algebraic equations are needed. A *basis* is a set of linearly independent vectors such that any vector can be expressed as a linear combination of the basis vectors; the coefficients are the *components*, or *coordinates*, of the vector in the basis. If the basis vectors form a finite set with n elements, then the vector space is finite dimensional (dimension n); otherwise, it is infinite dimensional. With coordinates introduced, every n-dimensional vector space can be identified with the space of ordered n-tuples of scalars. In this chapter, we consider mainly finite-dimensional vector spaces, though we also treat function spaces as vector spaces in examples where dimensionality is not critical. The complications associated with infinite-dimensional spaces, especially Hilbert spaces, will appear in Chapters 6 and 7.

Here we consider only vector spaces in which nonzero vectors have a positive definite length (or *norm*); the distance between two vectors x and y is the length of the vector $x - y$. If a scalar product satisfying natural axioms can also be defined, the vector space is *unitary*. Any basis of a unitary vector space can be transformed into a set of mutually orthogonal unit vectors; vector components are then the projections of the vector onto the unit vectors.

A scalar-valued function defined on a vector space or function space is called a *functional*. Many functionals can be defined on function spaces, for example, the action integral of classical mechanics to be considered in Chapter 3. Linear functionals defined on a linear vector space $\mathcal{V}$ are of special interest; they form another linear vector space, the *dual space* $\mathcal{V}^*$ of $\mathcal{V}$. An important result is that a bounded linear functional Λ on a unitary vector space has the form of a scalar product, $\Lambda[x] = (u, x)$ with u a vector in the space, so that $\mathcal{V}$ and $\mathcal{V}^*$ are equivalent. Linear functionals defined on function spaces are also known as *distributions*; a notable example is the Dirac δ-function introduced in Chapter 1, defined here as a linear functional on the space of continuous functions.

Introduction to Mathematical Physics. Michael T. Vaughn
Copyright © 2007 WILEY-VCH Verlag GmbH & Co. KGaA, Weinheim
ISBN: 978-3-527-40627-2

Linear operators on a linear vector space arise in many contexts. A system of n linear algebraic equations

$$
\begin{aligned}
a_{11}x_1 + a_{12}x_2 + \cdots + a_{1n}x_n &= c_1 \\
a_{21}x_1 + a_{22}x_2 + \cdots + a_{2n}x_n &= c_2 \\
&\vdots \\
a_{n1}x_1 + a_{n2}x_2 + \cdots + a_{nn}x_n &= c_n
\end{aligned}
$$

can be written in the short form

$$\mathbf{A}x = c$$

in which x denotes a vector in an n-dimensional space (real or complex), c is a known vector, and $\mathbf{A}$ is an $n \times n$ matrix. If the determinant of the matrix $\mathbf{A}$ is not equal to zero, these equations have a unique solution for every c.

In general, a linear operator $\mathbf{A}$ on a linear vector space is defined by the property that if x_1 and x_2 are vectors and c_1 and c_2 are scalars, then

$$\mathbf{A}(c_1x_1 + c_2x_2) = c_1\mathbf{A}x_1 + c_2\mathbf{A}x_2$$

Then if we know how $\mathbf{A}$ acts on a set of basis vectors, we can define the action of $\mathbf{A}$ on any linear combination of the basis. A linear operator is also known as a *linear transformation*.

Important characteristics of a linear operator are its *domain* (the space on which it is defined) and its *image*, or *range* (the space into which it maps its domain). The domain of a linear operator $\mathbf{A}$ on a finite-dimensional vector space $\mathcal{V}^n$ can always be extended to include the entire space, and $\mathbf{A}$ can then be represented by an $n \times n$ matrix. The theory of linear operators on $\mathcal{V}^n$ is thus equivalent to the theory of finite-dimensional matrices. Questions of domain and image are more subtle in infinite-dimensional spaces (see Chapter 7).

For a linear operator $\mathbf{A}$ on a finite-dimensional space we have the alternatives

(i) $\mathbf{A}$ is *nonsingular*: $\mathbf{A}$ defines a one-to-one map of $\mathcal{V}^n$ onto itself, and the equation $\mathbf{A}x = y$ has a unique solution $x = \mathbf{A}^{-1}y$, or

(ii) $\mathbf{A}$ is *singular*: the homogeneous equation $\mathbf{A}x = 0$ has a nontrivial solution, and $\mathcal{V}^n$ is mapped into a smaller subspace of itself by $\mathbf{A}$.

The first alternative requires that the determinant of the matrix representing $\mathbf{A}$ be nonzero; the second, that the determinant vanishes.

The homogeneous equation

$$\mathbf{A}x = \lambda x$$

has solutions only for discrete values of λ, the *eigenvalues* (*characteristic values*) of $\mathbf{A}$; the corresponding solutions are the *eigenvectors* of $\mathbf{A}$. The eigenvalues of a linear operator $\mathbf{A}$ define the *spectrum* of $\mathbf{A}$; in many applications, determining the spectrum of an operator is critical, as we shall see. In a finite-dimensional space $\mathcal{V}^n$, the eigenvalues are obtained by solving the *characteristic equation*

$$\det \|\mathbf{A} - \lambda\mathbf{1}\| \equiv p_{\mathbf{A}}(\lambda) = 0$$

in which $p_{\mathbf{A}}(\lambda)$ is a polynomial of degree n in λ, the *characteristic polynomial* of $\mathbf{A}$.

The *adjoint* $\mathbf{A}^\dagger$ of a linear operator $\mathbf{A}$ on $\mathcal{V}$ can be defined as a linear operator on the dual space $\mathcal{V}^*$. If $\mathcal{V}$ is a unitary vector space, then $\mathbf{A}^\dagger$ is a linear operator on $\mathcal{V}$ itself; then $\mathbf{A}$ is *self-adjoint*, or *Hermitian*, if $\mathbf{A}^\dagger = \mathbf{A}$. A *projection operator* projects vectors onto a linear subspace. *Unitary operators* transform an orthonormal basis into another orthonormal basis; the length of a vector and the scalar product between two vectors are unchanged by a unitary transformation. A unitary operator on a real vector space is simply a rotation, or orthogonal transformation. Explicit matrix representations are constructed for rotations in two and three dimensions.

Self-adjoint operators play a special role in quantum mechanics. States of a quantum mechanical system are represented as vectors in a unitary vector space $\mathcal{V}$, and the physical observables of the system by self-adjoint operators on $\mathcal{V}$. The eigenvalues of an operator correspond to the allowed values of the observables in states of the system, and the eigenvectors of the operator to states in which the observable has the corresponding eigenvalue. Especially important is the operator corresponding to the Hamiltonian of a system; its eigenvalues correspond to the allowed energy levels of the system.

A question of general interest is whether an operator has eigenvalues and eigenvectors. It turns out that a general linear operator $\mathbf{A}$ on the finite-dimensional space $\mathcal{V}^n$ has the form

$$\mathbf{A} = \mathbf{D} + \mathbf{N}$$

where $\mathbf{D}$ is a diagonalizable operator (one with a diagonal matrix representation in some basis) whose eigenvalues are the eigenvalues of $\mathbf{A}$, and $\mathbf{N}$ is a *nilpotent* operator ($\mathbf{N}^p = \mathbf{0}$ for some integer p) and $\mathbf{DN} = \mathbf{ND}$. The spectral properties of operators on an infinite-dimensional space are more complicated in general, and are discussed in Chapter 7.

An important class of linear operators on a unitary vector space $\mathcal{V}$ is *normal* operators; $\mathbf{A}$ is a normal operator if

$$\mathbf{A}^\dagger \mathbf{A} = \mathbf{A} \mathbf{A}^\dagger$$

In a finite-dimensional space, it is true that every normal operator has a complete orthonormal system of eigenvectors, which leads to the *spectral representation*

$$\mathbf{A} = \sum_k \lambda_k \mathbf{P}_k$$

where the $\{\lambda_k\}$ are the eigenvalues of $\mathbf{A}$ and the $\{\mathbf{P}_k\}$ are the projection operators onto orthogonal subspaces such that

$$\sum_k \mathbf{P}_k = \mathbf{1}$$

$\mathbf{A}$ is diagonal in the basis defined by the eigenvectors; hence $\mathbf{N} = \mathbf{0}$ for normal operators. If $\mathbf{A}$ is normal, and λ is an eigenvalue of $\mathbf{A}$, then λ^* is an eigenvalue of $\mathbf{A}^\dagger$. Self-adjoint operators are normal, with real eigenvalues; unitary operators are normal, with eigenvalues lying on the unit circle in the complex λ-plane.

The eigenvalues of a self-adjoint operator have minimax properties that lead to useful approximation methods for determining the spectra. These methods are based on the fundamental result: if $\mathbf{A}$ is a self-adjoint operator on the n-dimensional space $\mathcal{V}^n$, with eigenvalues ordered so that $\lambda_1 \leq \lambda_2 \leq \cdots \leq \lambda_n$, and if $\mathbf{A}'$ is the restriction of $\mathbf{A}$ to a subspace of dimension $m < n$, with eigenvalues $\lambda_1' \leq \lambda_2' \leq \cdots \leq \lambda_m'$ similarly ordered, then

$$\lambda_h \leq \lambda_h' \leq \lambda_{h+n-m}$$

($h = 1, \ldots, m$). Thus the eigenvalues of the restricted operator provide bounds on the eigenvalues of the full operator. The inequalities $\lambda_h \leq \lambda_h'$ remain true in an infinite-dimensional space for self-adjoint operators with a discrete spectrum.

Many quantum mechanics textbooks use thus result to derive a variational method for estimating the ground state energy of a system, which is lower than the lowest eigenvalue of any restriction of the Hamiltonian operator of the system to a subspace of the state space of the system. Beyond that, however, the nth lowest eigenvalue of the restricted Hamiltonian is an upper bound to the nth lowest eigenvalue of the full Hamiltonian, so the method provides an upper bound for the energies of excited states as well.

Functions of operators can be defined in terms of power series expansions, and by other methods when the spectrum of the operator is known. One example is the exponential function, which is used to provide a formal solution

$$x(t) = e^{t\mathbf{A}} x(0)$$

to the linear differential equation

$$\frac{dx}{dt} = \mathbf{A}x$$

with constant coefficients, starting from $x(0)$ at $t = 0$.

This solution is applied in Section 2.5 to the study of linear dynamical systems, which are described exactly by this equation, with time as the independent variable. The qualitative behavior of the system is determined by the spectrum of $\mathbf{A}$. Components of $x(0)$ along eigenvectors belonging to eigenvalues with negative real part decrease exponentially with increasing t, and are irrelevant to the behavior of the system at large time. Eigenvalues of $\mathbf{A}$ that are purely imaginary correspond to oscillatory modes of the system, with closed orbits and definite frequencies. If any eigenvalue of $\mathbf{A}$ has a positive real part, any component of the initial state $x(0)$ along the associated eigenvector grows exponentially in time, and the system is unstable.

Real physical systems are rarely exactly linear, but the behavior near an equilibrium point of the system is often well approximated by a linear model. For an energy-conserving system, this gives small oscillations with well-defined characteristic frequencies that are analyzed in Appendix A. Instabilities can be present in systems that do not conserve energy, but exponential growth is a signal that the linear theory is no longer valid. There is some discussion of nonlinear systems in Chapter 8 as well as in books cited in the bibliography.

2.1   Linear Vector Spaces

2.1.1   Linear Vector Space Axioms

A *linear vector space* V is a collection of vectors such that if x and y are vectors in V, then so is any linear combination of the form

$$u = ax + by \tag{2.1}$$

where a and b are numbers (scalars). V is *real* or *complex*, depending on whether scalars are taken from the real numbers $\mathbf{R}$ or the complex numbers $\mathbf{C}$ (in mathematics, scalars can be taken from any field, but here we are only concerned with real and complex vector spaces).

Addition of vectors, and multiplication by scalars must satisfy a natural set of axioms. Addition of vectors must be commutative:

$$x + y = y + x \tag{2.2}$$

for all x and y. Multiplication by scalars must satisfy the associative law:

$$a(bx) = (ab)x \tag{2.3}$$

and the distributive laws:

$$a(x + y) = ax + ay \qquad (a + b)x = ax + bx \tag{2.4}$$

for all a, b, x, and y. There must also be a *zero* vector (denoted by θ when it is necessary to distinguish it from the scalar 0) such that

$$x + \theta = x \tag{2.5}$$

for all x, and for every vector x there must be a vector $-x$ (*negative x*) such that

$$x + (-x) = \theta \tag{2.6}$$

→ **Exercise 2.1.**   Show that the axioms (2.2)–(2.6) imply that

$$1 \cdot x = x \qquad 0 \cdot x = \theta$$

for all x. *Hint.* Use the distributive laws.   □

Definition 2.1.   A vector of the form $a_1 x_1 + \ldots + a_k x_k$ is a *linear combination* of the vectors $x_1, \ldots, x_k$ (here the *coefficients* $a_1, \ldots, a_k$ are scalars). The vectors $x_1, \ldots, x_k$ are *linearly dependent* if there is a nontrivial linear combination of the vectors that vanishes, that is, if there are coefficients $a_1, \ldots, a_k$, not all zero, such that

$$a_1 x_1 + \ldots + a_k x_k = \theta \tag{2.7}$$

If no such coefficients exist, then the vectors $x_1, \ldots, x_k$ are *linearly independent*.   ∎

If the vectors $x_1, \ldots, x_k$ are linearly dependent, then we can express at least one of them as a linear combination of the others. For suppose we have a linear combination of the vectors that vanishes, as in Eq. (2.7), with at least one nonvanishing coefficient. Then we can take $a_1 \neq 0$ (renumbering the vectors if necessary) and solve for x_1 to get

$$x_1 = -\frac{a_2}{a_1} x_2 - \cdots - \frac{a_n}{a_1} x_n \tag{2.8}$$

Definition 2.2. A *linear manifold* $\mathcal{M}$ in the linear vector space $\mathcal{V}$ is a set of elements of $\mathcal{V}$ such that if x, y are in $\mathcal{M}$, then so is the linear combination $ax + by$ for any scalars a and b. ∎

Thus $\mathcal{M}$ is itself a linear vector space, a *subspace* of $\mathcal{V}$. If we have a set $x_1, \ldots, x_k$ of elements of $\mathcal{V}$, then we can define the linear manifold $\mathcal{M} \equiv \mathcal{M}(x_1, \ldots, x_k)$, the manifold *spanned* by $x_1, \ldots, x_k$, as the set of all linear combinations of $x_1, \ldots, x_k$.

Definition 2.3. The vectors $x_1, \ldots, x_k$ form a *basis* of the linear manifold $\mathcal{M}$ if (i) the vectors $x_1, \ldots, x_k$ are linearly independent *and* (ii) the manifold $\mathcal{M}$ is spanned by the $x_1, \ldots, x_k$. If $\mathcal{M}$ has a basis with k elements, then k is the *dimension* of $\mathcal{M}$. If $\mathcal{M}$ has no basis with a finite number of elements, then $\mathcal{M}$ is *infinite dimensional*. ∎

❑ **Example 2.1.** The set of continuous functions on an interval $a \leq x \leq b$ forms a linear vector space $C(a, b)$, with addition and multiplication by scalars defined in the natural way. The monomials $1, x, x^2, x^3, \ldots$ are linearly independent; hence $C(a, b)$ is infinite dimensional. ∎

The concept of dimension is important, the more so because it uniquely characterizes a linear vector space: if $x_1, \ldots, x_k$ and $y_1, \ldots, y_m$ are two bases of $\mathcal{M}$, then $k = m$. To show this, suppose that $m > k$. Then each of the vectors $y_1, \ldots, y_m$ can be expressed in terms of the basis $x_1, \ldots, x_k$ as

$$y_p = a_{p1}x_1 + \cdots + a_{pk}x_k \tag{2.9}$$

($p = 1, \ldots, m$). But the first k of these equations can be solved to express the $x_1, \ldots, x_k$ in terms of $y_1, \ldots, y_k$ (if the determinant of the coefficients were zero, then a linear combination of $y_1, \ldots, y_k$ would be zero), and the remaining $y_{k+1}, \ldots, y_m$ can be expressed in terms of the $x_1, \ldots, x_k$. Since this is inconsistent with the assumption that the $y_1, \ldots, y_m$ are linearly independent, we must have $m = k$.

It follows that if we have an n-dimensional vector space $\mathcal{V}^n$, then any set of n linearly independent vectors $x_1, \ldots, x_n$ forms a basis of $\mathcal{V}^n$, and any vector x in $\mathcal{V}^n$ can be expressed as a (unique) linear combination

$$x = a_1x_1 + \ldots + a_nx_n \tag{2.10}$$

The coefficients $a_1, \ldots, a_n$ in this expansion are the *components* of the vector x (with respect to the basis $x_1, \ldots, x_n$); we have a one-to-one correspondence between vectors and ordered n-tuples $(a_1, \ldots, a_n)$ of scalars. Evidently this correspondence depends on the basis chosen; the components of a vector depend on the coordinate system.

2.1.2 Vector Norm; Scalar Product

We introduce the *length*, or *norm* of a vector x, to be denoted by $\|x\|$. The norm should be positive definite, that is, $\|x\| \geq 0$ for any vector x, and $\|x\| = 0$ if and only if $x = \theta$. Once length is defined, the *distance* between two vectors x and y is defined by $\|x - y\|$. This distance should satisfy as an axiom the *triangle inequality*:

$$\left| \|x\| - \|y\| \right| \leq \|x - y\| \leq \|x\| + \|y\| \tag{2.11}$$

with equality if and only if $y = ax$ for some real a; these inequalities are satisfied by the sides of any triangle in the Euclidean plane. With this norm, definitions of open, closed, and bounded sets, and limit points of sets and sequences in a vector space can be made by analogy with the definitions in Chapter 1 for real and complex numbers.

A *scalar product* of two vectors x, y must satisfy axioms of *Hermitian symmetry*:

$$(y, x) = (x, y)^* \tag{2.12}$$

and *bilinearity:*

$$(y, a_1 x_1 + a_2 x_2) = a_1(y, x_1) + a_2(y, x_2) \tag{2.13}$$

$$(b_1 y_1 + b_2 y_2, x) = b_1^*(y_1, x) + b_2^*(y_2, x) \tag{2.14}$$

If such a scalar product exists, then a standard norm is defined by

$$\|x\|^2 \equiv (x, x) \tag{2.15}$$

A vector space with scalar product satisfying axioms (2.12)–(2.14) is *unitary*. These axioms correspond to standard properties of three-dimensional vectors, apart from the presence of the complex conjugate. The complex conjugate is needed to make the norm positive definite in a complex vector space, since if z is a complex number, then $|z|^2 = z^*z$ is positive definite, while z^2 is not.

Remark. We are concerned here mainly with unitary spaces, but there are spaces with positive definite norm that satisfy the triangle inequality with no scalar product. For example, the length of a vector $x = \{\xi_1, \ldots, \xi_n\}$ can be defined for any $p \geq 1$ by

$$\|x\| \equiv \left(\sum_{k=1}^{n} |\xi_k|^p \right)^{\frac{1}{p}} \tag{2.16}$$

(Check that this satisfies the triangle inequality.) This space is denoted by ℓ^p; only for $p = 2$ can a scalar product satisfying the axioms be defined on ℓ^p.

Remark. A physically interesting space with a length defined that is *not* positive definite is the four-dimensional spacetime in which we live, with its standard (Minkowski) metric. Spacelike and timelike vectors have norm-squared of opposite sign, and there are nonzero vectors with zero norm (lightlike vectors). Spaces with such indefinite metrics will be discussed later. □

Definition 2.4. The vectors x and y in a unitary vector space are *orthogonal* if and only if

$$(x, y) = 0 \tag{2.17}$$

The vectors $x_1, \ldots, x_m$ form an *orthogonal system* if

$$(x_k, x_l) = 0 \tag{2.18}$$

for all $k \neq l$. The vector ϕ is a *unit vector* if $\|\phi\| = 1$. The vectors $\phi_1, \ldots, \phi_m$ form an *orthonormal system* if

$$(\phi_k, \phi_l) = \delta_{kl} = \begin{cases} 1, & k = l \\ 0, & k \neq l \end{cases} \tag{2.19}$$

The vectors in an orthonormal system are mutually orthogonal unit vectors; they are especially convenient to use as a basis for a coordinate system. If the vectors $\phi_1, \ldots, \phi_m$ form an orthonormal system, then the components $a_1, \ldots, a_m$ of a vector

$$x = a_1\phi_1 + \ldots + a_m\phi_m \tag{2.20}$$

are given simply by $a_k = (\phi_k, x)$, and we can write

$$x = (\phi_1, x)\phi_1 + \cdots + (\phi_m, x)\phi_m \tag{2.21}$$

It is clear from this that the vectors in an orthonormal system are linearly independent, since the only linear combination of $\phi_1, \ldots, \phi_m$ that vanishes necessarily has vanishing coefficients. Even if x cannot be expressed as a linear combination of $\phi_1, \ldots, \phi_m$, it is still true that

$$\sum_{k=1}^{m} |(\phi_k, x)|^2 \leq \|x\|^2 \tag{2.22}$$

(*Bessel's inequality*), since it follows from the positivity condition that

$$\| x - \sum_{k=1}^{m} (\phi_k, x)\phi_k \|^2 = \|x\|^2 - \sum_{k=1}^{m} |(\phi_k, x)|^2 \geq 0 \tag{2.23}$$

Note that the equality holds if and only if

$$x = \sum_{k=1}^{m} (\phi_k, x)\phi_k \tag{2.24}$$

Bessel's inequality leads directly to the *Schwarz inequality*: if x, y are any two vectors, then

$$|(x, y)| \leq \|x\| \, \|y\| \tag{2.25}$$

with equality if and only if y is a scalar multiple of x. To show this, simply consider the unit vector $\phi \equiv y/\|y\|$ and apply Bessel's inequality.

Remark. The Schwarz inequality corresponds to the bound $|\cos\vartheta| \leq 1$ for real angles ϑ, since for real vectors it is generally true that

$$(x, y) = \|x\| \, \|y\| \, \cos\vartheta \tag{2.26}$$

where ϑ is the angle between x and y. □

An orthonormal system $\phi_1, \ldots, \phi_m$ can be constructed from linearly independent vectors $x_1, \ldots, x_m$ by a standard method known as the *Gram–Schmidt orthogonalization process*: let

$$\phi_1 = \frac{x_1}{\|x_1\|} \tag{2.27}$$

and then let

$$\psi_k \equiv x_k - \sum_{p=1}^{k-1} (\phi_p, x_k)\phi_p \qquad \phi_k \equiv \frac{\psi_k}{\|\psi_k\|} \tag{2.28}$$

$(k = 2, \ldots, m)$. It is easy to see that the $\phi_1, \ldots, \phi_m$ form an orthonormal system, since none of the ψ_k can vanish due to the assumed linear independence of the x_k. Furthermore, the linear manifolds defined are the same at each stage of the process, that is,

$$
\begin{aligned}
\mathcal{M}(\phi_1) &= \mathcal{M}(x_1) \\
\mathcal{M}(\phi_1, \phi_2) &= \mathcal{M}(x_1, x_2) \\
&\vdots \\
\mathcal{M}(\phi_1, \phi_2, \ldots, \phi_m) &= \mathcal{M}(x_1, x_2, \ldots, x_m)
\end{aligned}
\tag{2.29}
$$

We can now introduce the concept of a *complete orthonormal system*.

Definition 2.5. The orthonormal system $\phi_1, \phi_2, \ldots$ is *complete* if and only if the only vector orthogonal to every ϕ_k is the zero vector; i.e., $(\phi_k, x) = 0$ for all k if and only if $x = 0$. ∎

This definition has been carefully stated to remain valid in an infinite-dimensional space, though the existence of a basis in the infinite-dimensional case requires an additional axiom. If a basis exists, as it must in a finite-dimensional space, then it can be made into a complete orthonormal system by the Gram–Schmidt process. Properties of a vector can be expressed in terms of its components with respect to this complete orthonormal system in a standard way.

Theorem 2.1. The orthonormal system $\phi_1, \ldots, \phi_m$ in the n-dimensional linear vector space $\mathcal{V}^n$ is complete if and only if

(i) $m = n$,

(ii) $\phi_1, \ldots, \phi_m$ is a basis of $\mathcal{V}^n$,

(iii) for every vector x in $\mathcal{V}^n$, we have

$$\|x\|^2 = \sum_{k=1}^m |(\phi_k, x)|^2 \tag{2.30}$$

(iv) for every pair of vectors x and y in $\mathcal{V}^n$, we have

$$(x, y) = \sum_{k=1}^m (x, \phi_k)(\phi_k, y) \tag{2.31}$$

Any one of these conditions is necessary and sufficient for all of them. The conditions not already been dealt with explicitly can be verified from expansion (2.21).

❑ **Example 2.2.** The space $\mathbf{R}^n$ of ordered n-tuples of real numbers of the form $(x_1, \ldots, x_n)$ is a real n-dimensional vector space. Addition and multiplication by scalars are defined in the obvious way: if $x = (x_1, \ldots, x_n)$ and $y = (y_1, \ldots, y_n)$, then

$$ax + by = (ax_1 + by_1, \ldots, ax_n + by_n) \tag{2.32}$$

The scalar product can be defined by

$$(x, y) = \sum_{k=1}^{n} x_k y_k \tag{2.33}$$

(this definition is not unique—see Exercise 2.2). The vectors

$$
\begin{aligned}
\phi_1 &= (1, 0, \ldots, 0) \\
\phi_2 &= (0, 1, \ldots, 0) \\
&\;\;\vdots \\
\phi_n &= (0, 0, \ldots, 1)
\end{aligned}
\tag{2.34}
$$

define a complete orthonormal system, and the x_k and y_k are just the components of x and y with respect to this complete orthonormal system. ∎

❑ **Example 2.3.** The space $\mathbf{C}^n$ of ordered n-tuples of complex numbers of the form $(\xi_1, \ldots, \xi_n)$ is an n-dimensional complex vector space. Addition and multiplication by scalars are defined in the obvious way: if $x = (\xi_1, \ldots, \xi_n)$ and $y = (\eta_1, \ldots, \eta_n)$, then

$$ax + by = (a\xi_1 + b\eta_1, \ldots, a\xi_n + b\eta_n) \tag{2.35}$$

The scalar product can be defined by

$$(x, y) = \sum_{k=1}^{n} \xi_k^* \eta_k \tag{2.36}$$

Note the presence of the complex conjugate of the components of the vector on the left side of the scalar product; this is necessary to ensure positive definiteness of the norm, which is here given by

$$\|x\|^2 = \sum_{k=1}^{n} |\xi_k|^2 \tag{2.37}$$

The complete orthonormal system $\phi_1, \phi_2, \ldots, \phi_n$ defined in the previous example is also a complete orthonormal system in $\mathbf{C}^n$. In $\mathbf{C}^n$, however, the scalars (and hence the components) are complex. ∎

In a sense, the preceding two examples are the only possibilities for real or complex n-dimensional vector spaces. If there is a one-to-one map between two linear vector spaces $\mathcal{U}$ and $\mathcal{V}$ such that if $u_1 \leftrightarrow v_1$ and $u_2 \leftrightarrow v_2$ under the map, then

$$au_1 + bu_2 \leftrightarrow av_1 + bv_2 \tag{2.38}$$

for any scalars a and b ($\mathcal{U}$ and $\mathcal{V}$ must have the same scalars), then $\mathcal{U}$ and $\mathcal{V}$ are isomorphic, and the corresponding map is an *isomorphism*. If in addition, the vector spaces are unitary and

$$(u_1, u_2) = (v_1, v_2) \tag{2.39}$$

whenever $u_1 \leftrightarrow v_1$ and $u_2 \leftrightarrow v_2$ under the map, then the map is an *isometry*, and the spaces are *isometric*. It is clear that any n-dimensional real vector space is isomorphic to $\mathbf{R}^n$, and any n-dimensional complex vector space is isomorphic to $\mathbf{C}^n$. If we identify a vector in the n-dimensional vector space $\mathcal{V}^n$ with the n-tuple defined by its components with respect to some fixed basis, then we have an isomorphism between $\mathcal{V}^n$ and $\mathbf{R}^n$ (or between $\mathcal{V}^n$ and $\mathbf{C}^n$ if $\mathcal{V}^n$ is complex). If $\mathcal{V}^n$ is unitary, then the isomorphism can be extended to an isometry.

→ **Exercise 2.2.** Consider the linear vector space $\mathbf{C}^n$ with the scalar product of $x = (\xi_1, \dots, \xi_n)$ and $y = (\eta_1, \dots, \eta_n)$, defined by

$$(x, y) = \sum_{k=1}^{n} \xi_k^* w_k \eta_k$$

What restrictions must be put on the w_k in order that $\mathbf{C}^n$, with this scalar product, be a unitary vector space? If these restrictions are satisfied, denote the corresponding unitary vector space by $\mathbf{C}^n(w_1, \dots, w_n)$. Construct an explicit map between $\mathbf{C}^n(w_1, \dots, w_n)$ and the standard $\mathbf{C}^n$ that is both an isomorphism and an isometry. □

2.1.3 Sum and Product Spaces

We can join two linear vector spaces $\mathcal{U}$ and $\mathcal{V}$ together to make a new vector space $\mathcal{W} \equiv \mathcal{U} \oplus \mathcal{V}$, the *direct sum* of $\mathcal{U}$ and $\mathcal{V}$. Take ordered pairs $\langle u, v \rangle$ of vectors with u from $\mathcal{U}$ and v from $\mathcal{V}$, and define

$$a \langle u_1, v_1 \rangle + b \langle u_2, v_2 \rangle \equiv \langle au_1 + bu_2, av_1 + bv_2 \rangle \tag{2.40}$$

This defines $\mathcal{U} \oplus \mathcal{V}$ as a linear vector space. If $\mathcal{U}$ and $\mathcal{V}$ are unitary, define the scalar product in $\mathcal{U} \oplus \mathcal{V}$ by

$$(\langle u_1, v_1 \rangle, \langle u_2, v_2 \rangle) \equiv (u_1, u_2) + (v_1, v_2) \tag{2.41}$$

where the first scalar product on the right-hand side is taken in $\mathcal{U}$ and the second is taken in $\mathcal{V}$. This defines $\mathcal{U} \oplus \mathcal{V}$ as a unitary vector space.

→ **Exercise 2.3.** Let $\mathcal{U}$ and $\mathcal{V}$ be a linear vector spaces of dimensions m and n, respectively. Show that the dimension of the direct sum $\mathcal{U} \oplus \mathcal{V}$ is given by $\dim(\mathcal{U} \oplus \mathcal{V}) = m + n$. □

→ **Exercise 2.4.** If $\mathcal{U}$ and $\mathcal{V}$ are linear vector spaces with the same scalars, then the *tensor product* $W \equiv \mathcal{U} \otimes \mathcal{V}$ is defined as follows: start with ordered pairs $\langle u, v \rangle \equiv u \otimes v$ of vectors u from $\mathcal{U}$ and v from $\mathcal{V}$, and include all linear combinations of the form

$$\sum_{k,l} a_{kl} u_k \otimes v_l \qquad (*)$$

with the $\{u_k\}$ from $\mathcal{U}$ and the $\{v_l\}$ from $\mathcal{V}$. If $\mathcal{U}$, $\mathcal{V}$ are unitary, define a scalar product on $\mathcal{U} \otimes \mathcal{V}$ by

$$(u_1 \otimes v_1, u_2 \otimes v_2) \equiv (u_1, u_2)(v_1, v_2)$$

and by bilinearity for linear combinations of the type $(*)$.

(i) Show that if $\phi_1, \phi_2, \ldots$ is an orthonormal system in $\mathcal{U}$ and $\psi_1, \psi_2, \ldots$ is an orthonormal system in $\mathcal{V}$, then the vectors

$$\Phi_{kl} \equiv \phi_k \otimes \psi_l$$

form an orthonormal system in $\mathcal{U} \otimes \mathcal{V}$. The $\{\Phi_{kl}\}$ form a complete orthonormal system in $\mathcal{U} \otimes \mathcal{V}$ if and only if the $\{\phi_k\}$ are complete in $\mathcal{U}$ and the $\{\psi_l\}$ are complete in $\mathcal{V}$.

(ii) Show that if $\dim \mathcal{U} = m$ and $\dim \mathcal{V} = n$, then $\dim(\mathcal{U} \otimes \mathcal{V}) = mn$.

(iii) Does every vector in $\mathcal{U} \otimes \mathcal{V}$ have the form $u \otimes v$ with u from $\mathcal{U}$ and v from $\mathcal{V}$? □

If $\mathcal{M}$ is a linear manifold in the unitary vector space $\mathcal{V}$, the *orthogonal complement* of $\mathcal{M}$, denoted by $\mathcal{M}^\perp$, is the set of all vectors x in $\mathcal{V}$ such that $(y, x) = 0$ for all y in $\mathcal{M}$. Thus $\mathcal{M}^\perp$ contains the vectors that are orthogonal to every vector in $\mathcal{M}$. If $\mathcal{M}$ is finite dimensional, then every vector x in $\mathcal{V}$ can be decomposed according to

$$x = x' + x'' \qquad (2.42)$$

where x' is in $\mathcal{M}$ (x' is the *projection* of x onto $\mathcal{M}$), and x'' is in $\mathcal{M}^\perp$. This result is known as the *projection theorem*; it is equivalent to expressing $\mathcal{V}$ as the direct sum

$$\mathcal{V} = \mathcal{M} \oplus \mathcal{M}^\perp \qquad (2.43)$$

To prove the theorem, let $\phi_1, \ldots, \phi_m$ be an orthonormal basis of $\mathcal{M}$ (such a basis exists since $\mathcal{M}$ is finite dimensional). Then define

$$x' \equiv \sum_{k=1}^{m} (\phi_k, x)\phi_k \qquad x'' \equiv x - x' \qquad (2.44)$$

This provides the required decomposition.

→ **Exercise 2.5.** Let $\mathcal{M}$ be a finite-dimensional linear manifold in the unitary vector space ($\mathcal{M}$ need not be finite dimensional here). Then $(\mathcal{M}^\perp)^\perp = \mathcal{M}$. □

2.1.4 Sequences of Vectors

We are also interested in sequences of vectors and their convergence properties. In a finite-dimensional vector space $\mathcal{V}^n$ with a norm, these properties are directly related to those of sequences in the field of scalars. We define a *Cauchy sequence* to be a sequence $\{x_1, x_2, \ldots\}$ of vectors such that for every $\varepsilon > 0$ there is an integer N such that $\|x_p - x_q\| < \varepsilon$ whenever $p, q > N$. The sequence $\{x_1, x_2, \ldots\}$ *converges* if there is a vector x such that for every $\varepsilon > 0$ there is an integer N such that $\|x_p - x\| < \varepsilon$ whenever $p > N$. These definitions parallel the definitions for sequences of real and complex numbers given in Chapter 1.

If we represent a vector in $\mathcal{V}^n$ by its components $\xi_1, \ldots, \xi_n$ with respect to some basis, then the sequence $\{x_1, x_2, \ldots\}$ of vectors is a Cauchy sequence if and only if each of the sequences $\{\xi_{1k}\}, \{\xi_{2k}\}, \ldots, \{\xi_{nk}\}$ is a Cauchy sequence. Here ξ_{pk} denotes the pth component of the vector x_k. To see this in a unitary space with a complete orthonormal system given by $\phi_1, \ldots, \phi_n$, suppose

$$x_k = \sum_{p=1}^{n} \xi_{kp} \phi_p \tag{2.45}$$

Then, since

$$\|x_k - x_m\|^2 = \sum_{p=1}^{n} |\xi_{kp} - \xi_{mp}|^2 \tag{2.46}$$

it follows that the sequence of vectors is a Cauchy sequence if and only if each of the sequences of components is a Cauchy sequence. In $\mathcal{V}^n$, every Cauchy sequence of vectors converges to a limit vector because the corresponding Cauchy sequences of the components must converge (recall that this is how the real and complex number systems were constructed from the rationals).

It is not so simple in an infinite-dimensional space; a sequence $\{\phi_1, \phi_2, \ldots\}$ of orthonormal vectors is not a Cauchy sequence, yet each sequence of components of these vectors converges to zero, which leads to the concept of weak convergence in an infinite-dimensional space. This and other subtleties will be discussed in Chapter 6.

2.1.5 Linear Functionals and Dual Spaces

Definition 2.6. A function defined on a linear vector space $\mathcal{V}$ that takes on scalar values is called a *functional*, or *distribution*, on $\mathcal{V}$. The functional Λ is a *linear functional* if

$$\Lambda[ax + by] = a\Lambda[x] + b\Lambda[y] \tag{2.47}$$

for all vectors x, y and scalars a, b. The functional Λ is *bounded* if there is some positive constant C such that

$$|\Lambda[x]| \leq C\|x\| \tag{2.48}$$

for every vector x in $\mathcal{V}$. The smallest C for which this is true is the *bound* $|\Lambda|$ of Λ. ∎

Theorem 2.2. If Λ is a linear functional on the finite-dimensional unitary vector space $\mathcal{V}^n$, then there is a vector u in $\mathcal{V}^n$ such that Λ has the form of a scalar product

$$\Lambda[x] = (u, x) \equiv \Lambda_u[x] \tag{2.49}$$

Proof. If $\{\phi_1, \ldots, \phi_n\}$ is a complete orthonormal system in $\mathcal{V}^n$ and $\Lambda[\phi_k] = a_k$, then

$$\Lambda\phi = \Lambda \sum_k \xi_k \phi_k = \sum_k a_k \xi_k = (u, \phi) \tag{2.50}$$

with (note the complex conjugation here)

$$u = \sum_k a_k^* \phi_k \tag{2.51}$$

It follows that every linear functional on $\mathcal{V}^n$ is bounded, with $|\Lambda_u| = \|u\|$, since

$$|(u, x)| \leq \|u\| \|x\| \tag{2.52}$$

for all x by the Schwarz inequality (2.25). This need not be true in an infinite-dimensional space. For example, the Dirac δ-function $\delta(x)$ introduced in Section 1.3 is a linear functional defined on the space $\mathcal{C}^1$ of functions $f(x)$ that are continuous at $x = 0$; we have

$$\delta[f] = f(0) \tag{2.53}$$

This is not of the form (u, f) with u in $\mathcal{C}^1$. However, it is not required to have that form, as a function with unit norm can have an arbitrarily large value at a single point. To see this, consider the sequence of functions $\{\phi_n\}$ defined by

$$\phi_n(x) = \begin{cases} \sqrt{\dfrac{n}{2}} & -\dfrac{1}{n} \leq x \leq \dfrac{1}{n} \\ 0 & \text{otherwise} \end{cases} \tag{2.54}$$

Then $\|\phi_n\| = 1$ and $\delta[\phi_n] = \sqrt{n/2}$. Hence the Dirac δ-function is not bounded.

The linear functionals on $\mathcal{V}$ form a linear vector space $\mathcal{V}^*$, the *dual space* of $\mathcal{V}$. If $x_1, x_2, \ldots$ is a basis of $\mathcal{V}$, then the *dual basis* $u_1, u_2, \ldots$ of $\mathcal{V}^*$ is defined by the relations

$$u_k[x_\ell] = \delta_{k\ell} \tag{2.55}$$

Equation (2.55) resembles the relations that define an orthonormal system, except that here the $u_1, u_2, \ldots$ need not belong to the same vector space as the $x_1, x_2, \ldots$. In a unitary space $\mathcal{V}$, relations (2.55) define a dual basis even if the original basis $x_1, x_2, \ldots$ is not orthogonal. A simple two-dimensional illustration of this is given in Fig. 2.1, in which we start from a nonorthogonal pair of $\mathbf{X}$ and $\mathbf{Y}$ axes, and construct the $\mathbf{X}^*$-axis orthogonal to the $\mathbf{Y}$-axis and the $\mathbf{Y}^*$-axis orthogonal to the $\mathbf{X}$-axis. Nevertheless, the general theorem provides a natural one-to-one correspondence between bounded linear functionals in $\mathcal{V}^*$ and vectors in $\mathcal{V}$. This relation is not quite linear in a complex vector space, since Eq. (2.51) shows that

$$\Lambda_{au+bv} = a^* \Lambda_u + b^* \Lambda_v, \tag{2.56}$$

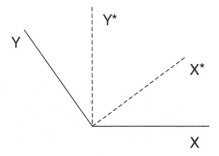

Figure 2.1: An example of a dual basis. If unit vectors in the **X** and **Y** directions define a basis in $\mathbf{R}^2$, then unit vectors in the $\mathbf{X}^*$ and $\mathbf{Y}^*$ directions define the dual basis.

A relation of this type is called *conjugate linear*, or *antilinear*; here it means that we can think of V^* as a "complex conjugate" of V in some sense. A notation due to Dirac emphasizes the duality between vectors and linear functionals. A vector is denoted by $|x\rangle$ (a *ket*) and a linear functional by $\langle u|$ (a *bra*). Then the value of the linear functional u at the vector x is denoted by

$$u(x) = \langle u, x \rangle = \langle u|x \rangle \tag{2.57}$$

where the last form on the right-hand side is known as a *Dirac bra(c)ket*. In a unitary vector space, the Dirac bracket is equivalent to the scalar product with the natural identification between linear functionals and vectors. The relations

$$\langle u, ax + by \rangle = a\langle u, x \rangle + b\langle u, y \rangle \tag{2.58}$$

$$\langle au + bv, x \rangle = a^*\langle u, x \rangle + b^*\langle v, x \rangle \tag{2.59}$$

are equivalent to the bilinearity of the scalar product in Eqs. (2.13)–(2.14).

2.2 Linear Operators

2.2.1 Linear Operators; Domain and Image; Bounded Operators

A *linear operator* **A** on a linear vector space V is a map of V into itself that satisfies

$$\mathbf{A}(ax + by) = a\mathbf{A}x + b\mathbf{A}y \tag{2.60}$$

for all vectors x, y and scalars a, b. The operator **A** need not be defined on all of V, but it must be defined on a linear manifold $\mathcal{D}_\mathbf{A}$, the *domain* of **A**. The domain can always be extended to the entire space in a finite-dimensional space V^n, but not so in an infinite-dimensional Hilbert space, as explained in Chapter 7. In this chapter, we deal mainly with properties of linear operators on finite-dimensional vector spaces.

The set of all vectors y such that $y = \mathbf{A}x$ for some x in the domain of $\mathbf{A}$ is the *image* of $\mathbf{A}$ (im $\mathbf{A}$)[1] or the *range* of $\mathbf{A}$ ($\mathcal{R}_\mathbf{A}$), as indicated schematically in Fig. 2.2. It is easy to see that im $\mathbf{A}$ is a linear manifold; its dimension is the *rank* $\rho(\mathbf{A})$ of $\mathbf{A}$. The set of vectors x such that

$$\mathbf{A}x = \theta \tag{2.61}$$

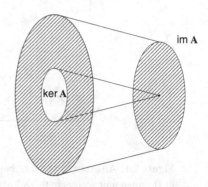

defines a linear manifold ker $\mathbf{A}$, the *kernel* (or *null space*) of $\mathbf{A}$. Figure 2.2 is a schematic illustration of ker $\mathbf{A}$ and im $\mathbf{A}$. If the only solution to Eq. (2.61) is the zero vector (ker $\mathbf{A} = \{\theta\}$), then the operator $\mathbf{A}$ is *nonsingular*. If there are nonzero solutions, then $\mathbf{A}$ is *singular*. In a vector space of finite dimension n, we have

Figure 2.2:   Schematic illustration of the kernel and image of an operator $\mathbf{A}$. The kernel (ker $\mathbf{A}$) is the manifold mapped into the zero vector by $\mathbf{A}$, while the image (im $\mathbf{A}$) is the map of the entire space $\mathcal{V}$ under $\mathbf{A}$.

$$\rho(\mathbf{A}) + \dim(\ker \mathbf{A}) = n \tag{2.62}$$

Equation (2.62) says that the dimension of im $\mathbf{A}$ is the original dimension of the space reduced by the dimension of the subspace which $\mathbf{A}$ *annihilates* (transforms into the zero vector).

To prove this intuitively plausible relation, let $m = \dim(\text{im }\mathbf{A})$ and let $y_1, \dots, y_m$ be a basis of im $\mathbf{A}$, with

$$y_k = \mathbf{A}x_k \tag{2.63}$$

($k = 1, \dots, m$). The vectors $x_1, \dots, x_m$ are linearly independent; if $\sum c_k x_k = \theta$ then $\sum c_k y_k = \theta$ by linearity, but $y_1, \dots, y_m$ are linearly independent so all the c_k must vanish. We can then choose $x_{m+1}, \dots, x_n$ to complete a basis of $\mathcal{V}^n$. Since the $y_1, \dots, y_m$ form a basis of im $\mathbf{A}$, the action of $\mathbf{A}$ on the $x_{m+1}, \dots, x_n$ can be expressed as

$$\mathbf{A}x_k = \sum_{\ell=1}^{m} c_\ell^{(k)} y_\ell \tag{2.64}$$

($k = m+1, \dots, n$). Now define vectors $z_{m+1}, \dots, z_n$ by

$$z_k \equiv x_k - \sum_{\ell=1}^{m} c_\ell^{(k)} x_\ell \tag{2.65}$$

The $z_{m+1}, \dots, z_n$ are linearly independent (show this) and $\mathbf{A}z_k = \theta$ ($k = m+1, \dots, n$). Hence they define a basis of ker $\mathbf{A}$ (ker $\mathbf{A} = \mathcal{M}(z_{m+1}, \dots, z_n)$) and $\mathcal{V}^n$ can be expressed as

$$\mathcal{V}^n = \mathcal{M}(x_1, \dots, x_m) \oplus \ker \mathbf{A} \tag{2.66}$$

from which Eq. (2.62) follows.

---

[1]More precisely, the image of $\mathcal{D}_\mathbf{A}$ under $\mathbf{A}$, but the shorter terminology has become standard.

The inhomogeneous equation

$$\mathbf{A}x = y \tag{2.67}$$

can have a solution only if y is in im $\mathbf{A}$. If $\mathbf{A}$ is nonsingular, then im $\mathbf{A}$ is the entire vector space $\mathcal{V}$, and Eq. (2.67) has a unique solution for every y in $\mathcal{V}$. If $\mathbf{A}$ is singular, then Eq. (2.67) has a solution only if y is in im $\mathbf{A}$, and the solution is not unique, since any solution of the homogeneous equation (2.61) (i.e., any vector in ker $\mathbf{A}$) can be added to a solution of Eq. (2.67) to obtain a new solution. If Eq. (2.67) has a unique solution for every x in $\mathcal{V}$, then define the operator $\mathbf{A}^{-1}$, the *inverse* of $\mathbf{A}$, by

$$\mathbf{A}^{-1}y = x \tag{2.68}$$

➜ **Exercise 2.6.** $\mathbf{A}$ and $\mathbf{B}$ are linear operators on the finite-dimensional vector space $\mathcal{V}^n$.
   (i) The product $\mathbf{AB}$ is nonsingular if and only if both $\mathbf{A}$ and $\mathbf{B}$ are nonsingular.
   (ii) If $\mathbf{A}$ and $\mathbf{B}$ are nonsingular, then

$$(\mathbf{AB})^{-1} = \mathbf{B}^{-1}\mathbf{A}^{-1}$$

   (iii) If $\mathbf{A}$ is nonsingular and $\mathbf{BA} = 1$, then $\mathbf{AB} = 1$.   □

Remark.   Statement (iii) means that if $\mathbf{B}$ is a left inverse of $\mathbf{A}$, then $\mathbf{B}$ is also a right inverse. This is true only in a finite-dimensional space; a counterexample in infinite dimensions can be found in Chapter 7.   □

Definition 2.7.   The linear operator $\mathbf{A}$ is *bounded* if there is a constant $C > 0$ such that $\|\mathbf{A}x\| \le C\|x\|$ for every vector x. The smallest C for which this inequality is true is the *bound* of $\mathbf{A}$, denoted by $|\mathbf{A}|$.   ∎

Theorem 2.3.   Every linear operator on a finite-dimensional unitary vector space $\mathcal{V}^n$ is bounded.

Proof.   To show this, suppose $\phi_1, \ldots, \phi_n$ is a complete orthonormal system and let

$$\phi = \sum_{k=1}^{n} \xi_k \phi_k \tag{2.69}$$

be a unit vector. Then

$$\mathbf{A}\phi = \sum_{k=1}^{n} \xi_k \mathbf{A}\phi_k \tag{2.70}$$

and

$$\|\mathbf{A}\phi\|^2 \le \left(\sum_{k=1}^{n} |\xi_k| \, \|\mathbf{A}\phi_k\| \right)^2 \le \sum_{k=1}^{n} |\xi_k|^2 \sum_{k=1}^{n} \|\mathbf{A}\phi_k\|^2 \le C^2\|\phi\|^2 \tag{2.71}$$

where

$$C^2 \equiv \sum_{k=1}^{n} \|\mathbf{A}\phi_k\|^2 \tag{2.72}$$

Remark.   The theorem is even if the space $\mathcal{V}^n$ is not unitary, but the proof is slightly more subtle, and we do not give it here.   □

2.2.2   Matrix Representation; Multiplication of Linear Operators

Linear operators can be given an explicit realization in terms of matrices. If the linear operator $\mathbf{A}$ is defined on a basis $x_1, x_2, \ldots$ of $\mathcal{V}$, then we have

$$\mathbf{A}x_k = \sum_j A_{jk}x_j \tag{2.73}$$

and the action of $\mathbf{A}$ on a vector $x = \sum_k \xi_k x_k$ is given by

$$\mathbf{A}x = \sum_j \left(\sum_k A_{jk}\xi_k \right) x_j \equiv \sum_j \eta_j x_j \tag{2.74}$$

This corresponds to the standard matrix multiplication rule

$$\begin{pmatrix} \eta_1 \\ \eta_2 \\ \vdots \end{pmatrix} = \begin{pmatrix} A_{11} & A_{12} & \cdots \\ A_{21} & A_{22} & \cdots \\ \vdots & \vdots & \ddots \end{pmatrix} \begin{pmatrix} \xi_1 \\ \xi_2 \\ \vdots \end{pmatrix} \tag{2.75}$$

if we identify x with the column vector (ξ_k) formed by its components and $\mathbf{A}$ with the matrix (A_{jk}). The A_{jk} are the *matrix elements* of $\mathbf{A}$ in the basis $x_1, x_2, \ldots$. The kth column of the matrix contains the components of the vector $\mathbf{A}x_k$.

→ **Exercise 2.7.**   Show that if $\mathcal{V}$ is a unitary vector space with complex scalars, then the operator $\mathbf{A} = \mathbf{0}$ if and only if $(x, \mathbf{A}x) = 0$ for every vector x in $\mathcal{V}$. Show that the condition $(x, \mathbf{A}x) = 0$ for every vector x in $\mathcal{V}$ is not sufficient to have $\mathbf{A} = \mathbf{0}$ in a real vector space $\mathcal{V}$ by constructing a nonsingular linear operator $\mathbf{A}$ on $\mathbf{R}^2$ such that

$$(x, \mathbf{A}x) = 0$$

for every x in $\mathbf{R}^2$.   □

Remark.   The theme of this exercise will reappear in several contexts, as certain algebraic properties are relevant only in complex vector spaces.   □

The *product* of two linear operators $\mathbf{A}$ and $\mathbf{B}$ is defined in a natural way as

$$(\mathbf{AB})x = \mathbf{A}(\mathbf{B}x) \tag{2.76}$$

Acting on a basis $x_1, x_2, \ldots$ this gives

$$(\mathbf{AB})x_k = \mathbf{A}\sum_j B_{jk}x_j = \sum_{j,\ell} A_{\ell j}B_{jk}x_j \tag{2.77}$$

so that

$$(\mathbf{AB})_{\ell k} = \sum_j A_{\ell j}B_{jk} \tag{2.78}$$

This explains the standard rule for matrix multiplication.

The order of the operators in the product is important; multiplication of operators (or their corresponding matrices) is not commutative in general.

Definition 2.8.   The *commutator* $[\mathbf{A}, \mathbf{B}]$ of two operators $\mathbf{A}$ and $\mathbf{B}$ is defined by

$$[\mathbf{A}, \mathbf{B}] \equiv \mathbf{AB} - \mathbf{BA} = -[\mathbf{B}, \mathbf{A}] \tag{2.79}$$

and the *anticommutator* $\{\mathbf{A}, \mathbf{B}\}$ by

$$\{\mathbf{A}, \mathbf{B}\} \equiv \mathbf{AB} + \mathbf{BA} = \{\mathbf{B}, \mathbf{A}\} \tag{2.80}$$

The operators $\mathbf{A}$, $\mathbf{B}$ *commute* if $[\mathbf{A}, \mathbf{B}] = 0$ ($\Rightarrow \mathbf{AB} = \mathbf{BA}$) and *anticommute* if $\{\mathbf{A}, \mathbf{B}\} = 0$ ($\Rightarrow \mathbf{AB} = -\mathbf{BA}$).   ∎

❑ **Example 2.4.**   The 2×2 *Pauli matrices* $\sigma_x, \sigma_y, \sigma_z$ (sometimes called $\sigma_1, \sigma_2, \sigma_3$) are defined by

$$\sigma_x \equiv \begin{pmatrix} 0 & 1 \\ 1 & 0 \end{pmatrix} \qquad \sigma_y \equiv \begin{pmatrix} 0 & -i \\ i & 0 \end{pmatrix} \qquad \sigma_z \equiv \begin{pmatrix} 1 & 0 \\ 0 & -1 \end{pmatrix} \tag{2.81}$$

Some quick arithmetic shows that

$$\sigma_x^2 = \sigma_y^2 = \sigma_z^2 = 1 \tag{2.82}$$

and

$$\sigma_x \sigma_y = i\sigma_z = -\sigma_y \sigma_x \quad \sigma_y \sigma_z = i\sigma_x = -\sigma_z \sigma_y \quad \sigma_z \sigma_x = i\sigma_y = -\sigma_x \sigma_z \tag{2.83}$$

so that two different Pauli matrices do *not* commute under multiplication, but the minus signs show that they anticommute. Note that

$$\operatorname{tr} \sigma_a = 0 \qquad \operatorname{tr} \sigma_a \sigma_b = 2\delta_{ab} \tag{2.84}$$

(here tr denotes the *trace* defined in Eq. (2.103)) so that any 2×2 matrix $\mathbf{A}$ can be expressed in the form

$$\mathbf{A} = a_0 \mathbf{1} + \vec{a} \cdot \vec{\sigma} \tag{2.85}$$

with a_0 and $\vec{a} = (a_x, a_y, a_z)$ determined by

$$a_0 = \tfrac{1}{2} \operatorname{tr} \mathbf{A} \qquad \vec{a} = \tfrac{1}{2} \operatorname{tr} \vec{\sigma} \mathbf{A} \tag{2.86}$$

These results are useful in dealing with general 2×2 matrices.   ∎

→ **Exercise 2.8.**   (i) Show that if $\vec{a} = (a_x, a_y, a_z)$ and $\vec{b} = (b_x, b_y, b_z)$, then

$$\vec{\sigma} \cdot \vec{a}\, \vec{\sigma} \cdot \vec{b} = \vec{a} \cdot \vec{b} + i\vec{\sigma} \cdot \vec{a} \times \vec{b}$$

(ii) Show that if $\hat{n}$ is a three-dimensional unit vector, then

$$e^{i\vec{\sigma} \cdot \hat{n}\xi} = \cos \xi + i\vec{\sigma} \cdot \hat{n} \sin \xi$$

for any complex number ξ.   ☐

2.2.3   The Adjoint Operator

To every linear operator $\mathbf{A}$ on $\mathcal{V}$ corresponds a linear operator $\mathbf{A}^\dagger$ (the *adjoint* of $\mathbf{A}$) on the dual space $\mathcal{V}^*$ of linear functionals on $\mathcal{V}$ (see Section 2.1.5) defined by

$$\langle \mathbf{A}^\dagger u, x \rangle = \langle u, \mathbf{A}x \rangle \qquad (2.87)$$

for every vector x in $\mathcal{V}$ and every linear functional u in $\mathcal{V}^*$. If $\mathcal{V}$ is a unitary vector space, this is equivalent to

$$(\mathbf{A}^\dagger y, x) = (y, \mathbf{A}x) \qquad (2.88)$$

for every pair of vectors x, y in $\mathcal{V}$.

→ **Exercise 2.9.**   Show that

$$(\mathbf{A}^\dagger)^\dagger = \mathbf{A} \qquad (a\mathbf{A})^\dagger = a^* \mathbf{A}^\dagger \qquad (\mathbf{AB})^\dagger = \mathbf{B}^\dagger \mathbf{A}^\dagger$$

where a is a scalar, and $\mathbf{A}, \mathbf{B}$ are linear operators on the linear vector space $\mathcal{V}$.   □

→ **Exercise 2.10.**   If $\mathbf{A}$ is a linear operator on a unitary vector space $\mathcal{V}$, then

$$\ker \mathbf{A} = (\operatorname{im} \mathbf{A}^\dagger)^\perp \qquad \ker \mathbf{A}^\dagger = (\operatorname{im} \mathbf{A})^\perp$$

If $\mathcal{V}$ is finite dimensional, then

$$\dim(\operatorname{im} \mathbf{A}) = \rho(\mathbf{A}) = \rho(\mathbf{A}^\dagger) = \dim(\operatorname{im} \mathbf{A}^\dagger)$$

Remark. Thus, in a finite-dimensional vector space, $\mathbf{A}$ is nonsingular if and only if $\mathbf{A}^\dagger$ is nonsingular. An infinite-dimensional counterexample will be seen in Chapter 7.   □

In a unitary vector space a matrix element of an operator with respect to a complete orthonormal system can be expressed as a scalar product. If $\phi_1, \phi_2, \ldots$ is a complete orthonormal system and

$$\mathbf{A}\phi_k = \sum_j A_{jk} \phi_j \qquad (2.89)$$

then the orthonormality conditions give

$$A_{jk} = (\phi_j, \mathbf{A}\phi_k) \qquad (2.90)$$

The matrix elements of the adjoint operator are given by

$$(\mathbf{A}^\dagger)_{jk} = (\phi_j, \mathbf{A}^\dagger \phi_k) = (\mathbf{A}\phi_j, \phi_k) = A_{kj}^* \qquad (2.91)$$

so the matrix representing $\mathbf{A}^\dagger$ is obtained from the matrix representing $\mathbf{A}$ by complex conjugation and transposition (interchanging rows and columns, or reflecting about the main diagonal of the matrix).

Definition 2.9. The operator $\mathbf{A}$ is *self-adjoint*, or *Hermitian*, if

$$\mathbf{A}^\dagger = \mathbf{A} \tag{2.92}$$

In terms of the operator matrix elements, this requires

$$(y, \mathbf{A}x) = (x, \mathbf{A}y)^* \tag{2.93}$$

for every pair of vectors x, y. ∎

Remark. It follows that the diagonal matrix elements $(x, \mathbf{A}x)$ of a Hermitian operator $\mathbf{A}$ are real for every vector x. In a complex vector space, it is also sufficient for $\mathbf{A}$ to be Hermitian that $(x, \mathbf{A}x)$ be real for every vector x, but not so in a real vector space, where $(x, \mathbf{A}x)$ is real for all x for any linear operator $\mathbf{A}$ (see Exercise 2.7). □

Remark. Self-adjoint is defined by Eq. (2.92) and Hermitian by Eq. (2.93). In a finite-dimensional space, these two conditions are equivalent and the terms self-adjoint and Hermitian are often used interchangeably. In an infinite-dimensional space, some care is required, since the operators $\mathbf{A}$ and $\mathbf{A}^\dagger$ may not be defined on the same domain. Such subtle points will be discussed further in Chapter 7. □

2.2.4 Change of Basis; Rotations; Unitary Operators

The coordinates of a vector x, and elements of the matrix representing a linear operator $\mathbf{A}$, depend on the basis chosen in the vector space $\mathcal{V}$. Suppose $x_1, x_2, \ldots$ and $y_1, y_2, \ldots$ are two sets of basis vectors in $\mathcal{V}$. Then define a linear operator $\mathbf{S}$ by

$$\mathbf{S}x_k = y_k = \sum_j S_{jk} x_j \tag{2.94}$$

$(k = 1, 2, \ldots)$. $\mathbf{S}$ is nonsingular (show this), and

$$x_k = \mathbf{S}^{-1} y_k \equiv \sum_j \overline{S}_{jk} y_j \tag{2.95}$$

where the $\overline{S}_{jk} = (\mathbf{S}^{-1})_{jk}$ are the elements of the matrix inverse of the matrix (S_{jk}). The coordinates of a vector

$$x = \sum_k \xi_k x_k = \sum_\ell \eta_\ell y_\ell \tag{2.96}$$

in the two bases are related by

$$\xi_k = \sum_\ell S_{k\ell} \eta_\ell \qquad \eta_\ell = \sum_k \overline{S}_{\ell k} \xi_k \tag{2.97}$$

The operator $\mathbf{S}$ can also be viewed as a transformation of vectors:

$$x' \equiv \mathbf{S}x \tag{2.98}$$

in which x' is a vector whose coordinates in the basis $y_1, y_2, \ldots$ are the same as those of the original vector x in the basis $x_1, x_2, \ldots$, since

$$\mathbf{S}\left(\sum_k \xi_k x_k\right) = \sum_k \xi_k(\mathbf{S}x_k) = \sum_k \xi_k y_k \tag{2.99}$$

Also, note that

$$\mathbf{A}y_k = \sum_j S_{jk}\mathbf{A}x_j = \sum_{j,\ell} A_{\ell j}S_{jk}x_\ell = \sum_{j,\ell,m}(\overline{S}_{m\ell}A_{\ell j}S_{jk})y_m$$

$$= \sum_m (\mathbf{S}^{-1}\mathbf{A}\mathbf{S})_{mk}y_m \tag{2.100}$$

Thus the matrix elements of $\mathbf{A}$ in the basis $y_1, y_2, \ldots$ are given by

$$A'_{jk} = (\mathbf{S}^{-1}\mathbf{A}\mathbf{S})_{jk} \tag{2.101}$$

These are the same as the matrix elements of the operator

$$\mathbf{A}' \equiv \mathbf{S}^{-1}\mathbf{A}\mathbf{S} \tag{2.102}$$

in the basis $x_1, x_2, \ldots$. The transformation (2.102) with nonsingular operator $\mathbf{S}$ is a *similarity transformation* of $\mathbf{A}$. Equation (2.101) shows that a similarity transformation of an operator is equivalent to the change of basis defined by Eq. (2.94). The matrix elements of $\mathbf{A}'$ in the original basis $x_1, x_2, \ldots$ are the same as the matrix elements of the original operator $\mathbf{A}$ in the transformed basis.

There are certain characteristics associated with a matrix that are *invariant* under a similarity transformation. For example, the *trace* of $\mathbf{A}$ defined by

$$\text{tr}\,\mathbf{A} \equiv \sum_k A_{kk} \tag{2.103}$$

is invariant, since

$$\text{tr}\,\mathbf{S}^{-1}\mathbf{A}\mathbf{S} = \sum_{k,\ell,m} \overline{S}_{k\ell}A_{\ell m}S_{mk} = \sum_{k,\ell,m} A_{\ell m}S_{mk}\overline{S}_{k\ell} = \text{tr}\,\mathbf{A} \tag{2.104}$$

since $\mathbf{S}\mathbf{S}^{-1} = 1$. Further invariants are then the trace of any integer power m of $\mathbf{A}$ ($\text{tr}\,\mathbf{A}^m$) and the determinant ($\det\mathbf{A}$). The invariance of the determinant follows either from the standard result of linear algebra that $\det\mathbf{M}\mathbf{N} = \det\mathbf{M}\det\mathbf{N}$, or from expressing the $\det\mathbf{A}$ in terms of traces of various powers of $\mathbf{A}$ (such expressions depend on the dimension of the space—see Problem 27).

A change of basis corresponds to a nonsingular transformation $\mathbf{S}$ as shown in Eq. (2.94). If the length of vectors is unchanged by this transformation ($\|\mathbf{S}x\| = \|x\|$ for every x), then the transformation is an *isometry*. If in addition the scalar product is preserved,

$$(\mathbf{S}x, \mathbf{S}y) = (x, y) \tag{2.105}$$

for all vectors x, y ($\Rightarrow \mathbf{S}^\dagger\mathbf{S} = 1$), then the transformation is *unitary*, or *orthogonal* on a real vector space, where it is a rotation. In general, a unitary operator $\mathbf{U}$ is characterized by any one of the equivalent conditions:

(i) $\mathbf{U}^\dagger\mathbf{U} = 1 = \mathbf{U}\mathbf{U}^\dagger$, or simply $\mathbf{U}^\dagger = \mathbf{U}^{-1}$,

(ii) $(\mathbf{U}x, \mathbf{U}y) = (x, y) = (\mathbf{U}^\dagger x, \mathbf{U}^\dagger y)$ for every pair of vectors x, y.

(iii) $\|\mathbf{U}x\| = \|x\| = \|\mathbf{U}^\dagger x\|$ for every vector x.

(iv) if $\phi_1, \phi_2, \dots$ is a complete orthonormal system, then $\mathbf{U}\phi_1, \mathbf{U}\phi_2, \dots$ is also a complete orthonormal system.

❑ **Example 2.5.** Consider a rotation in two dimensions in which the coordinate axes are rotated by an angle θ (see Fig. 2.3). The unit vectors $u_{x'}, u_{y'}$ along the X', Y' axes are related to the unit vectors u_x, u_y along the X, Y axes by

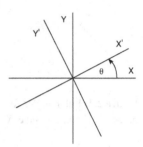

$$\begin{aligned} u_{x'} &= u_x \cos\theta + u_y \sin\theta \\ u_{y'} &= -u_x \sin\theta + u_y \cos\theta \end{aligned} \quad (2.106)$$

Thus the matrix corresponding to the rotation is

$$\mathbf{R}(\theta) = \begin{pmatrix} \cos\theta & -\sin\theta \\ \sin\theta & \cos\theta \end{pmatrix} \quad (2.107)$$

Figure 2.3: Rotation of coordinate axes in two dimensions.

Note that $\mathbf{R}^{-1}(\theta) = \mathbf{R}(-\theta)$. ∎

Remark. The rotation $\mathbf{R}(\theta)$ presented here is a rotation of coordinate axes that changes the components of vectors, which are viewed as intrinsic objects independent of the coordinate system. This viewpoint (the *passive point of view*) is emphasized further in Chapter 3. The alternative *active point of view* considers rotations as operations on the vectors themselves, so that a rotation $\overline{\mathbf{R}}(\theta)$ rotates vectors through an angle θ while keeping the coordinate axes fixed. Evidently the two views of rotations are inverse to each other:

$$\overline{\mathbf{R}}(\theta) = \mathbf{R}(-\theta) \quad (2.108)$$

since rotating the a vector through angle θ induces the same change in the components of a vector as rotating the coordinate axes through angle $-\theta$. ☐

❑ **Example 2.6.** Consider a rotation in three dimensions as illustrated in Fig. 2.4. The rotation is characterized by three angles (the *Euler angles*) ϕ, θ, ψ, which are defined as follows:

ϕ is the angle from the Y-axis to the line of nodes ($0 \le \phi < 2\pi$),

θ is the angle from the Z-axis to the Z'-axis ($0 \le \theta \le \pi$) and

ψ is the angle from the line of nodes to the Y'-axis ($0 \le \psi < 2\pi$)

where the angles are uniquely defined in the ranges given.[2]

---

[2]In some classical mechanics books, notably the book by Goldstein cited in the bibliography, the angle ϕ is defined from the X-axis to the line of nodes, and the angle ψ from the line of nodes to the X'-axis. It is technically more convenient in quantum mechanics to use the definition given here.

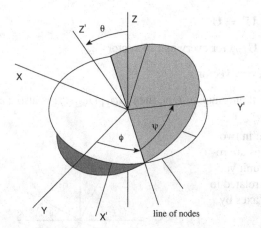

Figure 2.4: Euler angles for rotations in three dimensions. The *line of nodes* is the intersection of the X–Y plane and the X'–Y' plane, in the direction of the vector $\mathbf{u}_z \times \mathbf{u}'_z$.

Denote the associated rotation matrix by $\mathbf{R}(\phi, \theta, \psi)$. If $\mathbf{R_n}(\theta)$ denotes a rotation by angle θ about an axis along the unit vector $\mathbf{n}$, then

$$\mathbf{R}(\phi, \theta, \psi) = \mathbf{R}_{z'}(\psi)\mathbf{R_u}(\theta)\mathbf{R}_z(\phi) \qquad (2.109)$$

where $\mathbf{u}$ is a unit vector along the line of nodes, since the full rotation is a product of

(i) rotation by angle ϕ about the Z-axis,

(ii) rotation by angle θ about the line of nodes, and

(iii) rotation by angle ψ about the Z'-axis.

To compute the matrix $\mathbf{R}(\phi, \theta, \psi)$, it is better to express the rotation as a product of rotations about fixed axes (the X, Y, Z axes, say). It turns out that this can be done, with a strikingly simple result. Note first that the rotation about the line of nodes can be expressed as a rotation about the Y-axis if we first undo the rotation through ϕ about the Z-axis, rotate by angle θ, and then redo the rotation about the Z-axis, i.e.,

$$\mathbf{R_u}(\theta) = \mathbf{R}_z(\phi)\mathbf{R}_y(\theta)\mathbf{R}_z^{-1}(\phi) \qquad (2.110)$$

The rotation about the Z'-axis can be expressed as a rotation about the Z-axis if we undo the rotation about the line of nodes, rotate through ψ about the Z-axis, and redo the rotation about the line of nodes, so that

$$\mathbf{R}_{z'}(\psi) = \mathbf{R_u}(\theta)\mathbf{R}_z(\psi)\mathbf{R_u}^{-1}(\theta) \qquad (2.111)$$

Then

$$\mathbf{R}(\phi, \theta, \psi) = \mathbf{R}_z(\phi)\mathbf{R}_y(\theta)\mathbf{R}_z(\psi) \qquad (2.112)$$

and the rotation $\mathbf{R}(\phi, \theta, \psi)$ is obtained as the product of

(i) rotation by angle ψ about the Z-axis,

(ii) rotation by angle θ about the Y-axis, and

(iii) rotation by angle ϕ about the Z-axis.

Thus the rotations about the fixed axes have the same angles as the rotations about the moving axes, but they are done in reverse order! The final result for the matrix corresponding to a general rotation in three dimensions is then

$$\mathbf{R}(\phi, \theta, \psi) \tag{2.113}$$

$$= \begin{pmatrix} \cos\psi\cos\theta\cos\phi - \sin\psi\sin\phi & -\sin\psi\cos\theta\cos\phi - \cos\psi\sin\phi & \sin\theta\cos\phi \\ \cos\psi\cos\theta\sin\phi + \sin\psi\cos\phi & -\sin\psi\cos\theta\sin\phi + \cos\psi\cos\phi & \sin\theta\sin\phi \\ -\cos\psi\sin\theta & \sin\psi\sin\theta & \cos\theta \end{pmatrix}$$

Note that Eq. (2.112) is easier to remember than this result! ∎

2.2.5 Invariant Manifolds

A linear manifold $\mathcal{M}$ in $\mathcal{V}^n$ is an *invariant manifold* of the linear operator $\mathbf{A}$ if $\mathbf{A}x$ is in $\mathcal{M}$ for every vector x in $\mathcal{M}$. If $\mathcal{M}$ has dimension m and a basis $x_1, \ldots, x_m$, there are linearly independent vectors $x_{m+1}, \ldots, x_n$ that complete a basis of $\mathcal{V}^n$. Let $\mathcal{M}^*$ denote the $(n - m)$-dimensional manifold spanned by $x_{m+1}, \ldots, x_n$; then

$$\mathcal{V}^n = \mathcal{M} \oplus \mathcal{M}^* \tag{2.114}$$

This is similar to the split of a unitary space $\mathcal{V}$ into $\mathcal{M} \oplus \mathcal{M}^\perp$, but $\mathcal{M}^\perp$ is unique, while the manifold $\mathcal{M}^*$ is not, since to each of the basis vectors $x_{m+1}, \ldots, x_n$ can be added an arbitrary vector in $\mathcal{M}$. The matrix representation of $\mathbf{A}$ has the form

$$\mathbf{A} = \begin{pmatrix} \mathbf{A}_{\mathcal{M}} & \mathbf{B} \\ \mathbf{0} & \mathbf{A}_{\mathcal{M}^*} \end{pmatrix} \tag{2.115}$$

in this basis. $\mathbf{A}_{\mathcal{M}}$ is the *restriction* of $\mathbf{A}$ to $\mathcal{M}$, and $\mathbf{A}_{\mathcal{M}^*}$ the *restriction* of $\mathbf{A}$ to $\mathcal{M}^*$, Note that $\mathcal{M}^*$ is also an invariant manifold of $\mathbf{A}$ if and only if $\mathbf{B} = \mathbf{0}$. The properties of $\mathbf{A}$ that decide whether or not it is possible to find a basis in which $\mathbf{B} = \mathbf{0}$ are examined below.

Two invariant manifolds of the operator $\mathbf{A}$ are $\ker \mathbf{A}$ and $\mathrm{im}\, \mathbf{A}$. In fact, all the manifolds $\ker(\mathbf{A}^k)$ and $\mathrm{im}(\mathbf{A}^k)$ $(k = 0, 1, 2, \ldots)$ are invariant manifolds of $\mathbf{A}$; we have

$$\ker(\mathbf{A}^0) = \{\theta\} \subseteq \ker(\mathbf{A}) \subseteq \ker(\mathbf{A}^2) \subseteq \cdots \tag{2.116}$$

and

$$\mathrm{im}(\mathbf{A}^0) = \mathcal{V}^n \supseteq \mathrm{im}(\mathbf{A}) \supseteq \mathrm{im}(\mathbf{A}^2) \supseteq \cdots \tag{2.117}$$

At each stage, there is a linear manifold $\mathcal{M}_k$ such that

$$\mathcal{V}^n = \mathcal{M}_k \oplus \ker(\mathbf{A}^k) \tag{2.118}$$

constructed as in the derivation of Eq. (2.66). On a space $\mathcal{V}^n$ of finite dimension n, there must be an integer $p \leq n$ such that

$$\ker(\mathbf{A}^{p+1}) = \ker(\mathbf{A}^p) \tag{2.119}$$

(the dimension of $\ker(\mathbf{A}^p)$ cannot exceed n). Then also

$$\mathrm{im}(\mathbf{A}^{p+1}) = \mathrm{im}(\mathbf{A}^p) \tag{2.120}$$

and $\mathcal{V}^n$ can be expressed as

$$\mathcal{V}^n = \mathrm{im}(\mathbf{A}^p) \oplus \ker(\mathbf{A}^p) \tag{2.121}$$

With this split of $\mathcal{V}^n$, the matrix representation of $\mathbf{A}$ has the form

$$\mathbf{A} = \begin{pmatrix} \bar{\mathbf{A}} & \mathbf{0} \\ \mathbf{0} & \mathbf{N} \end{pmatrix} \tag{2.122}$$

where $\bar{\mathbf{A}}$ is a nonsingular operator on $\mathrm{im}(\mathbf{A}^p)$, and $\mathbf{N}^p = \mathbf{0}$.

Remark. A nonsingular operator maps the space $\mathcal{V}^n$ onto the whole space $\mathcal{V}^n$, while a singular operator maps $\mathcal{V}^n$ onto a subspace of smaller dimension. If $\mathbf{A}$ is singular, then repeated application of $\mathbf{A}$ (n times is enough on $\mathcal{V}^n$) leads to a subspace that cannot be reduced further by $\mathbf{A}$. This space is just the space $\mathrm{im}(\mathbf{A}^p)$. $\qquad\qquad\square$

Definition 2.10. An operator $\mathbf{N}$ is *nilpotent* if $\mathbf{N}^m = \mathbf{0}$ for some integer $m \geq 0$. $\qquad\blacksquare$

❏ **Example 2.7.** The linear operator σ^+ on $\mathbf{C}^2$ defined by

$$\sigma^+ = \frac{\sigma_x + i\sigma_y}{2} = \begin{pmatrix} 0 & 1 \\ 0 & 0 \end{pmatrix} \tag{2.123}$$

is nilpotent (see Eq. (2.81) for the definition of σ_x and σ_y). $\qquad\blacksquare$

➜ **Exercise 2.11.** Let $\mathbf{A}$ be a nilpotent operator on the n-dimensional vector space $\mathcal{V}^n$.
   (i) Show that $\mathbf{A}$ has a matrix representation

$$\mathbf{A} = \begin{pmatrix} 0 & a_{12} & \cdots & a_{1n} \\ 0 & 0 & \cdots & a_{2n} \\ \vdots & \vdots & \ddots & \vdots \\ 0 & 0 & \cdots & 0 \end{pmatrix}$$

with nonzero elements only above the main diagonal ($a_{k\ell} = 0$ if $k \geq \ell$).
   (ii) Show that if $\mathcal{V}^n$ is unitary, then there is an orthonormal basis on which this representation is valid. What is the matrix representation of $\mathbf{A}^\dagger$ in this basis? Is $\mathbf{A}^\dagger$ nilpotent?
   (iii) Suppose $\mathbf{A}$ and $\mathbf{B}$ are nilpotent. Is $\mathbf{AB}$ necessarily nilpotent? What about $[\mathbf{A}, \mathbf{B}]$? $\square$

2.2.6 Projection Operators

We have seen that if $\mathcal{M}$ is a linear manifold in a unitary vector space $\mathcal{V}$, then any vector x in $\mathcal{V}$ can be uniquely expressed as

$$x = x' + x'' \tag{2.124}$$

with x' in $\mathcal{M}$ and x'' in $\mathcal{M}^{\perp}$ (see Eq. (2.42)). The *(orthogonal) projection* operator (or *projector*) $\mathbf{P}_{\mathcal{M}}$ onto $\mathcal{M}$ is defined by

$$\mathbf{P}_{\mathcal{M}}x \equiv x' \tag{2.125}$$

for every vector x. If $\mathcal{M}$ is a one-dimensional manifold with a unit vector ϕ, then

$$\mathbf{P}_{\phi}x = (\phi, x)\phi \tag{2.126}$$

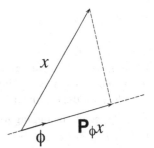

Figure 2.5: Projection $\mathbf{P}_{\phi}$ onto the unit vector ϕ.

selects the component of x in the direction of ϕ and eliminates the orthogonal components, as shown in Fig. 2.5.

More generally, if $\mathcal{M}$ is a manifold with an orthonormal basis $\phi_1, \phi_2, \ldots$, then

$$\mathbf{P}_{\mathcal{M}}x = \sum_k (\phi_k, x)\phi_k \tag{2.127}$$

Every projection operator $\mathbf{P}$ is Hermitian and *idempotent*

$$\mathbf{P}^2 = \mathbf{P} \tag{2.128}$$

(repeated application of a projection operator gives the same result as the first projection). The converse is also true: any Hermitian operator that is idempotent is a projection operator (see the exercise below). If $\mathbf{P}$ projects onto $\mathcal{M}$, then $1 - \mathbf{P}$ projects onto $\mathcal{M}^{\perp}$, since for every x expressed as in Eq. (2.124), we have

$$(1 - \mathbf{P}_{\mathcal{M}})x = x'' = (\mathbf{P}_{\mathcal{M}^{\perp}})x \tag{2.129}$$

→ **Exercise 2.12.** Show that if $\mathbf{P}$ is a Hermitian operator with $\mathbf{P}^2 = \mathbf{P}$, then $\mathbf{P}$ is a projection operator. What is the manifold onto which it projects? □

→ **Exercise 2.13.** Let $\mathcal{M}, \mathcal{N}$ be linear manifolds, and $\mathbf{P}, \mathbf{Q}$ the projection operators onto $\mathcal{M}, \mathcal{N}$. Under what conditions on $\mathcal{M}$ and $\mathcal{N}$ is it true that

(i) $\mathbf{P} + \mathbf{Q}$ is a projection operator,

(ii) $\mathbf{P} - \mathbf{Q}$ is a projection operator,

(iii) $\mathbf{PQ}$ is a projection operator, and

(iv) $\mathbf{PQ} = \mathbf{QP}$?

Give a geometrical description of these conditions. □

2.3   Eigenvectors and Eigenvalues

2.3.1   Eigenvalue Equation

Definition 2.11. If the linear operator $\mathbf{A}$ transforms a nonzero vector x into a scalar multiple of itself,

$$\mathbf{A}x = \lambda x \tag{2.130}$$

then x is an *eigenvector* of $\mathbf{A}$, and λ is the corresponding *eigenvalue*. The eigenvectors of $\mathbf{A}$ belonging to eigenvalue λ, together with the zero vector, form a linear manifold $\mathcal{M}_\lambda$, the *eigenmanifold* of $\mathbf{A}$ belonging to eigenvalue λ. The dimension m_λ of $\mathcal{M}_\lambda$ is the *geometric multiplicity* of the eigenvalue λ. The set of distinct eigenvalues of $\mathbf{A}$ is the *spectrum* of $\mathbf{A}$. ∎

Eigenvectors belonging to distinct eigenvalues are linearly independent. To see this, suppose $\mathbf{A}x_k = \lambda_k x_k$ ($k = 0, 1, \ldots, p$) and

$$x_0 = c_1 x_1 + \cdots + c_p x_p \tag{2.131}$$

with $x_1, \ldots, x_p$ linearly independent, and $c_1, \ldots, c_p$ nonvanishing. Then

$$\mathbf{A}x_0 = \lambda_0 x_0 = c_1 \lambda_1 x_1 + \cdots + c_p \lambda_p x_p \tag{2.132}$$

is consistent with linear independence of $x_1, \ldots, x_p$ only if

$$\lambda_0 = \lambda_1 = \cdots = \lambda_p \tag{2.133}$$

Equation (2.130) has a nonzero solution if and only if λ satisfies the *characteristic equation*

$$p_\mathbf{A}(\lambda) \equiv \det \|\mathbf{A} - \lambda\mathbf{1}\| = 0 \tag{2.134}$$

On a space $\mathcal{V}^n$ of finite dimension n, $\det \|\mathbf{A} - \lambda\mathbf{1}\|$ is a polynomial of degree n, the *characteristic polynomial* of $\mathbf{A}$, and the eigenvalues of $\mathbf{A}$ are the roots of this polynomial. From the fundamental theorem of algebra derived in Chapter 4, it follows that $p_\mathbf{A}(\lambda)$ can be expressed in the factorized form

$$p_\mathbf{A}(\lambda) = (-1)^n (\lambda - \lambda_1)^{p_1} \cdots (\lambda - \lambda_m)^{p_m} \tag{2.135}$$

with $\lambda_1, \ldots, \lambda_m$ the distinct roots of $p_\mathbf{A}(\lambda)$, and $p_1, \ldots, p_m$ the corresponding multiplicities ($p_1 + \cdots + p_m = n$). p_k is the *algebraic multiplicity* of the eigenvalue λ_k. It is important to note that the characteristic polynomial is independent of the coordinate system used to evaluate the determinant, since if $\mathbf{S}$ is a nonsingular matrix, then

$$\det \|\mathbf{S}^{-1}(\mathbf{A} - \lambda\mathbf{1})\mathbf{S}\| = \det(\mathbf{S}^{-1}) \det \|\mathbf{A} - \lambda\mathbf{1}\| \det(\mathbf{S}) = \det \|\mathbf{A} - \lambda\mathbf{1}\| \tag{2.136}$$

(it is a standard result of linear algebra that $\det \mathbf{MN} = \det \mathbf{M} \det \mathbf{N}$).

2.3.2   Diagonalization of a Linear Operator

Definition 2.12.   The operator $\mathbf{A}$ is *diagonalizable* if it has a basis of eigenvectors.   ∎

To understand this definition, suppose $x_1, \ldots, x_n$ is a basis with

$$\mathbf{A}x_k = \lambda_k x_k \tag{2.137}$$

$(k = 1, \ldots, n)$. In this basis, $\mathbf{A}$ has the matrix representation

$$\mathbf{A} = \begin{pmatrix} \lambda_1 & 0 & \cdots & 0 \\ 0 & \lambda_2 & \cdots & 0 \\ \vdots & \vdots & \ddots & \vdots \\ 0 & 0 & \cdots & \lambda_n \end{pmatrix} \equiv \mathrm{diag}(\lambda_1, \lambda_2, \ldots, \lambda_n) \tag{2.138}$$

with nonvanishing matrix elements only on the diagonal. In a general basis $y_1, \ldots, y_n$, there is then a nonsingular matrix $\mathbf{S}$ such that $\mathbf{S}^{-1}\mathbf{A}\mathbf{S}$ is diagonal in that basis.

The operator $\mathbf{A}$ is certainly diagonalizable if each root of the characteristic polynomial $p_{\mathbf{A}}(\lambda)$ is simple (multiplicity $= 1$), since we can then choose one eigenvector for each eigenvalue to create a set of n linearly independent eigenvectors of $\mathbf{A}$; these must form a basis of $\mathcal{V}^n$. On the other hand, suppose λ_0 is a multiple root of $p_{\mathbf{A}}(\lambda)$ (an eigenvalue of multiplicity > 1 is sometimes called *degenerate*). Then there may be fewer linearly independent eigenvectors belonging to λ_0 than the algebraic multiplicity allows, as shown by the following example.

❏ **Example 2.8.**   Consider the 2×2 matrix

$$\mathbf{A} = \begin{pmatrix} \mu & 1 \\ 0 & \mu \end{pmatrix} \tag{2.139}$$

that has $\lambda = \mu$ as a double root of its characteristic polynomial $p_{\mathbf{A}}(\lambda) = (\lambda - \mu)^2$. The only eigenvectors belonging to the eigenvalue μ are multiples of the unit vector

$$\phi = \begin{pmatrix} 1 \\ 0 \end{pmatrix} \tag{2.140}$$

and there is no second linearly independent eigenvector.   ∎

Remark.   Note that

$$(\mathbf{A} - \mu\mathbf{1})^2 = \mathbf{0} \tag{2.141}$$

so that $\mathbf{A}$ satisfies its own characteristic equation $p_{\mathbf{A}}(\mathbf{A}) = 0$. This is true in general.   ☐

The geometric multiplicity of an eigenvalue is always less than or equal to the algebraic multiplicity of the eigenvalue. To see this, suppose μ is an eigenvalue of $\mathbf{A}$, and $\mathcal{M}$ the corresponding eigenmanifold (of dimension m). If $x_1, \ldots, x_m$ is a basis of $\mathcal{M}$, and $x_{m+1}, \ldots, x_n$

a set of linearly independent vectors that complete a basis of $\mathcal{V}^n$, then $\mathbf{A}$ has the matrix representation

$$
\mathbf{A} = \begin{pmatrix}
\mu & 0 & \cdots & 0 & a_{1\,m+1} & \cdots & a_{1n} \\
0 & \mu & \cdots & 0 & a_{2\,m+1} & \cdots & a_{2n} \\
\vdots & \vdots & \ddots & \vdots & \vdots & & \vdots \\
0 & 0 & \cdots & \mu & a_{m\,m+1} & \cdots & a_{mn} \\
0 & 0 & \cdots & 0 & & & \\
\vdots & \vdots & & \vdots & & \bar{\mathbf{A}} & \\
0 & 0 & \cdots & 0 & & &
\end{pmatrix}
\tag{2.142}
$$

with $\bar{\mathbf{A}}$ an operator on the linear manifold $\mathcal{M}^* \equiv \mathcal{M}(x_{m+1}, \ldots, x_n)$ of dimension $n - m$. The characteristic polynomial of $\mathbf{A}$ can then be written as

$$
p_{\mathbf{A}}(\lambda) = (\mu - \lambda)^m \det \|\bar{\mathbf{A}} - \lambda \mathbf{1}\|
\tag{2.143}
$$

so the algebraic multiplicity of μ is at least m.

If the algebraic multiplicity p of μ is actually greater than m, then the operator $\bar{\mathbf{A}}$ defined on $\mathcal{M}^*$ must have an eigenvector belonging to eigenvalue μ. We can choose this eigenvector to be the basis vector x_{m+1}; then

$$
\mathbf{A} x_{m+1} = \mu x_{m+1} + y_{m+1}
\tag{2.144}
$$

where

$$
y_{m+1} = \sum_{k=1}^{m} a_{k\,m+1} x_k
\tag{2.145}
$$

is a vector in $\mathcal{M}$. Thus the manifold $\mathcal{M}_1 \equiv \mathcal{M}(x_1, \ldots, x_{m+1})$ of dimension $m + 1$ is an invariant manifold of $\mathbf{A}$. This procedure can be repeated until we have a manifold $\overline{\mathcal{M}}_\mu$ whose dimension is equal to the algebraic multiplicity p of the eigenvalue μ. The matrix representation of $\mathbf{A}$ on $\overline{\mathcal{M}}_\mu$ then has the form

$$
\mathbf{A}_\mu = \begin{pmatrix}
\mu & \cdots & 0 & a_{1\,m+1} & \cdots & a_{1p} \\
\vdots & \ddots & \vdots & \vdots & & \vdots \\
0 & \cdots & \mu & a_{m\,m+1} & \cdots & a_{mp} \\
0 & \cdots & 0 & \mu & \cdots & a_{m+1\,p} \\
\vdots & \vdots & \vdots & & \ddots & \vdots \\
0 & \cdots & 0 & 0 & \cdots & \mu
\end{pmatrix} \equiv \mu \mathbf{1} + \mathbf{N}_\mu
\tag{2.146}
$$

$\mathbf{N}_\mu = \mathbf{A}_\mu - \mu \mathbf{1}$ has nonzero elements only above the main diagonal, hence is nilpotent; in fact

$$
\mathbf{N}_\mu^{p-m+1} = \mathbf{0}
\tag{2.147}
$$

Thus if the characteristic polynomial $p_{\mathbf{A}}(\lambda)$ of $\mathbf{A}$ has roots $\lambda_1, \ldots, \lambda_m$ with algebraic multiplicities $p_1, \ldots, p_m$ as in the factorized form (2.135), then there are invariant manifolds $\overline{\mathcal{M}}_1, \ldots, \overline{\mathcal{M}}_m$ of dimensions $p_1, \ldots, p_m$, respectively, such that the restriction $\mathbf{A}_k$ of $\mathbf{A}$ to $\overline{\mathcal{M}}_k$ has the form

$$\mathbf{A}_k = \lambda_k \mathbf{1} + \mathbf{N}_k \tag{2.148}$$

with $\mathbf{N}_k$ nilpotent,

$$\mathbf{N}_k^{p_k - m_k + 1} = \mathbf{0} \tag{2.149}$$

(m_k is the geometric multiplicity of the eigenvalue λ_k). The eigenmanifold $\mathcal{M}_k$ is a subset of $\overline{\mathcal{M}}_k$; the two manifolds are the same if and only if $p_k = m_k$. The vector space $\mathcal{V}^n$ can be expressed as

$$\mathcal{V}^n = \overline{\mathcal{M}}_1 \oplus \cdots \oplus \overline{\mathcal{M}}_m \tag{2.150}$$

and $\mathbf{A}$ can be expressed in the form (the *Jordan canonical form*)

$$\mathbf{A} = \mathbf{D} + \mathbf{N} \tag{2.151}$$

where $\mathbf{D}$ is a diagonalizable matrix, $\mathbf{N}$ is nilpotent ($\mathbf{N} = \mathbf{0}$ if and only if $\mathbf{A}$ is diagonalizable), and $\mathbf{N}$ commutes with $\mathbf{D}$ (since $\mathbf{D}$ is just a multiple of $\mathbf{1}$ on each of the invariant manifolds $\overline{\mathcal{M}}_k$). Equation (2.151) is the unique split of a linear operator $\mathbf{A}$ into the sum of a diagonalizable operator $\mathbf{D}$ and a nilpotent operator $\mathbf{N}$ that commutes with $\mathbf{D}$.

Now suppose $\mathbf{B}$ is an operator that commutes with $\mathbf{A}$. Then $\mathbf{B}$ also commutes with any power of $\mathbf{A}$, $\mathbf{A}^p \mathbf{B} = \mathbf{B} \mathbf{A}^p$. If x is a vector in $\overline{\mathcal{M}}_k$, then

$$\mathbf{A}^{p_k} \mathbf{B} x = \mathbf{B} \mathbf{A}^{p_k} x = \lambda_k^{p_k} x \tag{2.152}$$

so that $\mathbf{B}x$ is also in $\overline{\mathcal{M}}_k$. Thus each of the manifolds $\overline{\mathcal{M}}_k$ is an invariant manifold of $\mathbf{B}$, on which $\mathbf{B}$ can be expressed as the sum of a diagonalizable matrix and a nilpotent operator as in Eq. (2.151). Thus $\mathbf{B}$ can be expressed in the form

$$\mathbf{B} = \mathbf{D}' + \mathbf{N}' \tag{2.153}$$

with $\mathbf{D}'$ diagonalizable, $\mathbf{N}'$ nilpotent and $\mathbf{D}'\mathbf{N}' = \mathbf{N}'\mathbf{D}'$. Furthermore, there is a basis in which both $\mathbf{D}$ and $\mathbf{D}'$ are diagonal, $\mathbf{N}$ and $\mathbf{N}'$ have nonzero elements only above the main diagonal, and $\mathbf{N}$ commutes with $\mathbf{N}'$. In particular, if $\mathbf{A}$ and $\mathbf{B}$ are each diagonalizable and $\mathbf{AB} = \mathbf{BA}$, then there is a basis of vectors that are simultaneous eigenvectors of $\mathbf{A}$ and $\mathbf{B}$.

2.3.3 Spectral Representation of Normal Operators

The study of the eigenvectors and eigenvalues of linear operators on $\mathcal{V}^n$ in the preceding section leads to the general structure expressed in Eq. (2.151). On a unitary vector space, however, there is a broad class of operators for which it is possible to go further and construct a complete orthonormal system of eigenvectors. This is the class of *normal* operators, for which we have the formal

Definition 2.13. The linear operator $\mathbf{A}$ on a unitary vector space is *normal* if

$$\mathbf{A}^\dagger \mathbf{A} = \mathbf{A}\mathbf{A}^\dagger \tag{2.154}$$

that is, $\mathbf{A}$ is normal if it commutes with its adjoint.   ∎

→ **Exercise 2.14.** Every self-adjoint operator is normal; every unitary operator is normal. □
Normal operators have several useful properties:
   (i) If $\mathbf{A}$ is normal, then

$$\|\mathbf{A}^\dagger x\| = \|\mathbf{A}x\| \tag{2.155}$$

for every vector x, since if $\mathbf{A}$ is normal, then

$$\|\mathbf{A}x\|^2 = (x, \mathbf{A}^\dagger \mathbf{A}x) = (x, \mathbf{A}\mathbf{A}^\dagger x) = \|\mathbf{A}^\dagger x\|^2 \tag{2.156}$$

It is also true in a complex vector space that $\mathbf{A}$ is normal if $\|\mathbf{A}^\dagger x\| = \|\mathbf{A}x\|$ for every x
(Exercise 2.7 shows why a complex vector space is needed).
   (ii) If $\mathbf{A}$ is normal and $\mathbf{A}x = \lambda x$, then

$$\mathbf{A}^\dagger x = \lambda^* x \tag{2.157}$$

since if $\mathbf{A}$ is normal and $\mathbf{A}x = \lambda x$, then

$$\|(\mathbf{A}^\dagger - \lambda^* \mathbf{1})x\| = \|(\mathbf{A} - \lambda \mathbf{1})x\| = 0 \tag{2.158}$$

   (iii) If $\mathbf{A}$ is normal, eigenvectors belonging to distinct eigenvalues are orthogonal. For if
$\mathbf{A}x_1 = \lambda_1 x_1$ and $\mathbf{A}x_2 = \lambda_2 x_2$, then

$$\lambda_1(x_2, x_1) = (x_2, \mathbf{A}x_1) = (\mathbf{A}^\dagger x_2, x_1) = \lambda_2(x_2, x_1) \tag{2.159}$$

so that $(x_2, x_1) = 0$ if $\lambda_1 \neq \lambda_2$.

Theorem 2.4. (Fundamental Theorem on Normal Operators) If the linear operator $\mathbf{A}$ on $\mathcal{V}^n$
is normal, then $\mathbf{A}$ has a complete orthonormal system of eigenvectors.

Proof. $\mathbf{A}$ has at least one eigenvalue λ_1; let $\mathcal{M}_1$ denote the corresponding eigenmanifold.
If x is in $\mathcal{M}_1$ and y is in $\mathcal{M}_1^\perp$, then

$$(x, \mathbf{A}y) = (\mathbf{A}^\dagger x, y) = \lambda_1(x, y) = 0 \tag{2.160}$$

so $\mathbf{A}y$ is also in $\mathcal{M}_1^\perp$; thus $\mathcal{M}_1^\perp$ is also an invariant manifold of $\mathbf{A}$. If $\dim(\mathcal{M}_1^\perp) > 0$,
then $\mathbf{A}$ has at least one eigenvalue λ_2 in $\mathcal{M}_1^\perp$, with corresponding eigenmanifold $\mathcal{M}_2$. It is
possible to proceed in this way to find eigenvalues $\lambda_1, \lambda_2, \ldots, \lambda_m$ with mutually orthogonal
eigenmanifolds $\mathcal{M}_1, \mathcal{M}_2, \ldots \mathcal{M}_m$ that span $\mathcal{V}^n$, so that

$$\mathcal{V}^n = \mathcal{M}_1 \oplus \mathcal{M}_2 \oplus \cdots \oplus \mathcal{M}_m \tag{2.161}$$

If we then choose a set of orthonormal elements $\{\phi_{11}, \ldots, \phi_{1p_1}\}$ spanning $\mathcal{M}_1$, a second
orthonormal system $\{\phi_{21}, \ldots, \phi_{2p_2}\}$ spanning $\mathcal{M}_2$, ..., and so on until we choose an orthonormal system $\{\phi_{m1}, \ldots, \phi_{mp_m}\}$ spanning $\mathcal{M}_m$ we will then have a complete orthonormal system of eigenvectors of $\mathbf{A}$.

The converse is also true; if $\mathbf{A}$ has a complete orthonormal system of eigenvectors, then $\mathbf{A}$ is normal. For if $\phi_1, \ldots, \phi_n$ is a complete orthonormal system of eigenvectors of $\mathbf{A}$ with eigenvalues $\lambda_1, \ldots, \lambda_n$ (not necessarily distinct), then $\mathbf{A}$ and $\mathbf{A}^\dagger$ are represented by diagonal matrices,

$$\mathbf{A} = \mathrm{diag}(\lambda_1, \ldots, \lambda_n) \qquad \mathbf{A}^\dagger = \mathrm{diag}(\lambda_1^*, \ldots, \lambda_n^*) \tag{2.162}$$

Then $\mathbf{A}$ is normal, since

$$\mathbf{A}^\dagger \mathbf{A} = \mathrm{diag}(\lambda_1^* \lambda_1, \ldots, \lambda_n^* \lambda_n) = \mathbf{A}\mathbf{A}^\dagger \tag{2.163}$$

If $\mathbf{A}$ is a normal operator with distinct eigenvalues $\lambda_1, \ldots, \lambda_m$ and corresponding eigenmanifolds $\mathcal{M}_1, \ldots \mathcal{M}_m$, introduce the projection operators

$$\mathbf{P}_1 \equiv \mathbf{P}_{\mathcal{M}_1}, \ldots, \mathbf{P}_m \equiv \mathbf{P}_{\mathcal{M}_m} \tag{2.164}$$

The operators $\mathbf{P}_1, \ldots, \mathbf{P}_m$ are the *eigenprojectors* of $\mathbf{A}$. They have the properties

$$\mathbf{P}_k \mathbf{P}_\ell = \mathbf{P}_\ell \mathbf{P}_k = 0 \quad k \neq \ell \tag{2.165}$$

$$\mathbf{P}_1 + \cdots + \mathbf{P}_m = 1 \tag{2.166}$$

and

$$\mathbf{A} = \sum_{k=1}^{m} \lambda_k \mathbf{P}_k = \lambda_1 \mathbf{P}_1 + \cdots + \lambda_m \mathbf{P}_m \tag{2.167}$$

Equation (2.167) is the *spectral representation* of the normal operator $\mathbf{A}$.

❏ **Example 2.9.**   A simple example is provided by the 2×2 matrix

$$\sigma_z = \begin{pmatrix} 1 & 0 \\ 0 & -1 \end{pmatrix} \tag{2.168}$$

introduced in Eq. (2.81). This matrix is already diagonal; its eigenvalues are 1 and -1. The corresponding eigenprojectors are given by

$$\mathbf{P}_+ = \begin{pmatrix} 1 & 0 \\ 0 & 0 \end{pmatrix} \qquad \mathbf{P}_- = \begin{pmatrix} 0 & 0 \\ 0 & 1 \end{pmatrix} \tag{2.169}$$

in terms of which the spectral representation of σ_z is

$$\sigma_z = \mathbf{P}_+ - \mathbf{P}_- \tag{2.170}$$

Note also that $\mathbf{P}_\pm = \frac{1}{2}(1 \pm \sigma_z)$ since $1 \pm \sigma_z$ gives zero when it acts on the eigenvector of σ_z with eigenvalue ∓ 1, but this formula works only because σ_z has just two distinct eigenvalues.   ∎

→ **Exercise 2.15.** Find the eigenvalues and eigenprojectors for the σ_x and σ_y in Eq. (2.81). ☐

→ **Exercise 2.16.** Suppose $\mathbf{A}$ is a normal operator with spectral representation

$$\mathbf{A} = \sum_{k=1}^{m} \lambda_k \mathbf{P}_k$$

Show that $\mathbf{A}$ is
  (i) self-adjoint if and only if all the eigenvalues $\{\lambda_k\}$ are real,
  (ii) unitary if and only if all the eigenvalues satisfy $|\lambda_k| = 1$, and
  (iii) nonsingular if and only if all the eigenvalues are nonzero.   □

Definition 2.14. A linear operator $\mathbf{A}$ on the unitary vector space $\mathcal{V}$ is *positive definite* ($\mathbf{A} > 0$) if $(x, \mathbf{A}x) > 0$ for every vector x, *negative definite* ($\mathbf{A} < 0$) if $(x, \mathbf{A}x) < 0$ for every x, and *definite* if it is either positive or negative definite.   ∎

→ **Exercise 2.17.** If $\mathcal{V}$ is a complex vector space, then the linear operator $\mathbf{A}$ is definite if and only if $\mathbf{A}$ is self-adjoint *and* all its eigenvalues are of real and of the same sign. What if $\mathcal{V}$ is a real vector space?   □

Definition 2.15. $\mathbf{A}$ is *nonnegative* ($\mathbf{A} \geq 0$) if $(x, \mathbf{A}x) \geq 0$ for every x, *nonpositive* ($\mathbf{A} \leq 0$) if $(x, \mathbf{A}x) \leq 0$ for every x, and *semidefinite* if it is either of these.   ∎

→ **Exercise 2.18.** What can you say about the eigenvalues of a semidefinite operator $\mathbf{A}$?   □

Suppose now that the normal operators $\mathbf{A}$ and $\mathbf{B}$ have the spectral representations

$$\mathbf{A} = \sum_k \lambda_k \mathbf{P}_k \qquad \mathbf{B} = \sum_\ell \mu_\ell \mathbf{Q}_l \tag{2.171}$$

and $\mathbf{A}\mathbf{B} = \mathbf{B}\mathbf{A}$ (that is, $\mathbf{A}$ and $\mathbf{B}$ commute). From the discussion in the previous section, it follows that every eigenmanifold of $\mathbf{A}$ is an eigenmanifold of $\mathbf{B}$ (here we know that $\mathbf{A}$ and $\mathbf{B}$ are diagonalizable) and that there is a basis of vectors that are simultaneous eigenvectors of $\mathbf{A}$ and $\mathbf{B}$. In terms of the eigenprojectors, this is equivalent to the statement that

$$\mathbf{P}_k \mathbf{Q}_\ell = \mathbf{Q}_\ell \mathbf{P}_k \quad \text{for every } k, \ell \tag{2.172}$$

Remark. States of a quantum mechanical system are represented as vectors in a unitary vector space $\mathcal{V}$, and the physical observables of the system by self-adjoint operators on $\mathcal{V}$. The eigenvalues of an operator correspond to allowed values of the observable in states of the system, and eigenvectors of the operator to states in which the observable has the corresponding eigenvalue. Linear combinations of the eigenvectors are states in which the observable has no definite value, but in which measurements of the observables will yield values with a probability distribution related to the amplitudes of the different eigenvectors in the state.

Measurement of an observable with a definite value projects the state of a system onto the eigenmanifold of the operator corresponding to that eigenvalue. Independent measurement of two observables at the same time is possible only if the eigenprojectors of the two operators commute, i.e., if the two operators commute. This will be discussed further in Chapter 7.   □

2.3.4 Minimax Properties of Eigenvalues of Self-Adjoint Operators

The natural frequencies of a vibrating system (a string, or a drum head) are the eigenvalues of a linear differential operator. Energy levels of a quantum mechanical system are eigenvalues of the Hamiltonian operator of the system. Theoretical understanding of such systems requires a knowledge of the spectra of these operators. In actual physical systems, there are often too many degrees of freedom to allow an exact computation of the eigenvalues, even with the power of modern computers. Hence we need to find relatively simple ways to estimate eigenvalues.

Powerful methods for estimating eigenvalues are based on extremum principles. If a real function $f(x)$ of a real variable x is known to have a relative minimum at some point x_*, then a very accurate estimate of the minimum value $f_* \equiv f(x_*)$ can be obtained with a moderately accurate knowledge of x_*. Since the behavior of $f(x)$ is parabolic near the minimum, we know that $f - f_*$ will be quadratic in $x - x_*$.

Here the (real) eigenvalues of a self-adjoint operator $\mathbf{A}$ on the linear vector space V are bounded by the eigenvalues of the restriction of the operator to any subspace V_0. If the subspace V_0 contains vectors close to the actual eigenvectors of $\mathbf{A}$, then estimates of the eigenvalues can be quite accurate, since the errors in the eigenvalues are of second order in the (presumed small) errors of the eigenvector.

Suppose $\mathbf{A}$ is a self-adjoint operator in V^n. Let the eigenvalues of $\mathbf{A}$ be arranged in non-decreasing order so that $\lambda_1 \leq \lambda_2 \leq \cdots \leq \lambda_n$, with corresponding orthonormal eigenvectors $\phi_1, \phi_2, \ldots, \phi_n$. Then for any vector x,

$$(x, \mathbf{A}x) = \sum_{k=1}^{n} \lambda_k |(\phi_k, x)|^2 \geq \lambda_1 \sum_{k=1}^{n} |(\phi_k, x)|^2 = \lambda_1 \|x\|^2 \tag{2.173}$$

(and equality if $x = c\phi_1$ for some scalar c). Similarly,

$$(x, \mathbf{A}x) \leq \lambda_n \sum_{k=1}^{n} |(\phi_k, x)|^2 = \lambda_n \|x\|^2 \tag{2.174}$$

for any vector x (and equality if $x = c\phi_n$ for some scalar c). Thus

$$\lambda_1 = \inf_{\|\phi\|=1} (\phi, \mathbf{A}\phi) \qquad \lambda_n = \sup_{\|\phi\|=1} (\phi, \mathbf{A}\phi) \tag{2.175}$$

Here "inf" (Latin: *infimum*) denotes the greatest lower bound (or simply minimum) and "sup" (Latin: *supremum*) the least upper bound (or maximum) (over the unit sphere $\|\phi\| = 1$).

Remark. Equation (2.175) is the basis of the standard variational method used in quantum mechanics to estimate the lowest (*ground state*) energy of a system with Hamiltonian $\mathbf{H}$. Choose a unit vector (*trial wave function*) $\phi = \phi_\alpha$, dependent on a set of (variational) parameters α, and minimize the energy

$$E(\alpha) = (\phi_\alpha, \mathbf{H}\phi_\alpha) \tag{2.176}$$

with respect to the parameters α. This leads to an upper bound for the ground state energy that is often an excellent approximation to its actual value. As explained below, this method can also be extended to generate upper bounds for excited state energies. $\square$

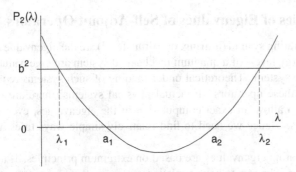

Figure 2.6: Roots of the characteristic polynomial of a 2×2 Hermitian matrix.

❑ **Example 2.10.** Consider a Hermitian operator **A** represented by the 2×2 matrix

$$\mathbf{A} = \begin{pmatrix} a_1 & b \\ b & a_2 \end{pmatrix} \tag{2.177}$$

(with $a_1 < a_2$ and b real). The eigenvalues are the roots of the characteristic polynomial

$$P_\mathbf{A}(\lambda) = (a_1 - \lambda)(a_2 - \lambda) - b^2 \tag{2.178}$$

The graph in Fig. 2.6 shows $(a_1 - \lambda)(a_2 - \lambda)$ as a function of λ; the eigenvalues of **A** are the intersections of this curve with the line $\lambda = b^2$. It is clear from the graph that

$$\lambda_1 < a_1 < a_2 < \lambda_2 \tag{2.179}$$

consistent with Eq. (2.175).   ∎

Equation (2.175) is especially useful, as it provides rigorous bounds for the extreme eigenvalues of a bounded operator. Formal bounds on other eigenvalues are obtained by restricting **A** to manifolds orthogonal to the eigenvectors already found. For example,

$$\lambda_2 = \inf_{\|\phi\|=1,(\phi_1,\phi)=0} (\phi, \mathbf{A}\phi) \qquad \lambda_{n-1} = \sup_{\|\phi\|=1,(\phi_n,\phi)=0} (\phi, \mathbf{A}\phi) \tag{2.180}$$

These results are not as useful as (2.175) in practice, however, since the orthogonality constraints cannot be satisfied exactly without knowing the eigenvectors ϕ_1 and ϕ_n. Bounds of more practical importance for other eigenvalues can be derived by suitably restricting the operator **A** to various subspaces of $\mathcal{V}^n$. Let

$$\mu_\mathbf{A}(\mathcal{M}) \equiv \inf_{\mathcal{M},\|\phi\|=1} (\phi, \mathbf{A}\phi) \tag{2.181}$$

be the minimum of $(\phi, \mathbf{A}\phi)$ on the intersection of the unit sphere with the linear manifold $\mathcal{M}$. $\mu_\mathbf{A}(\mathcal{M})$ is the smallest eigenvalue of the operator **A** restricted to $\mathcal{M}$. If we let $\mathcal{M}$ range over the $(n-1)$-dimensional subspaces of $\mathcal{V}$, we might expect that $\mu_\mathbf{A}(\mathcal{M})$ will vary between λ_1 and λ_2. To show this, suppose ϕ is a unit vector in $\mathcal{M}^\perp$ with the expansion

$$\phi = a_1 \phi_1 + a_2 \phi_2 + \cdots \tag{2.182}$$

If $a_1 = 0$, then ϕ_1 is in $\mathcal{M}$, and $\mu_{\mathbf{A}}(\mathcal{M}) = \lambda_1$. If $a_1 \neq 0$, let

$$a = \sqrt{z_1^2 + c_2^2} \qquad a_1 = a \sin \alpha \qquad a_2 = a \cos \alpha \qquad (2.183)$$

Then the unit vector

$$\psi \equiv \cos \alpha \, \phi_1 - \sin \alpha \, \phi_2 \qquad (2.184)$$

is in $\mathcal{M}$, since it is orthogonal to the vector ϕ that spans $\mathcal{M}^\perp$, and we have

$$\mu_{\mathbf{A}}(\mathcal{M}) = \cos^2 \alpha \, \lambda_1 + \sin^2 \alpha \, \lambda_2 \qquad (2.185)$$

Thus we have

$$\lambda_1 \leq \mu_{\mathbf{A}}(\mathcal{M}) \leq \lambda_2 \qquad (2.186)$$

for any subspace $\mathcal{M}$ of dimension $n - 1$, with equality for $\mathcal{M} = [\mathcal{M}(\phi_1)]^\perp$. This argument can be extended show that if $\mathcal{M}$ is a subspace of dimension $n - h$, then

$$\mu_{\mathbf{A}}(\mathcal{M}) \leq \lambda_{h+1} \qquad (2.187)$$

with equality for $\mathcal{M} = [\mathcal{M}(\phi_1, \ldots, \phi_h)]^\perp$. Thus the eigenvalue λ_{h+1} is larger than the smallest eigenvalue of any restriction of $\mathbf{A}$ to a manifold of dimension $n-h$ $(h = 0, \ldots, n-1)$.

To work from the other end, let

$$\nu_{\mathbf{A}}(\mathcal{M}) \equiv \sup_{\mathcal{M}, \|\phi\|=1} (\phi, \mathbf{A}\phi) \qquad (2.188)$$

$\nu_{\mathbf{A}}(\mathcal{M})$ is the largest eigenvalue of the operator $\mathbf{A}$ restricted to the manifold $\mathcal{M}$. Then

$$\lambda_h \leq \nu_{\mathbf{A}}(\mathcal{M}) \qquad (2.189)$$

for any manifold of dimension h, with equality for $\mathcal{M} = \mathcal{M}(\phi_1, \ldots, \phi_h)$; the eigenvalue λ_h is smaller than the largest eigenvalue of any restriction of $\mathbf{A}$ to a manifold of dimension h.

Inequalities (2.187) and (2.189) provide rigorous bounds for the eigenvalue λ_h. One further refinement is to let $\mathcal{M}$ be a linear manifold of dimension m with $\mathbf{P}_{\mathcal{M}}$ the projection operator onto $\mathcal{M}$, and let

$$\mathbf{A}' = \mathbf{A}_{\mathcal{M}} \equiv \mathbf{P}_{\mathcal{M}} \mathbf{A} \mathbf{P}_{\mathcal{M}} \qquad (2.190)$$

be the restriction of $\mathbf{A}$ to $\mathcal{M}$. $\mathbf{A}'$ is a self-adjoint operator on $\mathcal{M}$; its eigenvalues $\{\lambda_k'\}$ can be ordered so that

$$\lambda_1' \leq \lambda_2' \leq \cdots \leq \lambda_m' \qquad (2.191)$$

The preceding inequalities imply here that

$$\lambda_h \leq \lambda_h' \leq \lambda_{h+n-m} \qquad (2.192)$$

so that the eigenvalues of $\mathbf{A}'$ give direct bounds on the eigenvalues of $\mathbf{A}$.

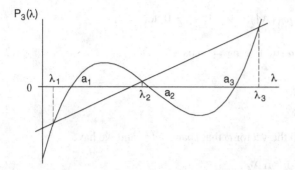

Figure 2.7: Roots of the characteristic polynomial of a 3×3 Hermitian matrix.

Remark. Thus if a linear operator $\mathbf{A}$ on $\mathcal{V}$ is restricted to an m-dimensional subspace of $\mathcal{V}$, the eigenvalues of the restricted operator are upper bounds on the corresponding eigenvalues of the full operator; the inequalities $\lambda_h \leq \lambda'_h$ give bounds for the lowest m eigenvalues, and not just the lowest. In quantum mechanical systems, this provides estimates of excited state energies as well as ground state energies.   □

❑ **Example 2.11.** Let $\mathbf{A}$ be a self-adjoint operator with a 3×3 matrix representation

$$\mathbf{A} = \begin{pmatrix} a_1 & 0 & b_1 \\ 0 & a_2 & b_2 \\ b_1^* & b_2^* & a_3 \end{pmatrix} \tag{2.193}$$

(with $a_1 < a_2 < a_3$ real). The characteristic polynomial of $\mathbf{A}$ is

$$P(\lambda) = (a_1 - \lambda)(a_2 - \lambda)(a_2 - \lambda) - (a_1 - \lambda)|b_2|^2 - (a_2 - \lambda)|b_1|^2 \tag{2.194}$$

The roots of the $P(\lambda)$ correspond to the solutions of

$$(\lambda - a_1)(\lambda - a_2)(\lambda - a_3) = (\lambda - a_1)|b_2|^2 + (\lambda - a_2)|b_1|^2 \tag{2.195}$$

Here the left-hand side is a cubic and the right-hand side a straight line as a function of λ as shown in Fig. 2.7. The eigenvalues are the values of λ at the intersections of the cubic and the straight line. These eigenvalues evidently satisfy the inequalities

$$\lambda_1 < a_1 < \lambda_2 < a_2 < \lambda_3 \tag{2.196}$$

consistent with the general inequalities (2.192). Note that it is essential to diagonalize the 2×2 matrix in the upper-left corner in order to conclude that $a_1 < \lambda_2 < a_2$, although the inequalities $\lambda_1 < a_1$ and $a_3 < \lambda_3$ are true in general (that is, even if the two zero elements in the matrix are replaced by a real parameter c).   ∎

2.4 Functions of Operators

If $\mathbf{A}$ is a linear operator, it is straightforward to define powers of $\mathbf{A}$, and then polynomials of $\mathbf{A}$ as linear combinations of powers of $\mathbf{A}$. If $\mathbf{A}$ is nonsingular, then negative integer powers of $\mathbf{A}$ are defined as powers of $\mathbf{A}^{-1}$. If a function $f(z)$ is an analytic function of a complex variable z in some neighborhood of $z = 0$, then a natural way to define $f(\mathbf{A})$ is through the power series expansion of $f(z)$. Thus, for example, we have the formal definitions

$$(1 - \alpha\mathbf{A})^{-1} = \sum_{n=0}^{\infty} \alpha^n \mathbf{A}^n \tag{2.197}$$

and

$$e^{t\mathbf{A}} = \sum_{n=0}^{\infty} \frac{t^n}{n!} \mathbf{A}^n \tag{2.198}$$

To give meaning to such series, we need to define convergence of sequences of operators.

Definition 2.16. Let $\{\mathbf{A}_k\}$ be a sequence of linear operators on the linear vector space $\mathcal{V}$. Then the sequence converges *uniformly* to $\mathbf{A}$ ($\{\mathbf{A}_k\} \Rightarrow \mathbf{A}$) if $\{|\mathbf{A}_k - \mathbf{A}|\} \to 0$. The sequence converges *strongly* to $\mathbf{A}$ ($\{\mathbf{A}_k\} \to \mathbf{A}$) if $\{\mathbf{A}_k x\} \to \mathbf{A}x$ for every vector x, and *weakly* to $\mathbf{A}$ ($\{\mathbf{A}_k\} \rightharpoonup \mathbf{A}$) if $\{(y, \mathbf{A}_k x)\} \to (y, \mathbf{A}x)$ for every pair of vectors x, y. ∎

As with sequences of vectors, strong and weak convergence of sequences of operators are equivalent in a finite-dimensional space $\mathcal{V}^n$. The equivalence of uniform convergence follows from the observation that if $\{\phi_1, \phi_2, \ldots, \phi_n\}$ is a complete orthonormal system, and if the sequences $\{\mathbf{A}_k \phi_1\}, \{\mathbf{A}_k \phi_2\}, \ldots, \{\mathbf{A}_k \phi_n\}$ all converge, then also $\{|\mathbf{A}_k - \mathbf{A}|\} \to 0$, since

$$|\mathbf{A}_k - \mathbf{A}|^2 \leq \sum_{m=1}^{n} \|(\mathbf{A}_k - \mathbf{A})\phi_m\|^2 \tag{2.199}$$

It is clear that an infinite series of operators on $\mathcal{V}^n$ converges if and only if the corresponding series of complex numbers obtained by replacing $\mathbf{A}$ by $|\mathbf{A}|$ is convergent. Thus, the infinite series (2.197) converges if $|\alpha|\,|\mathbf{A}| < 1$. It does not converge as an operator series if $|\alpha|\,|\mathbf{A}| \geq 1$, although the series $(\sum \alpha^n \mathbf{A}^n)x$ may converge as a series of vectors for some x. The infinite series (2.198) converges for all t (even uniformly) if $\mathbf{A}$ is bounded.

Suppose $\mathbf{A}$ is a normal operator with eigenvalues $\lambda_1, \ldots, \lambda_m$, corresponding eigenprojectors $\mathbf{P}_1, \ldots, \mathbf{P}_m$, and thus spectral representation given by Eq. (2.167). If $f(z)$ is any function defined on the spectrum of $\mathbf{A}$, then define

$$f(\mathbf{A}) \equiv \sum_{k=1}^{m} f(\lambda_k)\mathbf{P}_k \tag{2.200}$$

This definition is as general as possible, and coincides with other natural definitions. Furthermore, it provides the equivalent of analytic continuation beyond the circle of convergence of a power series. In the following examples, it is understood that $\mathbf{A}$ is a normal operator with spectral representation (2.167).

❑ **Example 2.12.** Suppose $f(z) = 1/z$. Then if $\mathbf{A}$ is nonsingular, we have

$$f(\mathbf{A}) = \sum_{k=1}^{m} \frac{1}{\lambda_k} \mathbf{P}_k = \mathbf{A}^{-1} \tag{2.201}$$

as expected (no eigenvalue vanishes if $\mathbf{A}$ is nonsingular).   ∎

❑ **Example 2.13.** Suppose $f(z) = z^*$. Then we have

$$f(\mathbf{A}) = \sum_{k=1}^{m} \lambda_k^* \mathbf{P}_k = \mathbf{A}^\dagger \tag{2.202}$$

so the adjoint operator is the equivalent of the complex conjugate of a complex number (at least for a normal operator).   ∎

❑ **Example 2.14.** Suppose $f(z) = 1/(1 - \alpha z)$. Then if $1/\alpha$ is not in the spectrum of $\mathbf{A}$, we have

$$f(\mathbf{A}) = \sum_{k=1}^{m} \frac{1}{1 - \alpha\lambda_k} \mathbf{P}_k = (1 - \alpha\mathbf{A})^{-1} \tag{2.203}$$

which gives the analytic continuation of the series (2.197) outside its circle of convergence in the complex α-plane.   ∎

❑ **Example 2.15.** Suppose

$$w = f(z) = \frac{z - i}{z + i} \tag{2.204}$$

which maps the real z-axis into the unit circle except the point $w = 1$ (show this). If $\mathbf{A}$ is self-adjoint (the λ_k are all real), then the operator

$$\mathbf{U} \equiv f(\mathbf{A}) = \sum_{k=1}^{m} \frac{\lambda_k - i}{\lambda_k + i} \mathbf{P}_k = \frac{\mathbf{A} - i\mathbf{1}}{\mathbf{A} + i\mathbf{1}} \tag{2.205}$$

is unitary. $\mathbf{U}$ is the *Cayley transform* of $\mathbf{A}$. Also, if $\mathbf{U}$ is unitary with no eigenvalue $= 1$, then

$$\mathbf{A} = f^{-1}(\mathbf{U}) = i\frac{1 + \mathbf{U}}{1 - \mathbf{U}} \tag{2.206}$$

is self-adjoint. $\mathbf{A}$ is the *inverse Cayley transform* of $\mathbf{U}$.   ∎

❑ **Example 2.16.** Suppose $f(z) = \sqrt{z}$, and suppose $\mathbf{A} \geq 0$. Then

$$f(\mathbf{A}) = \sum_{k=1}^{m} \sqrt{\lambda_k} \, \mathbf{P}_k = \sqrt{\mathbf{A}} \geq 0 \tag{2.207}$$

and $(\sqrt{\mathbf{A}})^2 = \mathbf{A}$.   ∎

2.5   Linear Dynamical Systems

A system whose coordinates $x_1, \ldots, x_n$ evolve in time according to the equations of motion

$$\frac{dx_k}{dt} = \sum_\ell A_{k\ell} x_\ell \tag{2.208}$$

is a *linear dynamical system*. The coordinates $x_1, \ldots, x_n$ characterize a vector in a linear vector space $\mathcal{V}^n$, which allows Eq. (2.208) to be written in the compact form

$$\frac{dx}{dt} = \mathbf{A}x \tag{2.209}$$

in which $\mathbf{A}$ is a linear operator on $\mathcal{V}^n$ with matrix elements $A_{k\ell}$.

The solution to Eq. (2.209) that starts from the vector $x(0)$ at time $t = 0$ is

$$x(t) = e^{t\mathbf{A}} x(0) \tag{2.210}$$

This defines a solution for all t, according to the discussion in Section 2.4. To describe the time dependence of the solution in more detail, recall the general decomposition

$$\mathbf{A} = \mathbf{D} + \mathbf{N} \tag{2.211}$$

with $\mathbf{D}$ diagonalizable and $\mathbf{N}$ nilpotent (Eq. (2.151)). Choose a basis $x_1, \ldots, x_n$ in which $\mathbf{D}$ is diagonal, with eigenvalues $\lambda_1, \ldots, \lambda_n$, and suppose that $x(0)$ is expressed as

$$x(0) = \sum_{k=1}^{n} \xi_{k0} x_k \tag{2.212}$$

Then the solution (2.210) has the form

$$x(t) = \sum_{k=1}^{n} \xi_k(t) x_k \tag{2.213}$$

with $\xi_k(0) = \xi_{k0}$. If $\mathbf{N} = 0$, this solution has the simple form

$$\xi_k(t) = e^{\lambda_k t} \xi_{k0} \tag{2.214}$$

❑ **Example 2.17.**   Consider a damped harmonic oscillator, with the equation of motion

$$\frac{d^2 x}{dt^2} + 2\gamma \frac{dx}{dt} + \omega_0^2 x = 0 \tag{2.215}$$

where $\gamma > 0$ to insure damping, and ω_0 is real. With the new variable

$$u \equiv \frac{dx}{dt} \tag{2.216}$$

the second-order equation (2.215) becomes the pair of equations

$$\frac{dx}{dt} = u \qquad \frac{du}{dt} = -2\gamma u - \omega_0^2 x \tag{2.217}$$

This has the form of Eq. (2.209) with

$$\mathbf{A} = \begin{pmatrix} 0 & 1 \\ -\omega_0^2 & -2\gamma \end{pmatrix} \tag{2.218}$$

The eigenvalues of $\mathbf{A}$ are evidently given by

$$\lambda_\pm \equiv -\gamma \pm \sqrt{-\omega_0^2 + \gamma^2} \tag{2.219}$$

and $\mathrm{Re}\,\lambda_\pm < 0$. If $\omega_0^2 > \gamma^2$, the eigenvalues are a complex conjugate pair, and the system undergoes damped oscillations; otherwise, both eigenvalues are real and the system is purely damped. The solutions of the system (2.217) are given explicitly by

$$x(t) = A_+ e^{\lambda_+ t} + A_- e^{\lambda_- t} \tag{2.220}$$

$$u(t) = \lambda_+ A_+ e^{\lambda_+ t} + \lambda_- A_- e^{\lambda_- t} \tag{2.221}$$

with constants $A_\pm$ chosen to satisfy the initial conditions (see also the discussion after Eq. (6.94)).   ∎

If $\mathbf{A}$ has one or more degenerate eigenvalues, then it is possible that $\mathbf{N} \neq \mathbf{0}$. Even in this case, the solution is straightforward. Since $\mathbf{N}^p = \mathbf{0}$ for some p, the power series expansion

$$e^{t\mathbf{N}} = \sum_{k=0}^{p-1} \frac{t^k}{k!} \mathbf{N}^k \tag{2.222}$$

has only a finite number of nonzero terms, so the exponential factors are simply modified by polynomials in t.

❑ **Example 2.18.** Consider a two-dimensional system, with

$$\mathbf{A} = \begin{pmatrix} \lambda & a \\ 0 & \lambda \end{pmatrix} \tag{2.223}$$

We have

$$e^{t\mathbf{A}} = \begin{pmatrix} e^{\lambda t} & at \\ 0 & e^{\lambda t} \end{pmatrix} \tag{2.224}$$

so the solution starting from the initial vector $x = (\xi_{10}, \xi_{20})$ is given by

$$\xi_1(t) = e^{\lambda t}(\xi_{10} + at\xi_{20}) \tag{2.225}$$

$$\xi_2(t) = e^{\lambda t}\xi_{20} \tag{2.226}$$

Note here that $x_1(t)$ is unbounded if $\mathrm{Re}\,\lambda = 0$.   ∎

In general, the behavior of the solution $\xi_k(t)$ for $t \to \infty$ depends on the corresponding eigenvalue λ_k of $\mathbf{A}$:

(i) if $\mathrm{Re}\,\lambda_k < 0$, then $\xi_k(t) \to 0$ for $t \to \infty$,

(ii) if $\mathrm{Re}\,\lambda_k > 0$, then $\xi_k(t) \to \infty$ for $t \to \infty$, and

(iii) if $\mathrm{Re}\,\lambda_k = 0$, then $\xi_k(t)$ moves around a circle of fixed radius $|\xi_{k0}|$ in the complex ξ-plane if $\mathbf{N} = \mathbf{0}$, unless $\lambda_k = 0$, in which case $\xi_k(t)$ is constant.

The behavior of $\xi_k(t)$ for $t \to \infty$ if $\mathrm{Re}\,\lambda_k \neq 0$ is qualitatively independent of whether or not $\mathbf{N} = \mathbf{0}$, as the exponential factor dominates the asymptotic behavior. If $\mathbf{N} \neq \mathbf{0}$, the $\xi_k(t)$ corresponding to $\mathrm{Re}\,\lambda_k = 0$ grows in magnitude like a power of t for $t \to \infty$ (the power may be zero if the corresponding vector x_k is actually an eigenvector of $\mathbf{A}$).

The vector $x = \theta$ is a *fixed point*, or *equilibrium solution*, of Eq. (2.209), since if $x(0) = \theta$, then $x(t) = \theta$ for all t. A fixed point x_* is *stable* if there is a neighborhood $\mathcal{N}(x_*)$ of x_* such that any solution starting in $\mathcal{N}(x_*)$ remains in $\mathcal{N}(x_*)$ for all t. x_* is *asymptotically stable* if there is a neighborhood of the x_* such that any solution starting in the neighborhood approaches x_* asymptotically for $t \to \infty$. Such a fixed point is a *sink*, or an *attractor*.

It is clear here that $x(0) = \theta$ is asymptotically stable if and only if $\mathrm{Re}\,\lambda_k < 0$ for every eigenvalue of $\mathbf{A}$, while stability requires only $\mathrm{Re}\,\lambda_k \leq 0$ and $\mathbf{N} = \mathbf{0}$ on the invariant manifolds belonging to the eigenvalues with $\mathrm{Re}\,\lambda_k = 0$. Physically, asymptotic stability means that small displacements from the equilibrium will eventually be exponentially damped, while simple stability allows bounded oscillations without damping, at least in some directions. A fixed point for which $\mathrm{Re}\,\lambda_k > 0$ for every eigenvalue of $\mathbf{A}$ is a *source*, or *repellor*. The solution curves move outward in all directions from a source.

A fixed point for which no eigenvalue has $\mathrm{Re}\,\lambda_k = 0$ is *hyperbolic*. In general, a hyperbolic fixed point will have a *stable manifold* $\mathcal{M}^s$ spanned by the invariant manifolds $\mathcal{M}_k$ for those eigenvalues with $\mathrm{Re}\,\lambda_k < 0$, and an *unstable manifold* $\mathcal{M}^u$ spanned by the invariant manifolds $\mathcal{M}_k$ for those eigenvalues with $\mathrm{Re}\,\lambda_k > 0$. As $t \to \infty$, the solution curves will be drawn very close to the unstable manifold as the components in the stable manifold are exponentially damped. The simple case of a two-dimensional system for which the operator $\mathbf{A}$ has one positive and one negative eigenvalue is illustrated in Fig. 2.8. Imaginary parts of the eigenvalues will give oscillatory behavior not shown in the figure.

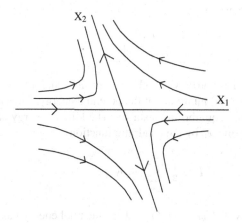

Figure 2.8: Schematic form of solutions to Eq. (2.208) in the hyperbolic case with one positive and one negative eigenvalue.

The types of behavior described here occur in nonlinear systems as well. As already noted, a nonlinear system can often be approximated by a linear system near a fixed point, and the linear properties qualitatively describe the system near the fixed point.

A   Small Oscillations

Energy conservation is a fundamental principle for many systems in classical mechanics (such systems are called *conservative*). For example, a harmonic oscillator of mass m and natural frequency ω_0 has a total energy

$$E = \tfrac{1}{2}mu^2 + \tfrac{1}{2}m\omega_0^2 x^2 \tag{2.A1}$$

($u = dx/dt$). This corresponds to the equation of motion (2.215) with $\gamma = 0$ (no damping). For a conservative system with coordinates $x_1, \ldots, x_n$ and corresponding velocities $u_1, \ldots, u_n$ ($u_k = dx_k/dt$), the energy often has the form

$$E = \tfrac{1}{2} \sum_{k,\ell=1}^{n} u_k T_{k\ell}(x) u_\ell + V(x_1, \ldots, x_n) \tag{2.A2}$$

Here the first term is the kinetic energy T of the system (hence the matrix $\mathbf{T}(x)$, which may depend on the coordinates, must be positive definite), and $V(x_1, \ldots, x_n)$ is the potential energy of the system.

An equilibrium of the system corresponds to a stationary point of the potential energy. If the system has an equilibrium at a point x_0 with coordinates $x_{10}, \ldots, x_{n0}$, then the first derivatives of the potential vanish at x_0, and we can assume that the expansion

$$V(x) \simeq V(x_0) + \tfrac{1}{2} \sum_{k,\ell=1}^{n} (x_k - x_{k0}) V_{k\ell}(x_\ell - x_{\ell0}) + \cdots \tag{2.A3}$$

exists, with

$$V_{k\ell} = \left. \frac{\partial^2 V}{\partial x_k \partial x_\ell} \right|_{x=x_0} \tag{2.A4}$$

The matrix $\mathbf{V} = (V_{k\ell})$ is positive definite at a stable equilibrium, which corresponds to a (local) minimum of the potential energy. Thus the eigenvalues of $\mathbf{V}$ determine whether or not the equilibrium is stable. The kinetic energy can also be approximated near the equilibrium point x_0 by the quadratic function

$$T \simeq \tfrac{1}{2} \sum_{k,\ell=1}^{n} u_k K_{k\ell} u_\ell \tag{2.A5}$$

with $K_{k\ell} = T_{k\ell}(x_0)$. Thus the total energy has the approximate quadratic form

$$E \simeq V(x_0) + \tfrac{1}{2} \sum_{k,\ell=1}^{n} u_k K_{k\ell} u_\ell + \tfrac{1}{2} \sum_{k,\ell=1}^{n} (x_k - x_{k0}) V_{k\ell}(x_\ell - x_{\ell0}) \tag{2.A6}$$

near the equilibrium point (the constant is not important for the motion near x_0 and will be dropped).

Since the matrix $\mathbf{K}$ is positive definite, we can choose a coordinate system in which it is diagonal, with positive eigenvalues, so that the kinetic energy has the form

$$K = \tfrac{1}{2} \sum_{k=1}^{n} \mu_k \left(\frac{d\xi_k}{dt} \right)^2 \tag{2.A7}$$

where the $\xi_k = x_k - x_{k0}$ are the coordinates of the displacement from equilibrium, and μ_k is the "mass" associated with the coordinate ξ_k. Introduce rescaled coordinates $y_k \equiv \sqrt{\mu_k}\,\xi_k$; the quadratic energy (2.A6) then has the form

$$E = \tfrac{1}{2} \sum_{k=1}^{n} \left(\frac{dy_k}{dt} \right)^2 + \tfrac{1}{2} \sum_{k,\ell=1}^{n} y_k \, \overline{V}_{k\ell} \, y_\ell \tag{2.A8}$$

where the $\overline{V}_{k\ell} = V_{k\ell}/\sqrt{\mu_k \mu_\ell}$ are the matrix elements of the potential energy matrix in terms of the rescaled coordinates. In the operator form,

$$\overline{\mathbf{V}} = \mathbf{K}^{-\frac{1}{2}} \mathbf{V} \mathbf{K}^{-\frac{1}{2}} \tag{2.A9}$$

The operator $\overline{\mathbf{V}}$ is self-adjoint (check this), so that we can diagonalize it by a unitary transformation, which does not affect the form of the kinetic energy in Eq. (2.A8). Denote the eigenvalues of $\overline{\mathbf{V}}$ by $a_1, \ldots, a_n$, and coordinates of a vector in the basis in which $\overline{\mathbf{V}}$ is diagonal by $\eta_1, \ldots, \eta_n$. Then the quadratic energy has the form

$$E = \tfrac{1}{2} \sum_{k=1}^{n} \left\{ \left(\frac{d\eta_k}{dt} \right)^2 + a_k \eta_k^2 \right\} \tag{2.A10}$$

and the equations of motion for the η_k are

$$\frac{d^2 \eta_k}{dt^2} = -a_k \eta_k \tag{2.A11}$$

Thus each of the η_k evolves independently of the others; the η_k define the *normal modes* of the system near the equilibrium x_0.

If $a_k > 0$, which will be the case for all the eigenvalues of $\overline{\mathbf{V}}$ at a stable equilibrium, let $a_k \equiv \omega_k^2$, and the coordinate η_k oscillates with (angular) frequency ω_k, which is a *natural frequency* of the system. If $a_k < 0$, then let $a_k \equiv -\alpha_k^2$, and the coordinate η_k is proportional to $\exp(\pm\alpha_k t)$, so it will in general have a component that grows exponentially for large t. In practice, this means that the coordinate η_k moves away from the equilibrium point, so that the approximation of a small displacement from equilibrium becomes invalid, and the equilibrium is unstable in the direction defined by the coordinate η_k. Finally, if $a_k = 0$, then the coordinate η_k evolves in time with constant velocity. Such a mode is called a *zero mode*; the approximate energy (2.A10) is independent of the corresponding coordinate. This zero mode indicates a symmetry of the approximate system, which may or may not be a symmetry of the original system as well.

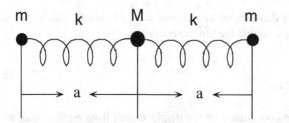

Figure 2.9: Three collinear masses joined by two springs, each with equilibrium length a.

❑ **Example 2.19.** Consider the linear system of three masses connected by two springs as shown in Fig. 2.9. Let ξ_1, ξ_2 denote the displacements of the masses m from their equilibrium positions, and ξ_3 the displacement of the mass M; let u_1, u_2, u_3 denote the corresponding velocities. The energy of the system is given by

$$E = \tfrac{1}{2}m(u_1^2 + u_2^2) + \tfrac{1}{2}Mu_3^2 + \tfrac{1}{2}k[(\xi_1 - \xi_3)^2 + (\xi_2 - \xi_3)^2] \tag{2.A12}$$

In terms of rescaled coordinates $y_1 \equiv \sqrt{m}\xi_1$, $y_2 \equiv \sqrt{m}\xi_2$, $y_3 \equiv \sqrt{M}\xi_3$, and velocities $v_k \equiv dy_k/dt$, the energy is

$$E = \tfrac{1}{2}(v_1^2 + v_2^2 + v_3^2) + \tfrac{1}{2}(y, \overline{\mathbf{V}}y) \tag{2.A13}$$

where $\overline{\mathbf{V}}$ is the 3×3 matrix

$$\overline{\mathbf{V}} \equiv \frac{k}{m}\begin{pmatrix} 1 & 0 & -\alpha \\ 0 & 1 & -\alpha \\ -\alpha & -\alpha & 2\alpha^2 \end{pmatrix} \tag{2.A14}$$

with $\alpha = \sqrt{m/M}$. The eigenvalues of $\overline{\mathbf{V}}$ are 0, k/m, and $(1 + 2\alpha^2)k/m$, with corresponding eigenvectors proportional to $(\alpha, \alpha, 1)$, $(1, -1, 0)$, and $(1, 1, -2\alpha)$, respectively. While these eigenvalues and eigenvectors can be obtained by straightforward algebra, it is also instructive to understand how they arise physically.

The zero mode has eigenvector $(\alpha, \alpha, 1)$ corresponding to the center of mass coordinate $m(\xi_1 + \xi_2) + M\xi_3$. Since the potential energy depends only on the relative coordinates of the masses, and not on the center of mass coordinate, the energy is unchanged if all three masses are shifted by the same displacement. The total momentum of the system is conserved, and the center of mass moves with constant velocity.

The mode with frequency $\sqrt{k/m}$ and eigenvector $(1, -1, 0)$ corresponds to an oscillation of the two masses m with the mass M fixed. The eigenvector corresponds to the two masses oscillating with opposite displacements; the frequency is just the frequency of the a single spring connecting the two masses, since the mass M does not oscillate in this mode.

The third mode, with frequency $\sqrt{(1 + 2\alpha^2)k/m}$ and eigenvector $(1, 1, -2\alpha)$, corresponds to an oscillation in which the two masses m oscillate in phase, such that the center of mass of these two has a displacement in a direction opposite to the displacement of the mass M (the center of mass must remain fixed). ∎

Bibliography and Notes

There are many modern textbooks on the subject of linear algebra and finite-dimensional vector spaces, to which no special references are given. A classic book on finite-dimensional vector spaces is

> Paul R. Halmos, *Finite Dimensional Vector Spaces* (2nd edition), van Nostrand (1958).

It is quite readable, though mathematically formal and rigorous, and contains many useful exercises.

A standard earlier textbook is

> Frederick W. Byron, Jr. and Robert W. Fuller, *Mathematics of Classical and Quantum Physics*, Dover (1992).

It is a reprint of a two-volume work first published in 1969–70. It has a thorough treatment of linear vector spaces and operators, with many physics examples.

An excellent introduction to linear vector spaces and linear operators that leads to the applications to dynamical systems is in

> Morris W. Hirsch and Stephen Smale, *Differential Equations, Dynamical Systems and Linear Algebra*, Academic Press (1974).

Starting from the analysis of linear systems, the book describes dynamical systems in terms of vector fields, and proceeds to a general description of the qualitative behavior of systems near critical points of the vector field. A major revision of this book that has more emphasis on nonlinear systems is

> Morris W. Hirsch, Stephen Smale and Robert Devaney, *Differential Equations, Dynamical Systems and an Introduction to Chaos* (2nd edition), Academic Press (2004).

In this new edition, the discussion of vector spaces has been drastically reduced, and more emphasis has been given to nonlinear systems; there are many new applications and examples, especially of systems with chaotic behavior. There are also many computer-generated pictures to illustrate qualitative behavior of systems.

A standard graduate textbook on classical mechanics is

> Herbert Goldstein, *Classical Mechanics* (2nd edition), Addison-Wesley (1980).

This book describes the theory of rotations in three dimensions and its application to the motion of a rigid body with one point held fixed. It also has a useful chapter on the theory of small oscillations. However, it does not make use of the geometrical ideas introduced in Chapter 3, in contrast to several books cited there.

An excellent alternative that also has a good discussion of small oscillations is

> L. D. Landau and E. M. Lifshitz, *Mechanics* (3rd edition), Butterworth (1976).

This was the first of the famous series by Landau and Lifshitz that covered the foundations of physics at a graduate level as viewed in the 1960s. It is still useful today.

Problems

1. Consider a real two-dimensional vector space with vectors x_1, x_2, and x_3 that satisfy

$$\|x_1\|^2 = \|x_3\|^2 = 1 \qquad \|x_2\|^2 = \tfrac{5}{4} \tag{$*$}$$

   and

$$x_3 = x_2 - (x_1, x_2)x_1 \tag{$**$}$$

   (i) Show that x_1, x_3 form a basis.

   (ii) How many linearly independent vectors x_2 satisfy ($*$) and ($**$)?

   (iii) For a vector x_2 satisfying ($*$) and ($**$), find a unit vector orthogonal to x_2. Express this vector as a linear combination of x_1 and x_3.

2. Let $\{\phi_1, \phi_2\}$ be an orthonormal system in $\mathbf{C}^2$ and consider the unit vectors u_1, u_2, u_3 defined by

$$u_1 = \phi_1 \quad u_2 = -\tfrac{1}{2}\phi_1 + \tfrac{\sqrt{3}}{2}\phi_2 \quad u_3 = -\tfrac{1}{2}\phi_1 - \tfrac{\sqrt{3}}{2}\phi_2$$

   (i) Draw a picture of the three unit vectors in the plane.

   (ii) Show that for any vector x in $\mathbf{C}^2$,

$$\|x\|^2 = \tfrac{2}{3}\sum_{k=1}^{3}|(u_k, x)|^2 \tag{$*$}$$

   (iii) Show that any vector x in $\mathbf{C}^2$ can be written as

$$x = \tfrac{2}{3}\sum_{k=1}^{3}(u_k, x)u_k \tag{$**$}$$

   Remark. Here ($**$) looks like the expansion of a vector along a complete orthonormal system, except for the factor $\tfrac{2}{3}$. In fact the vectors u_1, u_2, u_3 form a (*tight*) *frame*, which is characterized by a relation of the form ($*$), although the constant may be different. The vectors u_1, u_2, u_3 do *not* form a basis, since they are not linearly independent; they are "redundant," or "overcomplete." Such sets are nevertheless useful in various contexts; see the book by Daubechies cited in Chapter 6, where this example is given. □

3. Show that the polynomials $p(t)$ of degree $\leq N$ form a linear vector space if we define addition and multiplication by scalars (which may be real or complex) in the natural way. What is the dimension of this space? Write down a basis for this space. If we define the scalar product of two polynomials $p(t)$ and $q(t)$ by

$$(p, q) \equiv \int_0^1 p^*(t)q(t)\, dt$$

   then this space is unitary (verify this). Compute explicitly the first four elements of the complete orthonormal system obtained by the Gram–Schmidt process from the basis you have just given.

4. (i) Show that

$$\sum_{n=1}^{N} e^{\frac{2\pi i k n}{N}} e^{-\frac{2\pi i q n}{N}} = N\delta_{kq}$$

$(k, q = 1, \ldots, N)$.

(ii) Show that a set of numbers $f_1, \ldots, f_N$ can be expressed in the form

$$f_n = \sum_{k=1}^{N} c_k e^{\frac{2\pi i k n}{N}} \tag{$*$}$$

with coefficients c_k given by

$$c_k = \frac{1}{N} \sum_{m=1}^{N} f_m e^{-\frac{2\pi i k m}{N}}$$

(iii) Show that if the $f_1, \ldots, f_N$ are real, then $c_{N-k} = c_k^* \ (k = 1, \ldots, N)$.

Remark. Expansion $(*)$ is the *finite Fourier transform*. $\qquad\square$

5. Show that

$$\sum_{n=1}^{N-1} \sin \frac{n\pi k}{N} \sin \frac{n\pi q}{N} = \frac{N}{2}\delta_{kq}$$

$(k, q = 1, \ldots, N-1)$ and thus that the set of numbers $\{f_1, \ldots, f_{N-1}\}$ can be expressed as

$$f_n = \sum_{k=1}^{N-1} b_k \sin \frac{k\pi n}{N}$$

with coefficients b_k given by

$$b_k = \frac{2}{N} \sum_{m=1}^{N-1} f_m \sin \frac{m\pi k}{N}$$

Remark. This expansion is the *finite Fourier sine transform*. $\qquad\square$

6. A *lattice* is a set of vectors such that if x_1 and x_2 are lattice vectors, then any linear combination

$$x = n_1 x_1 + n_2 x_2$$

of x_1 and x_2 with integer coefficients n_1 and n_2 is also a lattice vector. The linearly independent vectors $e_1, \ldots, e_N$ are *generators* of an N-dimensional lattice if the lattice consists of all linear combinations

$$x = \sum_{k=1}^{N} n_k e_k$$

with integer coefficients $n_1, \ldots, n_N$. The *reciprocal lattice*, or *dual lattice*, is the set of vectors k for which $k \cdot x$ is an integer for any lattice vector x.

(i) Show that the reciprocal lattice is actually a lattice, that is, if k_1 and k_2 are in the reciprocal lattice, then so is any linear combination $n_1 k_1 + n_2 k_2$ with integer coefficients.

(ii) The reciprocal lattice is generated by the basis dual to $e_1, \ldots, e_N$ in $\mathbf{R}^N$.

7. (i) If $\mathbf{A}$, $\mathbf{B}$ and $\mathbf{C}$ are three linear operators, then

$$[[\mathbf{A}, \mathbf{B}], \mathbf{C}] + [[\mathbf{B}, \mathbf{C}], \mathbf{A}] + [[\mathbf{C}, \mathbf{A}], \mathbf{B}] = 0$$

Remark. This result is the *Jacobi identity*.   □

(ii) $\mathbf{A}$, $\mathbf{B}$ and $\mathbf{C}$ satisfy the further identities

$$\{\{\mathbf{A}, \mathbf{B}\}, \mathbf{C}\} - \{\mathbf{A}, \{\mathbf{B}, \mathbf{C}\}\} = [\mathbf{B}, [\mathbf{A}, \mathbf{C}]]$$

$$\{[\mathbf{A}, \mathbf{B}], \mathbf{C}\} - \{\mathbf{A}, [\mathbf{B}, \mathbf{C}]\} = [\{\mathbf{A}, \mathbf{C}\}, \mathbf{B}]$$

$$[\{\mathbf{A}, \mathbf{B}\}, \mathbf{C}] - [\mathbf{A}, \{\mathbf{B}, \mathbf{C}\}] = [\mathbf{B}, \{\mathbf{A}, \mathbf{C}\}]$$

8. Let $\mathbf{A}$ be a linear operator on a unitary vector space $\mathcal{V}$. If $\mathcal{M}$ is an invariant manifold of $\mathbf{A}$, then $\mathcal{M}^{\perp}$ is an invariant manifold of $\mathbf{A}^{\dagger}$.

9. Show that with addition and multiplication by scalars defined in the natural way, the linear operators on an n-dimensional vector space $\mathcal{V}^n$ themselves form a linear vector space. What is the dimension of the vector space? Give a basis for linear operators on $\mathbf{C}^2$ that includes the three Pauli matrices defined by Eq. (2.81).

10. Let $\mathbf{U}$ and $\mathbf{V}$ be unitary operators. Then

    (i) $\mathbf{UV}$ is unitary,

    (ii) $\mathbf{U} + \mathbf{V}$ is unitary if and only if

$$\mathbf{UV}^{\dagger} = \omega \mathbf{P} + \omega^2 (1 - \mathbf{P})$$

with $\mathbf{P}$ a projection operator and $\omega^3 = 1$.

11. What are the Euler angles $(\bar{\phi}, \bar{\theta}, \bar{\psi})$ of the rotation inverse to the rotation $\mathbf{R}(\phi, \theta, \psi)$, given the constraints $0 \leq \bar{\phi} < 2\pi$, $0 \leq \bar{\theta} \leq \pi$, and $0 \leq \bar{\psi} < 2\pi$?

12. Show that the rotation $\mathbf{R}(\phi, \theta, \psi)$ in three dimensions can be expressed as a product of
    (i) rotation through angle ψ about the Z'-axis,
    (ii) rotation through angle θ about the Y'-axis, and
    (iii) rotation through angle ϕ about the Z'-axis.

13. Let $\mathbf{n} = \mathbf{n}(\theta, \phi)$ be a unit vector in the direction defined by the usual spherical angles θ, ϕ. Show that the matrix corresponding to rotation through angle Φ about $\mathbf{n}$ can be expressed as

    $$\mathbf{R_n}(\Phi) = \mathbf{R}(\phi, \theta, \xi)\mathbf{R}_z(\Phi)\mathbf{R}^{-1}(\phi, \theta, \xi)$$

    with an arbitrary angle ξ.

14. Show that the eigenvalues of the proper rotation matrix $\mathbf{R_n}(\Phi)$ in Problem 13 are

    $$\lambda = 1, e^{\pm i\Phi}$$

    Remark. The eigenvector $\mathbf{n}$ belonging to eigenvalue 1 is evidently along the axis of rotation. □

    (ii) Explain the relation of the (complex) eigenvectors belonging to eigenvalues $\exp(\pm i\Phi)$ to real coordinate axes in the plane normal to $\mathbf{n}$. What is the form of the rotation matrix in a real basis?

15. Show that if $\mathbf{R}$ is a proper rotation matrix that is symmetric, then either $\mathbf{R}$ is the identity, or $\mathbf{R}$ corresponds to a rotation through angle π about some axis.

16. (i) Let $\mathbf{R}_x(\Phi), \mathbf{R}_y(\Phi)$, and $\mathbf{R}_z(\Phi)$ be the 3×3 matrices corresponding to rotations through angle Φ about the X, Y, Z axes, respectively. Write down explicit forms for these matrices.

    (ii) Define the matrices $\mathbf{L}_x, \mathbf{L}_y$, and $\mathbf{L}_z$ by

    $$\mathbf{L}_\alpha \equiv i\mathbf{R}'_\alpha(\Phi = 0)$$

    ($\alpha = x, y, z$) where the prime denotes differentiation with respect to Φ. Show that the commutator

    $$[\mathbf{L}_\alpha, \mathbf{L}_\beta] \equiv \mathbf{L}_\alpha \mathbf{L}_\beta - \mathbf{L}_\beta \mathbf{L}_\alpha = i \sum_\gamma \varepsilon_{\alpha\beta\gamma} \mathbf{L}_\gamma$$

    where $\varepsilon_{\alpha\beta\gamma}$ is the usual antisymmetric symbol on three indices.

    Remark. These commutation relations can be written informally as $\vec{\mathbf{L}} \times \vec{\mathbf{L}} = i\vec{\mathbf{L}}$. That the commutators of $\mathbf{L}_x, \mathbf{L}_y$, and $\mathbf{L}_z$ are expressed as linear combinations of themselves means that the $\mathbf{L}_x, \mathbf{L}_y$, and $\mathbf{L}_z$ span a *Lie algebra*. This and more general Lie algebras are studied in Chapter 10. □

    (iii) Show that

    $$\mathbf{R}_\alpha(\Phi) = \exp(-i\mathbf{L}_\alpha\Phi)$$

17. (i) Show that any 2×2 matrix $\mathbf{M}$ can be expressed as

$$\mathbf{M} = \exp\left\{ i\left(\xi_0 \mathbf{1} + \vec{\sigma} \cdot \vec{\xi} \right) \right\}$$

where $\vec{\sigma}$ denotes the usual three Pauli matrices, and $\mathbf{1}$ is the 2×2 unit matrix.

(ii) Show that $\mathbf{M}$ is unitary if and only if ξ_0 and $\vec{\xi}$ are real.

(iii) Show that

$$\operatorname{tr} \mathbf{M} = 2 \exp(i\xi_0) \cos | \vec{\xi} | \qquad \det \mathbf{M} = \exp(2i\xi_0)$$

18. Let $\mathbf{S}_\alpha \equiv \frac{1}{2}\sigma_\alpha$ where σ_α are the 2×2 Pauli matrices (Eq. (2.81)) ($\alpha = x, y, z$).

(i) Show that the $\mathbf{S}_\alpha$ satisfy the same commutation relations as the $\mathbf{L}_\alpha$ in Problem 16.

Remark. The 2×2 matrices $\mathbf{U}_\alpha(\Phi)$ defined by

$$\mathbf{U}_\alpha(\Phi) \equiv \exp(-i\mathbf{S}_\alpha\Phi)$$

then satisfy the same multiplication rules as the corresponding 3×3 matrices $\mathbf{R}_\alpha(\Phi)$ defined in Problem 16, and can thus be used to study geometrical properties of rotations. This is useful, since 2×2 matrices are easier to multiply than 3×3 matrices.   □

(ii) Find explicit 2×2 matrices $\mathbf{U}_x(\theta), \mathbf{U}_y(\theta), \mathbf{U}_z(\theta)$ corresponding to rotations through angle θ about each of the coordinate axes.

(iii) Construct a 2×2 matrix $\mathbf{U}(\phi, \theta, \psi)$ corresponding to a rotation characterized by Euler angles ϕ, θ, ψ.

19. Suppose $\mathbf{A}$ is a 2×2 matrix with $\det \mathbf{A} = 1$. Show that

$$\tfrac{1}{2} \operatorname{tr} \mathbf{A}\mathbf{A}^\dagger \geq 1$$

and the equality is true if and only if $\mathbf{A}$ is unitary.

20. A rotation in a real vector space is described by a real unitary matrix $\mathbf{R}$, also known as an *orthogonal* matrix. The matrix $\mathbf{R}$ can be diagonalized in a complex vector space, but not in a real vector space, since its eigenvalues are complex (see Example 2.5, where the eigenvalues of the two-dimensional rotation matrix are $\exp(\pm i\theta)$). Show that

(i) in $2n$ dimensions, the eigenvalues of a proper rotation matrix $\mathbf{R}$ ($\det \mathbf{R} = 1$) occur in complex conjugate pairs of the form

$$\lambda = e^{\pm i\theta_1}, e^{\pm i\theta_2}, \ldots, e^{\pm i\theta_n}$$

(ii) in $2n + 1$ dimensions, the eigenvalues have the same form with an additional eigenvalue $\lambda_{2n+1} = 1$. (*Note.* You may use the fact that the complex roots of a polynomial with real coefficients occur in complex conjugate pairs.)

(iii) a proper rotation $\mathbf{R}$ in $2n$ dimensions can be brought to the standard form

$$\mathbf{R} = \begin{pmatrix} \cos\theta_1 & -\sin\theta_1 & 0 & 0 & \cdots & 0 & 0 \\ \sin\theta_1 & \cos\theta_1 & 0 & 0 & \cdots & 0 & 0 \\ 0 & 0 & \cos\theta_2 & -\sin\theta_2 & \cdots & 0 & 0 \\ 0 & 0 & \sin\theta_2 & \cos\theta_2 & \cdots & 0 & 0 \\ \vdots & \vdots & \vdots & \vdots & \ddots & \vdots & \vdots \\ 0 & 0 & 0 & 0 & \cdots & \cos\theta_n & -\sin\theta_n \\ 0 & 0 & 0 & 0 & \cdots & \sin\theta_n & \cos\theta_n \end{pmatrix}$$

What is the corresponding form in $2n + 1$ dimensions? How do these results change for a rotation-reflection matrix (a real unitary matrix $\mathbf{R}$ with $\det \mathbf{R} = -1$)?

21. Consider the matrix

$$\mathbf{A} = \frac{1}{2}\begin{pmatrix} 3 & -1 \\ -1 & 3 \end{pmatrix}$$

Find the eigenvalues of $\mathbf{A}$, and find one eigenvector for each distinct eigenvalue.

22. Consider the matrices

$$\mathbf{A}_1 \equiv \frac{1}{2}\begin{pmatrix} 1 & \sqrt{3} \\ \sqrt{3} & -1 \end{pmatrix} \qquad \mathbf{A}_2 \equiv \frac{1}{2}\begin{pmatrix} 1 & -\sqrt{3} \\ -\sqrt{3} & -1 \end{pmatrix} \qquad \mathbf{A}_3 \equiv \begin{pmatrix} -1 & 0 \\ 0 & 1 \end{pmatrix}$$

(i) Show that

$$\mathbf{A}_k \mathbf{A}_l \mathbf{A}_k = \mathbf{A}_m$$

where (k, l, m) denotes a cyclic permutation of $(1, 2, 3)$.

(ii) Show that the operators $\mathbf{P}_k^{\pm} \equiv \frac{1}{2}(1 \pm \mathbf{A}_k)$ are projectors ($k = 1, 2, 3$), and find the projection manifolds.

(iii) Find a complete orthonormal set of eigenvectors of each $\mathbf{A}_k$, together with the corresponding eigenvalues.

23. The linear operator $\mathbf{A}$ on $\mathbf{C}^2$ has the matrix representation

$$\mathbf{A} = \begin{pmatrix} 1 & -i \\ i & -1 \end{pmatrix}$$

(i) Find the eigenvalues of $\mathbf{A}$, and express the corresponding eigenvectors in terms of the unit vectors

$$u_+ \equiv \begin{pmatrix} 1 \\ 0 \end{pmatrix} \qquad u_- \equiv \begin{pmatrix} 0 \\ 1 \end{pmatrix}$$

(ii) Find matrices representing the eigenprojectors of $\mathbf{A}$ in the u_+, u_- basis.

(iii) Find matrices representing the operators $\sqrt{\mathbf{A}}$ and $(1 + \mathbf{A}^2)^{-\frac{1}{2}}$ in the same basis.

24. A linear operator $\mathbf{A}$ on $\mathbf{C}^2$ has the matrix representation

$$\mathbf{A} = \begin{pmatrix} 1 & 1 \\ 0 & 1+\varepsilon \end{pmatrix}$$

(i) Find the eigenvalues of $\mathbf{A}$, and express the corresponding (normalized) eigenvectors in terms of the unit vectors $u_\pm$ defined above in Problem 23.

(ii) What is the angle between the eigenvectors belonging to the two eigenvalues?

(iii) What happens in the limit $\varepsilon \to 0$?

(iv) Repeat the analysis of parts (i)–(iii) for the operator $\mathbf{A}^\dagger$.

25. Let $\phi_1, \ldots \phi_n$ be a complete orthonormal system in $\mathcal{V}^n$. Define the linear operator $\mathbf{U}$ by

$$\mathbf{U}\phi_k = \phi_{k+1} \quad (k = 1, \ldots, \text{n-1}) \qquad \mathbf{U}\phi_n = \phi_1$$

(i) Show that $\mathbf{U}$ is unitary.

(ii) Find the eigenvalues of $\mathbf{U}$ and a corresponding orthonormal system of eigenvectors.

26. Consider a linear operator $\mathbf{A}$ on $\mathcal{V}^n$ whose matrix representation in one orthonormal basis has the form

$$\mathbf{A} = (\xi_k \eta_\ell^* + \eta_k \xi_\ell^*)$$

where $\xi \equiv (\xi_k)$ and $\eta \equiv (\eta_k)$ are two linearly independent vectors in $\mathcal{V}^n$. Find the eigenvalues and eigenvectors of $\mathbf{A}$.

27. (i) Show that any 2×2 matrix $\mathbf{A}$ satisfies

$$\det \mathbf{A} = \tfrac{1}{2}\left\{ (\operatorname{tr} \mathbf{A})^2 - \operatorname{tr} \mathbf{A}^2 \right\}$$

(ii) Show that any 3×3 matrix $\mathbf{A}$ satisfies

$$\det \mathbf{A} = \tfrac{1}{6}\left\{ (\operatorname{tr} \mathbf{A})^3 - 3(\operatorname{tr} \mathbf{A})(\operatorname{tr} \mathbf{A}^2) + 2\operatorname{tr} \mathbf{A}^3 \right\}$$

(iii) For extra credit, derive the corresponding relation between $\det \mathbf{A}$ and the traces of powers of $\mathbf{A}$ for a 4×4 matrix.

28. Show that

$$\det \mathbf{A} = e^{\operatorname{tr}(\ln \mathbf{A})}$$

for any linear operator $\mathbf{A}$ on $\mathcal{V}^n$.

29. Suppose $\mathbf{A}$ is a positive definite linear operator on the real vector space $\mathcal{V}^n$. Show that

$$G(\mathbf{A}) \equiv \int_\infty^\infty \cdots \int_\infty^\infty e^{-(x, \mathbf{A}x)} \, dx_1 \cdots dx_n = \sqrt{\frac{\pi^n}{\det \mathbf{A}}}$$

where $(x, \mathbf{A}x)$ is the usual scalar product

$$(x, \mathbf{A}x) = \sum_{j,k=1}^{n} x_j A_{jk} x_k$$

30. A linear operator $\mathbf{A}$ is represented by the 4×4 matrix

$$\mathbf{A} = \begin{pmatrix} a_1 & 0 & 0 & b_1 \\ 0 & a_2 & 0 & b_2 \\ 0 & 0 & a_3 & b_3 \\ b_1^* & b_2^* & b_3^* & a_4 \end{pmatrix}$$

with $a_1 < a_2 < a_3 < a_4$ real, and $b_k \neq 0$ ($k = 1, 2, 3$). By sketching the characteristic polynomial $p_{\mathbf{A}}(\lambda)$, show that the eigenvalues of $\mathbf{A}$ (suitably ordered) satisfy

$$\lambda_1 < a_1 < \lambda_2 < a_2 < \lambda_3 < a_3 < \lambda_4$$

31. The linear operator $\mathbf{A}$ can be expressed as

$$\mathbf{A} = \mathbf{X} + i\mathbf{Y}$$

where $\mathbf{X}$ and $\mathbf{Y}$ are Hermitian operators defined by

$$\mathbf{X} = \frac{\mathbf{A} + \mathbf{A}^\dagger}{2} \qquad \mathbf{Y} = \frac{\mathbf{A} - \mathbf{A}^\dagger}{2i}$$

(this is the *Cartesian decomposition* of $\mathbf{A}$).

(i) Show that $\mathbf{A}$ is normal if and only if

$$[\mathbf{X}, \mathbf{Y}] = 0$$

(ii) Show that the nonsingular linear operator $\mathbf{A}$ has a unique *left polar decomposition* of the form

$$\mathbf{A} = \mathbf{R}\mathbf{U}$$

with $\mathbf{R} > 0$ and $\mathbf{U}$ unitary, and express $\mathbf{R}$ and $\mathbf{U}$ in terms of $\mathbf{A}$. Also find the *right polar decomposition*

$$\mathbf{A} = \mathbf{U}'\mathbf{R}'$$

with $\mathbf{R}' > 0$ and $\mathbf{U}'$ unitary.

(iii) Under what conditions on the polar decompositions in part (ii) is $\mathbf{A}$ normal? State and prove both necessary and sufficient conditions.

32. The linear operator $\mathbf{A}$ on $\mathbf{C}^2$ has the matrix representation

$$\mathbf{A} = \begin{pmatrix} 1 & \xi \\ 0 & 1 \end{pmatrix}$$

with ξ a complex number. What are the eigenvalues of $\mathbf{A}$? Is $\mathbf{A}$ positive definite? Find a matrix representation for each of the operators

$$\mathbf{S} \equiv \sqrt{\mathbf{A}} \quad \text{and} \quad \mathbf{H}(t) \equiv e^{t\mathbf{A}}$$

33. Linear operators $\mathbf{A}$ and $\mathbf{B}$ on $\mathbf{C}^2$ are represented by the matrices

$$\mathbf{A} = \begin{pmatrix} 0 & 1 \\ 0 & 0 \end{pmatrix} \quad \text{and} \quad \mathbf{B} = \begin{pmatrix} 0 & 0 \\ 1 & 0 \end{pmatrix}$$

Find the matrices representing the operators

$$\mathbf{C} = e^{t\mathbf{A}}e^{t\mathbf{B}} \quad \mathbf{D} = e^{t\mathbf{B}}e^{t\mathbf{A}} \quad \text{and} \quad \mathbf{F} = e^{t(\mathbf{A}+\mathbf{B})}$$

34. Let $\mathbf{A}$ be a 3×3 matrix given by

$$\mathbf{A} \equiv \begin{pmatrix} a & b & 0 \\ 0 & a & b \\ 0 & 0 & a \end{pmatrix}$$

Evaluate the matrix

$$\mathbf{B} \equiv e^{-at}e^{t\mathbf{A}}$$

35. Linear operators $\mathbf{A}$ and $\mathbf{B}$ on $\mathbf{C}^3$ are represented by the matrices

$$\mathbf{A} = \begin{pmatrix} 0 & 1 & 0 \\ 0 & 0 & 1 \\ 0 & 0 & 0 \end{pmatrix} \quad \text{and} \quad \mathbf{B} = \begin{pmatrix} 0 & 0 & 0 \\ 1 & 0 & 0 \\ 0 & 1 & 0 \end{pmatrix}$$

Find the matrices representing the operators

$$\mathbf{C} = e^{t\mathbf{A}}e^{t\mathbf{B}}, \quad \mathbf{D} = e^{t\mathbf{B}}e^{t\mathbf{A}} \quad \text{and} \quad \mathbf{F} = e^{t(\mathbf{A}+\mathbf{B})}$$

36. Consider the linear nth-order differential equation

$$u^{(n)}(t) + \alpha_1 u^{(n-1)}(t) + \cdots + \alpha_{n-1}u'(t) + \alpha_n u(t) = 0$$

with constant coefficients $\alpha_1, \ldots, \alpha_n$. Express this equation in the matrix form

$$\frac{dx}{dt} = \mathbf{A}x$$

by introducing the vector

$$x = (u, u', \ldots, u^{(n-1)})^T$$

(where x^T denotes the transpose of x). Find an explicit representation of the matrix $\mathbf{A}$. Show that the eigenvalues of $\mathbf{A}$ are the roots of the polynomial

$$p(\lambda) = \lambda^n + \alpha_1\lambda^{n-1} + \cdots + \alpha_{n-1}\lambda + \alpha_n$$

3 Geometry in Physics

Analytic and algebraic methods are essential both for deriving exact results and for obtaining useful approximations to the behavior of complex systems. However, it is important to augment these traditional methods with geometrical ideas, which are often easy to visualize, and which can provide useful insights into the qualitative behavior of a system. Such qualitative understanding is especially relevant when analyzing lengthy numerical computations, both for evaluating the accuracy and validity of the computations, and for extracting general conclusions from numerical results.

Moreover, there is a deep connection between geometry and gravitation at the classical level, first formalized in Einstein's theory of gravity (*general relativity*), which has led to the prediction of exotic phenomena such as black holes and gravitational radiation that have only been confirmed in the last 30 years or so. Beyond that, one of the fundamental problems of contemporary physics is to find a theory of gravity that incorporates quantum physics. This has stimulated the development and application of even deeper geometrical ideas to various string theories.

Thus we introduce here some basic elements of differential geometry. We begin with the concept of a *manifold*, which is simply a collection of points labeled by coordinates. The number of coordinates required to identify a point is the *dimension* of the manifold. An *atlas* on a manifold is defined by a set of overlapping regions (*coordinate patches*) in each of which points are characterized by smoothly varying n-tuples $\{x_1, \ldots, x_n\}$ of numbers such that in the regions of overlap, the different sets of coordinates are smooth functions of each other. Mathematical properties such as continuity and differentiability of functions on the manifold are defined in terms of the corresponding properties of functions of the coordinates. Assumed smoothness properties of the relations between coordinates ensure that these properties do not depend on which particular set of equivalent coordinate systems is used to describe the manifold; there is no preferred coordinate system on the manifold.

For example, the space and time in which we live appears to be a four-dimensional manifold, *spacetime*, though string theorists suggest that there may be extra dimensions that have not yet been observed. The spatial coordinates of a classical system of N particles define a $3N$-dimensional manifold, the *configuration space* of the system; the coordinates together with the particle momenta define a $6N$-dimensional manifold, the *phase space* of the system. These variables are drastically reduced in number when a system is considered in thermodynamics, where the states of a system are described by a small number of thermodynamic variables such as temperature, pressure, volume, etc. These variables also define a manifold, the (thermodynamic) *state space* of the system.

Introduction to Mathematical Physics. Michael T. Vaughn
Copyright © 2007 WILEY-VCH Verlag GmbH & Co. KGaA, Weinheim
ISBN: 978-3-527-40627-2

With each point P in a manifold $\mathcal{M}$ is associated a linear vector space $\mathcal{T}_P$, the *tangent space* at P, whose dimension is the same as the dimension of $\mathcal{M}$. A vector $\vec{v}$ in $\mathcal{T}_P$ corresponds to an equivalence class of curves with the same tangent vector at P, and is represented by the differential operator $\vec{v} \cdot \vec{\nabla}$, the *directional derivative*. The collection of the tangent spaces at all the points of $\mathcal{M}$, together with $\mathcal{M}$ itself, defines the *tangent bundle* $\mathcal{T}(\mathcal{M})$ of $\mathcal{M}$.

A *differential form*, or simply *form*, at a point P is an element of vector space $\mathcal{T}_P^*$ dual to the tangent space at P. A form $\vec{\alpha}$ is represented naturally by an expression $\vec{\alpha} \cdot d\vec{x}$, which corresponds to a line element normal to a surface of dimension $n - 1$. The collection of the cotangent spaces at all the points of a manifold $\mathcal{M}$, together with the manifold itself, defines the *cotangent bundle* $\mathcal{T}^*(\mathcal{M})$ of $\mathcal{M}$. If $\mathcal{M}$ is the configuration space of a classical mechanical system, for example, then $\mathcal{T}(\mathcal{M})$ is a manifold characterized by the coordinates and the velocities, while $\mathcal{T}^*(\mathcal{M})$ is the phase space manifold defined by the coordinates of the system and their conjugate momenta.

Tensors of rank greater than 1 are defined in the usual way as products of vectors and differential forms. A special type of tensor, a *p-form*, is obtained from the antisymmetric product (also known as *wedge product*, or *exterior product*) of p elementary differential forms. This is a generalization of the cross-product of three-dimensional vectors. A p-form defines an element of an oriented p-dimensional surface in the manifold, and provides the integrand of a surface integral. The volume element in an n-dimensional manifold is an n-form. The set of all p-forms $(p = 1, \ldots, n)$ on a manifold $\mathcal{M}$ defines the *exterior algebra* $\mathcal{E}(\mathcal{M})$ of $\mathcal{M}$.

A *vector field* is introduced as a collection of vectors, one for each point of a manifold, such that the components of the vectors are smooth functions of the coordinates. A vector field is defined by a set of first-order differential equations

$$\frac{dx^k}{dt} = v^k(x)$$

that have a unique solution passing through a given point if the $v^k(x)$ are smooth. The solutions to these equations define the *integral curves*, or *lines of flow*, of the field; they fill the manifold without intersecting each other. Familiar examples are the lines of flow of electric and magnetic fields, and of the velocity flow of a fluid. Associated with a vector field v is a differential operator $\mathcal{L}_v$, the *Lie derivative*, that drags the functions on which it acts along the lines of flow of the field.

A generalization of the $\vec{\nabla}$ operator of three-dimensional vector calculus is the *exterior derivative d*, which transforms a p-form into a $(p + 1)$-form. This operator appears in the generalized Stokes' theorem, which relates the integral of a p-form σ over the boundary $\partial\mathbf{R}$ of a closed region $\mathbf{R}$ to the integral of the exterior derivative $d\sigma$ over the entire region $\mathbf{R}$,

$$\int_{\mathbf{R}} d\sigma = \int_{\partial\mathbf{R}} \sigma$$

When $\mathbf{R}$ is a closed region in the manifold, this formula is equivalent to the divergence theorem of elementary calculus; here it can be used to define the divergence of a vector field.

A *metric tensor* $\mathbf{g} = (g_{jk})$ on a manifold defines a distance ds between nearby points according to the standard rule

$$ds^2 = g_{jk}(x)dx^j dx^k$$

Also associated with the metric tensor is a natural n-form volume element

$$\Omega = \sqrt{|\det \mathbf{g}|}\, dx^1 \wedge dx^2 \wedge \cdots \wedge dx^n \equiv \rho(x)\, dx^1 \wedge dx^2 \wedge \cdots \wedge dx^n$$

The line element ds^2 is supposed to be independent of the coordinates on the manifold. This leads to a transformation law for the components of the metric tensor under a change of coordinates that also leads to an easy calculation of the transformation of the volume element. These transformation laws are useful even in mundane problems involving curvilinear coordinates in three dimensions.

There is also a natural definition of the length of a curve joining two points a and b:

$$\ell_{ab}(C) = \int_a^b \sqrt{g_{jk}(x)\dot{x}^j \dot{x}^k}\, d\lambda \equiv \int_a^b \sqrt{\sigma}\, d\lambda$$

where λ is a parameter that varies smoothly along the curve and $\dot{x} = dx/d\lambda$. A *geodesic* is a curve for which the path length between two points is an extremum relative to nearby curves. Finding the extrema of integrals such as $\ell_{ab}(C)$ is a fundamental problem of the calculus of variations, which is briefly reviewed in Appendix A. There it is shown that an extremal curve must satisfy a set of differential equations, the Euler–Lagrange equations. These lead to the standard *geodesic equations* if we make a natural choice of parameter along the curve, as seen in Section 3.4.6.

Apart from defining a measure of distance, the metric tensor provides a natural duality between vectors and forms, in which the metric tensor is used for raising and lowering indices on tensor components, transforming components of a vector into components of a form, and vice versa. One important use of this duality is to give a general definition of the *Laplacian* Δ, a differential operator that acts on scalar functions in an n-dimensional manifold according to

$$\Delta f = \mathrm{div}\,(\mathrm{grad}\, f)$$

Since the gradient df of a scalar function is a 1-form, the metric tensor needs to be applied to convert the gradient into a vector field before taking the divergence. This leads to

$$\Delta f = \frac{1}{\rho(x)} \sum_{k,\ell=1}^n \frac{\partial}{\partial x^k} \left\{ g^{k\ell}(x)\rho(x) \frac{\partial f}{\partial x^\ell} \right\}$$

where the $g^{k\ell}(x)$ are elements of the matrix $\mathbf{g}^{-1}$. Further refinements are needed to define a Laplacian operator on vector fields or forms, but those are beyond the scope of this book.

There has been a massive body of research in the last 40 years or so analyzing the behavior of dynamical systems, which can be represented as vector fields on a manifold. Here we introduce a simple two-dimensional model from ecology that illustrates some of the methods available to analyze complex systems. The differential equations of the model have fixed points that fix the qualitative behavior of the solutions in the plane, and allow a sketch of the lines of flow of the associated vector field even without detailed calculations.

The Lagrangian and Hamiltonian formulations of the dynamics of classical mechanical systems can also be described in geometric terms. The *configuration space* of a system is a manifold $\mathcal{M}$ with (generalized) coordinates $q = \{q^k\}$; the tangent bundle $\mathcal{T}(\mathcal{M})$ of $\mathcal{M}$ is characterized by the $\{q^k\}$ and the corresponding velocities $\dot{q} = \{\dot{q}^k\}$. The Lagrangian of the system is a function defined on the tangent bundle that serves to define the *action integral*

$$S[q(t)] = \int_a^b L(q, \dot{q}, t)\, dt$$

associated with a path $q(t)$ of the system joining points a and b in $\mathcal{M}$. *Hamilton's principle* requires that the action integral for the actual trajectory of the system be an extremum relative to nearby paths. This leads to *Lagrange's equations of motion* using the methods of the calculus of variations.

In the Hamiltonian formulation, the conjugate momenta $p = \{p_k\}$ of the system are introduced as components of a 1-form field on the configuration space; the *phase space* defined by the coordinates $\{q^k\}$ and the conjugate momenta $\{p_k\}$ is identified with the cotangent bundle $\mathcal{T}^*(\mathcal{M})$ of $\mathcal{M}$. The Hamiltonian of the system is defined by

$$H(q, p, t) = \sum_k p_k q^k - L(q, \dot{q}, t)$$

Hamilton's equations of motion, which can also be derived from Hamilton's principle, are

$$\frac{dq^k}{dt} = \frac{\partial H}{\partial p_k} \qquad \frac{dp_k}{dt} = \frac{\partial H}{\partial q^k}$$

These equations define the trajectories of the system in phase space, which are also the integral curves of a vector field on $\mathcal{T}^*(\mathcal{M})$ that is closely related to the Hamiltonian.

There is a *canonical 2-form*

$$\omega \equiv \sum_k dp_k \wedge dq^k$$

defined on the phase space (the dp_k form a basis of 1-forms in momentum space). This form is used to associate vectors and 1-forms on the phase space. *Canonical transformations* (also known as *symplectic transformations*) are introduced as transformations of the phase space coordinates that leave the canonical form invariant. For a Hamiltonian independent of time, transformations that translate the system by a time τ are canonical, and define a one-parameter group of canonical transformations of the system. Other canonical transformations can make explicit other constants of motion that a system may, or may not, have.

An ideal fluid can be described in terms of a few smooth functions such as density, velocity, pressure, and temperature defined on the space occupied by the fluid. The integral curves of the velocity field provide a nice picture of the fluid behavior, and conservation laws of the fluid are also naturally expressed in the geometric language of this chapter. A brief description of the basic equations governing fluids is found in Section 3.6.

The description of many-particle systems by a few thermodynamic variables is one of the conceptual highlights of theoretical physics. Visualizing the space of states as a manifold whose coordinates are the thermodynamic variables, and the first law (energy conservation) as a relation between forms on this manifold, clarifies many relations between thermodynamic variables and their derivatives, as we explain in Appendix B.

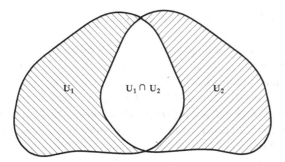

Figure 3.1: Two coordinate patches $\mathcal{U}_1$ and $\mathcal{U}_2$ in a manifold $\mathcal{M}$. The two sets of coordinates in the overlap region $\mathcal{U}_1 \cap \mathcal{U}_2$ are required to be smooth one-to-one functions of each other.

3.1   Manifolds and Coordinates

3.1.1   Coordinates on Manifolds

A *manifold* is characterized as a set of points that can be identified by n-tuples $(x_1, \ldots, x_n)$ of real numbers (*coordinates*), such that the coordinates vary smoothly in some neighborhood of each point in the manifold. In general, these coordinates are defined only over some bounded region of the manifold (a *coordinate patch*); more than one coordinate patch will be needed to cover the whole manifold. There will be regions in which two or more coordinate patches overlap as shown schematically in Fig. 3.1. In these regions, points will have more than one set of coordinates, and the coordinates of one set will be functions of the other set as we move through the overlap region. If these functions are one to one and smooth (here *smooth* means differentiable to whatever order is needed), then the manifold is a *differentiable manifold*. Here all manifolds are assumed to be differentiable unless otherwise stated.[1]

Definition 3.1.   A set $\mathcal{M}$ of points is a *manifold* if for every point p in $\mathcal{M}$ there is an (open) neighborhood $\mathcal{N}_p$ of p on which there is a continuous one-to-one mapping onto a bounded open subset of $\mathbf{R}^n$ for some n. The smallest n for which such a mapping exists is the *dimension* of $\mathcal{M}$. We thus have a collection of open sets $\mathcal{U}_1, \ldots, \mathcal{U}_N$ that taken together cover the manifold, and a set of functions (*maps*) $\phi_1, \ldots, \phi_N$ such that for every point p in $\mathcal{U}_k$, there is an element $(\phi_k^1(p), \ldots, \phi_k^n(p))$ of $\mathbf{R}^n$, the *coordinates* of p under the map ϕ_k (here we use

---

[1]Mathematicians often start from the concept of a topological space on which open and closed sets are defined in some abstract way; continuity of functions (or mappings) is defined in terms of the images of open sets. Here we always deal with manifolds defined concretely in terms of coordinates, so that we can use the standard notions of open and closed sets and of continuity in $\mathbf{R}^n$ introduced in elementary calculus. In particular, for discussions of limits, continuity, closure, etc., we can define an ε-neighborhood of a point P with coordinates $x_0 = (x_{10}, \ldots, x_{n0})$ in some coordinate system to be the set of points x such that

$$\|x - x_0\| < \varepsilon$$

where $\|x - x_0\|$ is the Euclidean norm on $\mathbf{R}^n$. This use of the norm is independent of the possible existence of a metric on the manifold as described below, but serves simply to give a concrete definition of neighborhood. Since coordinate transformations are required to be smooth, it does not matter which particular coordinates are used to define neighborhoods.

superscripts to label the coordinates). The pair $(\mathcal{U}_k, \phi_k)$ is a *chart*; a collection of charts that covers the manifold is an *atlas*.

If the open sets $\mathcal{U}_k$ and $\mathcal{U}_l$ overlap, then the map $\phi_{kl} \equiv \phi_k \circ \phi_l^{-1}$ defines a function (the *transition function*) that maps the image of $\mathcal{U}_l$ in $\mathbf{R}^n$ onto the image of $\mathcal{U}_k$ in $\mathbf{R}^n$; that is, it defines the coordinates ϕ_k as functions of the coordinates ϕ_l. More simply stated, the function $\phi_k \circ \phi_l^{-1}$ defines a *coordinate transformation* in the overlap region. If all the transition functions in an atlas are differentiable, then $\mathcal{M}$ is a *differentiable manifold*. If the transition functions are all analytic (differentiable to all orders), then $\mathcal{M}$ is an *analytic manifold*. A manifold on which the coordinates are complex, and the mappings between coordinate systems are analytic as functions of the complex coordinates, is a *complex manifold*.   ∎

❑ **Example 3.1.** $\mathbf{R}^n$ is a manifold of dimension n. The space $\mathbf{C}^n$ of n-tuples of complex numbers is a manifold of dimension $2n$, which can also be viewed as a complex manifold of dimension n. The linear manifolds introduced in Chapter 2 are manifolds in the sense used here, with the additional restriction that only linear transformations of the coordinates are allowed.   ∎

3.1.2   Some Elementary Manifolds

❑ **Example 3.2.** The circle $\mathbf{S}^1$ defined in $\mathbf{R}^2$ by

$$x_1^2 + x_2^2 = 1 \tag{3.1}$$

is a one-dimensional manifold. However, there is no single coordinate that covers the entire circle, since the angle θ shown on the left-hand circle in Fig. 3.2 is not continuous around the entire circle, whether we choose the range of θ to be $-\pi \leq \theta < \pi$ or $0 \leq \theta < 2\pi$. One solution is to introduce two overlapping arcs $\mathcal{A}_1$ and $\mathcal{A}_2$ as shown on the right of the figure. On $\mathcal{A}_1$ use the angle θ_1 measured counterclockwise from the horizontal axis, on $\mathcal{A}_2$ the angle θ_2 measured clockwise from the axis. A suitable range for θ_1 and θ_2 is

$$-\tfrac{\pi}{2} - \delta_1 \leq \theta_1 \leq \tfrac{\pi}{2} + \delta_1 \qquad -\tfrac{\pi}{2} - \delta_2 \leq \theta_2 \leq \tfrac{\pi}{2} + \delta_2 \tag{3.2}$$

where δ_1, δ_2 are small positive angles (so that the two arcs overlap). Evidently

$$\theta_2 = \pi - \theta_1 \quad (\theta_1, \theta_2 > 0) \qquad \theta_2 = -\pi - \theta_1 \quad (\theta_1, \theta_2 < 0) \tag{3.3}$$

The first relation covers the upper interval of overlap; the second, the lower. Note that there is no requirement that the relation between the two coordinates has the same form on the disjoint intervals of overlap.

There are mappings of the circle $\mathbf{S}^1$ into the real line $\mathbf{R}^1$ that illustrate the topological distinction between the two. Consider, for example, the mapping

$$\theta \to x \equiv \frac{\theta}{\pi^2 - \theta^2} \tag{3.4}$$

that maps the interval $-\pi < \theta < \pi$ (an open arc on the circle) onto the entire real line. In order to map the point on the circle corresponding to $\theta = \pm\pi$, we must add to the real line a single "point" at infinity (∞).   ∎

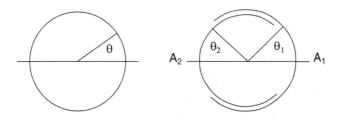

Figure 3.2: The manifold $\mathbf{S}^1$ (the circle). The usual polar angle θ is shown on the left. To have coordinates that are everywhere smooth, it is necessary to split the circle into two overlapping arcs $\mathcal{A}_1$ and $\mathcal{A}_2$, with angles θ_1 (measured counterclockwise from the horizontal axis) and θ_2 (measured clockwise from the axis).

❑ **Example 3.3.** The unit sphere $\mathbf{S}^2$ defined in $\mathbf{R}^3$ by

$$x_1^2 + x_2^2 + x_3^2 = 1 \tag{3.5}$$

is a two-dimensional manifold. The usual spherical coordinates θ, ϕ do not cover the whole manifold, since ϕ is undefined at the poles ($\theta = 0, \pi$), and the range $0 \leq \phi < 2\pi$ leaves a line of discontinuity of ϕ that joins the two poles (think of the International Date Line on the surface of the Earth, for example). A coordinate system that covers all but one point of the sphere is the *stereographic projection* illustrated in Fig. 3.3, which maps the sphere minus the North pole into the entire plane $\mathbf{R}^2$ (details of the projection are left to Problem 1). Note that the complex plane, which includes a point at ∞ (see Chapter 4), is mapped into the entire sphere, as the North pole is the image of the point ∞. ∎

❑ **Example 3.4.** The *unit ball* $\mathbf{B}^n$ is defined in $\mathbf{R}^n$ as the set of points that satisfy

$$x_1^2 + \cdots + x_n^2 < 1 \tag{3.6}$$

$\mathbf{B}^n$ is a submanifold of $\mathbf{R}^n$, and yet it can be mapped into all of $\mathbf{R}^n$ by the map

$$r \rightarrow r' \equiv \frac{r}{1 - r} \tag{3.7}$$

which transforms the open interval $0 < r < 1$ into the positive real axis, thus stretching the ball to an infinite radius. ∎

❑ **Example 3.5.** The (unit) *sphere* $\mathbf{S}^n$ is defined in $\mathbf{R}^{n+1}$ as the set of points that satisfy

$$x_1^2 + \cdots + x_{n+1}^2 = 1 \tag{3.8}$$

It is also the *boundary* of the unit ball $\mathbf{B}^{n+1}$; note that the sphere is not included in the definition of the ball. Again, there is no single coordinate system that covers the whole sphere smoothly. A generalization of the stereographic projection described above for $\mathbf{S}^2$ maps the entire sphere $\mathbf{S}^n$ except for one point onto $\mathbf{R}^n$. To include an image for the last

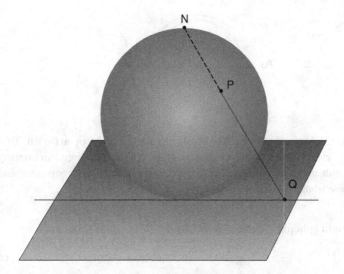

Figure 3.3: Schematic illustration of the stereographic projection of the sphere S^2 onto the plane R^2. The stereographic projection of the point P on the sphere is the point Q where the line from the North pole N through P intersects the plane tangent to the South pole.

point of S^n, it is necessary to add a single point at infinity to R^n. This addition of a single point at ∞ to turn R^n into a manifold equivalent to the compact manifold S^n is called the *one-point compactification* of R^n.   ∎

❏ **Example 3.6.** A monatomic gas is described by thermodynamic variables p (the pressure), V (the volume), and T (the temperature); each of these variables must be positive. The states of thermal equilibrium must satisfy an equation of state of the form

$$f(p, V, T) = 0 \tag{3.9}$$

so that these states form a two-dimensional manifold. Note that there is no natural notion of distance on the thermodynamic manifold, in contrast to the other examples, where the Euclidean metric of the R^n in which the manifolds are embedded leads to a natural definition of distance on the manifold.   ∎

❏ **Example 3.7.** The coordinates of a classical system of N particles define a manifold, the *configuration space* of the system. The dimension n of this manifold is $3N$ if the particles can move freely in three dimensions, but is smaller if the motion is subject to constraints. Each coordinate on the manifold is a *degree of freedom* of the system; the number of degrees of freedom is equal to the dimension of the configuration space. For a Hamiltonian system with conjugate momenta for each of the coordinates, the coordinates and momenta together form the $2n$-dimensional *phase space* of the system, to be described in detail later in Section 3.5.3.   ∎

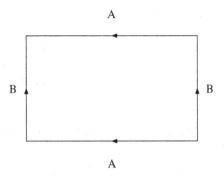

Figure 3.4: Representation of the torus on the plane. The two sides marked A are to be identified, as are the two sides marked B. Imagine the rectangle to be rolled up. Then glue the two sides marked A together (in the direction of the arrows) to form a hollow cylinder. Finally, glue the two sides marked B together (in the direction of the arrows).

3.1.3 Elementary Properties of Manifolds

If $\mathcal{M}$ and $\mathcal{N}$ are manifolds, the *product manifold* $\mathcal{M} \times \mathcal{N}$ is defined as the set of ordered pairs (x, y) where x is a point from $\mathcal{M}$ and y is a point from $\mathcal{N}$. If x has coordinates $(x_1, \ldots, x_m)$ and y has coordinates $(y_1, \ldots, y_n)$, then (x, y) has coordinates $(x_1, \ldots, x_m, y_1, \ldots, y_n)$. Evidently the dimension of $\mathcal{M} \times \mathcal{N}$ is the sum of the dimensions of $\mathcal{M}$ and $\mathcal{N}$. Note that if $\mathcal{M}$ and $\mathcal{N}$ are linear vector spaces, the product manifold is actually the direct sum $\mathcal{M} \oplus \mathcal{N}$ of the two vector spaces, *not* the tensor product.

❏ **Example 3.8.** It is easy to see that $\mathbf{R}^m \times \mathbf{R}^n = \mathbf{R}^{m+n}$. ∎

❏ **Example 3.9.** The *torus* $\mathcal{T}$ is the product $\mathbf{S}^1 \times \mathbf{S}^1$ of two circles. It can be represented in the plane as a rectangle in which opposite sides are identified; that is, a point on the upper side A is the same as the point opposite on the lower side A in Fig. 3.4, and a point on the left side B is the same as the opposite point on the right side B. The torus is also equivalent to the unit cell in the complex plane shown in Fig. 4.5. ∎

❏ **Example 3.10.** The *n-torus* $\mathcal{T}^n$ is the product $\mathbf{S}^1 \times \cdots \times \mathbf{S}^1$ (n factors) of n circles. ∎

Coordinates are introduced to give a concrete description of a manifold. However, many properties of a manifold are independent of the specific coordinates assigned to points of the manifold. Indeed, it is a principle of physics that all observable properties of a physical system should be independent of the coordinates used to describe the system, at least within a restricted class of coordinate systems.[2] For example, the distance between two points in a manifold is defined in terms of coordinates in a region containing the two points. Once defined, however, it must be the same for any equivalent set of coordinates on the region.

---

[2]This is the essence of principles of relativity from Galileo to Einstein. Einstein's principle of general relativity is the broadest, since it insists that the equations of physics be unchanged by diffeomorphisms of the spacetime manifold, which are defined shortly.

Equivalence of two differentiable manifolds $\mathcal{M}_1$ and $\mathcal{M}_2$ means that there is a one-to-one map between the manifolds for which the coordinates in any region in $\mathcal{M}_2$ are smooth functions of the coordinates in the corresponding region of $\mathcal{M}_1$. Such a map is a *diffeomorphism*. If such a map exists, then the two manifolds are *diffeomorphic*, or simply *equivalent*. If only continuity of the map is required, then the map is a *homeomorphism*, and the manifolds are *homeomorphic*.

❑ **Example 3.11.**   The unit ball $\mathbf{B}^n$ is diffeomorphic to all of $\mathbf{R}^n$, since the radial rescaling in Eq. (3.7) is differentiable for $0 \le r < 1$.   ∎

❑ **Example 3.12.**   Consider the group $SU(2)$ of unitary 2×2 matrices with determinant $+1$. Every matrix in this group has the form

$$\mathbf{U} = \exp\left(\pi i \vec{\sigma} \cdot \vec{a}\right) \tag{3.10}$$

where $\vec{\sigma} = (\sigma_1, \sigma_2, \sigma_3)$ denote the three Pauli matrices, and $\vec{a}$ is a vector with $|\vec{a}| \le 1$ (if $|\vec{a}| = 1$, then $\mathbf{U} = -1$ independent of the direction of $\vec{a}$). The map

$$\vec{a} \to \vec{r} = \frac{\vec{a}}{1 - |\vec{a}|} \tag{3.11}$$

takes the group manifold into $\mathbf{R}^3$ augmented by a single point at ∞ that corresponds to $\mathbf{U} = -1$. As seen above, this is the one-point compactification of $\mathbf{R}^3$ to the sphere $\mathbf{S}^3$. Thus the group $SU(2)$ is diffeomorphic to $\mathbf{S}^3$.   ∎

Convergence properties of sequences of points in a manifold are defined in terms of the corresponding properties of the sequences in $\mathbf{R}^n$ of the coordinates of the points. Because we require all coordinate transformations to be smooth, these convergence properties will not depend on which particular coordinate system we use to test the convergence. Thus, for example, the sequence $\{x_1, x_2, \ldots\}$ of points in an n-dimensional manifold is a Cauchy sequence if the sequence of coordinates is a Cauchy sequence in $\mathbf{R}^n$, using any positive definite norm on $\mathbf{R}^n$.

A function $f(x)$ defined on a manifold $\mathcal{M}$ maps the points of the manifold to real (or complex) numbers; there may be exceptional points, or *singular points* at which the function is not defined. The function does not depend on the coordinate system introduced on the manifold; hence is also called a *scalar field*. The field is continuous (differentiable, smooth, ...) if $f(x)$ is a continuous (differentiable, smooth, ...) function of the coordinates. Again, these properties of the field are coordinate independent because of our restriction to smooth coordinate transformations.

A manifold is *closed* if every Cauchy sequence of points in the manifold converges to a limit in the manifold. If a manifold is $\mathcal{M}$ is not closed, then new points can be defined as limit points of Cauchy sequences in the manifold. The manifold obtained from $\mathcal{M}$ by adjoining the limits of all Cauchy sequences in $\mathcal{M}$ is the *closure* of $\mathcal{M}$, denoted by $\mathcal{Cl}\,\mathcal{M}$, or by $\overline{\mathcal{M}}$. Recall that a region in $\mathbf{R}^n$ is *compact* if it is closed (contains all its limit points) and bounded (it can be contained inside a sufficiently large sphere). A manifold $\mathcal{M}$ is *compact* if it is equivalent to the union of a finite number of compact regions in some $\mathbf{R}^n$.

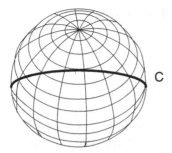
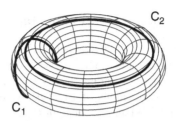

Figure 3.5: Illustrations of the sphere $\mathbf{S}^2$ and torus T^2. Any closed curve C on $\mathbf{S}^2$ can be smoothly contracted to a single point. On the other hand, the curves labeled C_1 and C_2 on T^2 cannot be smoothly contracted to a point, nor deformed into each other.

❏ **Example 3.13.** The sphere $\mathbf{S}^n$ is compact; by its definition it is a closed bounded region in $\mathbf{R}^{n+1}$. The unit ball $\mathbf{B}^n$ is not compact even though it is bounded, since it is not closed. There are sequences of points in $\mathbf{B}^n$ that converge to a point on the boundary sphere $\mathbf{S}^{n-1}$, which is not part of $\mathbf{B}^n$. ∎

❏ **Example 3.14.** $\mathbf{R}^n$ is not compact but, as just noted, addition of a single point at ∞ turns it into a compact manifold, the one-point compactification of $\mathbf{R}^n$, equivalent to $\mathbf{S}^n$. ∎

Two manifolds of the same dimension are always equivalent locally, (i.e., in some neighborhood of any point) since they are represented by coordinates in $\mathbf{R}^n$. However, even with the limited structure we have introduced so far, there are manifolds of the same dimension that are not globally equivalent, i.e., there is no global diffeomorphism that takes one manifold into the other. A compact manifold cannot be equivalent to a noncompact manifold ($\mathbf{B}^n$ and $\mathbf{S}^n$ are not globally equivalent, for example). Manifolds can be classified according to the kinds of distinct closed curves and surfaces that can be constructed in them. This classification is one of the main problems of *algebraic topology*.

❏ **Example 3.15.** The sphere $\mathbf{S}^2$ has no nontrivial closed loops; every closed loop on the sphere can be contracted smoothly to a single point. However, on the two-dimensional torus $T^2 \sim \mathbf{S}^1 \times \mathbf{S}^1$, closed curves corresponding to one circuit around either of the circles cannot be contracted to a point (see Fig. 3.5). Closed curves in T^2 are characterized by a pair of integers (n_1, n_2) (the *winding numbers*), depending on how many times they wrap around each of the two circles. Curves with different winding numbers cannot be continuously deformed into each other. Since the winding numbers are independent of the coordinates, they are *topological invariants* of the manifold. Equivalence classes of closed curves in a manifold define a group, the *homotopy group* of the manifold. ∎

❏ **Example 3.16.** The group $SO(3)$ of rotations in three dimensions looks like a ball $\mathbf{B}^3$ of radius π, since every rotation can be characterized by an axis $\mathbf{n}$ and an angle θ ($0 \leq \theta \leq \pi$). But, unlike in $SU(2)$, a rotation about $\mathbf{n}$ through angle π is the same as a rotation about $-\mathbf{n}$ through angle π. Thus the opposite ends of a diameter of the ball correspond to the same rotation, and any diameter is a closed path that cannot be shrunk to a point. ∎

3.2   Vectors, Differential Forms, and Tensors

3.2.1   Smooth Curves and Tangent Vectors

The concept of a *curve* in a manifold is intuitive. A formal statement of this intuitive concept is that a curve on the manifold $\mathcal{M}$ is a map from an open interval on the real axis into $\mathcal{M}$. A curve is defined parametrically by the coordinates $\{x^1(\lambda), \ldots, x^k(\lambda)\}$ of the points on the curve[3] as functions of a real parameter λ over some range $\lambda_1 < \lambda < \lambda_2$. The curve is differentiable (smooth, analytic) if these coordinates are differentiable (smooth, analytic) functions of the parameter λ.

If $f(x)$ is a differentiable function defined on $\mathcal{M}$, and C is a smooth curve in $\mathcal{M}$, then the function

$$g(\lambda) \equiv f[x(\lambda)] \tag{3.12}$$

defined on the curve C is a differentiable function of λ, and

$$\frac{dg}{d\lambda} = \sum_{k=1}^{n} \frac{dx^k}{d\lambda} \frac{\partial f}{\partial x^k} = \sum_{k=1}^{n} \xi^k(\lambda) \frac{\partial f}{\partial x^k} \tag{3.13}$$

by the chain rule. The $\{\xi^k(\lambda) = dx^k/d\lambda\}$ define the components of a vector tangent to the curve at the point, and the derivative operator

$$\frac{d}{d\lambda} \equiv \sum_{k=1}^{n} \xi^k(\lambda) \frac{\partial}{\partial x^k} = \xi^k(\lambda) \frac{\partial}{\partial x^k} = \vec{\xi} \cdot \vec{\nabla} \tag{3.14}$$

itself is identified with the *tangent vector*.

This is more abstract than the traditional physicist's view of a vector. However, if we recall the geometrical interpretation of the derivative dy/dx as the tangent to a curve $y = y(x)$ in two dimensions, then we can recognize Eq. (3.14) as a generalization of this interpretation to curves in higher dimensional spaces. The derivative operator in Eq. (3.14) is the *directional derivative* in the direction of the tangent. It is also known as the *Lie derivative* when it is applied to vectors and tensors. The form $\vec{\xi} \cdot \vec{\nabla}$ emphasizes the view of this derivative as an intrinsic (i.e., coordinate-independent) entity, although the components of $\vec{\xi}$ depend on the coordinate system.

Remark.   With the notation $(\xi^k \, \partial/\partial x^k)$ we introduce the (Einstein) *summation convention*, which instructs us to sum over a pair of repeated indices, unless explicitly instructed not to. In this chapter, we use the summation convention without further comment. However, we insist that the pair of indices must consist of one subscript and one superscript, understanding that a subscript (superscript) in the denominator of a partial derivative is equivalent to a superscript (subscript) in the numerator.   □

---

[3]We use here superscripts to identify the coordinates $\{x^1, \ldots, x^k\}$ in keeping with historical usage. The reader is alerted to be aware of the distinction between these superscripts and exponents.

3.2.2 Tangent Spaces and the Tangent Bundle $T(\mathcal{M})$

The tangent vectors at a point P in the manifold $\mathcal{M}$ define a linear vector space T_P, the *tangent space* to $\mathcal{M}$ at P. A vector V in T_P corresponds to the class of smooth curves through P with tangent vector V.[4].

If $\{x^1, \ldots, x^n\}$ provides a coordinate system in a neighborhood of P, a natural set of basis vectors in the tangent space T_P is given by the partial derivatives

$$\mathbf{e}_k \equiv \frac{\partial}{\partial x^k} \tag{3.15}$$

This is the *coordinate basis* associated with the $\{x^1, \ldots, x^n\}$.

If $\{y^1, \ldots, y^n\}$ is another set of coordinates, smooth functions of the old coordinates $\{x^1, \ldots, x^n\}$, then we have the corresponding coordinate basis

$$\mathbf{f}_k \equiv \frac{\partial}{\partial y^k} = \frac{\partial x^m}{\partial y^k} \frac{\partial}{\partial x^m} = \frac{\partial x^m}{\partial y^k} \mathbf{e}_m \tag{3.16}$$

(by the chain rule again), so long as the Jacobian determinant[5]

$$J \equiv \frac{\partial(x^1, \ldots, x^n)}{\partial(y^1, \ldots, y^n)} \equiv \det \left\| \frac{\partial x^m}{\partial y^k} \right\| \neq 0 \tag{3.17}$$

This leads to the transformation law for the components of vectors. If

$$V = \xi^k \frac{\partial}{\partial x^k} = \eta^k \frac{\partial}{\partial y^k} \tag{3.18}$$

then we have

$$\xi^k = \eta^m \frac{\partial x^k}{\partial y^m} \quad and \quad \eta^m = \xi^k \frac{\partial y^m}{\partial x^k} \tag{3.19}$$

to relate the components of the vector in the two coordinate systems. This follows yet again from the application of the chain rule, here to the components of the tangent vector:

$$\frac{dx^k}{d\lambda} = \frac{\partial x^k}{\partial y^m} \frac{dy^m}{d\lambda} \tag{3.20}$$

The following exercise illustrates how the relations in Eq. (3.19) are generalizations of the relations for changing basis in a linear vector space.

→ **Exercise 3.1.** Suppose $\{\phi_1, \ldots, \phi_n\}$ and $\{\psi_1, \ldots, \psi_n\}$ are two bases on the linear vector space $\mathcal{V}^n$, related by

$$\psi_k = \sum_{j=1}^n S_{jk} \phi_j$$

---

[4]Note that there is an arbitrary positive scale factor in the definition of the tangent vector
[5]A geometrical interpretation of the Jacobian will be given later.

with constants S_{jk}. Find relations between the components of a vector in the two bases. That is, if

$$x = \sum_{j=1}^{n} \xi_j \phi_j = \sum_{k=1}^{n} \eta_k \psi_k$$

express the $\{\xi_j\}$ in terms of the $\{\eta_k\}$ and vice versa. Show that these relations are special cases of Eq. (3.19) for a linear coordinate transformation.   □

The manifold $\mathcal{M}$ together with the collection of tangent spaces at each point of the manifold forms a structure known as the *tangent bundle* $\mathcal{T}(\mathcal{M})$.

❑ **Example 3.17.**   If $\mathcal{M}$ is the n-dimensional configuration space of a mechanical system, then the tangent space at a point of $\mathcal{M}$ is defined by the velocities associated with trajectories that pass through the point. The tangent bundle $\mathcal{T}(\mathcal{M})$ is the $2n$-dimensional space defined by the coordinates and velocities of the system.   ∎

Remark.   The concept of a *bundle*, or *fiber bundle*, over a manifold is important in differential geometry and its applications to physics. In general, a fiber bundle consists of a manifold $\mathcal{M}$, the *base space*, and another space $\mathcal{F}$, such as a linear vector space or a Lie group, the *fiber space*. A copy of $\mathcal{F}$ is attached to each point of the manifold. A function on the manifold that has values in $\mathcal{F}$ is a *(cross)-section* of the bundle. If the space $\mathcal{F}$ is a linear vector space, such as the tangent space $\mathcal{T}(\mathcal{M})$ introduced here, then the bundle is a *vector bundle*. To complete the definition of a fiber bundle, it is necessary to also specify a group $\mathcal{G}$ (the *structure group*) of diffeomorphisms of the fiber space. If $\mathcal{F}$ is a real n-dimensional linear vector space, the structure group might be the group $SO(n)$ of rotations of $\mathbf{R}^n$, or even the group $GL(n, \mathbf{R})$ of nonsingular linear operators on $\mathbf{R}^n$. In this book, we consider only the tangent bundle and the closely related cotangent bundle introduced below. For more general bundles, see the book by Frankel cited in the bibliography.   □

The slice of the tangent bundle corresponding to the points from $\mathcal{M}$ in some neighborhood $\mathcal{U}$ of a point P is locally equivalent to a product $\mathcal{U} \times \mathcal{V}^n$. However, the complete tangent bundle $\mathcal{T}(\mathcal{M})$ may, or may not, be globally equivalent to the product $\mathcal{M} \times \mathcal{V}^n$. This depends how vectors at different points in the manifold are related. Can we define parallel curves, or parallel vectors at different points, for example? These important questions are important for understanding general relativity, but we must refer the reader to the bibliography for answers.

3.2.3   Differential Forms

The tangent space $\mathcal{T}_P$ to the manifold $\mathcal{M}$ at a point P is a linear vector space. As in Section 2.1.5, we can introduce the dual space $\mathcal{T}_P^*$ of linear functionals on $\mathcal{T}_P$. An element ω of $\mathcal{T}_P^*$ maps $\mathcal{T}_P$ into the real (or perhaps complex) numbers,

$$\omega : v \mapsto \omega(v)$$

such that

$$\omega(a_1 v_1 + a_2 v_2) = a_1 \omega(v_1) + a_2 \omega(v_2) \tag{3.21}$$

for any pair of vectors v_1, v_2 and any pair of numbers a_1, a_2. Such a linear functional ω is a (differential) *1-form*, or simply *form*. The dual relation between forms and vectors is emphasized in the various notations

$$\omega(v) = (\omega, v) = \langle \omega, v \rangle = \langle \omega | v \rangle = v(\omega) \tag{3.22}$$

for the number (scalar) associated with the pair ω, v. The space $\mathcal{T}_P^*$ is the *cotangent space* of $\mathcal{M}$ at P, and a 1-form is sometimes called a *cotangent vector*.

A basis $\{\mathbf{e}_k\}$ on $\mathcal{T}_P$ induces a preferred basis $\{\mu^k\}$ on $\mathcal{T}_P^*$, the *dual basis*, with

$$\langle \mu^k, \mathbf{e}_m \rangle = \delta_m^k \tag{3.23}$$

so that if $v = v^k\,\mathbf{e}_k$, then $\mu^k(v) = v^k$, and if $\omega = \omega_k \mu^k$, then

$$(\omega, v) = \omega_k\,v^k \tag{3.24}$$

A coordinate system x^k on a region $\mathcal{U}$ in $\mathcal{M}$ provides a natural (coordinate) basis $\{\partial/\partial x^k\}$ of vectors and a corresponding coordinate basis $\{dx^k\}$ of 1-forms; we have

$$\left\langle dx^k, \frac{\partial}{\partial x^m} \right\rangle = \delta_m^k \tag{3.25}$$

In a coordinate basis, a form ω is expressed as

$$\omega = \alpha_k dx^k = \vec{\alpha} \cdot \vec{dx} \tag{3.26}$$

The expression $\vec{\alpha} \cdot \vec{dx}$ again emphasizes the intrinsic, coordinate-independent nature of a form. Under a smooth coordinate transformation from $\{x^1, \ldots, x^n\}$ to $\{y^1, \ldots, y^n\}$, we have

$$dx^k = \frac{\partial x^k}{\partial y^m}\,dy^m \tag{3.27}$$

which leads to the transformation law for the components of a 1-form

$$\omega = \alpha_k\,dx^k = \beta_m\,dy^m \tag{3.28}$$

We have

$$\beta_m = \alpha_k \frac{\partial x^k}{\partial y^m} \quad \text{and} \quad \alpha_k = \beta_m \frac{\partial y^m}{\partial x^k} \tag{3.29}$$

Note that the role of the partial derivatives is reversed here, compared to the transformation law (3.19) for the components of a vector. That is, if

$$\eta^m \equiv \mathbf{S}^m_{\ k}\,\xi^k \quad \text{and} \quad \beta_m \equiv \bar{\mathbf{S}}^k_{\ m}\,\alpha_k \tag{3.30}$$

then the matrices $\mathbf{S}$ and $\bar{\mathbf{S}}$, with elements

$$\mathbf{S}^m_{\ k} = \frac{\partial y^m}{\partial x^k} \quad \text{and} \quad \bar{\mathbf{S}}^k_{\ m} = \frac{\partial x^k}{\partial y^m} \tag{3.31}$$

are inverse to each other.

The manifold $\mathcal{M}$ together with the collection of cotangent spaces at each point of the manifold forms a structure known as the *cotangent bundle* $T^*(\mathcal{M})$, just as the manifold together with the collection of tangent spaces at each point forms the tangent bundle.

❑ **Example 3.18.** If $\mathcal{M}$ is the n-dimensional configuration space of a Hamiltonian system of classical mechanics, then the cotangent bundle $T^*(\mathcal{M})$ is the $2n$-dimensional *phase space* defined by the coordinates and conjugate momenta of the system, as discussed in more detail in Section 3.5.3.   ∎

Remark. We introduced the tangent space T_P of vectors at a point P and then defined 1-forms as linear functionals on T_P. Alternatively, we could have introduced the cotangent space T_P^* of forms as a linear vector space spanned by the coordinate differentials and then defined vectors as linear functionals on T_P^*. This duality is emphasized by Eq. (3.22), as well as by the discussion of dual spaces in Section 2.1.5.   ☐

Remark. A change of basis on the tangent space T_P or cotangent space T_P^* need not be a coordinate transformation, since there is no necessary relation between the transformations at neighboring points. However, the transformation laws for the components of vectors and forms are still related if we insist that the transformations on T_P and T_P^* change a pair of dual bases into another pair of dual bases. See Problem 3 for details.   ☐

Remark. In an older terminology, the components of a 1-form were known as the *covariant* components of a vector. They transform in the same way as the components of the gradient, which was treated as the prototype of a vector, perhaps since so many fields are expressed as the gradient of some scalar potential. The components of a (tangent) vector were known as the *contravariant* components of a vector, since they transform according to the inverse (*contra*) transformation law. This emphasis on transformation laws led to a proliferation of indices, and tended to obscure the point that vectors and forms are geometrical and physical objects that do *not* depend on the particular coordinate system used to describe the manifold in which they live. The modern view emphasizes the intrinsic properties of forms and vectors, and the duality relation between them.   ☐

Remark. In a manifold with a metric, there is a natural mapping between vectors and forms induced by a metric tensor, as will appear soon in Section 3.4. Here we note that in a linear vector space $\mathbf{R}^n$ with the (Euclidean) scalar product as defined in Chapter 2, the distinction between vectors and forms is purely formal, and we can identify the vector

$$V = V^k \frac{\partial}{\partial x^k} \tag{3.32}$$

and the form

$$\tilde{V} = V_k \, dx^k \tag{3.33}$$

with the same numerical components $V_k = V^k$ in any Cartesian coordinate basis.   ☐

→ **Exercise 3.2.** Show that if these numerical components are the same in one coordinate basis, they are the same in any basis obtained by a rotation from the original basis, i.e., show that these components have the same transformation law under rotations.   ☐

3.2.4 Tensors

Tensors arise when we consider the products of vectors and forms. The tensor product of linear vector spaces has been introduced in Section 2.1.3. Here we introduce the tensor spaces

$$T_P^{(M,N)} \equiv \underbrace{T_P \otimes \cdots \otimes T_P}_{M \text{ times}} \otimes \underbrace{T_P^* \otimes \cdots \otimes T_P^*}_{N \text{ times}} \tag{3.34}$$

at a point P in a manifold. Elements of $T_P^{(M,N)}$ are formed from linear combinations of products of M vectors from T_P and N 1-forms from T_P^*. Elements of this space are $\binom{M}{N}$ tensors. If $N = 0$, we have a tensor of *rank M*. If $M = 0$, we have a *dual* tensor of rank N. Otherwise, we have a *mixed* tensor of rank $M + N$.

❑ **Example 3.19.** A linear operator on a linear vector space is a $\binom{1}{1}$ tensor, since the equation $y = \mathbf{A}x$ becomes

$$y^j = \mathbf{A}_j^{\ k} x_k \tag{3.35}$$

in terms of components.   ∎

❑ **Example 3.20.** The inertia tensor of a rigid body about some point is a (symmetric) tensor of rank 2. Since a rigid body is defined only in a Euclidean space, we need not distinguish between vector and form indices, at least for Cartesian components.   ∎

Tensor components are defined with respect to some basis. Corresponding to coordinates $\{x^1, \ldots, x^n\}$ in some neighborhood of P is the coordinate basis

$$\mathbf{e}_{i_1 \ldots i_M}{}^{j_1 \ldots j_N} \equiv \frac{\partial}{\partial x^{i_1}} \otimes \cdots \otimes \frac{\partial}{\partial x^{i_M}} \otimes dx^{j_1} \otimes \cdots \otimes dx^{j_N} \tag{3.36}$$

on the tensor space $T_P^{(M,N)}$. A tensor $\mathbf{T}$ is expressed in terms of this basis as

$$\mathbf{T} = \mathbf{T}_{j_1 \ldots j_N}{}^{i_1 \ldots i_M} \mathbf{e}_{i_1 \ldots i_M}{}^{j_1 \ldots j_N} \tag{3.37}$$

(summation convention in effect). The $\mathbf{T}_{j_1 \ldots j_N}{}^{i_1 \ldots i_M}$ are the *components* of $\mathbf{T}$.

A tensor can also be introduced as a functional: an $\binom{M}{N}$ tensor is a linear functional on the linear vector space $T_P^{(N,M)}$,

$$\mathbf{T} = \mathbf{T} \, (\underbrace{., \ldots, .}_{N \text{ arguments}} \, | \, \underbrace{., \ldots, .}_{M \text{ arguments}}) \tag{3.38}$$

in which the first N arguments require vectors, and the last M arguments need forms, in order to reduce the functional to a scalar. This is equivalent to the statement

$$T_P^{(M,N)*} = T_P^{(N,M)} \tag{3.39}$$

that generalizes the duality of vectors and 1-forms. Again, note that in a manifold with a metric, there is a natural mapping between a tensor and its dual induced by the metric tensor.

Inserting arguments in the empty slots in the functional (3.38) reduces the rank of the tensor. Thus if v is a vector, then

$$\mathbf{T}\left(., \ldots, v \,|\, ., \ldots, .\right) \tag{3.40}$$

is an $\binom{M}{N-1}$ tensor with components

$$\mathbf{T}_{j_1,\ldots,j_{N-1}j}^{\quad i_1,\ldots,i_M}\, v^j \tag{3.41}$$

Similarly, if ω is a (1-)form, then

$$\mathbf{T}\left(., \ldots, . \,|\, ., \ldots, \omega\right) \tag{3.42}$$

is an $\binom{M-1}{N}$ tensor with components

$$\mathbf{T}_{j_1\ldots j_N}^{\quad i_1,\ldots,i_{M-1}i}\, \omega_i \tag{3.43}$$

The reduction of rank by inserting arguments in the functional is called *contraction*.

A mixed $\binom{M}{N}$ tensor can also be contracted with itself to give an $\binom{M-1}{N-1}$ tensor $\tilde{\mathbf{T}}$ with components

$$\tilde{\mathbf{T}}_{j_1,\ldots,j_{N-1}}^{\quad i_1,\ldots,i_{M-1}} \equiv \mathbf{T}_{j_1,\ldots,j_{N-1}m}^{\quad i_1,\ldots,i_{M-1}m} \tag{3.44}$$

❑ **Example 3.21.** For a linear operator $\mathbf{A}$ on $\mathcal{T}_P$, with components $\mathbf{A}_j^{\;k}$ in some basis, the contraction

$$A \equiv \mathbf{A}_k^{\;k} \tag{3.45}$$

is a scalar already identified as the *trace* of $\mathbf{A}$. It is indeed a scalar, as its value is independent of the coordinate system. ∎

3.2.5 Vector and Tensor Fields

A *vector field* v on a manifold $\mathcal{M}$ is a map from the manifold into the tangent bundle $\mathcal{T}(\mathcal{M})$ such that at each point x in $\mathcal{M}$, there is attached a vector $v(x)$ in the tangent space $\mathcal{T}_x$, whose components are smooth functions of the coordinates. Simple fluid flow is described by a *velocity field* $u(x)$, the velocity of the fluid element at x. The electric field $E(x)$ and the magnetic field $B(x)$ due to a system of charges and currents are vector fields. As already noted, a vector field $v(x)$ can be identified with the directional derivative

$$v = v^k(x)\,\frac{\partial}{\partial x^k} = \overrightarrow{v} \cdot \overrightarrow{\nabla} \tag{3.46}$$

More intuitive is the visualization of a vector field by drawing some of its *integral curves*, which are also known as *streamlines* (fluid flow), *lines of force* (electric fields), or *flux lines* (magnetic fields). Two examples are shown in Fig. 3.6. The integral curve of the vector field $v(x)$ passing through a point P is generated by starting at P, moving in the direction of

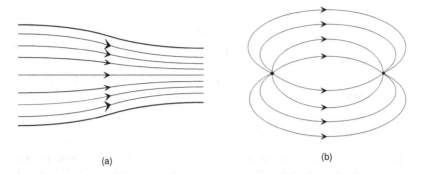

(a)   (b)

Figure 3.6: Examples of the integral curves of a vector field: (a) the streamlines of a fluid flowing through a pipe with a nonconstant cross-section, and (b) schematic representation of the lines of force of the electric field due to equal and opposite charges (an extended dipole).

the field at that point and continuing in the direction of the field at each point. This can be accomplished by solving the system of first-order differential equations

$$\frac{dx^k}{d\lambda} = v^k(x) \tag{3.47}$$

with initial condition $x = x(P)$ at $\lambda = 0$. This system has a unique solution provided the $v^k(x)$ are well-behaved functions, not all zero, at the point P.[6] Two integral curves cannot intersect, except perhaps at points where $v(x) = 0$ or where $v(x)$ is undefined. Such points are *singular points* of the vector field.

Remark.   The electric field is undefined at the location of a point charge, for example. Also, a point where $v(x) = 0$ is a *fixed point* of Eq. (3.47). The solution to Eq. (3.47) starting at such a point P does not generate a curve but remains fixed at P. However, integral curves near the point P may approach P asymptotically, diverge from P, or exhibit hyperbolic behavior as illustrated in Fig. 2.8. Understanding the generic behavior of solutions near a fixed point is the object of the stability analysis in Section 2.5.   □

The vector field v in Eq. (3.46) defines a tangent vector at each point on an integral curve of the field. To generate a finite displacement along the curve, say from λ_0 to $\lambda_0 + \alpha$, we can formally exponentiate the differential operator to give

$$x^k(\lambda_0 + \alpha) = \exp\left(\alpha\frac{d}{d\lambda}\right)x^k(\lambda_0) = x^k(\lambda_0) + \sum_{n=1}^{\infty}\left[\frac{d^n}{d\lambda^n}x^k(\lambda)\right]_{\lambda=\lambda_0}\frac{\alpha^n}{n!} \tag{3.48}$$

This formal Taylor series expansion actually converges only if the coordinates $x^k(\lambda)$ are analytic functions of the parameter λ along the curve. This may, or may not, be the case.

---

[6]It is sufficient that there is some constant $M > 0$ such that

$$|v^k(x) - v^k(x')| < M\|x - x'\|$$

for all k and all x, x' in some neighborhood of P (*Lipschitz condition*). For proofs, consult a textbook on differential equations. Note that the Lipschitz condition includes the case in which the $v^k(x)$ are differentiable at P, but is more restrictive than continuity of the $v^k(x)$ at P.

The partial derivatives $\partial/\partial x^k$ with respect to the coordinates x^k define a basis for the tangent space $\mathcal{T}_x$ at each point x. But at each x, we can choose *any* set of n linearly independent vectors as a basis for $\mathcal{T}_x$, and in some open region $\mathcal{U}$ around x, we can choose any n linearly independent vector fields as a basis.

Can we use these fields to define a coordinate system? To answer this question, suppose we have two vector fields

$$v = \frac{d}{d\lambda} = v^k \frac{\partial}{\partial x^k} \quad \text{and} \quad w = \frac{d}{d\mu} = w^k \frac{\partial}{\partial x^k} \tag{3.49}$$

Let us carry out displacements from a point $\mathcal{O}$, as shown in Fig. 3.7. First move a distance $d\lambda$ along the integral curve of v through $\mathcal{O}$ to the point A. The coordinates then change according to

$$x^k(A) - x^k(\mathcal{O}) \simeq v^k(\mathcal{O})d\lambda \tag{3.50}$$

Then move a distance $d\mu$ along the integral curve of w through A, arriving at the point P whose coordinates are given by

$$x^k(P) - x^k(A) \simeq w^k(A)d\mu \tag{3.51}$$

Alternatively, move first $d\mu$ along the integral curve of w through $\mathcal{O}$ to Q, with

$$x^k(B) - x^k(\mathcal{O}) \simeq w^k(\mathcal{O})d\mu \tag{3.52}$$

and then $d\lambda$ along the integral curve of v through Q, arriving at the point P', with

$$x^k(P') - x^k(B) \simeq v^k(B)d\lambda \tag{3.53}$$

If v and w are to generate a coordinate system, either order of carrying out the two displacements should lead to the same final coordinates, and hence the same final point, as in Fig. 3.7(b). But we can compute

$$w^k(A) \simeq w^k(\mathcal{O}) + \left[v^m \frac{\partial w^k}{\partial x^m} \right]_{\mathcal{O}} d\lambda \tag{3.54}$$

and

$$v^k(B) \simeq v^k(\mathcal{O}) + \left[w^m \frac{\partial v^k}{\partial w^m} \right]_{\mathcal{O}} d\mu \tag{3.55}$$

Then

$$x^k(P) - x^k(P') \simeq \left[v^m \frac{\partial w^k}{\partial x^m} - w^m \frac{\partial v^k}{\partial x^m} \right]_{\mathcal{O}} d\lambda \, d\mu \tag{3.56}$$

The term in brackets must vanish if P and P' are the same points. Now the commutator of the two operators v and w is defined by

$$[v, w] = \left[\frac{d}{d\lambda}, \frac{d}{d\mu} \right] = \left[v^m \frac{\partial w^k}{\partial x^m} - w^m \frac{\partial v^k}{\partial x^m} \right] \frac{\partial}{\partial x^k} \tag{3.57}$$

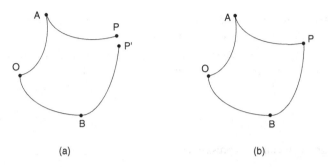

Figure 3.7: (a) A general displacement from point $\mathcal{O}$ along the integral curves of two vector fields v and w. A displacement $d\lambda$ along the integral curve of v carries us from $\mathcal{O}$ to A, and a further displacement $d\mu$ along the integral curve of w through A carries us to P. A displacement $d\mu$ along the integral curve of w through $\mathcal{O}$ to B, followed by a displacement $d\lambda$ along the integral curve of v through B takes us to P'. In general, the points P and P' are not the same, but if the vector fields v and w commute (have vanishing Lie bracket), as in (b), then the two paths lead to the same endpoint, and the vector fields v and w can be used as a part of a coordinate basis.

This commutator is also called the *Lie bracket* of the two vector fields. In order for λ and μ to be coordinates, this commutator must vanish. More generally, a *necessary* condition for a set of vector fields to form the basis of a coordinate system is that the Lie bracket, or commutator, of any pair of the fields must vanish.

Remark.   This condition should not be a surprise—we expect mixed partial derivatives with respect to a pair of coordinates to be independent of the ordering of the derivatives.   □

The vanishing of the commutator (3.57) is also sufficient. A formal way to see this is to note the formal expression

$$x^k(\lambda_0 + \alpha, \mu_0 + \beta) = \exp\left(\alpha \frac{d}{d\lambda} + \beta \frac{d}{d\mu}\right) x^k(\lambda_0, \mu_0) \tag{3.58}$$

based on Eq. (3.48). The exponential factorizes into a displacement α of the coordinate λ and a displacement β of the coordinate μ if (and only if) the derivative operators commute.

→ **Exercise 3.3.**   Suppose r, θ are the standard polar coordinates in the plane defined by

$$x = r\cos\theta \qquad y = r\sin\theta$$

Consider the radial and tangential unit vectors

$$\mathbf{e}_r \equiv \cos\theta\,\mathbf{e}_x + \sin\theta\,\mathbf{e}_y \qquad \mathbf{e}_\tau \equiv -\sin\theta\,\mathbf{e}_x + \cos\theta\,\mathbf{e}_y$$

Corresponding to these are the vector fields

$$\frac{\partial}{\partial r} \equiv \cos\theta \frac{\partial}{\partial x} + \sin\theta \frac{\partial}{\partial y} \qquad \frac{\partial}{\partial \tau} \equiv -\sin\theta \frac{\partial}{\partial x} + \cos\theta \frac{\partial}{\partial y}$$

Do these vector fields form a coordinate basis? If not, find a suitable coordinate basis for polar coordinates, and express this basis in terms of $\partial/\partial x$ and $\partial/\partial y$.   □

A *1-form field* ω on a manifold $\mathcal{M}$ is a map that takes each point x in $\mathcal{M}$ to a 1-form $\omega(x)$ in the cotangent space $\mathcal{T}_x^*$. If $f(x)$ is a differentiable function (scalar field) on $\mathcal{M}$ the *gradient* df of f is the 1-form field defined by

$$\left\langle df, \frac{d}{d\lambda} \right\rangle = \frac{df}{d\lambda} \tag{3.59}$$

In a coordinate basis, the gradient can be expressed as

$$df = \frac{\partial f}{\partial x^1} \, dx^1 + \cdots + \frac{\partial f}{\partial x^n} \, dx^n = \frac{\partial f}{\partial x^k} \, dx^k = \vec{\nabla} f \cdot \vec{dx} \tag{3.60}$$

Thus the components of df in this basis are just the partial derivatives of f, as introduced in elementary calculus. We can also note that df is the element of $\mathcal{T}_P^*$ whose value on v in $\mathcal{T}_P$ is the directional derivative of f along v. Since this derivative vanishes in any direction on a surface $f = $ constant, df is in some sense *normal* to such a surface. If a metric is defined on $\mathcal{M}$, this statement can be made more precise, as will be seen soon.

The tensor space $\mathcal{T}_P^{(M,N)}$ is introduced at each point P of a manifold. A *tensor field* of type $\binom{M}{N}$ is a collection of $\binom{M}{N}$ tensors defined at each point on the manifold (or in some region of the manifold). The field is continuous (differentiable, smooth,...) if its components in some coordinate basis are continuous (differentiable, smooth,...).

❑ **Example 3.22.**   Stress and strain in an elastic medium are (symmetric) tensor fields of rank 2.   ∎

❑ **Example 3.23.**   The electromagnetic field $\mathbf{F} = (F_{\mu\nu})$ is an antisymmetric tensor field of rank 2. The electromagnetic stress–energy tensor $\mathbf{T} = (T_{\mu\nu})$ is a symmetric tensor field of rank 2.   ∎

Remark.   The examples given here all have rank 2. Tensors of higher rank are less common in macroscopic physics, appearing mainly in the theories of elasticity in deformable solids, and also in Einstein's theory of gravity (general relativity). However, they are relevant in quantum theories with higher symmetries, as in the examples discussed in Chapter 10. It is useful to realize that conceptually, at least, higher rank tensors are no more complicated than those of rank 2.   □

3.2.6   The Lie Derivative

The *Lie derivative* $\mathcal{L}_v$ associated with a vector field v is an operator that measures the rate of change along integral curves of the vector field. The Lie derivative of a scalar function $f(x)$ is the directional derivative

$$\mathcal{L}_v f(x) = \vec{v} \cdot \vec{\nabla} f(x) \tag{3.61}$$

The Lie derivative $\mathcal{L}_v$ of a vector field w has been introduced in Eq. (3.57) as the commutator

$$\mathcal{L}_v w = [v, w] \tag{3.62}$$

It measures the change in the coordinate distance along integral curves of w between intersections with two nearby integral curves of v, as we move along the integral curves of v. If $\mathcal{L}_v w = 0$, then this coordinate distance does not change as we move along the integral curves, and the fields v and w can be a part of a coordinate basis.

Expressions for the Lie derivative of other types of tensor fields can be generated using the Leibniz rule

$$\mathcal{L}_v (ab) = (\mathcal{L}_v a) b + a (\mathcal{L}_v b) \tag{3.63}$$

❏ **Example 3.24.** To obtain a rule for the Lie derivative of a 1-form w, note that the Leibniz rule gives

$$(\mathcal{L}_v w, u) = \mathcal{L}_v (w, u) - (w, \mathcal{L}_v u) \tag{3.64}$$

for any vector u. Now

$$\mathcal{L}_v (w, u) = v^k \frac{\partial}{\partial x^k} (w_\ell u^\ell) = v^k \left(\frac{\partial w_\ell}{\partial x^k} u^\ell + \frac{\partial u^\ell}{\partial x^k} w_\ell \right) \tag{3.65}$$

and

$$(w, \mathcal{L}_v u) = w_\ell \left(v^k \frac{\partial u^\ell}{\partial x^k} - u^k \frac{\partial v^\ell}{\partial x^k} \right) \tag{3.66}$$

so that

$$(\mathcal{L}_v w)_\ell = v^k \frac{\partial w_\ell}{\partial x^k} + w_k \frac{\partial v^k}{\partial x^\ell} \tag{3.67}$$

Note the positive sign here, in contrast to the negative sign in Eq. (3.57). ∎

Remark. Note that the Lie derivative does not change the rank of a tensor. ❏

❏ **Example 3.25.** The Lie derivative appears naturally in the description of the flow of a fluid with velocity field U. The *co-moving derivative*, which measures the rate of change of a property of a fixed drop of the fluid moving with the flow, rather than at a fixed point in space, is expressed as

$$\frac{d}{dt} = \frac{\partial}{\partial t} + \mathcal{L}_U \tag{3.68}$$

Fluids are discussed at greater length in Section 3.6. ∎

Another important use of the Lie derivative is to identify invariants. If $\mathbf{T}$ is a tensor such that

$$\mathcal{L}_v \mathbf{T} = 0 \tag{3.69}$$

then $\mathbf{T}$ is constant along any integral curve of v; $\mathbf{T}$ is an *invariant* of the vector field v.

3.3   Calculus on Manifolds

3.3.1   Wedge Product: p-Forms and p-Vectors

A general tensor of type $\binom{0}{N}$ has components $\alpha_{i_1,\ldots,i_N}$ with no particular symmetry under permutations of the indices. However, such a tensor can be split into sets of components that have definite symmetry, i.e., they transform according to irreducible representationsof the group $\mathcal{S}_N$ that permutes the indices of the tensor; such sets are invariant under change of basis. General irreducible representations are studied in Chapter 10; for now we are interested only in completely symmetric or completely antisymmetric tensors.

If σ and τ are 1-forms, and

$$\omega = \sigma \otimes \tau \tag{3.70}$$

then introduce the symmetric and antisymmetric forms

$$\omega^S \equiv \tfrac{1}{2}\left(\sigma \otimes \tau + \tau \otimes \sigma\right) \qquad \omega^A \equiv \tfrac{1}{2}\left(\sigma \otimes \tau - \tau \otimes \sigma\right) \tag{3.71}$$

If $\{\mu^k\}$ is a basis of 1-forms and

$$\sigma = \alpha_k \mu^k \qquad \tau = \beta_k \mu^k \tag{3.72}$$

then

$$\omega^S \;=\; \tfrac{1}{2}\left(\alpha_k\beta_\ell + \alpha_\ell\beta_k\right)\mu^k \otimes \mu^\ell \equiv \tfrac{1}{2}\,\omega^S_{k\ell}\,\mu^k \otimes \mu^\ell \tag{3.73}$$

$$\omega^A \;=\; \tfrac{1}{2}\left(\alpha_k\beta_\ell - \alpha_\ell\beta_k\right)\mu^k \otimes \mu^\ell \equiv \tfrac{1}{2}\,\omega^A_{k\ell}\,\mu^k \otimes \mu^\ell \tag{3.74}$$

where the factor $\tfrac{1}{2}$ is to compensate for double counting in the sum over k and ℓ.

→ **Exercise 3.4.**   Show that $\omega^S_{k\ell}$ ($\omega^A_{k\ell}$) remain symmetric (antisymmetric) under any change of basis on the tensor space $T_P^{(0,2)}$ on which the forms are defined.   □

The antisymmetric part of the product of two 1-forms σ and τ is important enough that it has a special name, the *wedge product* (also *exterior product* or *Grassmann product* in the mathematical literature). It is denoted by

$$\sigma \wedge \tau \equiv \tfrac{1}{2}\left(\sigma \otimes \tau - \tau \otimes \sigma\right) = -\tau \wedge \sigma \tag{3.75}$$

In terms of components of σ and τ with respect to a basis $\{\mu^k\}$, we have

$$\sigma \wedge \tau = \tfrac{1}{2}\left(\alpha_k\beta_\ell - \alpha_\ell\beta_k\right)\mu^k \wedge \mu^\ell \tag{3.76}$$

→ **Exercise 3.5.**   Show that if σ and τ are 1-forms, and v is a vector, then

$$\langle \sigma \wedge \tau, v \rangle = \langle \sigma, v \rangle \tau - \sigma \langle \tau, v \rangle$$

The wedge product generalizes the three-dimensional cross product $\mathbf{A} \times \mathbf{B}$ of vectors. Indeed, this cross product is often denoted by $\mathbf{A} \wedge \mathbf{B}$ in the European literature.   □

The wedge product of two 1-forms is a *2-form*. The antisymmetric product of p 1-forms is a *p-form*. If $\sigma_1, \ldots, \sigma_p$ are 1-forms, we have

$$\sigma_1 \wedge \cdots \wedge \sigma_p \equiv \frac{1}{p!} \sum_{i_1, \ldots, i_p} \varepsilon_{i_1, \ldots, i_p} \, \sigma_1 \otimes \cdots \otimes \sigma_p \tag{3.77}$$

where $\varepsilon_{i_1, \ldots, i_p}$ is the usual antisymmetric (Levi-Civita) symbol with p indices defined as by

$$\varepsilon_{i_1, \ldots, i_p} = \varepsilon^{i_1, \ldots, i_p} = \begin{cases} +1, & \text{if } i_1, \ldots, i_p \text{ is an even permutation of } 1 \ldots p \\ -1, & \text{if } i_1, \ldots, i_p \text{ is an odd permutation of } 1 \ldots p \\ 0, & \text{if any two indices are equal} \end{cases} \tag{3.78}$$

and the division by $p!$ is needed to compensate for multiple counting in the sum over $i_1, \ldots, i_p$. A p-form σ can be expressed in terms of components as

$$\sigma = \frac{1}{p!} \sigma_{i_1, \ldots, i_p} \, \mu^{i_1} \wedge \cdots \wedge \mu^{i_p} \tag{3.79}$$

→ **Exercise 3.6.** Show that the wedge product of a p-form σ_p and a q-form τ_q is a $(p+q)$-form, with

$$\sigma_p \wedge \tau_q = (-1)^{pq} \, \tau_q \wedge \sigma_p$$

Evidently this generalizes the antisymmetry of the wedge product of two 1-forms. □

The appearance of coordinate differentials as a basis for forms suggests that forms can be integrated. For example, a 1-form $\sigma = \sigma_k \, dx^k$ appears naturally as the integrand of a line integral

$$\int_C \sigma = \int_C \sigma_k \, dx^k \tag{3.80}$$

along a curve C, and the dx^k in the integral are the usual coordinate differentials of elementary calculus. Note that it may be necessary to use more than one coordinate patch to integrate along the entire curve C.

Similarly, a 2-form $\sigma = \frac{1}{2}\sigma_{k\ell} \, dx^k \wedge dx^\ell$ appears naturally as the integrand of a surface integral

$$\int_S \sigma = \frac{1}{2} \int_S \sigma_{k\ell} \, dx^k \wedge dx^\ell \tag{3.81}$$

over a surface S. In this context, the product $dx^k \wedge dx^\ell$ represents an oriented surface element defined by the line elements dx^k and dx^ℓ; the orientation arises from the antisymmetry of the wedge product.

Remark. Expressing a surface element as a 2-form $dx \wedge dy$ leads directly to understanding the appearance of the Jacobian determinant when integration variables are changed. If

$$u = u(x, y) \qquad v = v(x, y) \tag{3.82}$$

then

$$du = \frac{\partial u}{\partial x} \, dx + \frac{\partial u}{\partial y} \, dy \qquad dv = \frac{\partial v}{\partial x} \, dx + \frac{\partial v}{\partial y} \, dy \tag{3.83}$$

whence

$$du \wedge dv = \left(\frac{\partial u}{\partial x} \frac{\partial v}{\partial y} - \frac{\partial u}{\partial y} \frac{\partial v}{\partial x} \right) dx \wedge dy \tag{3.84}$$

The minus sign and the absence of $dx \wedge dx$ and $dy \wedge dy$ terms follow from the antisymmetry of the wedge product that is associated with the orientation of the surface element.   □

In general, a p-form appears as an integrand in a p-dimensional integral. If σ is a smooth p-form defined on a manifold $\mathcal{M}$, and $\mathbf{R}$ is a p-dimensional submanifold of $\mathcal{M}$, then we denote such an integral simply by

$$I \equiv \int_{\mathbf{R}} \sigma \tag{3.85}$$

where I represents a number, the *value* of the integral. To relate this integral to the usual integral of elementary calculus, and to apply the usual methods to evaluate the integral, we need to introduce a coordinate system (or a covering set of coordinate patches) on $\mathbf{R}$. If $x^1, \ldots, x^p$ is such a coordinate system, then the form σ can be expressed as

$$\sigma = f_\sigma(x^1, \ldots, x^p) \, dx^1 \wedge \cdots \wedge dx^p \tag{3.86}$$

(locally on each coordinate patch in $\mathbf{R}$, if necessary). The integral I is then given by

$$I = \int_{\mathbf{R}} f_\sigma(x^1, \ldots, x^p) \, dx^1, \ldots, dx^p \tag{3.87}$$

where now the $dx^1, \ldots, dx^p$ are the usual coordinate differentials of elementary calculus, rather than a set of abstract basis elements for forms.

Remark.   Note that the antisymmetry of the underlying p-form leads to the relation

$$dy^1 \wedge \cdots \wedge dy^p = \det \left\| \frac{\partial y^k}{\partial x^\ell} \right\| dx^1 \wedge \cdots \wedge dx^p \tag{3.88}$$

for changing variables from $x^1, \ldots, x^p$ to $y^1, \ldots, y^p$ in a p-dimensional integral, where the determinant is the usual Jacobian. This result leads to a natural definition of volume element on a manifold with a metric tensor, as will be seen in the next section.   □

In an n-dimensional manifold, there are no p-forms at all with $p > n$, and there is exactly one independent n-form, which in a coordinate basis is proportional to

$$\Omega \equiv dx^1 \wedge \cdots \wedge dx^n = \frac{1}{n!} \, \varepsilon_{i_1, \ldots, i_n} dx^{i_1} \wedge \cdots \wedge dx^{i_n} \tag{3.89}$$

where $\varepsilon_{i_1, \ldots, i_n}$ is the Levi-Civita symbol defined in Eq. (3.78).

→ **Exercise 3.7.**   Show that the dimension of the linear vector space $\mathcal{E}^p(P)$ of p-forms at a point P is given by

$$d_p = \binom{n}{p}$$

Note that this implies $d_{n-p} = d_p$.   □

Definition 3.2.   The collection of all p-forms ($p = 1, \ldots, n$) at a point P of an n-dimensional manifold $\mathcal{M}$ defined the *exterior algebra* $\mathcal{E}(P)$ at P, with product of forms defined as the wedge product.   ∎

→ **Exercise 3.8.**   Show that the dimension of $\mathcal{E}(P)$ is 2^n.   □

The p-forms have been introduced as antisymmetric tensor products of 1-forms at a point P. These can be extended to p-form fields defined on a region of a manifold—even on the entire manifold. One important field on a manifold is the *fundamental volume form*

$$\Omega(x) \equiv \rho(x)\, dx^1 \wedge \cdots \wedge dx^n \tag{3.90}$$

where $\rho(x)$ is a somewhat arbitrary density function that is modified by a Jacobian determinant under a change of coordinates. However, in a manifold with a metric tensor, there is a natural definition of the fundamental volume form (see Section 3.4). If it is possible to define a continuous n-form on the n-dimensional manifold $\mathcal{M}$ with $\rho(x)$ nowhere vanishing, then $\mathcal{M}$ is (internally) *orientable*, and we can take $\rho(x)$ to be positive everywhere on $\mathcal{M}$.

The volume form serves to define another duality between vectors and forms. If σ is a p-form, then the *(Hodge)*-dual* of σ is the antisymmetric $\binom{n-p}{0}$ tensor V defined by

$$V^{i_1,\ldots,i_{n-p}} = (^*\sigma)^{i_1,\ldots,i_{n-p}} \equiv \frac{1}{p!} \frac{1}{\rho(x)} \varepsilon^{i_1,\ldots,i_n} \sigma_{i_{n-p+1},\ldots,i_n} \tag{3.91}$$

Conversely, if V is an antisymmetric $\binom{n-p}{0}$ tensor, then the (Hodge)*-dual of V is the p-form σ defined by

$$\sigma_{i_1,\ldots,i_p} = (^*V)_{i_1,\ldots,i_p} \equiv \frac{1}{(n-p)!} \rho(x) \varepsilon_{i_1,\ldots,i_n} V^{i_{p+1},\ldots,i_n} \tag{3.92}$$

❑ **Example 3.26.**   Consider the real linear vector space $\mathbf{R}^3$ with the usual volume form

$$\Omega = dx^1 \wedge dx^2 \wedge dx^3 \tag{3.93}$$

A 2-form

$$\sigma = \tfrac{1}{2}\, \sigma_{jk} dx^j \wedge dx^k \tag{3.94}$$

is dual to the vector

$$^*\sigma = \tfrac{1}{2}\varepsilon^{jk\ell}\sigma_{jk}\frac{\partial}{\partial x^\ell} \equiv {}^*\sigma^\ell \frac{\partial}{\partial x^\ell} \tag{3.95}$$

Conversely, a vector

$$V = V^\ell \frac{\partial}{\partial x^\ell} \tag{3.96}$$

corresponds to the dual 2-form

$$*V = \tfrac{1}{2}\, \varepsilon_{jkl}\, V^\ell\, dx^j \wedge dx^k \equiv \tfrac{1}{2}\, {}^*V_{jk}\, dx^j \wedge dx^k \tag{3.97}$$

Thus in three dimensions (only), the wedge product, or *vector product*, of two vectors can be identified with another vector as in the usual expression $\mathbf{C} = \mathbf{A} \times \mathbf{B}$. ∎

→ **Exercise 3.9.**   Show that if σ is a p-form, then $*(*\sigma) = (-1)^{p(n-p)}\sigma$.   □

→ **Exercise 3.10.**   Show that the dual of the volume form is a constant scalar function (a 0-vector), and, in fact, $*\Omega = 1$.   □

3.3.2   Exterior Derivative

The derivative operator $\overrightarrow{\nabla}$ of elementary vector calculus can be generalized to a linear operator d, the *exterior derivative*, which transforms a p-form field into a $(p + 1)$-form field. The operator d is defined by the following rules:

1. For a scalar field f, df is simply the gradient of f, defined in a coordinate basis by the usual rule, Eq. (3.59),

$$df = \frac{\partial f}{\partial x^k}\, dx^k \tag{3.98}$$

so that if

$$v = v^k \frac{\partial}{\partial x^k} \tag{3.99}$$

is a vector field, then

$$(df, v) = v^k \frac{\partial f}{\partial x^k} = \overrightarrow{v} \cdot \overrightarrow{\nabla} f(x) \tag{3.100}$$

which is the directional derivative of f in the direction of v.

2. The operator d is linear, so that if σ and τ are p-forms, then

$$d(\sigma + \tau) = d\sigma + d\tau \tag{3.101}$$

3. If σ is a p-form and τ is a q-form, then

$$d(\sigma \wedge \tau) = d\sigma \wedge \tau + (-1)^p \sigma \wedge d\tau \tag{3.102}$$

4. *Poincaré's lemma.* For any form σ,

$$d(d\sigma) = 0 \tag{3.103}$$

Remark. It is actually enough to assume that this is true for 0-forms (scalars) only, but we do not want to go into details of the proof of the generalized formula, Eq. (3.103). □

The combination of rules 1–4 allows us to construct explicit representations for the d operator in terms of components. If

$$\sigma = \sigma_k \, dx^k \tag{3.104}$$

is a 1-form, then $d\sigma$ is given by

$$d\sigma = d\sigma_k \wedge dx^k = \frac{\partial \sigma_k}{\partial x^j} \, dx^j \wedge dx^k = \frac{1}{2} \left(\frac{\partial \sigma_k}{\partial x^j} - \frac{\partial \sigma_j}{\partial x^k} \right) dx^j \wedge dx^k \tag{3.105}$$

where Poincaré's lemma has been used to set $d(dx^k) = 0$. Thus $d\sigma$ is a generalization of $\operatorname{curl} \sigma = \vec{\nabla} \times \vec{\sigma}$ of elementary calculus. Note that if $\sigma = df$, with f being a scalar function, then

$$d\sigma = \frac{\partial^2 f}{\partial x^j \partial x^k} \, dx^j \wedge dx^k = 0 \tag{3.106}$$

since the second derivative is symmetric in j and k, while the wedge product is antisymmetric. This is consistent with Poincaré's lemma $d(df) = 0$, and corresponds to the elementary result

$$\operatorname{curl} \operatorname{grad} f = \vec{\nabla} \times \vec{\nabla} f = 0 \tag{3.107}$$

Similarly, if

$$\sigma = \tfrac{1}{2} \sigma_{k\ell} \, dx^k \wedge dx^\ell \tag{3.108}$$

is a 2-form, then

$$d\sigma = \frac{1}{2} \frac{\partial \sigma_{k\ell}}{\partial x^m} \, dx^k \wedge dx^\ell \wedge dx^m \tag{3.109}$$

$$= \frac{1}{6} \left(\frac{\partial \sigma_{k\ell}}{\partial x^m} + \frac{\partial \sigma_{mk}}{\partial x^\ell} + \frac{\partial \sigma_{\ell m}}{\partial x^k} \right) dx^k \wedge dx^\ell \wedge dx^m \tag{3.110}$$

with obvious generalization to higher forms.

❑ **Example 3.27.** In three dimensions, we have the standard volume element

$$\Omega = \tfrac{1}{6} \varepsilon_{jkl} \, dx^j \wedge dx^k \wedge dx^\ell \tag{3.111}$$

so that

$$dx^j \wedge dx^k \wedge dx^\ell = \varepsilon^{jkl} \, \Omega \tag{3.112}$$

The 2-form σ is the dual of a vector V, $\sigma = {}^*V$, with

$$\sigma_{jk} = \varepsilon_{jkm} V^m \tag{3.113}$$

Thus

$$d\sigma = \left(\frac{\partial V^m}{\partial x^m}\right)\Omega = (\text{div } V)\,\Omega \tag{3.114}$$

and the divergence of a vector field V can be expressed as

$$(\text{div } V)\,\Omega = d\,{}^*V \tag{3.115}$$

In view of Exercise 3.10, we also have

$$\text{div } V = \,{}^*d({}^*V) \tag{3.116}$$

This definition serves to define the divergence of a vector field in any dimension.   ∎

→ **Exercise 3.11.**   Show that in a space with a general volume form defined by Eq. (3.90), the divergence of the vector field V is given by

$$\text{div } V = \frac{1}{\rho(x)}\,\frac{\partial}{\partial x^k}\left[\rho(x)V^k\right]$$

Show also that

$$\mathcal{L}_V\Omega = (\text{div } V)\Omega$$

This is yet another geometrical property of the Lie derivative.   □

→ **Exercise 3.12.**   Show that Poincaré's lemma also gives the identity

$$\text{div curl } \overrightarrow{V} = 0$$

of elementary calculus.   □

Definition 3.3.   If σ is a p-form and V is a vector field, the *contraction* of σ with V defines a $(p-1)$-form that we denote by

$$i_V\sigma \equiv \sigma(V,\,.\,) \tag{3.117}$$

or, in terms of components,

$$(i_V\sigma)_{i_1,\dots,i_{p-1}} = \sigma_{k\,i_1,\dots,i_{p-1}}V^k \tag{3.118}$$

as in Eq. (3.44).

Remark. $i_V\sigma$ is sometimes called an *interior derivative*, or *interior product*.   □

→ **Exercise 3.13.**   Show that if σ is a p-form and τ is a q-form, then

$$i_V(\sigma \wedge \tau) = \sigma \wedge (i_V\tau) + (-1)^q(i_V\sigma) \wedge \tau$$

Note that we define $i_V\sigma$ by contracting the first index of σ with the vector index of V. The alternative of contraction with the pth index leads to more awkward minus signs.   □

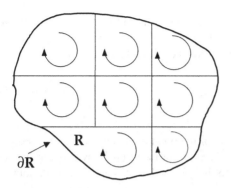

Figure 3.8: A closed region **R** with boundary $\partial\mathbf{R}$. **R** is orientable if it can be subdivided into subregions with boundaries, such that each internal boundary bounds two subregions with opposite orientation, while the external boundaries of the subregions join smoothly to form a consistent orientation on all of $\partial\mathbf{R}$.

An important relation between the Lie derivative and the exterior derivative is the

Theorem 3.1. If V is a vector field and σ a p-form, then

$$\mathcal{L}_V\sigma = d(i_V\sigma) + i_V(d\sigma) \tag{3.119}$$

Proof. If σ is a 1-form, then

$$d(i_V\sigma)_k = \frac{\partial}{\partial x^k}(\sigma_\ell V^\ell) = \frac{\partial\sigma_\ell}{\partial x^k}V^\ell + \sigma_\ell\frac{\partial V^\ell}{\partial x^k} \tag{3.120}$$

and

$$[i_V(d\sigma)]_k = \left(\frac{\partial\sigma_k}{\partial x_\ell} - \frac{\partial\sigma_\ell}{\partial x^k}\right)V^\ell \tag{3.121}$$

The two terms combine to give the Lie derivative $\mathcal{L}_V\sigma$ as obtained in Eq. (3.67). See Problem 5 for a general p-form.

3.3.3   Stokes' Theorem and its Generalizations

A fundamental result of integral calculus is that if $f(x)$ is a differentiable function, then

$$\int_{x=a}^{x=b} df(x) = f(b) - f(a) \tag{3.122}$$

Thus the integral of the derivative of a function (0-form) over an interval depends only on the values of the function at the endpoints of the interval. This result can be generalized to the integral of an exterior derivative over a closed region in an arbitrary manifold. Suppose **R** is a closed p-dimensional region in a manifold $\mathcal{M}$, with boundary $\partial\mathbf{R}$ as shown schematically in

Fig. 3.8. Assume the boundary is orientable, which means that $\mathbf{R}$ can be divided into subregions with a consistent orientation on the boundary of each of the subregions, as indicated in the figure. Then if σ is a smooth $(p-1)$-form defined on $\mathbf{R}$ and $\partial\mathbf{R}$, we have

$$\int_{\mathbf{R}} d\sigma = \int_{\partial\mathbf{R}} \sigma \qquad (3.123)$$

Equation (3.123) is *Stokes' theorem*.

To derive this result, consider a segment $\mathbf{S}_0$ of the boundary characterized by coordinates $x^2, \ldots, x^p$, as indicated in Fig. 3.9. We can drag this surface into $\mathbf{R}$ by a small distance $\delta\xi$ along the integral curves of a coordinate ξ that is independent of the $x^2, \ldots, x^p$, obtaining a new surface $\mathbf{S}_1$.

Let $\delta\mathbf{S}$ denote the small region bounded by $\mathbf{S}_0$, $\mathbf{S}_1$ and the edges traced out by the boundary of $\mathbf{S}_0$ as it is dragged. Then in the region $\delta\mathbf{S}$, we have

$$d\sigma = \frac{\partial\sigma(\xi, x^2, \ldots, x^p)}{\partial\xi} \, d\xi \wedge dx^2, \ldots, \wedge dx^n \quad (3.124)$$

and we can write

$$\int_{\delta\mathbf{S}} d\sigma = \int_{\mathbf{S}_0} \sigma - \int_{\mathbf{S}_1} \sigma + \cdots \text{ (edge terms)} \quad (3.125)$$

up to a sign that is fixed by choosing a suitable orientation of the region $\delta\mathbf{S}$.

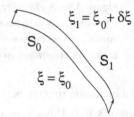

Fig. 3.9: Integration over a region $\delta\mathbf{R}$ between two surfaces $\mathbf{S}_0$ and $\mathbf{S}_1$ separated by a small coordinate displacement $\delta\xi$.

The surface $\mathbf{S}_0$ can be extended to cover the entire original boundary surface $\partial\mathbf{R}$, perhaps using more than one coordinate patch in covering $\partial\mathbf{R}$. The region $\mathbf{R}$ is reduced to a smaller region $\mathbf{R}_1$ with boundary $\partial\mathbf{R}_1$. Denote the region between $\partial\mathbf{R}$ and $\partial\mathbf{R}_1$ by $\delta\mathbf{R}$. Then

$$\int_{\delta\mathbf{R}} d\sigma = \int_{\partial\mathbf{R}} \sigma - \int_{\partial\mathbf{R}_1} \sigma \qquad (3.126)$$

and the edge terms cancel since they have opposite orientation when they appear from adjacent regions. If the region $\mathbf{R}$ is simply connected, this procedure can be repeated until the inner region is reduced to a single point. Then $\delta\mathbf{R}$ includes the entire region $\mathbf{R}$, and $\partial\mathbf{R}_1$ disappears, so that Eq. (3.126) is reduced to Eq. (3.123). If $\mathbf{R}$ is not simply connected, we can still apply the procedure starting with the outer boundary, which we call $\partial\mathbf{R}_0$. When we shrink this boundary, we eventually reach an inner boundary $\partial\mathbf{R}_1$ that can be contracted no further ($\partial\mathbf{R}_1$ may even have several disconnected components). Then Eq. (3.126) becomes

$$\int_{\mathbf{R}} d\sigma = \int_{\partial\mathbf{R}_0} \sigma - \int_{\partial\mathbf{R}_1} \sigma \qquad (3.127)$$

To see that this is equivalent to Stokes' theorem, note that if the orientation of $\partial\mathbf{R}_0$ is "outward" from the region $\mathbf{R}$, then the orientation of $\partial\mathbf{R}_1$ points *into* $\mathbf{R}$. To obtain the consistent "outward" orientation of the boundary $\partial\mathbf{R}$ in Eq. (3.123), we need to reverse the orientation of $\partial \mathbf{R}_1$. The integral on the right-hand side of Eq. (3.127) is then an integral over the entire boundary of $\mathbf{R}$.

The general Stokes' theorem (Eq. (3.123)) includes the elementary version

$$\oint_C \vec{V} \cdot \vec{dl} = \int_S \operatorname{curl} \vec{V} \cdot \vec{dS} \tag{3.128}$$

that relates the integral of a vector $\vec{V}$ around a closed curve C to the integral of curl $\vec{V}$ over a surface S bounded by C.

Remark. Equation (3.128) shows that any surface bounded by C can be chosen, since the integral on the right-hand side depends only on the boundary curve C. □

❑ **Example 3.28.** A magnetic field $\vec{B}$ can be expressed in terms of a vector potential $\vec{A}$ as

$$\vec{B} = \operatorname{curl} \vec{A} \tag{3.129}$$

The magnetic flux Φ through a surface S bounded by the closed curve C can then be expressed as

$$\Phi = \int_S \vec{B} \cdot \vec{dS} = \oint_C \vec{A} \cdot \vec{dl} \tag{3.130}$$

Note that this is invariant under a gauge transformation $\vec{A} \rightarrow \vec{A} + \vec{\nabla}\chi$ for any single-valued gauge function χ. ∎

→ **Exercise 3.14.** The vector potential at the point $\vec{r}$ due to a point magnetic dipole $\vec{m}$ at the origin is given by

$$\vec{A}(\vec{r}) = k\, \frac{\vec{m} \times \vec{r}}{r^3} \tag{3.131}$$

where k is a constant. Find the magnetic flux through a circular loop of radius R with center at a distance d from the dipole, oriented perpendicular to the line from the dipole to the center (i) by using the result of the previous example, and (ii) by integrating the magnetic field over a spherical cap with center at the dipole, and bounded by the circular loop. □

Stokes' theorem also includes the divergence theorem in three dimensions,

$$\oint_S \vec{V} \cdot \vec{dS} = \int_{\mathbf{R}} (\operatorname{div} \vec{V})\, \Omega \tag{3.132}$$

that relates the flow of a vector $\vec{V}$ across a closed surface S to the integral of the divergence of $\vec{V}$ over the volume $\mathbf{R}$ bounded by S. The divergence theorem leads to relations between global conservation laws and local equations of continuity. For example, the total electric charge $Q_{\mathbf{R}}$ in a region $\mathbf{R}$ is given in terms of the charge density ρ by

$$Q_{\mathbf{R}} = \int_{\mathbf{R}} \rho\, \Omega \tag{3.133}$$

The current I across a surface S is expressed in terms of the current density $\vec{j}$ as

$$I = \int_S \vec{j} \cdot \vec{dS} \tag{3.134}$$

Conservation of charge means that the total charge in a closed region changes only if charge flows across the boundary of the region. This is expressed formally as

$$\frac{dQ_\mathbf{R}}{dt} = \frac{d}{dt} \int_\mathbf{R} \rho\, \Omega = \int_\mathbf{R} \frac{\partial \rho}{\partial t}\, \Omega = -\int_{\partial \mathbf{R}} \vec{j} \cdot \vec{dS} \tag{3.135}$$

where as above, $\partial \mathbf{R}$ denotes the boundary of $\mathbf{R}$. Hence, using Eq. (3.132), we have

$$\int_{\partial \mathbf{R}} \vec{j} \cdot \vec{dS} = \int_\mathbf{R} \left(\operatorname{div} \vec{j} \right)\, \Omega \tag{3.136}$$

for any region $\mathbf{R}$, and then

$$\int_\mathbf{R} \left(\frac{\partial \rho}{\partial t} + \operatorname{div} \vec{j} \right)\, \Omega \tag{3.137}$$

Since this is true for any $\mathbf{R}$, the integrand must vanish everywhere, so that

$$\frac{\partial \rho}{\partial t} + \operatorname{div} \vec{j} = 0 \tag{3.138}$$

Remark. Equation (3.138) is the *equation of continuity*. It is also true for matter flow in a fluid, where ρ is the mass density and $\vec{j} = \rho \vec{u}$ is the mass flow density, and the total mass of the fluid is conserved. See Section 3.6 for further discussion of fluids. $\square$

Stokes' theorem also serves to show the equivalence between differential and integral forms of Maxwell's equations for electric and magnetic fields. Gauss' Law

$$\int_S \vec{E} \cdot \vec{dS} = 4\pi k Q = 4\pi k \int_\mathbf{R} \rho\, \Omega \tag{3.139}$$

relates the flux of the electric field $\vec{E}$ through a closed surface S to the total charge Q in the $\mathbf{R}$ bounded by S (k is a constant that relates the unit of charge and the unit of electric field). The divergence theorem allows us to write this in the form

$$\int_\mathbf{R} \left(\operatorname{div} \vec{E} - 4\pi k \rho \right)\, \Omega \tag{3.140}$$

(ρ is again the charge density) for any region $\mathbf{R}$, and hence

$$\operatorname{div} \vec{E} = 4\pi k \rho \tag{3.141}$$

In the absence of magnetic charge, the total magnetic flux through any closed surface vanishes, and thus the magnetic field satisfies

$$\operatorname{div} \vec{B} = 0 \tag{3.142}$$

Faraday's Law

$$\oint_C \vec{E} \cdot \vec{dl} = -\frac{d}{dt} \int_S \vec{B} \cdot \vec{dS} \tag{3.143}$$

relates the integral of the electric field around a closed curve C to the rate of change of the magnetic flux through a surface S bounded by C (in view of Eq. (3.130), any surface S bounded by C will do here). Stokes' theorem allows us to write this equation as

$$\int_S \left(\operatorname{curl} \vec{E} + \frac{\partial \vec{B}}{\partial t} \right) \cdot \vec{dS} = 0 \tag{3.144}$$

for any surface S, and then

$$\operatorname{curl} \vec{E} = -\frac{\partial \vec{B}}{\partial t} \tag{3.145}$$

Finally, the original version of Ampère's Law

$$\int_C \vec{B} \cdot \vec{dl} = 4\pi\alpha I = 4\pi\alpha \int_S \vec{j} \cdot \vec{dS} \tag{3.146}$$

relates the integral of the magnetic field around a closed curve C to the current I flowing through a surface S bounded by the curve (α is a constant that relates the unit of current to the unit of magnetic field). Ampère's Law corresponds to the differential law

$$\operatorname{curl} \vec{B} = 4\pi\alpha \, \vec{j} \tag{3.147}$$

But this equation as it stands is inconsistent with the equation of continuity, Eq (3.138), which is derived from conservation of electric charge, since div curl $\vec{B} = 0$ always, while div $\vec{j} = 0$ only if the charge density is time independent.

   Maxwell's great step forward was to recognize that adding a term proportional to the rate of change of the electric field to the right-hand side of Eq. (3.147) would restore consistency with charge conservation, and he proposed the modified form

$$\operatorname{curl} \vec{B} = 4\pi\alpha \left(\vec{j} + \frac{1}{4\pi k} \frac{\partial \vec{E}}{\partial t} \right) \tag{3.148}$$

The term proportional to $\partial \vec{E}/\partial t$ on the right-hand side of Eq. (3.148) is sometimes called the (Maxwell) *displacement current density*. It is analogous to the time derivative of the magnetic field in Faraday's Law (apart from a significant minus sign), and leads to a consistent classical theory of electromagnetic radiation.

Remark.   Equations (3.141), (3.142), (3.145), and (3.148) are *Maxwell's equations*. Note that in SI units, the constants k and α are given by $k = 1/4\pi\varepsilon_0$ and $\alpha = \mu_0/4\pi$.   □

3.3.4   Closed and Exact Forms

Poincaré's lemma 3.103 states that $d(d\sigma) = 0$ for any form σ, and we have noted the special cases

$$\text{curl grad } f = 0 \qquad \text{and} \qquad \text{div curl } \vec{V} = 0$$

from elementary vector calculus. Is the converse true? That is, if ω is a form such that $d\omega = 0$, is there always a form σ such that $\omega = d\sigma$? To answer this question, we introduce the

Definition 3.4. The form ω is *closed* if

$$d\omega = 0 \tag{3.149}$$

The form ω is *exact* if there is a form σ such that

$$\omega = d\sigma \tag{3.150}$$

Poincaré's lemma tells us that every exact form is closed. Here we ask: is every closed form exact? If ω is a closed form, then there is always a form σ such that $\omega = d\sigma$ locally, but it may not be possible to extend this relation to the entire manifold. In fact the existence of closed forms that are not exact is an important topological characteristic of a manifold. Here we simply illustrate this with some examples.

❑ **Example 3.29.**   On the circle, with coordinate angle θ, the form $d\theta$ is closed. But it is not exact, since there is no single-valued function on the circle whose gradient is $d\theta$.   ∎

❑ **Example 3.30.**   If $\vec{V}$ is a vector in a two-dimensional manifold $\mathcal{M}$, such that curl $\vec{V} = 0$, then we can define a function f(x) by

$$f(x) = \int_{x_0}^{x} \vec{V} \cdot \vec{dl} \tag{3.151}$$

If the manifold $\mathcal{M}$ is simply connected (this means that any closed curve in the manifold can be smoothly deformed to a circle of arbitrarily small radius), the integral defining $f(x)$ is independent of the path of integration from x_0 to x. For if C_1 and C_2 are two smooth curves joining x_0 and x, then Stokes' theorem insures that

$$\int_{C_1} \vec{V} \cdot \vec{dl} - \int_{C_2} \vec{V} \cdot \vec{dl} = \int_{\mathcal{R}} \text{curl } \vec{V} \cdot \vec{dS} = 0 \tag{3.152}$$

where $\mathcal{R}$ is a region (any region) bounded by the curves C_1 and C_2, such that $\mathcal{R}$ lies entirely in the manifold $\mathcal{M}$. Then also

$$\text{grad } f = \text{grad} \int_{x_0}^{x} \vec{V} \cdot \vec{dl} = \vec{V} \tag{3.153}$$

It should be clear that the crucial point is that the manifold be simply connected. Otherwise, there could be two paths joining x_0 and x with no region between them lying entirely inside the manifold, and then we could not apply Stokes' theorem in Eq. (3.152).   ∎

❑ **Example 3.31.** Consider the magnetic field $\vec{B}$ due to a hypothetical magnetic charge (monopole) of strength g located at the origin,

$$\vec{B} = \kappa g \frac{\vec{r}}{r^3} \tag{3.154}$$

where κ is a constant to set the physical units of g. $\vec{B}$ satisfies div $\vec{B} = 0$ everywhere except at the location of the charge, yet there is no single vector potential $\vec{A}$ such that

$$\vec{B} = \text{curl } \vec{A}$$

everywhere except at the origin. As Dirac discovered in the early 1930s, any such vector potential $\vec{A}$ will be singular on some curve extending from the origin to ∞. He called this singularity a "string" (this Dirac string has no direct connection to modern string theory).

For example, consider two vector potentials

$$\begin{aligned} \vec{A}_N &= \frac{\kappa g}{r(z+r)}(x\hat{e}_y - y\hat{e}_x) \\ \vec{A}_S &= \frac{\kappa g}{r(z-r)}(x\hat{e}_y - y\hat{e}_x) \end{aligned} \tag{3.155}$$

$\vec{A}_N$ is singular along the negative Z-axis (the "string" for $\vec{A}_N$), while $\vec{A}_S$ is singular along the positive Z-axis. Nonetheless, we have

$$\vec{B} = \text{curl } \vec{A}_N = \text{curl } \vec{A}_S \tag{3.156}$$

except along the Z-axis (show this). Thus, just as we often need more than a single coordinate patch to cover a manifold (as here in the case of $\mathbf{R}^3$ with the origin removed), we may need more than a single function to describe the vector potential of a magnetic field, as explained by T. T. Wu and C. N. Yang [Phys. Rev. **D12**, 3843 (1975)]. ∎

→ **Exercise 3.15.** Find a function Λ such that

$$\vec{A}_N - \vec{A}_S = \vec{\nabla}\Lambda$$

except along the Z-axis. ☐

Remark. Thus $\vec{A}_N$ and $\vec{A}_S$ are gauge equivalent, as we already know from Eq. (3.156). This gauge equivalence, together with the gauge transformation properties of quantum mechanical wave functions for charged particles, leads to the *Dirac quantization condition* that the product of the fundamental unit e of electric charge and the corresponding unit g of magnetic charge must be an integer multiple of a fundamental constant. Though many searches have been made, a magnetic monopole has never been observed. ☐

3.4   Metric Tensor and Distance

3.4.1   Metric Tensor of a Linear Vector Space

Many general properties of a manifold do not require the concept of a metric, or measure of distance, on the manifold. However, metric is especially important in our physical spacetime, so we want to consider the added structure a metric brings to a manifold.

Definition 3.5.   A *metric tensor* $\mathbf{g}$ on a (real) linear vector space is a symmetric $\binom{0}{2}$ tensor that maps each pair (u, v) of vectors into a real number

$$\mathbf{g}(u, v) = \mathbf{g}(v, u) = u \cdot v \tag{3.157}$$

(the *scalar product* of u and v). The components of the metric tensor

$$g_{jk} \equiv \mathbf{g}(\mathbf{e}_j, \mathbf{e}_k) = \mathbf{e}_j \cdot \mathbf{e}_k \tag{3.158}$$

in a basis $\{\mathbf{e}_m\}$ define a real symmetric matrix $\mathbf{g}$. We require $\mathbf{g}$ to be nonsingular (or *nondegenerate*), i.e., if $\mathbf{g}(u, v) = 0$ for every v, then $u = 0$, which is equivalent to the matrix condition $\det \mathbf{g} \neq 0$. In terms of the components of u and v, the scalar product (3.157) has the form

$$u \cdot v = g_{jk} u^j v^k \tag{3.159}$$

The matrix $\mathbf{g}$ can be diagonalized, and the basis vectors rescaled so that

$$\mathbf{g} = \mathrm{diag}\,(\underbrace{1, \dots, 1}_{p \text{ times}}, \underbrace{-1, \dots, -1}_{q \text{ times}}) \tag{3.160}$$

(there are no zero eigenvalues since $\mathbf{g}$ is nonsingular). The pair (p, q) is the *signature* of the metric. The pair (q, p) is equivalent to the pair (p, q), since the overall sign of the metric is in general arbitrary. If the rescaled metric in Eq. (3.160) has diagonal elements all of the same sign (that is, the signature is either $(p, 0)$ or $(0, q)$), then the metric is *definite*; otherwise, it is *indefinite*. An indefinite metric is sometimes called a *pseudometric*.

Remark.   In the discussion of linear vector spaces in Chapter 2, we insisted that the metric be positive definite. This is often appropriate, but not in describing relativistic physics of the spacetime manifold in which we live. The spacetime manifold of special relativity is described by coordinates $\{x^0 = ct, x^1, x^2, x^3\}$ in $\mathbf{R}^4$ (t is the usual time coordinate, x^1, x^2, x^3 are the usual Cartesian space coordinates, and c is the speed of light) with a metric tensor

$$\mathbf{g} = \mathrm{diag}\,(1, -1, -1, -1) \tag{3.161}$$

the *Minkowski metric*, with signature $(1, 3)$. This metric is not definite, but is evidently nonsingular. In some contexts, it is more convenient to use a metric $\mathbf{g} = \mathrm{diag}\,(-1, 1, 1, 1)$ with signature $(3, 1)$. The reader should recognize either version.   □

3.4.2 Raising and Lowering Indices

The metric tensor can be used to define a direct correspondence between vectors and 1-forms. If v is a vector, then

$$\tilde{v} \equiv \mathbf{g}(\,.\,,v) \tag{3.162}$$

is a 1-form *associated* with v. Conversely, if the matrix **g** is nonsingular, as required, then the inverse matrix $\bar{\mathbf{g}}$ defines a symmetric $\binom{2}{0}$ tensor that maps pairs (σ, τ) of forms into real numbers:

$$\bar{\mathbf{g}}(\sigma, \tau) = \bar{\mathbf{g}}(\tau, \sigma) = \sigma \cdot \tau \tag{3.163}$$

Then $\bar{\mathbf{g}}$ maps the 1-form ω into the associated vector $\tilde{\omega}$, with

$$\tilde{\omega} = \bar{\mathbf{g}}(\,.\,,\omega) \tag{3.164}$$

Since $\bar{\mathbf{g}}$ is the matrix inverse of **g**, we have

$$\bar{\mathbf{g}}(\,.\,,\mathbf{g}(\,.\,,v)) = v \tag{3.165}$$

The correspondence between vectors and 1-forms using the metric tensor can be described in terms of components. If

$$\mathbf{g} = (g_{jk}) \quad \text{and} \quad \bar{\mathbf{g}} = (g^{jk}) \tag{3.166}$$

in some coordinate system, then we have the relations

$$\tilde{v}_j = g_{jk}v^k \quad \text{and} \quad \tilde{\omega}^k = g^{kj}\omega_j \tag{3.167}$$

between components of associated vectors and 1-forms. Thus the metric tensor is used for *raising* and *lowering* of indices between components of associated vectors and 1-forms. Thus the metric tensor is used for *raising* and *lowering* of indices.

Raising and lowering of indices with the metric tensor can also be applied to tensors of higher rank. Thus a tensor of rank N can be expressed as a $\binom{p}{q}$ tensor with any p, q so long as $p + q = N$. Starting from a pure $\binom{N}{0}$ tensor **T** of rank N with components $T^{k_1,\ldots,k_N}$, we can form an $\binom{N-q}{q}$ tensor with components $\mathbf{T}_{j_1,\ldots,j_q}{}^{k_{q+1},\ldots,k_N}$ by lowering indices using the metric tensor,

$$\mathbf{T}_{j_1,\ldots,j_q}{}^{k_{q+1},\ldots,k_N} = g_{j_1 k_1}, \ldots, g_{j_q k_q} T^{k_1,\ldots,k_N} \tag{3.168}$$

Remark. Here we lowered the first q indices of the original tensor **T**. This is simple, but not necessary, and in general we get different mixed tensor components by lowering a different set of q indices. If the original tensor **T** is either symmetric or completely antisymmetric, however, then the alternatives will differ by at most an overall sign. $\qquad\square$

3.4.3 Metric Tensor of a Manifold

A *metric tensor* $\mathbf{g}(x)$ on a manifold $\mathcal{M}$ is a metric defined on the tangent space $\mathcal{T}_P$ at each point P of $\mathcal{M}$ in such a way that the components of $\mathbf{g}$ in some coordinate basis are continuous (even differentiable) to some order. Continuity guarantees that the signature of the metric is a global property of the manifold, since the signature can change along a curve only if $\det(\mathbf{g})$ vanishes at some point on the curve, contrary to the assumption that $\mathbf{g}$ is nonsingular. $\mathbf{g}(x)$ will often be described simply as the *metric* of $\mathcal{M}$. A manifold with a metric is *Riemannian*, and its geometry is *Riemannian geometry*.

The metric tensor provides a measure of distance along smooth curves in the manifold, in particular along lines of flow of a vector field. Suppose $v = d/d\lambda$ is a vector field with coordinates $x^k = x^k(\lambda)$ along an integral curve of the field. Then the distance ds corresponding to an infinitesimal parameter change $d\lambda$ along the curve is given by

$$ds^2 = \mathbf{g}\left(\frac{d}{d\lambda}, \frac{d}{d\lambda}\right) d\lambda^2 = g_{jk}\frac{dx^j}{d\lambda}\frac{dx^k}{d\lambda}\, d\lambda^2 = g_{jk}\dot{x}^j\dot{x}^k d\lambda^2 \tag{3.169}$$

where we introduce the notation

$$\dot{x}^m \equiv \frac{dx^m}{d\lambda} \tag{3.170}$$

for the components of the tangent vector along the curve.

Remark.   A metric tensor $\mathbf{g}(x)$ is *Euclidean* if $\mathbf{g}(x) = \mathbf{1}$ for all x; the corresponding coordinate system is *Cartesian*. A metric tensor is *pseudo-Euclidean* if it is constant over the entire manifold, (e.g., the Minkowski metric of relativistic spacetime). A manifold is *flat* if it has a metric tensor that is (pseudo)-Euclidean in some coordinate system. It is important to determine whether a manifold with a given metric tensor $\mathbf{g}(x)$ is flat, especially in Einstein's theory of gravity (general relativity) where gravity appears as a deviation of the metric of spacetime from the flat Minkowski metric. To provide a definitive answer, we need the concept of *curvature*, which is described in the books on general relativity cited at the end of the chapter, but the study of geodesics (see Section 3.4.6) can provide some hints. □

❑ **Example 3.32.**   For the standard polar coordinates r, θ in the plane (see Exercise 3.3), the Euclidean distance corresponding to an infinitesimal coordinate displacement is

$$ds^2 = dx^2 + dy^2 = dr^2 + r^2 d\theta^2 \tag{3.171}$$

corresponding to nonvanishing elements

$$g_{rr} = \mathbf{g}\left(\frac{\partial}{\partial r}, \frac{\partial}{\partial r}\right) = 1 \qquad g_{\theta\theta} = \mathbf{g}\left(\frac{\partial}{\partial\theta}, \frac{\partial}{\partial\theta}\right) = r^2 \tag{3.172}$$

of the metric tensor.   ∎

❑ **Example 3.33.**   Consider the standard spherical coordinates r, θ, ϕ on $\mathbf{R}^3$ defined in terms of Cartesian coordinates x, y, z by

$$x = r\sin\theta\cos\phi \qquad y = r\sin\theta\sin\phi \qquad z = r\cos\theta \tag{3.173}$$

The Euclidean distance ds corresponding to a displacement $(dr, d\theta, d\phi)$ is again obtained by transformation from Cartesian coordinates, and is given by

$$ds^2 = dr^2 + r^2 d\theta^2 + r^2 \sin^2 \theta d\phi^2 \tag{3.174}$$

The metric tensor then has nonvanishing elements

$$g_{rr} = 1 \quad g_{\theta\theta} = r^2 \qquad g_{\phi\phi} = r^2 \sin^2 \theta \tag{3.175}$$

The spherical volume element is derived from this metric tensor in Section 3.4.4. ∎

❑ **Example 3.34.** From the metric tensor of $\mathbf{R}^3$ expressed in spherical coordinates, we also obtain a metric tensor on the 2-sphere $\mathbf{S}^2$, the metric *induced* on $\mathbf{S}^2$ by the Euclidean metric of the $\mathbf{R}^3$ in which the sphere is embedded. With coordinates θ, ϕ on $\mathbf{S}^2$, the metric tensor is given by

$$g_{\theta\theta} = a^2 \qquad g_{\phi\phi} = a^2 \sin^2 \theta \quad g_{\theta\phi} = 0 \tag{3.176}$$

with a being an arbitrary scale factor. ∎

These examples show that the metric tensor need not be constant, even if the underlying manifold is flat, as is any $\mathbf{R}^n$ with its standard Euclidean metric—the form of the metric is coordinate dependent. In the examples of polar and spherical coordinates, the Euclidean metric is recovered by transforming back to Cartesian coordinates. However, the induced metric (Eq. (3.176)) on $\mathbf{S}^2$ cannot be transformed into a flat metric. One indication of this is the fact that geodesics through a point can intersect away from that point, as explained in Section 3.4.6.

3.4.4 Metric Tensor and Volume

In a manifold with a metric tensor, there is a natural volume element. As already noted, changing coordinates from $x^1, \ldots, x^n$ to $y^1, \ldots, y^n$ leads to the transformation

$$dy^1 \wedge \cdots \wedge dy^n = |\det \mathbf{J}| \, dx^1 \wedge \cdots \wedge x^n \tag{3.177}$$

where $\mathbf{J}$ is the Jacobian matrix

$$\mathbf{J} = (\mathbf{J}_{k\ell}) = \left(\frac{\partial y^k}{\partial x^\ell} \right) \tag{3.178}$$

and the determinant is the usual Jacobian.

If $\mathbf{g}$ denotes the metric tensor in the x coordinate system, $\mathbf{h}$ the metric tensor in the y coordinate system, then the components of the two tensors are related by

$$g_{jk} = \frac{\partial y^m}{\partial x^j} \frac{\partial y^n}{\partial x^k} h_{mn} \tag{3.179}$$

which we can write in the matrix form as

$$\mathbf{g} = \mathbf{J}^\dagger \mathbf{h} \mathbf{J} \tag{3.180}$$

Evidently

$$\det \mathbf{g} = (\det \mathbf{J})^* \det \mathbf{h} \, (\det \mathbf{J}) \tag{3.181}$$

and then

$$\sqrt{|\det \mathbf{g}|} \, dx^1 \wedge \cdots \wedge x^n = \sqrt{|\det \mathbf{h}|} \, dy^1 \wedge \cdots \wedge dy^n \tag{3.182}$$

in view of Eq. (3.177). Thus we have a natural volume form on the manifold,

$$\Omega = \sqrt{|\det \mathbf{g}|} \, dx^1 \wedge \cdots \wedge dx^n \tag{3.183}$$

Ω is invariant under (smooth) coordinate transformations.

❑ **Example 3.35.** For the standard spherical coordinates (Eq. (3.173)), we have

$$\det \mathbf{g} = r^4 \sin^2 \theta \tag{3.184}$$

which gives the standard volume form

$$\Omega = r^2 \sin \theta \, dr \wedge d\theta \wedge d\phi \tag{3.185}$$

Further examples are given in the problems. ∎

3.4.5　The Laplacian Operator

A important differential operator in many branches of physics is the *Laplacian* operator Δ, defined in Cartesian coordinates by

$$\Delta = \nabla^2 = \frac{\partial^2}{\partial x^2} + \frac{\partial^2}{\partial y^2} + \frac{\partial^2}{\partial z^2} = \mathrm{div}(\mathrm{grad}) \tag{3.186}$$

To define the Laplacian on a general manifold $\mathcal{M}$, we note that the divergence of a vector field is defined by Eq. (3.116) in Example 3.27, with an explicit form given in Exercise 3.11 for a manifold with volume form given by

$$\Omega(x) \equiv \rho(x) \, dx^1 \wedge \cdots \wedge dx^n \tag{3.187}$$

(Eq. (3.90)), with

$$\rho(x) = \sqrt{|\det \mathbf{g}|} \tag{3.188}$$

However, the gradient df of a scalar function f is a 1-form, that we need to convert to a vector field before using Eq. (3.116) to define the divergence. If $\mathcal{M}$ has a metric tensor $\mathbf{g}$ with inverse $\bar{\mathbf{g}}$, then we can convert the gradient df into a vector field

$$\delta f = \bar{\mathbf{g}}(\cdot, df) \tag{3.189}$$

The Laplacian of the scalar f can then be defined by

$$\Delta f = {}^*d\,{}^*\delta f \tag{3.190}$$

or, in terms of components,

$$\Delta f = \frac{1}{\sqrt{|\det g|}} \frac{\partial}{\partial x^j} \left(\sqrt{|\det g|}\, g^{jk} \frac{\partial f}{\partial x^k} \right) \tag{3.191}$$

❑ **Example 3.36.**   For the standard spherical coordinates (Eq. (3.173)), we have

$$g^{rr} = 1 \qquad g^{\theta\theta} = \frac{1}{r^2} \qquad g_{\phi\phi} = \frac{1}{r^2 \sin^2\theta} \tag{3.192}$$

and then

$$\Delta = \frac{1}{r^2} \frac{\partial}{\partial r} \left(r^2 \frac{\partial}{\partial r} \right) + \frac{1}{r^2 \sin\theta} \frac{\partial}{\partial \theta} \left(\sin\theta \frac{\partial}{\partial \theta} \right) + \frac{1}{r^2 \sin^2\theta} \frac{\partial^2}{\partial \phi^2} \tag{3.193}$$

This result will no doubt be familiar to many readers.   ∎

→ **Exercise 3.16.**   Standard *cylindrical coordinates* ρ, ϕ, z are defined in terms of Cartesian coordinates x, y, z by

$$x = \rho \cos\phi \qquad y = \rho \sin\phi \qquad z = z \tag{3.194}$$

(i) Find an expression for the infinitesimal line element ds in terms of $d\rho$, dz, and $d\phi$, and thus the components of the metric tensor.
(ii) Find an expression for the volume element Ω in terms of $d\rho$, dz, and $d\phi$.
(iii) Find the Laplacian Δ in terms of partial derivatives with respect to ρ, z, and ϕ.   ☐

3.4.6   Geodesic Curves on a Manifold

In a manifold with a positive definite metric, the path length along a curve $C : x = x(\lambda)$ joining two points a and b can be expressed as

$$\ell_{ab}(C) = \int_a^b \sqrt{g_{jk}(x)\dot{x}^j \dot{x}^k}\, d\lambda \equiv \int_a^b \sqrt{\sigma}\, d\lambda \tag{3.195}$$

where we define

$$\sigma \equiv g_{jk}(x)\dot{x}^j \dot{x}^k \tag{3.196}$$

The *distance* between the two points a and b is the minimum value of $\ell_{ab}(C)$ for any curve C joining the two points. A curve on which this minimum is achieved is a *geodesic* connecting a and b, the analog of a straight line in a Euclidean space.

Finding the extrema (maxima or minima) of an integral such as $\ell_{ab}(C)$, a functional of the curve C, is a standard problem in the calculus of variations, which we review in Appendix A. From this, it follows that a geodesic curve must satisfy the Euler–Lagrange equations (3.A10)

$$\frac{d}{d\lambda} \frac{\partial \sqrt{\sigma}}{\partial \dot{x}^j} - \frac{\partial \sqrt{\sigma}}{\partial x^j} = 0 \tag{3.197}$$

To express these in a standard form, note first that

$$\frac{\partial \sqrt{\sigma}}{\partial \dot{x}^j} = \frac{1}{\sqrt{\sigma}} g_{jk} \dot{x}^k \qquad \frac{\partial \sqrt{\sigma}}{\partial x^j} = \frac{1}{2\sqrt{\sigma}} \frac{\partial g_{k\ell}}{\partial x^j} \dot{x}^k \dot{x}^\ell \tag{3.198}$$

Then also

$$\frac{d}{d\lambda} \frac{\partial \sqrt{\sigma}}{\partial \dot{x}^j} = \frac{1}{\sqrt{\sigma}} \left\{ g_{jk} \ddot{x}^k + \frac{\partial g_{jk}}{\partial x^\ell} \dot{x}^k \dot{x}^\ell - \frac{1}{2\sigma} g_{jk} \dot{x}^k \frac{d\sigma}{d\lambda} \right\} \tag{3.199}$$

and

$$\frac{d\sigma}{d\lambda} = \dot{x}^\ell \frac{\partial \sigma}{\partial x^\ell} + \ddot{x}^\ell \frac{\partial \sigma}{\partial \dot{x}^\ell} = \frac{\partial g_{mn}}{\partial x^\ell} \dot{x}^\ell \dot{x}^m \dot{x}^n + 2 g_{\ell m} \dot{x}^\ell \ddot{x}^m$$

$$= 2 \dot{x}^\ell \left(g_{\ell m} \ddot{x}^m + \Gamma_{\ell,mn} \dot{x}^m \dot{x}^n \right) \tag{3.200}$$

where we have introduced the *Christoffel symbols* (of the *first kind*)

$$\Gamma_{j,k\ell} \equiv \frac{1}{2} \left(\frac{\partial g_{jk}}{\partial x^\ell} + \frac{\partial g_{j\ell}}{\partial x^k} - \frac{\partial g_{k\ell}}{\partial x^j} \right) = \Gamma_{j,\ell k} \tag{3.201}$$

After some algebra, we can write the Euler–Lagrange equations as

$$\frac{d}{d\lambda} \frac{\partial \sqrt{\sigma}}{\partial \dot{x}^j} - \frac{\partial \sqrt{\sigma}}{\partial x^j} = \frac{1}{\sqrt{\sigma}} \left(\delta_j^{\,\ell} - \frac{1}{\sigma} g_{jk} \dot{x}^k \dot{x}^\ell \right) \left(g_{\ell m} \ddot{x}^m + \Gamma_{\ell,mn} \dot{x}^m \dot{x}^n \right) \tag{3.202}$$

$$\equiv M_j^{\,k}(\dot{x}) g_{k\ell} D^\ell = 0$$

where

$$M_j^{\,k}(\dot{x}) = \left(\sigma \delta_j^{\,k} - g_{j\ell} \dot{x}^k \dot{x}^\ell \right) / \sigma^{3/2} \tag{3.203}$$

and

$$D^\ell \equiv \ddot{x}^\ell + \Gamma_{mn}^\ell \dot{x}^m \dot{x}^n \tag{3.204}$$

Here we have used the metric tensor to raise the free index, and

$$\Gamma_{mn}^\ell = g^{j\ell} \Gamma_{\ell,mn} = \frac{1}{2} g^{j\ell} \left(\frac{\partial g_{jm}}{\partial x^n} + \frac{\partial g_{jn}}{\partial x^m} - \frac{\partial g_{mn}}{\partial x^j} \right) \tag{3.205}$$

The Γ_{mn}^ℓ are also *Christoffel symbols* (of the *second kind*).

The matrix factor $\mathbf{M}(\dot{x}) = (M_j{}^k(\dot{x}))$ is singular, since

$$\mathbf{M}(\dot{x})_j{}^k \dot{x}_k = 0 \tag{3.206}$$

Thus the component of D^ℓ along the tangent vector $\dot{x}$ is not constrained by the Euler–Lagrange equations. However, the transverse components of D^ℓ must satisfy the *geodesic equation*

$$D^\ell = \ddot{x}^\ell + \Gamma^\ell_{mn} \dot{x}^m \dot{x}^n = 0 \tag{3.207}$$

To deal with the tangential component of D^ℓ, note that Eq. (3.200) tells us that

$$2\dot{x}^k g_{k\ell} D^\ell = 2\dot{x}^k g_{k\ell} (\ddot{x}^\ell + \Gamma^\ell_{mn} \dot{x}^m \dot{x}^n) = \frac{d\sigma}{d\lambda} \tag{3.208}$$

The parameter λ along the curve is arbitrary, so long as it increases continuously as we move forward along the curve. It is natural to choose λ to be proportional to the length s along the geodesic. Then, since $ds = \sqrt{\sigma}\, d\lambda$, $\sqrt{\sigma}$ is constant along the curve, the right-hand side of Eq. (3.208) vanishes, and the tangential component of D^ℓ also satisfies the geodesic equation.

→ **Exercise 3.17.** Derive transformation laws for $\Gamma_{j,k\ell}$ (Eq. (3.201)) and Γ^ℓ_{mn} (Eq. (3.205)) under coordinate transformations. Are the Christoffel symbols components of a tensor? Explain. □

Remark. The Christoffel symbols are important in the geometry of the manifold, as they provide a connection between the tangent spaces at different points of a manifold that is essential to the discussion of curvature, parallel transport of vectors, and covariant differentiation in the books noted at the end of the chapter. □

❑ **Example 3.37.** For the sphere $\mathbf{S}^2$ with metric (3.176), we have nonzero Christoffel symbols

$$\Gamma_{\theta,\phi\phi} = -a^2 \sin\theta \cos\theta \qquad \Gamma_{\phi,\theta\phi} = \Gamma_{\phi,\phi\theta} = a^2 \sin\theta \cos\theta \tag{3.209}$$

and

$$\Gamma^\theta_{\phi\phi} = -\sin\theta \cos\theta \qquad \Gamma^\phi_{\theta\phi} = \Gamma^\phi_{\phi\theta} = \cot\theta \tag{3.210}$$

The geodesic equations then read

$$\ddot{\theta} - \sin\theta \cos\theta\, \dot{\phi}^2 = 0 \qquad \ddot{\phi} + 2\cot\theta\, \dot{\phi}\, \dot{\theta} = 0 \tag{3.211}$$

The second equation can be written as

$$\frac{d}{d\lambda} \ln\dot{\phi} = \frac{\ddot{\phi}}{\dot{\phi}} = -2\cot\theta\, \dot{\theta} = -2\frac{d}{d\lambda} \ln(\sin\theta) \tag{3.212}$$

from which we have

$$\dot{\phi} = \frac{C}{\sin^2\theta} \tag{3.213}$$

where $C = \sin^2 \theta_0 \, \dot{\phi}_0$ is given by the initial conditions $(\theta_0, \dot{\phi}_0)$. Then the first equation is

$$\ddot{\theta} = C^2 \, \frac{\cos \theta}{\sin^3 \theta} \tag{3.214}$$

which gives

$$\dot{\theta}^2 = A - \frac{C^2}{\sin^2 \theta} \tag{3.215}$$

where $A = \dot{\theta}_0^2 + \sin^2 \theta_0 \, \dot{\phi}_0^2$ is another constant of integration. Note that

$$\dot{\theta}^2 + \sin^2 \theta \, \dot{\phi}^2 = A \tag{3.216}$$

so that σ is constant along the geodesic, consistent with the parameter λ being proportional to the arc length along the curve. The remaining integrations can also be done in closed form; we leave that to the reader. Problems 11 and 12 deal with two questions of interest.

It is instructive to use the symmetry of the sphere to obtain the geodesics more simply. To find the geodesic between two points on the sphere, choose a coordinate system with one point on the equator at longitude $\phi = \phi_0$, and the other is at a latitude θ and the same longitude. Then a solution of the geodesic equations is $\dot{\phi} = 0$ and $\dot{\theta} = \dot{\theta}_0$, a constant ω, and the geodesics are arcs along the line of constant longitude joining the two points. It can be expressed in any coordinate system on the sphere using the rotation matrices of Eq. (2.113). Viewed in three dimensions, the line of constant longitude is a circle about the center of the sphere. Such a circle is called a *great circle*—it is a circle of maximum radius that can be drawn on the sphere. Knowledge of the great circle geodesics is of practical importance to seafarers and airplane pilots, although these are fixed, while wind and ocean currents are more variable and less predictable currents

Note that both the short and long arcs of a great circle joining two points satisfy the geodesic equations; one has minimum, the other maximum, length relative to nearby curves. Joining two diametrically opposed points, which we can take to be the poles of the sphere, there are infinitely many geodesics, since any line of constant longitude joining two poles is a great circle, and hence a geodesic. These facts are an indication, though not yet a complete proof, that a sphere is not flat. ∎

The derivation of the geodesic equation above assumed that the metric was definite. In a manifold with an indefinite metric, the direction of a tangent vector can be classified as *positive* ($\sigma > 0$), *negative* ($\sigma < 0$), or *null* ($\sigma = 0$). In any case, the geodesic equation leads to curves that can be called geodesics, but for negative directions, we must define a real distance by

$$ds = \sqrt{-\sigma} \, d\lambda \tag{3.217}$$

For null geodesics, we need to provide a suitable parametrization along the curve, since $\sigma = 0$ by definition. The geodesics may also correspond to relative maxima rather than relative minima of the "distance."

❑ **Example 3.38.** In Minkowski spacetime, coordinates are denoted by

$$x = \{x^\mu\} = (x^0 = ct, x^1, x^2, x^3) = (x^0, \vec{x}) \tag{3.218}$$

using the standard range $\mu = 0, 1, 2, 3$ for Greek indices. The metric tensor is given by $\mathbf{g} = \text{diag}(1, -1, -1, -1)$ everywhere, as already noted (see Eq. (3.161)); hence

$$\sigma = (\dot{x}^0)^2 - \dot{\vec{x}} \cdot \dot{\vec{x}} \tag{3.219}$$

Tangent vectors with $\sigma > 0$ are *timelike*, those with $\sigma < 0$ are *spacelike*, and those with $\sigma = 0$ are *null*, or *lightlike*. The geodesic equations are $\ddot{x}^\mu = 0$, with the straight lines

$$x^\mu = a^\mu + b^\mu \lambda \quad \Rightarrow \quad x = a + b\lambda \tag{3.220}$$

as solutions (notation as defined in Eq. (3.218)). Evidently $\sigma = (b^0)^2 - \vec{b} \cdot \vec{b}$, and we have the cases:

1. $\sigma > 0$: The geodesic corresponds to the trajectory of a massive particle moving with velocity $\vec{v} = (\vec{b}/b^0)c = \vec{\beta}c$. If we choose $\sigma = 1$, then we must have

$$b^0 = 1/\sqrt{1 - \beta^2} \equiv \gamma \qquad \vec{b} = b^0 \vec{\beta} = \vec{\beta}/\sqrt{1 - \beta^2} \tag{3.221}$$

and the parameter λ is the *proper time* along the trajectory, often denoted by τ. It corresponds physically to the time measured by an observer moving along the trajectory.

2. $\sigma < 0$: The geodesic corresponds to a rod moving with velocity $\vec{v} = \beta c \mathbf{n}$ in the direction of the unit vector $\mathbf{n} = \vec{b}/|\vec{b}|$, where $\beta = b^0/|\vec{b}|$. If we choose $\sigma = -1$, then

$$\vec{b} = \gamma \mathbf{n} \qquad b^0 = \beta\gamma \tag{3.222}$$

where again $\gamma = 1/\sqrt{1 - \beta^2}$. Here the parameter λ is the *proper length* along the rod. It corresponds physically to the length measured by an observer moving along with the rod.

3. $\sigma = 0$: The geodesic corresponds to a light ray moving in the direction of the unit vector $\mathbf{n} = \vec{b}/|\vec{b}|$. The scale of the parameter λ is arbitrary.

Minkowski space and the Lorentz transformations that leave the metric $\mathbf{g}$ invariant are discussed at length in Appendix B of Chapter 10. ∎

3.5 Dynamical Systems and Vector Fields

3.5.1 What is a Dynamical System?

We have discussed several examples of dynamical systems without explaining exactly what *is* a dynamical system. The broadest concept of a dynamical systems is a set of variables, usually but not always defining a manifold, the *state space* of the system, together with a set of rules for generating the evolution in time of a system, the *equations of motion* of the system.

The equations of motion may be expressed as maps that generate the states of the system at discrete time intervals, as in Appendix A of Chapter 1. They may be a finite set of ordinary first-order differential equations, generating a vector field as described in this chapter. Or they may be a set of partial differential equations for a set of dynamical variables that are themselves defined as functions (fields) on some manifold—for example, the electromagnetic fields in space for which the equations of motion are Maxwell's equations, or the velocity field of a fluid, for which the equations of motion are derived in Section 3.6.

In this section, we give two examples. The first is a simple model from ecology in which the two variables are the populations of a predator species and a prey species. These variables satisfy relatively simple nonlinear differential equations that allow a straightforward analysis of the qualitative behavior of the solutions. The second is a study of the geometrical properties of Hamiltonian systems, which include energy-conserving systems of classical mechanics.

3.5.2   A Model from Ecology

A simple model that illustrates a two-dimensional vector field as a dynamical system is the Lotka–Volterra model of predator–prey relations. The populations of the predator (x_1) and prey (x_2) are supposed to satisfy the differential equations

$$\dot{x}_1 = \frac{dx_1}{dt} = -\lambda x_1 + a x_1 x_2$$

$$\dot{x}_2 = \frac{dx_2}{dt} = \mu x_2 - b x_1 x_2 \tag{3.223}$$

where λ, μ, a, and b are positive constants.

Remark.   The signs of these constants are based on a set of simple but realistic assumptions about the nature of the system. It is supposed that the predator population will decline in the absence of prey, so that $\lambda > 0$. However, the predators can survive if sufficient prey is available ($a > 0$). It is also supposed that the prey population will grow exponentially ($\mu > 0$) in the absence of predators, who serve to limit the prey population ($b > 0$).   ☐

There are two fixed points where the vector field vanishes:

$$\mathcal{O}: x_1 = x_2 = 0 \quad \text{and} \quad \mathcal{P}: x_1 = \frac{\mu}{b} \quad x_2 = \frac{\lambda}{a} \tag{3.224}$$

The character of each fixed point is established by linearizing the equations near the fixed point. The linear equations can then be analyzed as in Section 2.5.

To analyze the equations further, introduce the scaled variables y_1 and y_2 by

$$x_1 = \frac{\mu}{b} y_1 \qquad x_2 = \frac{\lambda}{a} y_2 \tag{3.225}$$

Then the Lotka–Volterra equations (3.223) become

$$\dot{y}_1 = -\lambda y_1 (1 - y_2) \qquad \dot{y}_2 = \mu y_2 (1 - y_1) \tag{3.226}$$

Near the origin $\mathcal{O}$, we have

$$\dot{y}_1 \simeq -\lambda y_1 \qquad \dot{y}_2 \simeq \mu y_2 \tag{3.227}$$

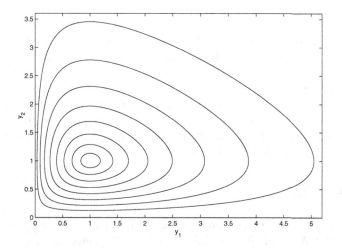

Figure 3.10: Flow lines of the two-dimensional vector field defined by the Lotka–Volterra equations (3.223) with $\lambda = 1.0$, $\mu = 2.0$. Here y_1, y_2 are the scaled variables introduced in Eq. (3.225).

so the origin is a hyperbolic fixed point, with stable manifold consisting of the line $y_2 = 0$ and unstable manifold consisting of the line $y_1 = 0$. Near the fixed point $\mathcal{P}$, which is at $y_1 = y_2 = 1$ in the scaled variables, let $\xi_1 = y_1 - 1$, $\xi_2 = y_2 - 1$. Then we have the linear equations

$$\dot{\xi}_1 \simeq \lambda\xi_2 \qquad \dot{\xi}_2 \simeq -\mu\xi_1 \tag{3.228}$$

whose solutions correspond to periodic motion with angular frequency ω given by

$$\omega = \sqrt{\lambda\mu} \tag{3.229}$$

For larger amplitudes, the motion is no longer elliptical, but the integral curves cannot run to ∞ in the first quadrant $y_1 > 0$ and $y_2 > 0$, and they cannot cross the axes. Hence they must form a set of closed curves around the fixed point $\mathcal{P}$, which is called a *center*. The motion along any curve is periodic, with period depending on the curve. A set of integral curves of the scaled equations (3.226) is shown in Fig. (3.10) for parameters $\lambda = 1.0$, $\mu = 2.0$.

Remark.   The vector field defined here is *structurally unstable*, since introducing a small quadratic term in either equation introduces new fixed points that change the qualitative behavior of the system (see Problem 14). The modified equations are perhaps slightly more realistic, but the reader is invited to look closely at the differences between those equations and the system (3.223). $\square$

3.5.3   Lagrangian and Hamiltonian Systems

A classical dynamical system is defined on an n-dimensional configuration space $\mathcal{C}$ with co-ordinates $q = \{q^1, \ldots, q^n\}$ on each coordinate patch in the atlas of $\mathcal{C}$. These are the *generalized coordinates* of the system, and we follow tradition in denoting them by $\{q^k\}$ rather than $\{x^k\}$ in this context. At each point of $\mathcal{C}$, we have the tangent space with coordinates $\dot{q} = \{\dot{q}^1, \ldots, \dot{q}^n\}$, the *velocities* of the system. The Lagrangian dynamics of the system is based on the introduction of a *Lagrangian* $L(q, \dot{q}, t)$, which depends on the coordinates and velocities, and perhaps explicitly on the time t. The dynamical trajectories of the system are determined from *Hamilton's principle*, which requires that the *action integral*

$$S[q(t)] = \int_a^b L(q, \dot{q}, t)\, dt \tag{3.230}$$

from point a to point b in $\mathcal{C}$ be an extremum along the actual trajectory, compared to nearby trajectories. As explained in Appendix A, this leads to the Euler–Lagrange equations of motion

$$\frac{d}{dt} \frac{\partial L}{\partial \dot{q}^k} - \frac{\partial L}{\partial q^k} = 0 \tag{3.231}$$

$(k = 1, \ldots, n)$.

❏ **Example 3.39.**   For a particle of mass m moving in a one-dimensional potential $V(x)$, the Lagrangian is

$$L = \tfrac{1}{2} m \dot{x}^2 - V(x) \tag{3.232}$$

leading to the Euler–Lagrange equation

$$m \ddot{x} + \frac{dV}{dx} = 0 \tag{3.233}$$

This is the same as Newton's second law with force $F = -dV/dx$.   ∎

❏ **Example 3.40.**   For a simple harmonic oscillator in one dimension ($V = \tfrac{1}{2} m \omega^2 x^2$), we have the equation of motion

$$\ddot{x}(t) + \omega^2 x(t) = 0 \tag{3.234}$$

The solution corresponding to initial position x_0 and initial velocity $\dot{x}_0$ is

$$x(t) = x_0 \cos \omega t - (\dot{x}_0/\omega) \sin \omega t = A \cos(\omega t + \phi) \tag{3.235}$$

The *amplitude* A and *phase* ϕ are expressed as

$$A = \sqrt{x_0^2 + (\dot{x}_0/\omega)^2} \qquad \phi = \tan^{-1}(\dot{x}_0/\omega x_0) \tag{3.236}$$

in terms of the initial position and velocity.   ∎

Hamiltonian dynamics is expressed in terms of the coordinates $q = \{q^1, \ldots, q^n\}$ and the *conjugate momenta* $p = \{p_1, \ldots, p_n\}$ introduced by

$$p_k = \frac{\partial L}{\partial \dot{q}^k} \tag{3.237}$$

$(k = 1, \ldots, n)$. The *Hamiltonian* is defined by

$$H(q, p, t) \equiv p_k \dot{q}^k - L(q, \dot{q}, t) \tag{3.238}$$

The equations of motion for the coordinates and momenta are

$$\dot{q}^k = \frac{dq^k}{dt} = \frac{\partial H}{\partial p_k} \qquad \dot{p}_k = \frac{dp_k}{dt} = -\frac{\partial H}{\partial q^k} \tag{3.239}$$

$(k = 1, \ldots, n)$. These are *Hamilton's equations* of motion.

Remark. Hamilton's equations (3.239) are equivalent to the Lagrange equations (3.231). However, a careful derivation of this equivalence needs to note that in taking partial derivatives, the Lagrangian is expressed as a function of the coordinates and the velocities, while the Hamiltonian is expressed as a function of the coordinates and momenta. See Problem 17. □

❑ **Example 3.41.** For a particle of mass m in a central potential $V(r)$, the Lagrangian is

$$L = \tfrac{1}{2}m\left(\dot{x}^2 + \dot{y}^2 + \dot{z}^2\right) - V(r) = \tfrac{1}{2}m\left(\dot{r}^2 + r^2\dot{\theta}^2 + r^2\sin^2\theta\,\dot{\phi}^2\right) - V(r) \tag{3.240}$$

where the second expression is in terms of the usual spherical coordinates defined in Eq. (3.173). The Lagrange equations of motion in spherical coordinates are

$$\frac{d}{dt}\frac{\partial L}{\partial \dot{r}} = m\ddot{r} = \frac{\partial L}{\partial r} = mr\left(\dot{\theta}^2 + \sin^2\theta\,\dot{\phi}^2\right) - \frac{dV}{dr} \tag{3.241}$$

$$\frac{d}{dt}\frac{\partial L}{\partial \dot{\theta}} = mr^2\ddot{\theta} + 2mr\dot{r}\dot{\theta} = \frac{\partial L}{\partial \theta} = mr^2\sin\theta\cos\theta\,\dot{\phi}^2 \tag{3.242}$$

$$\frac{d}{dt}\frac{\partial L}{\partial \dot{\phi}} = mr^2\sin^2\theta\,\ddot{\phi} + 2mr\sin^2\theta\,\dot{r}\dot{\phi} + 2mr^2\sin\theta\cos\theta\,\dot{\theta}\dot{\phi} = \frac{\partial L}{\partial \phi} = 0 \tag{3.243}$$

Note that for a particle constrained to move on the surface of a sphere ($\dot{r} = 0$), the equations of motion for θ and ϕ are the same as the geodesic equations (3.211).

The conjugate momenta in spherical coordinates are

$$p_r = m\dot{r} \qquad p_\theta = mr^2\dot{\theta} \qquad p_\phi = mr^2\sin^2\theta\,\dot{\phi} \tag{3.244}$$

and the Hamiltonian is expressed as

$$H = \frac{1}{2m}\left(p_r^2 + \frac{p_\theta^2}{r^2} + \frac{p_\phi^2}{r^2\sin^2\theta}\right) + V(r) \tag{3.245}$$

in terms of these coordinates and momenta. ∎

→ **Exercise 3.18.** Write down explicitly Hamilton's equations of motion for the Hamiltonian (3.245). Show that these equations of motion are equivalent to Eqs. (3.241)–(3.243).   □

If the Hamiltonian is independent of the coordinate q^a, then q^a is a *cyclic coordinate*, or an *ignorable coordinate*. The corresponding momentum p_a then satisfies

$$\dot{p}_a = -\frac{\partial H}{\partial q^a} = 0 \qquad (3.246)$$

so that p_a is conserved on any trajectory of the system—p_a is a *constant of the motion*.

❑ **Example 3.42.** The Hamiltonian equation (3.245) is independent of the spherical coordinate ϕ. Hence the conjugate momentum $p_\phi = mr^2 \sin^2 \theta \, \dot{\phi}$ is a constant of the motion. Note that p_ϕ is the Z-component of the angular momentum of the particle.   ∎

We have implicitly assumed that the momenta $\{p_k\}$ defined by Eq. (3.237) are independent. For this to be the case, it is necessary that the matrix $\mathbf{M}$ defined by

$$\mathbf{M} = (M_{jk}) = \left(\frac{\partial^2 L}{\partial \dot{q}^j \partial \dot{q}^k} \right) \qquad (3.247)$$

is nonsingular everywhere in $\mathcal{C}$. For a Lagrangian with the standard form of a kinetic energy quadratic in the velocities minus a potential energy $V(q)$, we have

$$L = \tfrac{1}{2} \dot{q}^j \, \mathbf{M}_{jk}(q) \, \dot{q}^k - V(q) \qquad (3.248)$$

and the momenta will be independent if the matrix $\mathbf{M}(q)$ is nonsingular everywhere in $\mathcal{C}$. Then also $\mathbf{M}(q)$ can serve as a metric tensor on the tangent space at each point, and we have

$$p_j = \frac{\partial L}{\partial \dot{q}^j} = \mathbf{M}_{jk}(q) \, \dot{q}^k \qquad (3.249)$$

Thus the momentum space and the velocity space at each point are dual to each other.

The Hamiltonian is related to a vector field $\mathbf{X}$ defined on the phase space of the dynamical system by

$$\mathbf{X} = \frac{d}{dt} = \dot{q}^k \frac{\partial}{\partial q^k} + \dot{p}^k \frac{\partial}{\partial p^k} \equiv \mathbf{X}_H \qquad (3.250)$$

From Hamilton's equations of motion, we have

$$\mathbf{X} = \frac{\partial H}{\partial p_k} \frac{\partial}{\partial q^k} - \frac{\partial H}{\partial q_k} \frac{\partial}{\partial p^k} \qquad (3.251)$$

The integral curves of $\mathbf{X}$ are the phase space trajectories of the system if H is time independent. The Lie derivative $\mathcal{L}_{\mathbf{X}}$ transports vectors and forms along the trajectories; in particular,

$$\mathbf{X}H = \mathcal{L}_{\mathbf{X}}H = \frac{\partial H}{\partial p_k} \frac{\partial H}{\partial q^k} - \frac{\partial H}{\partial q^k} \frac{\partial H}{\partial p_k} = 0 \qquad (3.252)$$

Thus the Hamiltonian is constant along the integral curves of $\mathbf{X}$. When H is identified with the total energy of the system, Eq. (3.252) is a statement of conservation of energy.

Remark.   For a general scalar function f, we have

$$\mathbf{X}f = \mathcal{L}_{\mathbf{X}}f = \frac{\partial H}{\partial p_k}\frac{\partial f}{\partial q^k} - \frac{\partial H}{\partial q^k}\frac{\partial f}{\partial p_k} \tag{3.253}$$

The right-hand side of this equation is the *Poisson bracket* of H and f, denoted by $\{H, f\}$. See Problem 19 for a general definition of the Poisson bracket.   □

The velocities $\dot{q} = \{\dot{q}^1, \ldots, \dot{q}^n\}$ transform as components of a vector on $\mathcal{C}$, since under a coordinate transformation $q = \{q^1, \ldots, q^n\} \to Q = \{Q^1, \ldots, Q^n\}$, we have

$$\dot{q}^k \to \dot{Q}^k = \left(\frac{\partial Q^k}{\partial q^\ell}\right)\dot{q}^\ell \tag{3.254}$$

On the other hand, the momenta $p = \{p_1, \ldots, p_n\}$ transform according to $p \to P$, with

$$P_k = \frac{\partial L}{\partial \dot{Q}^k} = \frac{\partial \dot{q}^\ell}{\partial \dot{Q}^k}\frac{\partial L}{\partial \dot{q}^\ell} = \left(\frac{\partial q^\ell}{\partial Q^k}\right)p_\ell \tag{3.255}$$

Hence the momenta are the components of a 1-form σ,

$$\sigma \equiv p_k\, dq^k \tag{3.256}$$

at each point of $\mathcal{C}$; σ is known as the *Poincaré 1-form*. Note that σ is *not* a 1-form field on $\mathcal{C}$ since the p_k are not functions of the coordinates. However, it *is* a 1-form field on the phase space $T^*(\mathcal{C})$, and its exterior derivative is the *canonical 2-form*

$$\omega = d\sigma = dp_k \wedge dq^k \tag{3.257}$$

The canonical 2-form is nonsingular on the phase space $T^*(\mathcal{C})$, and it provides a unique correspondence between vectors and forms on the phase space; if $\mathbf{V}$ is a vector field in the phase space, then (recall the notation $i_{\mathbf{V}}$ introduced in Eq. (3.117))

$$i_{\mathbf{V}}\omega = \omega(\mathbf{V}, \,.\,) \tag{3.258}$$

is the 1-form field associated with $\mathbf{V}$.

Remark.   The canonical 2-form ω is often denoted by Ω in classical mechanics textbooks; here we use ω to avoid confusion with the volume form Ω.   □

→ **Exercise 3.19.**   (i) Show that the 2-form ω is antisymmetric,

$$\omega(\mathbf{U}, \mathbf{V}) = -\omega(\mathbf{V}, \mathbf{U}) \tag{3.259}$$

for every pair of vector fields $\mathbf{U}$, $\mathbf{V}$ defined on $T^*(\mathcal{C})$. Then also show that

$$(i_{\mathbf{V}}\omega, \mathbf{V}) = \omega(\mathbf{V}, \mathbf{V}) = 0 \tag{3.260}$$

for every vector field $\mathbf{V}$ on $T^*(\mathcal{C})$.

(ii) Show that for every vector $\mathbf{V}$, there is a vector $\mathbf{U}$ such that

$$(i_{\mathbf{V}}\omega, \mathbf{U}) = \omega(\mathbf{V}, \mathbf{U}) > 0 \tag{3.261}$$

so that ω is nonsingular. In particular, suppose

$$\mathbf{V} = a^k \frac{\partial}{\partial q^k} + b_k \frac{\partial}{\partial p_k}$$

is defined on the tangent space of some point in the phase space. Find an explicit representation for the form $i_{\mathbf{V}}\omega$, and find a vector $\mathbf{U}$ such that $\omega(\mathbf{U}, \mathbf{V}) > 0$. $\square$

Remark.   In Exercise 2.7, it was stated that if $\mathbf{A}$ is a linear operator on a complex vector space such that $(x, \mathbf{A}x) = 0$ for every vector x, then $\mathbf{A} = \mathbf{0}$. The form ω shows why the statement is not true in a real vector space. $\square$

Definition 3.6. A form ω satisfying Eq. (3.260) is *symplectic*. A manifold on which such a 2-form field exists on the entire manifold is a *symplectic manifold*; the 2-form field defines a *symplectic structure* on the manifold.

➜ **Exercise 3.20.**   If ω is a 2-form such that $\omega(\mathbf{V}, \mathbf{V}) = 0$ for every vector $\mathbf{V}$, then

$$\omega(\mathbf{U}, \mathbf{V}) = -\omega(\mathbf{V}, \mathbf{U})$$

for every pair of vectors $\mathbf{U}, \mathbf{V}$. $\square$

➜ **Exercise 3.21.**   If σ is a 1-form *on the phase space*, then there is a unique vector $\mathbf{V}_\sigma$ such that

$$\sigma = \omega(\mathbf{V}_\sigma, \,.\,) = i_{\mathbf{V}_\sigma}\omega \tag{3.262}$$

These exercises show that ω can serve as a (pseudo-)metric to relate vectors and forms on phase space, as stated. By contrast, the kinetic energy matrix introduced in Eq. (3.247) serves as a metric on the original configuration space $\mathcal{C}$ (see Eq. (3.249)). $\square$

The vector field $\mathbf{X}$ (Eq. (3.250)) that defines the trajectories of the system is related to the Hamiltonian by

$$i_{\mathbf{X}}\omega \equiv \omega(\mathbf{X}, \,.\,) = -\frac{\partial H}{\partial q^k}\,dq^k - \frac{\partial H}{\partial p_k}\,dp_k \tag{3.263}$$

In general, we have

$$dH = \frac{\partial H}{\partial q^k}\,dq^k + \frac{\partial H}{\partial p_k}\,dp_k + \frac{\partial H}{\partial t} \tag{3.264}$$

Hence if H does not depend explicitly on time, we have

$$i_{\mathbf{X}}\omega = -\,dH \tag{3.265}$$

From Theorem 3.3.1, it then follows that

$$\mathcal{L}_{\mathbf{X}}\omega = i_{\mathbf{X}}(d\omega) + d(i_{\mathbf{X}}\omega) = 0 \tag{3.266}$$

since

$$dw = d(d\sigma) = 0 \qquad d(i_{\mathbf{X}}w) = -d(dH) = 0 \tag{3.267}$$

by Poincaré's lemma. Thus the canonical 2-form w is invariant along the trajectories of the system in phase space, as are the successive powers of w,

$$w_2 = w \wedge w , \ \ldots , \ w_n = \underbrace{w \wedge \cdots \wedge w}_{n \text{ factors}} \tag{3.268}$$

The forms $w, w_2, \ldots w_n$ are the *Poincaré invariants* of the system. The form w_n is the volume element in phase space. Invariance of w_n along the trajectories of the system means that if we start with a small region in phase space and let it develop in time according to Hamilton's equations of motion, the volume of the region will remain constant as the region evolves in time. This result is known as *Liouville's theorem*; it is an important and useful property of the phase space flow of a dynamical system.

The canonical 2-form w is invariant under coordinate transformations $q \to Q$, with the corresponding transformation $p \to P$ of the momenta given by Eq. (3.255), in the sense that

$$w = dp_k \wedge dq^k = dP_k \wedge dQ^k \tag{3.269}$$

However there are more general transformations $(q, p) \to (Q, P)$ of the phase space that leave w invariant as in Eq. (3.269). Such transformations are called *canonical transformations*, or *symplectic transformations*.

One use of canonical transformations is to attempt to reduce a Hamiltonian to a form that depends on as few coordinates as possible, so that the momenta corresponding to the remaining coordinates are constants of the motion. For example, the use of spherical coordinates for a spherically symmetric system shows explicitly the conservation of angular momentum, which is not so obvious in Cartesian coordinates, for example.

Relatively few systems can be completely solved in this way; a system for which the Hamiltonian can be completely expressed in terms of conserved momenta is called *integrable*. However, many systems are "nearly" integrable, in the sense that an integrable system can be used as a starting point for a systematic approximation scheme. One such scheme is classical perturbation theory, a simple example of which appears in Problem 20.

❏ **Example 3.43.**   The Hamiltonian of the one-dimensional harmonic oscillator is

$$H = \frac{p^2}{2m} + \frac{1}{2}mw^2x^2 \tag{3.270}$$

Rescaling the variables from (x, p) to (X, P) defined by

$$X \equiv \sqrt{mw}\, x \qquad P \equiv \frac{p}{\sqrt{mw}} \tag{3.271}$$

is a canonical transformation. In terms of (X, P), the Hamiltonian is

$$H = \tfrac{1}{2}w(P^2 + X^2) \tag{3.272}$$

We can further introduce variables J, α by

$$X \equiv \sqrt{2J} \sin \alpha \qquad P \equiv \sqrt{2J} \cos \alpha \tag{3.273}$$

corresponding to

$$J = \tfrac{1}{2}(P^2 + X^2) \qquad \tan\alpha = \frac{P}{X} \tag{3.274}$$

In terms of these variables, the Hamiltonian is given simply by

$$H = \omega J \tag{3.275}$$

The variables J, α are *action-angle* variables. Since the Hamiltonian (3.275) is independent of the *angle variable* α, the conjugate momentum J (the *action variable*) is a constant of the motion, and

$$\dot{\alpha} = \frac{\partial H}{\partial J} = \omega \tag{3.276}$$

Thus the motion in the phase space defined by the variables (X, P) is a circle of radius $\sqrt{2J}$, with angular velocity given by $\dot{\alpha} = \omega$. Note that for the special case of simple harmonic oscillator, the angular velocity $\dot{\alpha}$ is independent of the action variable. This is not true in general (see Problem 20). ∎

→ **Exercise 3.22.** Show that the transformation to action-angle variables is canonical, i.e., show that

$$dJ \wedge d\alpha = dP \wedge dX$$

Then explain the choice of $\sqrt{2J}$, rather than some arbitrary function of J, as the "radius" variable in Eq. (3.273). □

3.6   Fluid Mechanics

A real fluid consists of a large number of atoms or molecules whose interactions are sufficiently strong that the motion of the fluid on a macroscopic scale appears to be smooth flow superimposed on the thermal motion of the individual atoms or molecules, the thermal motion being generally unobservable except through the Brownian motion of particles introduced into the fluid.

An ideal fluid is characterized by a mass density $\rho = \rho(x,t)$ and a velocity field $\vec{u} = \vec{u}(x,t)$, as well as thermodynamic variables such as pressure $p = p(x,t)$ and temperature $T = T(x,t)$. If the fluid is a gas, then it is often important to consider the equation of state relating ρ, p, and T. For a liquid, on the other hand, it is usually a good approximation to treat the density as constant (*incompressible flow*).

Remark. It is implicitly assumed that the time scales associated with the fluid flow are long enough for local thermodynamic equilibrium to be established, though the temperature may vary within the fluid. □

Conservation of matter means that the total mass in a region **R** can only change if matter flows across the boundary of the region. Thus we have

$$\frac{d}{dt} \int_{\mathbf{R}} \rho \Omega = - \int_{\partial \mathbf{R}} \rho \vec{u} \cdot \vec{dS} \tag{3.277}$$

where Ω is the volume form on **R**, and $\vec{dS}$ is an outward normal to the boundary surface $\partial \mathbf{R}$. The integral on the left is the rate of change of the total mass within **R**, while the integral on the right is the rate at which matter flows out across the boundary surface. Using Stokes' theorem, we then have

$$\int_{\mathbf{R}} \left\{ \frac{\partial \rho}{\partial t} + \vec{\nabla} \cdot (\rho \vec{u}) \right\} \Omega = 0 \tag{3.278}$$

Since this is true for any region **R**, the integrand must vanish[7], and we then have the *equation of continuity*

$$\frac{\partial \rho}{\partial t} + \operatorname{div}(\rho \vec{u}) = 0 \tag{3.279}$$

Associated with the velocity field $\vec{u}$ are a vector field U, and a 1-form field $\tilde{U}$,

$$U = u^k \frac{\partial}{\partial x^k} \qquad \tilde{U} = u_k dx^k \tag{3.280}$$

where the components u^k and u_k are related by the metric tensor **g** in the usual way. In a Cartesian coordinate system in a flat space, these components are equal ($u^k = u_k$), but we want to be able to consider both curvilinear coordinate systems and the nonflat metrics associated with very massive stars, for example. The integral curves of the vector field U are the *lines of flow*, or *streamlines* of the fluid flow.

→ **Exercise 3.23.** Show that

$$\mathcal{L}_U \tilde{U} = (\vec{u} \cdot \vec{\nabla}) \vec{u} + \tfrac{1}{2} \vec{\nabla} u^2 \tag{3.281}$$

where $u^2 = \vec{u} \cdot \vec{u} = \langle \tilde{U}, U \rangle$. Then show also that

$$\langle \mathcal{L}_U \tilde{U}, U \rangle = (\vec{u} \cdot \vec{\nabla}) u^2 \tag{3.282}$$

(recall Eq. (3.67)). □

The equation of continuity (3.279) can be expressed as

$$\left(\frac{\partial}{\partial t} + \mathcal{L}_U \right) \rho \Omega = 0 \tag{3.283}$$

using Exercise 3.11, where $\mathcal{L}_U$ is the Lie derivative associated with the velocity field U.

---

[7]Technically, the integrand in Eq. (3.278) must be continuous. For a mass of water flowing through a pipe, there is a discontinuity at the leading edge of the water, which produces δ-function singularities in the partial derivatives in the integrand. However, the δ-function singularities must and do cancel in the end.

Newton's second law for a moving element of fluid is

$$\rho \frac{d\vec{u}}{dt} = \rho \left\{ \frac{\partial \vec{u}}{\partial t} + (\vec{u} \cdot \vec{\nabla})\vec{u} \right\} = \vec{f} \tag{3.284}$$

where $\vec{f}$ is the force per unit volume on the fluid element. If p is the pressure in the fluid, and Φ is the gravitational potential, then

$$\vec{f} = -\vec{\nabla}p - \rho\vec{\nabla}\Phi \tag{3.285}$$

and Eq. (3.284) becomes *Euler's equation*

$$\frac{\partial \vec{u}}{\partial t} + (\vec{u} \cdot \vec{\nabla})\vec{u} = -\frac{1}{\rho}\vec{\nabla}p - \vec{\nabla}\Phi \tag{3.286}$$

In the language of forms, Euler's equation takes the form

$$\left(\frac{\partial}{\partial t} + \mathcal{L}_U \right) \tilde{U} = -\frac{1}{\rho} dp + d\left(\tfrac{1}{2}u^2 - \Phi \right) \tag{3.287}$$

using the results of Exercise 3.23. Then also

$$\frac{1}{2}\left(\frac{\partial}{\partial t} + \mathcal{L}_U \right) u^2 + \frac{1}{\rho}\mathcal{L}_U\, p + \mathcal{L}_U\, \Phi = 0 \tag{3.288}$$

For steady (time-independent) flow of an incompressible fluid, this becomes

$$\mathcal{L}_U(\tfrac{1}{2}\rho u^2 + p + \rho\Phi) = 0 \tag{3.289}$$

so that the quantity $\frac{1}{2}\rho u^2 + p + \rho\Phi$ is constant along each streamline of the fluid flow. This is *Bernoulli's principle*, which is equivalent to conservation of energy of a drop of fluid as it moves along a streamline.

The momentum density Π of the fluid is the vector field

$$\Pi = \rho U \tag{3.290}$$

From the equation of continuity and Euler's equation, we have

$$\frac{\partial \Pi^k}{\partial t} = -u^k \frac{\partial}{\partial x^\ell}(\rho U^\ell) - \frac{\partial}{\partial x^k}(p + \rho\Phi) \equiv -\frac{\partial \mathbf{T}^{k\ell}}{\partial x^\ell} \tag{3.291}$$

where the *stress tensor* $\mathbf{T}$ is a symmetric $\binom{2}{0}$ tensor defined by

$$\mathbf{T} = (p + \rho\Phi)\bar{\mathbf{g}} + \rho U \otimes U \tag{3.292}$$

and $\bar{\mathbf{g}} = \mathbf{g}^{-1}$ is the dual of the metric tensor.

Remark. The concept of a stress tensor is more general; it appears in electromagnetic theory, as well as in the general theory of elastic media. It is always related to a rate of change of momentum density by the local form, Eq. (3.291), of Newton's second law.

The force $\mathbf{f}$ exerted by the fluid across a surface σ is then expressed in terms of components of the stress tensor by

$$\mathbf{f}^k = \mathbf{T}^{k\ell} g_{\ell m} (^*\sigma)^m \tag{3.293}$$

where $^*\sigma^m = \varepsilon^{jkm} \sigma_{jk}$ is a vector that can be identified with the usual normal to the surface. This local interpretation of the stress tensor is valid in any coordinate system, and can also be generalized to nonflat spaces.   □

In a Euclidean space with a Cartesian coordinate system (and only in such a space), the momentum density can be integrated to define a total momentum. If $\mathbf{R}$ is a region fixed in space (*not* co-moving with the fluid), then the total fluid momentum P in $\mathbf{R}$ has components

$$P^k = \int_{\mathbf{R}} \Pi^k \Omega \tag{3.294}$$

Then also

$$\frac{dP^k}{dt} = \int_{\mathbf{R}} \frac{\partial \Pi^k}{\partial t} \, \Omega = -\int_{\mathbf{R}} \frac{\partial \mathbf{T}^{k\ell}}{\partial x^\ell} \, \Omega = -\int_{\partial \mathbf{R}} \mathbf{T}^{k\ell} g_{\ell m} {}^*\sigma^m \tag{3.295}$$

where the last equality follows from Stokes' theorem. Here $\partial \mathbf{R}$ is the boundary of $\mathbf{R}$ and $^*\sigma_\ell$ is the (outward) normal to $\partial \mathbf{R}$ introduced above.

The *vorticity* α of the flow is a 2-form defined by

$$\alpha = d\tilde{U} \tag{3.296}$$

($\vec{\alpha} = \mathrm{curl}\ \vec{u}$ in ordinary vector notation). If S is a two-dimensional surface bounded by the closed curve C, then

$$\int_S \alpha = \int_C \tilde{U} = \int_C \vec{u} \cdot \vec{d\ell} \tag{3.297}$$

by Stokes' theorem. Thus α is a measure of the average tangential component of the velocity field around a closed curve. If $\alpha = 0$ throughout a simply connected region, then this integral vanishes for any closed curve in the region, and the flow is called *irrotational*. In this case, there is a scalar function ϕ (the *velocity potential*) such that

$$\tilde{U} = d\phi \tag{3.298}$$

in the region. For steady state flow of a incompressible fluid, the velocity potential must satisfy *Laplace's equation*

$$\nabla^2 \phi = 0 \tag{3.299}$$

Methods of finding solutions to Laplace's equation will be discussed in Chapter 8.

A   Calculus of Variations

Consider an integral of the form

$$S[x(\tau)] \equiv \int_a^b F(x, \dot{x}, \tau) \, d\tau \qquad (3.A1)$$

from point a to point b along a set of smooth curves $C : x = x(\tau)$ in a one-dimensional manifold, where here, as in Section 3.4,

$$\dot{x} = \frac{dx}{d\tau} \qquad (3.A2)$$

Higher derivatives may also be present, but they are absent from the examples we consider here, so we do not discuss them further (see, however, Problem 22). An integral of the type, Eq. (3.A1), is a *functional* of the curve $x(\tau)$. We have encountered linear functionals in Section 2.1.5, and quadratic functionals in Section 2.3.4, but here we are concerned with more general functionals, such as the one encountered in the study of geodesics on a manifold in Section 3.4.6.

The problem of interest is to find a curve $x = x_*(\tau)$ for which the integral $S[x(\tau)]$ is an extremum. By analogy with the extremum conditions for functions of n variables, we expect the condition for an extremum to have the form

$$\frac{\delta S}{\delta x(\tau)} = 0 \qquad (3.A3)$$

but we need a definition of $\delta S / \delta x(\tau)$. To provide such a definition, consider a curve

$$x(\tau) = x_*(\tau) + \varepsilon \eta(\tau) \qquad (3.A4)$$

near the curve $x = x_*(\tau)$. The variation of $S[x(\tau)]$ is given by

$$\delta S[x(\tau)] = \varepsilon \int_a^b \left(\frac{\partial F}{\partial x} \eta(\tau) + \frac{\partial F}{\partial \dot{x}} \dot{\eta}(\tau) \right) d\tau \qquad (3.A5)$$

The second term can be integrated by parts to give

$$\delta S[x(\tau)] = \varepsilon \int_a^b \left(\frac{\partial F}{\partial x} - \frac{d}{d\tau} \frac{\partial F}{\partial \dot{x}} \right) \eta(\tau) d\tau + \varepsilon \eta(\tau) F(x, \dot{x}, \tau) \Big|_{\tau=a}^{\tau=b} \qquad (3.A6)$$

The endpoint term in the integration by parts vanishes, since $S[x(\tau)]$ is defined by an integral between fixed endpoints, so that we must have

$$\eta(a) = \eta(b) = 0 \qquad (3.A7)$$

and

$$\delta S[x(\tau)] = \varepsilon \int_a^b \left(\frac{\partial F}{\partial x} - \frac{d}{d\tau} \frac{\partial F}{\partial \dot{x}} \right) \eta(\tau) d\tau \qquad (3.A8)$$

From Eq. (3.A4), we have the identification $\delta x(\tau) = \varepsilon \eta(\tau)$, and thus

$$\frac{\delta S}{\delta x(\tau)} = \frac{1}{\varepsilon} \frac{\delta S}{\delta \eta(\tau)} = \frac{\partial F}{\partial x} - \frac{d}{d\tau} \frac{\partial F}{\partial \dot{x}} \tag{3.A9}$$

The extremum condition, Eq. (3.A3), then becomes

$$\frac{d}{d\tau} \frac{\partial F}{\partial \dot{x}} - \frac{\partial F}{\partial x} = 0 \tag{3.A10}$$

This differential equation is the *Euler–Lagrange equation* for the functional $S[x(\tau)]$. In the present context, the Euler–Lagrange equation is typically a second-order differential equation whose solution is required to pass through two particular points. The question of existence and uniqueness of the solutions satisfying endpoint conditions is not so simple as for a system of first-order equations with fixed initial conditions, and there may be no solutions, one solution, or more than one solution satisfying the endpoint conditions.

Remark.   Beyond geodesics, functionals of the type, Eq. (3.A1), form the basis of the Lagrangian formulation of classical mechanics, in which the function $F(x, \dot{x}, \tau)$ is the Lagrangian of a system, and the Euler–Lagrange equations are the classical Lagrange equations of motion for the system, as discussed at length in Section 3.5.3.   □

If the integral (3.A1) is over a curve in an n-dimensional manifold with coordinates given by $x^1, \ldots, x^n$, then minimizing with respect to each of the coordinates leads to an Euler–Lagrange equation in each variable, so that we have the n conditions

$$\frac{d}{d\tau} \frac{\partial F}{\partial \dot{x}^k} - \frac{\partial F}{\partial x^k} = 0 \tag{3.A11}$$

($k = 1, \ldots, n$).

B   Thermodynamics

Consider a simple thermodynamic system, such as a one-component gas, described by the variables T (temperature), p (pressure), V (volume), U (internal energy), and S (entropy). The system satisfies the first law in the form

$$dU = T\, dS - p\, dV \tag{3.B12}$$

where the $T\, dS$ term represents the heat absorbed by the system, and the term $p\, dV$ the work done *by* the system on its environment. The system is supposed to satisfy an *equation of state* of the form

$$f(p, V, T) = 0 \tag{3.B13}$$

that allows any of the variables to be expressed in terms of two independent variables. The precise form of the equation of state is not important here; the essential point is that the equation of state defines a two-dimensional manifold in the space of the three variables p, V, T, the *thermodynamic state space* of the system.

The properties of forms allow us to derive some completely general relations between the derivatives of the thermodynamic variables in a relatively simple way. For example, the first law (Eq. (3.B12)) implies

$$T = \left(\frac{\partial U}{\partial S}\right)_V \qquad p = -\left(\frac{\partial U}{\partial V}\right)_S \qquad (3.B14)$$

Here we use the standard thermodynamic notation

$$\left(\frac{\partial u}{\partial x}\right)_y$$

to denote the partial derivative of u with respect to x, holding y fixed. In other words, we are treating u as a function of the variables x and y, so that $u = u(x, y)$ and

$$du = \left(\frac{\partial u}{\partial x}\right)_y dx + \left(\frac{\partial u}{\partial y}\right)_x dy \qquad (3.B15)$$

If instead we want to treat u as a function of the variables x and y, with $y = y(x, z)$, then

$$dy = \left(\frac{\partial y}{\partial x}\right)_z dx + \left(\frac{\partial y}{\partial z}\right)_x dz \qquad (3.B16)$$

so that

$$\left(\frac{\partial u}{\partial z}\right)_x = \left(\frac{\partial u}{\partial y}\right)_x \left(\frac{\partial y}{\partial z}\right)_x \qquad (3.B17)$$

and

$$\left(\frac{\partial u}{\partial x}\right)_z = \left(\frac{\partial u}{\partial x}\right)_y + \left(\frac{\partial u}{\partial y}\right)_x \left(\frac{\partial y}{\partial x}\right)_z \qquad (3.B18)$$

Equality of mixed second partial derivatives then gives the relation

$$\left(\frac{\partial T}{\partial V}\right)_S = -\left(\frac{\partial p}{\partial S}\right)_V \qquad (3.B19)$$

Since $d(dU) = 0$, we have

$$dT \wedge dS = dp \wedge dV \qquad (3.B20)$$

With

$$dS = \left(\frac{\partial S}{\partial T}\right)_V dT + \left(\frac{\partial S}{\partial V}\right)_T dV \qquad dp = \left(\frac{\partial p}{\partial T}\right)_V dT + \left(\frac{\partial p}{\partial V}\right)_T dV \qquad (3.B21)$$

we have

$$\left(\frac{\partial S}{\partial V}\right)_T dT \wedge dV = \left(\frac{\partial p}{\partial T}\right)_V dT \wedge dV \qquad (3.B22)$$

so that

$$\left(\frac{\partial S}{\partial V}\right)_T = \left(\frac{\partial p}{\partial T}\right)_V \tag{3.B23}$$

Taking other combinations of independent variables leads to the further relations

$$\left(\frac{\partial T}{\partial p}\right)_S = \left(\frac{\partial V}{\partial S}\right)_p \qquad \left(\frac{\partial V}{\partial T}\right)_p = -\left(\frac{\partial S}{\partial p}\right)_T \tag{3.B24}$$

Equations (3.B19), (3.B23), and (3.B24) are the (thermodynamic) *Maxwell relations*. They are completely general, and do not depend on the specific form of the equation of state.

Changing independent variables for U from S and V to T and V in Eq. (3.B12) gives

$$dU = T\left(\frac{\partial S}{\partial T}\right)_V dT + \left[T\left(\frac{\partial S}{\partial V}\right)_T - p\right]dV \tag{3.B25}$$

so that

$$\left(\frac{\partial U}{\partial V}\right)_T = T\left(\frac{\partial S}{\partial V}\right)_T - p = T\left(\frac{\partial p}{\partial T}\right)_V - p \tag{3.B26}$$

This relates the dependence of the energy on volume at constant temperature, which is a measure of the interaction between the particles in the gas, to the dependence of the pressure on temperature at constant volume. Note that for an ideal gas, in which the interactions between the particles are negligible, the equation of state

$$pV = nRT \tag{3.B27}$$

(R is the gas constant, n is the number of moles of the gas) implies that both sides of Eq. (3.B26) vanish, and the internal energy of the ideal gas depends only on temperature.

→ **Exercise 3.B1.**   The (Helmholtz) *free energy* F of a thermodynamic system is related to the internal energy by

$$F \equiv U - TS$$

(i) Show that

$$dF = -p\,dV - S\,dT$$

(ii) Use this to derive directly the Maxwell relation (3.B23),

$$\left(\frac{\partial p}{\partial T}\right)_V = \left(\frac{\partial S}{\partial V}\right)_T$$

(iii) Show that

$$\left(\frac{\partial F}{\partial T}\right)_p = -S + p\left(\frac{\partial S}{\partial p}\right)_T$$

□

→ **Exercise 3.B2.** The *enthalpy* H of a thermodynamic system is defined by

$$H \equiv U + pV$$

(i) Show that

$$dH = T\, dS + V\, dp$$

(ii) Use this to derive the Maxwell relation

$$\left(\frac{\partial T}{\partial p} \right)_S = \left(\frac{\partial V}{\partial S} \right)_p$$

stated above.   □

Other general identities can be derived using the general properties of forms. For example, if we view y as a function of x and u in Eq. (3.B15), so that

$$dy = \left(\frac{\partial y}{\partial x} \right)_u dx + \left(\frac{\partial y}{\partial u} \right)_x dz \tag{3.B28}$$

then we have

$$\left[1 - \left(\frac{\partial u}{\partial y} \right)_x \left(\frac{\partial y}{\partial u} \right)_x \right] du = \left[\left(\frac{\partial y}{\partial x} \right)_u + \left(\frac{\partial y}{\partial u} \right)_x \left(\frac{\partial u}{\partial x} \right)_y \right] dx \tag{3.B29}$$

Since

$$\left(\frac{\partial u}{\partial y} \right)_x \left(\frac{\partial y}{\partial u} \right)_x = 1 \tag{3.B30}$$

the left-hand side of Eq. (3.B29) vanishes, and thus

$$\left(\frac{\partial y}{\partial x} \right)_u = - \left(\frac{\partial y}{\partial u} \right)_x \left(\frac{\partial u}{\partial x} \right)_y \tag{3.B31}$$

or

$$\left(\frac{\partial x}{\partial y} \right)_u \left(\frac{\partial y}{\partial u} \right)_x \left(\frac{\partial u}{\partial x} \right)_y = -1 \tag{3.B32}$$

Note the minus sign on the right-hand side of Eqs. (3.B31) and (3.B32). Naive cancelation of partial derivatives here would be an error.

To illustrate the use of these results, consider the *heat capacity* C of a thermodynamic system, the rate at which heat must be added to raise the temperature of the system. If thermal energy δQ is added to the system, then we have

$$\delta Q = T\, dS \equiv C\, dT \tag{3.B33}$$

The heat capacity depends on the conditions under which heat is added to the system. For example, the volume of the system may be held fixed (imagine a gas in a rigid container), in which case the heat capacity C_V at constant volume is relevant. On the other hand, the heat capacity C_p at constant pressure is appropriate if the system is held at constant pressure (imagine a gas in a balloon at atmospheric pressure). The difference between C_p and C_V is due to the fact that the system at constant pressure will expand as heat is added, doing work on its environment, so less energy will be converted to internal energy of the system.

We have

$$C_V = \left(\frac{\partial U}{\partial T}\right)_V = T\left(\frac{\partial S}{\partial T}\right)_V \qquad (3.B34)$$

$$C_p = \left(\frac{\partial U}{\partial T}\right)_p + p\left(\frac{\partial V}{\partial T}\right)_p = T\left(\frac{\partial S}{\partial T}\right)_p \qquad (3.B35)$$

➡ **Exercise 3.B3.** Show that the heat capacity at constant pressure of a system is

$$C_p = \left(\frac{\partial H}{\partial T}\right)_p$$

where H is the enthalpy introduced above. ☐

Now

$$\left(\frac{\partial S}{\partial T}\right)_p = \left(\frac{\partial S}{\partial T}\right)_V + \left(\frac{\partial S}{\partial V}\right)_T \left(\frac{\partial V}{\partial T}\right)_p \qquad (3.B36)$$

so that

$$C_p - C_V = T\left(\frac{\partial S}{\partial V}\right)_T \left(\frac{\partial V}{\partial T}\right)_p = T\left(\frac{\partial p}{\partial T}\right)_V \left(\frac{\partial V}{\partial T}\right)_p \qquad (3.B37)$$

where the second equality follows from the Maxwell relation (3.B23).

The *thermal expansion coefficient* β and the *isothermal compressibility* k are defined by

$$\beta \equiv \frac{1}{V}\left(\frac{\partial V}{\partial T}\right)_p \qquad k_T \equiv -\frac{1}{V}\left(\frac{\partial V}{\partial p}\right)_T \qquad (3.B38)$$

and from Eq. (3.B32), we have

$$\left(\frac{\partial p}{\partial T}\right)_V = -\left(\frac{\partial p}{\partial V}\right)_T \left(\frac{\partial V}{\partial T}\right)_p = \frac{\beta}{k_T} \qquad (3.B39)$$

It follows that

$$C_p - C_V = \frac{\beta^2 V T}{k_T} \qquad (3.B40)$$

which can be tested experimentally. Note that this relation implies $C_p \geq C_V$, since k_T is always positive—an increase in pressure at fixed temperature must always lead to a decrease in volume for a stable system.

Bibliography and Notes

An excellent elementary introduction to the concepts of this chapter from a physics point of view is

> Bernard F. Schutz, *Geometrical Methods of Mathematical Physics*, Cambridge University Press (1980).

It is well written, with many useful diagrams and examples. A modern comprehensive introduction aimed at theoretical physicists is

> Theodore Frankel, *The Geometry of Physics* (2nd edition), Cambridge University Press (2004).

A modern advanced undergraduate textbook on classical mechanics that introduces geometrical concepts at various stages is

> Tom W. B. Kibble and Frank H. Berkshire, *Classical Mechanics* (5th edition), Imperial College Press (2004).

The final two chapters are a nice elementary introduction to the general area of dynamical systems and chaos.

Dynamical systems and vector fields are closely related. An early text that emphasizes geometry in classical mechanics is

> V. I. Arnold, *Mathematical Methods of Classical Mechanics*, Springer (1974).

The book by Hirsch, Smale and Devaney cited in Chapter 2 is also oriented to a geometrical analysis of dynamical systems.

Two recent classical mechanics textbooks in a similar spirit are

> Jorge V. José and Eugene J. Saletan, *Classical Dynamics: A Contemporary Approach*, Cambridge University Press (1998), and
> Joseph L. McCauley, *Classical Mechanics*, Cambridge University Press (1997),

These books emphasize the view of trajectories of Hamiltonian systems as flows in phase space along which the canonical 2-form is invariant, in addition to treating standard topics.

Two classic elementary introductions to thermodynamics are

> Enrico Fermi, *Thermodynamics*, Dover (1956), and
> A. B. Pippard, *The Elements of Classical Thermodynamics*, Cambridge University Press (1957).

Enrico Fermi was arguably the greatest experimental physicist of the 20th century, as well as a major theorist. He was also a magnificent teacher and writer, and his book on thermodynamics is relevant even now. Pippard's book is another clear introduction by a distinguished physicist.

Many books on general relativity give an introduction to the geometrical ideas discussed here as a prelude to general relativity. Two relatively elementary books are

> Bernard F. Schutz, *A First Course in General Relativity*, Cambridge University Press (1985). and
> Sean M. Carroll, *Spacetime Geometry: An Introduction to General Relativity*, Addison-Wesley (2004)

Problems

1. The stereographic projection of the unit sphere $\mathbf{S}^2$ from the North pole onto the plane tangent to the South pole is obtained from the figure at the right by rotating the entire figure about the vertical line through the poles (compare with Fig. 3.3).

   (i) Show that the coordinates (x_N, y_N) of the image Q in the plane of the point P with usual spherical angles (θ, ϕ) is given by (note that $2\alpha = \pi - \theta$)

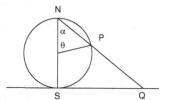

$$x_N = 2\cot\frac{\theta}{2}\cos\phi \qquad y_N = 2\cot\frac{\theta}{2}\sin\phi$$

   (ii) Find the corresponding image (x_S, y_S) of the stereographic projection of P from the South pole onto a plane tangent to the North pole.

   (iii) Express (x_S, y_S) in terms of (x_N, y_N).

   Remark. These functions are differentiable to all orders in the region of the sphere defined by $\delta \leq \theta \leq \pi - \delta$ ($0 < \delta < \frac{\pi}{2}$). Thus the two stereographic projections form an atlas that covers the entire sphere. □

2. (i) Find a diffeomorphism mapping the open square $(-1 < x < 1, -1 < y < 1)$ onto the entire plane $\mathbf{R}^2$.

   (ii) Find a diffeomorphism mapping the open disk D $(x^2 + y^2 < 1)$ (also defined as the unit ball $\mathbf{B}^2$) onto the entire plane $\mathbf{R}^2$.

   (iii) Combine these two results to find a diffeomorphism that maps the open square onto the open disk D. Comment on the fact that the sharp corners of the square have been smoothed out in the disk.

3. Let $\{\mathbf{e}_k\}$ be a basis of the tangent space T_P and $\mathbf{S} = (\mathbf{S}^k{}_m)$ a nonsingular matrix. Show that

   (i) the vectors $\{\bar{\mathbf{e}}_m\}$ defined by

   $$\bar{\mathbf{e}}_m = \mathbf{S}^k{}_m \, \mathbf{e}_k$$

   also form a basis.

   (ii) the basis $\{\omega^k\}$ of T_P^* dual to $\{\mathbf{e}_k\}$ and the basis $\{\bar{\omega}^m\}$ dual to $\{\bar{\mathbf{e}}_m\}$ are related by

   $$\bar{\omega}^m = \bar{\mathbf{S}}^m{}_k \, \omega^k$$

   where $\bar{\mathbf{S}} = (\bar{\mathbf{S}}^m{}_k)$ is the matrix inverse of $\mathbf{S}$.

   (iii) if $\mathbf{S} = \mathbf{S}(x)$ defines a change of basis on the tangent spaces T_x for x in some neighborhood of P, then $\mathbf{S}$ defines a coordinate transformation only if

   $$\frac{\partial}{\partial x^\ell}\,\bar{\mathbf{S}}^m{}_k(x) = \frac{\partial}{\partial x^k}\,\bar{\mathbf{S}}^m{}_\ell(x)$$

   where $\bar{\mathbf{S}} = \mathbf{S}^{-1}$ as in part (ii).

4. Show that the exterior derivative commutes with the Lie derivative. First, show that if V is a vector field and σ is a 1-form field, then

$$\mathcal{L}_V(d\sigma) = d(\mathcal{L}_V\sigma)$$

Then show that this relation is true for a general p-form σ. (Hint: use induction together with the Leibniz rule for differentiation of a product.)

5. Show that if V is a vector field and σ a p-form, then

$$\mathcal{L}_V\sigma = i_V(d\sigma) + d(i_V\sigma)$$

6. *Prolate spheroidal* coordinates ξ, η, ϕ on $\mathbf{R}^3$ are defined in terms of Cartesian coordinates x, y, z by

$$x = c \sinh\xi \sin\eta \cos\phi \qquad y = c \sinh\xi \sin\eta \sin\phi \qquad z = c \cosh\xi \cos\eta$$

where c is a constant with dimensions of length.

(i) Describe the surfaces of constant ξ, η, ϕ. What are the ranges of ξ, η, ϕ? What subsets of $\mathbf{R}^3$ correspond to coordinate singularities with this range of coordinates?

(ii) An alternative version of the coordinate set is u, v, ϕ with $u \equiv c \cosh\xi$ and $v \equiv \cos\eta$. What are the ranges of u, v, ϕ? What subsets of $\mathbf{R}^3$ correspond to coordinate singularities with this range of coordinates?

(iii) Find the metric tensor and the preferred volume form in these coordinates. Consider both the set ξ, η, ϕ and the set u, v, ϕ.

(iv) Express the Laplacian in terms of partial derivatives with respect to these coordinates.

Remark. These coordinates are useful for both the classical and quantum mechanical problem of a single charge q moving in the Coulomb field of two fixed point charges Q_1 and Q_2 (the fixed charges are placed at the foci of the spheroids of constant u), as well as scattering problems with spheroidal scatterers. They also simplify the problem of computing the electrostatic potential of a charged conducting prolate spheroid, and the limiting case of a long thin needle with a rounded edge.   □

7. Consider a system with two fixed charges q_1 and q_2, separated by a distance R, and a charge q of mass M that is free to move in the Coulomb field of these charges. Express the Coulomb potential energy of the system in terms of the position of the moving charge in prolate spheroidal coordinates with the fixed charges at the foci of the spheroid.

8. *Oblate spheroidal* coordinates ξ, η, ϕ on $\mathbf{R}^3$ are defined in terms of Cartesian coordinates x, y, z by

$$x = c \cosh\xi \sin\eta \cos\phi \qquad y = c \cosh\xi \sin\eta \sin\phi \qquad z = c \sinh\xi \cos\eta$$

(again c is a constant with dimensions of length).

(i) Describe the surfaces of constant ξ, η, ϕ. What are the ranges of ξ, η, ϕ? What subsets of $\mathbf{R}^3$ correspond to coordinate singularities with this range of coordinates?

(ii) An alternative version of the coordinate set is u, v, ϕ with $u \equiv c \sinh \xi$ and $v \equiv \cos \eta$. What are the ranges of u, v, ϕ? What subsets of $\mathbf{R}^3$ correspond to coordinate singularities with this range of coordinates?

(iii) Compute the metric tensor and the preferred volume form in these coordinates. Consider both the set ξ, η, ϕ and the set u, v, ϕ.

(iv) Express the Laplacian in terms of partial derivatives with respect to these coordinates.

Remark. These coordinates are useful for scattering problems with oblate spheroidal scatterers. They also simplify the computation of the electrostatic potential of a charged conducting oblate spheroid and its limiting case of a circular disk. □

9. *Parabolic* coordinates ξ, η, ϕ on $\mathbf{R}^3$ are defined in terms of Cartesian coordinates x, y, z by

$$x = \xi\eta \cos\phi \qquad y = \xi\eta \sin\phi \qquad z = \tfrac{1}{2}\left(\xi^2 - \eta^2\right).$$

(i) Describe the surfaces of constant ξ, η, ϕ. What are the ranges of ξ, η, ϕ? What subsets of $\mathbf{R}^3$ correspond to coordinate singularities with this range of coordinates?

(ii) Compute the metric tensor and preferred volume form in these coordinates.

(iii) Express the Laplacian in terms of partial derivatives with respect to these coordinates.

Remark. These coordinates are useful for scattering problems with parabolic scatterers. They are also useful in the quantum mechanical problem of the Coulomb scattering of a charge (an electron, for example) by a fixed point charge (an atomic nucleus, for example). □

10. Four-dimensional spherical coordinates r, α, θ, ϕ are defined in terms of Cartesian coordinates w, x, y, z by

$$x = r \sin\alpha \, \sin\theta \, \cos\phi \qquad y = r \sin\alpha \, \sin\theta \, \sin\phi$$

$$w = r \cos\alpha \qquad z = r \sin\alpha \, \cos\theta$$

(i) Express the four-dimensional Euclidean metric in terms of the coordinates r, α, θ, ϕ.

(ii) Express the four-dimensional volume element in terms of r, α, θ, ϕ.

(iii) Use this volume element to compute the four-volume of a ball of radius R.

(iv) Compute the surface "area" of the sphere $\mathbf{S}^3$ bounding the ball of radius R.

(v) Express the four-dimensional Laplacian □ in terms of partial derivatives with respect to these coordinates.

11. The geographic latitude α on a spherical surface is related to the spherical angle θ by $\alpha = \pi/2 - \theta$ (and is usually given in degrees, minutes and seconds, rather than in radians). Find an expression for the maximum latitude reached by the great circle starting from initial conditions $\theta_0, \dot{\theta}_0, \dot{\phi}_0$.

12. (i) What is the shortest distance between points (θ_1, ϕ_1) and (θ_2, ϕ_2) on the sphere $\mathbf{S}^2$? *Hint.* Let R be the radius of the sphere, and use the result that the geodesic is an arc of a great circle.

    (ii) Find the tangent vector to this geodesic at the initial point (θ_1, ϕ_1).

    (iii) Find the geodesic curve joining the points (θ_1, ϕ_1) and (θ_2, ϕ_2) on the sphere $\mathbf{S}^2$ using the arc length along the curve as parameter.

13. How does Stokes' theorem

$$\int_{\mathbf{R}} d\sigma = \int_{\partial \mathbf{R}} \sigma$$

    work if $\mathbf{R}$ is a region in a one-dimensional manifold? Explain this in the language of elementary calculus.

14. Consider the extension of the two-dimensional dynamical system introduced in Section 3.5.2 defined by

$$\dot{x}_1 = \frac{dx_1}{dt} = (-\lambda + ax_2 + px_1)\, x_1$$

$$\dot{x}_2 = \frac{dx_2}{dt} = (\mu - bx_1 + qx_2)\, x_2$$

    where the parameters λ, μ, a, b are all positive, as in Eq. (3.223). The signs of p, q are not specified, though one might expect them to be negative in the context of the Lotka–Volterra model.

    (i) Find the fixed points of this dynamical system.

    (ii) Investigate the stability of each of these fixed points. Note especially under what conditions, if any, on the parameters p, q each of these fixed points will be stable.

    (iii) Under what conditions on the parameters p, q do all the fixed points lie in the quadrant $x_1 \geq 0$, $x_2 \geq 0$?

    (iv) Explain why one might expect p, q to be negative. Under what conditions might one or the other of them be positive?

15. An alternative action for a relativistic free particle of mass m is

$$S = \frac{1}{2} \int \left\{ \frac{\dot{x}^\mu g_{\mu\nu}(x)\dot{x}^\nu}{\xi(\lambda)} + m^2 \xi(\lambda) \right\} d\lambda$$

    where $\xi(\lambda)$ is a positive and monotonic increasing function of λ; it plays the role of a one-dimensional "metric" along the particle trajectory. Changing the parameter from λ to $\mu = \mu(\lambda)$ leaves the action invariant if $\xi(\lambda)$ is replaced by $\eta(\mu)$ such that

$$\eta(\mu)d\mu = \xi(\lambda)d\lambda$$

(i) Show that if we choose $\xi(\lambda) = 1$, then the equations of motion for the $x^\mu(\lambda)$ are exactly the geodesic equations (3.207).

(ii) Show that if we fix $\xi(\lambda)$ by requiring $\delta S/\delta\xi(\lambda) = 0$, the action reduces to the standard geodesic action.

16. The classical trajectory of a relativistic free particle is an extremum of the relativistic action, but it is in fact a maximum. To see why, consider first the geodesic from the point $(0, \vec{0})$ to the point $(2T, \vec{0})$, which is simply the trajectory $x = (ct, \vec{0})$ of a particle at rest at the origin. Next consider the trajectory of a particle starting at the origin at $t = 0$. It is with constant velocity $\vec{v}$ until time $t = (1 - \varepsilon) * T$ $(0 < \varepsilon < 1)$. It is then accelerated with constant acceleration $-\vec{v}/\varepsilon T$ until time $t = (1 + \varepsilon) * T$, after which it moves with constant velocity $-\vec{v}$ until it returns to the origin at time $t = 2T$.

(i) Write an explicit formula for this trajectory, using the time of the observer at rest as the parameter along the curve.

(ii) Integrate the standard Lagrangian to show that the action S for this trajectory compared to the action S_0 for the observer at rest satisfies

$$\sqrt{1 - \frac{v^2}{c^2}} < \frac{S}{S_0} < (1 - \varepsilon)\sqrt{1 - \frac{v^2}{c^2}} + \varepsilon$$

Remark. This suggests that the action for the particle at rest is a maximum, though this calculation actually shows only that it is not a minimum. This result also resolves to the so-called *twin paradox*, which is the nonintuitive statement that a twin who moves along the trajectory described above appears younger than the other twin who remains at rest at the origin. Since the action is proportional to the proper time along the particle trajectory, more proper time has elapsed for the twin who remains at rest, assuming that physical clocks run at a rate proportional to the proper time, even in an accelerated system. This last assumption is Einstein's *equivalence principle*, which has been experimentally tested. □

17. The partial derivatives of the Lagrangian in Eq. (3.231) are taken with L as a function of the x^k and $\dot{x}^k$, while the partial derivatives of the Hamiltonian in Eq. (3.239) are taken with H as a function of the x^k and p_k. Use the relations in Appendix B between partial derivatives with different variables held fixed to show that the two sets of equations are equivalent.

18. (i) Express the kinetic energy for a free particle of mass m in terms of the prolate spheroidal coordinates ξ, η, ϕ and the alternate set u, v, ϕ introduced in Problem 6.

(ii) Find the momenta conjugate to each of these coordinates, and express the Hamiltonian in terms of the momenta and coordinates.

(iii) Find the Lagrangian and Hamiltonian equations of motion in each of these coordinate systems.

19. If f is a function on the phase space of a Hamiltonian system, let $\mathbf{X}_f$ be a vector field such that

$$i_{\mathbf{X}_f}\omega = -df$$

where ω is the canonical 2-form (Eq. (3.257)) on the phase space.

(i) Show that $\mathbf{X}_f$ is given uniquely by

$$\mathbf{X}_f = \frac{\partial f}{\partial p_k}\frac{\partial}{\partial q^k} - \frac{\partial f}{\partial q^k}\frac{\partial}{\partial p_k}$$

(ii) The *Poisson bracket* of two scalar functions f, g is defined by

$$\{f,g\} \equiv \omega(\mathbf{X}_f, \mathbf{X}_g)$$

Show that

$$\{f,g\} = \frac{\partial f}{\partial p_k}\frac{\partial g}{\partial q^k} - \frac{\partial f}{\partial q^k}\frac{\partial g}{\partial p_k}$$

and that

$$[\mathbf{X}_f, \mathbf{X}_g] = \mathbf{X}_{\{f,g\}}$$

(iii) If H is the Hamiltonian of the system, and f is a scalar function of the coordinates, momenta, and possibly time, then

$$\frac{df}{dt} = \{H, f\} + \frac{\partial f}{\partial t}$$

where the derivative on the left-hand side is the time derivative of f along a trajectory of the system.

Remark.   The Poisson bracket is a classical analog of the quantum mechanical commutator, as discussed in many quantum mechanics textbooks. □

20. Consider a nonlinear oscillator, with Hamiltonian

$$H = \tfrac{1}{2}\omega(P^2 + X^2) + \tfrac{1}{4}\lambda X^4$$

in terms of the variables introduced in the example at the end of Section 3.5.3.

(i) Express this Hamiltonian in terms of the action-angle variables J, α introduced in Eq. (3.273).

(ii) Write down Hamilton's equations of motion for J, α.

(iii) From the equation for $\dot{\alpha}$, find an approximate expression for the period as a function of J by averaging the right-hand side of the equation over a complete period, assuming that $\dot{\alpha}$ is constant (which it actually is not, but nearly so for small λ).

Remark.   This problem is a prototype for classical perturbation theory. □

21. Use Euler's equation (3.287) to show that the vorticity α defined by Eq. (3.296) satisfies

$$\left(\frac{\partial}{\partial t} + \mathcal{L}_U\right)\alpha = d\left(\frac{\partial}{\partial t} + \mathcal{L}_U\right)\tilde{U} = \frac{1}{\rho^2}\,d\rho \wedge dp$$

Remark. If the fluid is incompressible, or if it satisfies an equation of state $p = f(\rho)$, then the right-hand side vanishes, and vorticity is carried along with the fluid flow. □

22. Consider an integral of the form

$$S[x(\tau)] \equiv \int_a^b F\left(x, \dot{x}, \ddot{x}, \tau\right) d\tau$$

from point a to point b along a set of smooth curves $C : x = x(\tau)$ in a one-dimensional manifold, where here

$$\dot{x} = \frac{dx}{d\tau} \qquad \ddot{x} = \frac{d^2 x}{d\tau^2}$$

Derive the conditions that must be satisfied by a curve $x = x_*(\tau)$ for which $S[x(\tau)]$ is an extremum relative to nearby curves? Find differential equation(s) of the Euler–Lagrange type, and boundary conditions that must be satisfied at the endpoints of the interval $[a, b]$.

23. The *Gibbs function* G of a thermodynamic system is related to the internal energy by

$$G \equiv U - TS + pV$$

(i) Show that for the thermodynamic system in Appendix B,

$$dG = V\,dp - S\,dT$$

(ii) Show that

$$\left(\frac{\partial G}{\partial T}\right)_V = V\left(\frac{\partial S}{\partial V}\right)_T - S$$

24. Show that for an ideal gas,

(i) the entropy is given by

$$S = S_0(T) + nR\ln\left(\frac{V}{V_0}\right)$$

where $S_0(T)$ is the entropy at volume V_0 and temperature T,

(ii) if the internal energy of the gas is given by $U = \alpha nRT$, with α constant, then

$$S_0(T) = \sigma_0 + \alpha nR\ln\left(\frac{T}{T_0}\right)$$

where σ_0 is the entropy at temperature T_0, volume V_0, and

(iii) the heat capacities of the gas are related by

$$C_p - C_V = nR$$

4 Functions of a Complex Variable

There are many functions $f(x)$ of a real variable x whose definition contains a natural extension of the function to complex values of its argument. For example, functions defined by a convergent power series, or by an integral representation such as the Laplace integral introduced in Chapter 1, are already defined in some regions of the complex plane. Extending the definition of functions into the complex plane leads to new analytical tools that can be used to study these functions. In this chapter we survey some of these tools and present a collection of detailed examples.

Analytic functions of a complex variable are functions that are differentiable in a region of the complex plane. This definition is not quite so straightforward as for functions defined only for real argument; when the derivative is defined as a limit

$$f'(z) = \lim_{z \to z_0} \frac{f(z) - f(z_0)}{z - z_0}$$

in the complex plane, the limit must exist independent of the direction from which $z \to z_0$ in the complex plane. This requires special relations (the *Cauchy–Riemann conditions*) between the partial derivatives of the real and imaginary parts of an analytic function; these relations further imply that the real and imaginary parts of an analytic function satisfy Laplace's equation in two dimensions. Analytic functions define mappings of one complex region into another that are conformal (angle-preserving) except at singular points where the function or its inverse is not differentiable. These conformal mappings are described and some elementary mappings worked out in detail.

If a function $f(z)$ is not single valued in a neighborhood of some point z_0 (for example, $\sqrt{z}$ near $z = 0$), then $f(z)$ has a branch point at z_0. Branch points generally come in pairs, and can be connected by a branch cut such that the function is single valued in the complex plane excluding the cut. The domain of a function with branch points is a multisheeted surface (a Riemann surface) in which crossing a branch cut leads from one sheet of the surface to the next. Some important Riemann surfaces are described.

Integrals of analytic functions in the complex plane have many useful properties that follow from Cauchy's theorem: the integral of an analytic function around a closed curve C vanishes if there are no singularities of the function inside the curve. This leads to the Cauchy integral formula

$$\oint_C \frac{f(\xi)}{\xi - z} \, d\xi = 2\pi i f(z)$$

which expresses the values of a function $f(z)$ analytic within the region bounded by the closed curve C in terms of its values on the boundary. It also leads to the Cauchy residue theorem,

Introduction to Mathematical Physics. Michael T. Vaughn
Copyright © 2007 WILEY-VCH Verlag GmbH & Co. KGaA, Weinheim
ISBN: 978-3-527-40627-2

which expresses the integral of an analytic function with isolated singularities inside a contour in terms of the behavior of the function near these singularities. Several examples are given to show how this can be used to evaluate some definite integrals that are important in physics.

The Cauchy integral formula is also used to obtain power series expansions of an analytic function about a regular point and about an isolated singular point. The formal process of analytic continuation that leads to the global concept of an analytic function is explained with the use of the power series expansions.

The singular points of an analytic function are characterized as poles, essential singularities or branch points, depending on the behavior of the function near the singularity. If $f(z) \simeq A/(z - z_0)^n$ for $z \to z_0$, with A a constant and n a positive integer, then $f(z)$ has a pole of order n at z_0. If $f(z)$ has a power series expansion around z_0 that includes an infinite number of negative integer powers of $(z - z_0)$, then $f(z)$ has an essential singularity at z_0. Functions with branch points are not single valued and require the introduction of branch cuts and a Riemann surface to provide a maximal analytic continuation.

We show that every nonconstant analytic function has at least one singular point (possibly at ∞). This leads to a simple proof of the fundamental theorem of algebra that every nonconstant polynomial has at least one root. It follows that every polynomial of degree n has exactly n roots (counted according to multiplicity), and then can be expressed as a constant multiple of a product of factors $(z - z_k)$, where the z_k are the zeros of the polynomial.

The factorization of polynomials has several important consequences: (i) a function whose only singularities are poles can be expressed as a ratio of two polynomials, (ii) an entire function (a function whose only singularity is at ∞) can be expressed as an entire function with no zeros times a product of factors determined by its zeros (Weierstrass factorization theorem), and (iii) a function whose only singularities in the finite plane are poles can be expressed as a sum over terms determined by the singular parts of the function at the poles plus an entire function (Mittag–Leffler theorem).

Periodic functions of a real variable are familiar from the study of oscillatory systems. A general expansion of periodic functions in terms of trigonometric functions is the Fourier series expansion. Here we derive this expansion for periodic analytic functions by relating it to a Laurent expansion in the complex variable $w \equiv e^{2\pi i z/\alpha}$, where α is the period. Further properties of Fourier series will appear in the context of linear vector space theory in Chapter 6.

In addition to simply periodic functions, there are also analytic functions that have two independent periods in the complex plane. If $f(z + \alpha) = f(z)$ and $f(z + \beta) = f(z)$ with β/α a complex number, not real, then $f(z)$ is a doubly periodic function, known for historical reasons as an elliptic function if its only singularities are poles. Some general properties of these functions are described here. A standard set of elliptic functions will be studied further in Appendix A of Chapter 5.

The Γ-function is an extension to the complex plane of the factorial function $n!$ defined for integer n. Many properties of the Γ-function and the related beta function, including Stirling's formula for the asymptotic behavior of $\Gamma(x)$ for large positive x, are derived in Appendix A as an important illustration of the methods that can be used to study functions in the complex plane.

4.1 Elementary Properties of Analytic Functions

4.1.1 Cauchy–Riemann Conditions

Consider a function $f(z)$ of the complex variable $z = x + iy$; write

$$f(z) = w = u + iv = u(x, y) + iv(x, y) \tag{4.1}$$

with $u(x, y)$ and $v(x, y)$ real. $f(z)$ is *continuous* at z_0 if

$$\lim_{z \to z_0} f(z) = f(z_0) \tag{4.2}$$

This looks like the definition of continuity for a function of a real variable, except that the limit must exist as $z \to z_0$ from any direction in the complex z-plane.

Similarly, $f(z)$ is *differentiable* at z_0 if the limit

$$\lim_{z \to z_0} \frac{f(z) - f(z_0)}{z - z_0}$$

exists from any direction in the complex plane. If the limit does exist, then the derivative of $f(z)$ at z_0 is equal to the limit; we have

$$\left. \frac{df}{dz} \right|_{z_0} \equiv \lim_{z \to z_0} \frac{f(z) - f(z_0)}{z - z_0} \equiv f'(z_0) \tag{4.3}$$

If $f(z) = u(x, y) + iv(x, y)$ is differentiable at $z_0 = x_0 + iy_0$, then

$$f'(z_0) = \left[\frac{\partial u}{\partial x} + i \frac{\partial v}{\partial x} \right]_{(x_0, y_0)} = \left[\frac{\partial v}{\partial y} - i \frac{\partial u}{\partial y} \right]_{(x_0, y_0)} \tag{4.4}$$

where the two expressions on the right-hand side are obtained by taking the limit $z \to z_0$ first parallel to the real axis, then parallel to the imaginary axis. Hence the partial derivatives of $u(x, y)$ and $v(x, y)$ must exist at (x_0, y_0). Moreover, the two limits must be equal, so that the partial derivatives must satisfy the *Cauchy–Riemann conditions*

$$\frac{\partial u}{\partial x} = \frac{\partial v}{\partial y}, \quad \frac{\partial v}{\partial x} = -\frac{\partial u}{\partial y} \tag{4.5}$$

The converse is also true: If $u(x, y)$ and $v(x, y)$ are real functions with continuous first partial derivatives in some neighborhood of (x_0, y_0), and if the conditions (4.5) are satisfied at (x_0, y_0), then $f(z) = u(x, y) + iv(x, y)$ is differentiable at $z_0 = x_0 + iy_0$, with derivative given by Eq. (4.4). Thus the Cauchy–Riemann conditions, together with the continuity conditions, are both necessary and sufficient for differentiability of $f(z)$.

Definition 4.1. The function $f(z)$ of the complex variable z is *analytic* (*regular, holomorphic*) at the point z_0 if $f(z)$ is differentiable at z_0 *and* in some neighborhood of z_0.

Remark. To understand why analyticity requires $f(z)$ to be differentiable in a neighborhood of z_0, and not just at z_0 itself, consider the function

$$f(z) \equiv (\tfrac{1}{2}|z|^2 - 1) \, z^* \tag{4.6}$$

$f(z)$ is continuous everywhere in the complex plane, and is actually differentiable on the unit circle $|z| = 1$. But it is not differentiable anywhere off the unit circle, and hence not in the neighborhood of any point on the circle, so it is nowhere analytic. Note that the condition (4.13) is satisfied by $f(z)$ only on the unit circle. Similarly, it is important that the Cauchy–Riemann conditions are satisfied in some neighborhood of a point. The functions

$$u(x, y) = x^2 \qquad v(x, y) = y^2 \tag{4.7}$$

satisfy the Cauchy–Riemann conditions at the origin ($x = y = 0$), but nowhere else. Hence the function $f(z) = x^2 + iy^2$ is nowhere analytic. □

If $f(z)$ is analytic in a region $\mathcal{R}$, then the Cauchy–Riemann conditions imply that

$$\frac{\partial^2 u}{\partial x^2} + \frac{\partial^2 u}{\partial y^2} = 0 = \frac{\partial^2 v}{\partial x^2} + \frac{\partial^2 v}{\partial y^2} \tag{4.8}$$

(*Laplace's equation*) in $\mathcal{R}$. Thus both the real and imaginary parts of an analytic function satisfy the two-dimensional Laplace's equation. The Cauchy–Riemann conditions also give

$$\frac{\partial u}{\partial x}\frac{\partial v}{\partial x} + \frac{\partial u}{\partial y}\frac{\partial v}{\partial y} = \vec{\nabla}u \cdot \vec{\nabla}v = 0 \tag{4.9}$$

Now $\vec{\nabla}u$ ($\vec{\nabla}v$) is orthogonal to the curve $u(x, y) = $ constant ($v(x, y) = $ constant) at every point, and Eq. (4.9) shows that $\vec{\nabla}u$ and $\vec{\nabla}v$ are orthogonal. Hence any curve of constant u is orthogonal to any curve of constant v at any point where the two curves intersect. Thus an analytic function generates two families of mutually orthogonal curves. More generally, the analytic function generates a *conformal map* from the complex z-plane to the complex w-plane, as explained in the next section.

Another view of the Cauchy–Riemann conditions is obtained by treating $z \equiv x + iy$ and $z^* \equiv x - iy$ as the independent variables. Then

$$\frac{\partial}{\partial z} = \frac{1}{2}\left(\frac{\partial}{\partial x} - i\frac{\partial}{\partial y}\right) \quad \text{and} \quad \frac{\partial}{\partial z^*} = \frac{1}{2}\left(\frac{\partial}{\partial x} + i\frac{\partial}{\partial y}\right) \tag{4.10}$$

whence, with $f(z, z^*) = u(z, z^*) + iv(z, z^*)$,

$$\frac{\partial f}{\partial z} = \frac{1}{2}\left(\frac{\partial u}{\partial x} + \frac{\partial v}{\partial y}\right) + \frac{1}{2}i\left(\frac{\partial v}{\partial x} - \frac{\partial u}{\partial y}\right) \tag{4.11}$$

$$\frac{\partial f}{\partial z^*} = \frac{1}{2}\left(\frac{\partial u}{\partial x} - \frac{\partial v}{\partial y}\right) + \frac{1}{2}i\left(\frac{\partial v}{\partial x} + \frac{\partial u}{\partial y}\right) \tag{4.12}$$

The Cauchy–Riemann conditions are then equivalent to the condition

$$\frac{\partial f}{\partial z^*} = 0 \tag{4.13}$$

so that $f(z, z^*)$ is an analytic function of z if and only if it is both differentiable with respect to z *and* independent of z^*.

4.1.2 Conformal Mappings

The function $w = u + iv$ defines a mapping of the complex z-plane into the complex w-plane. The mapping is *conformal* if angle and sense of rotation are preserved by the mapping; that is, if C_1, C_2 are curves in the z-plane that intersect at an angle α, then the corresponding image curves C_1', C_2' in the w-plane intersect at the same angle α, with the sense of rotation preserved.

If $w = f(z)$ is analytic in a region $\mathcal{R}$ of the z-plane, then the mapping of $\mathcal{R}$ onto its image in the w-plane is conformal, except at points where $f'(z) = 0$. To show this, let z_0 be a point in $\mathcal{R}$, and C be a curve through z_0 that makes an angle ξ with respect to the real z axis. Then the image curve C' in the w-plane passes through $w_0 = f(z_0)$ at an angle ξ' with respect to the real w axis, where

$$\xi' - \xi = \lim_{\substack{w \to w_0 \\ \text{on } C'}} \arg(w - w_0) - \lim_{\substack{z \to z_0 \\ \text{on } C}} \arg(z - z_0)$$

$$= \lim_{\substack{z \to z_0 \\ \text{on } C}} \arg\left[\frac{f(z) - f(z_0)}{z - z_0}\right] = \arg f'(z_0) \tag{4.14}$$

independent of the curve C if $f(z)$ is analytic at z_0 and $f'(z_0) \neq 0$.

Examples. (i) *Linear transformations*

- $w = z + b$ is a *(rigid) translation* by b of the whole z-plane.

- $w = \varrho z$ (ϱ real) is a *scale transformation* by scale factor ϱ.

- $w = e^{i\varphi} z$ (φ real) is a *(rigid) rotation* through angle φ of the whole z-plane.

The general linear transformation

$$w = az + b \tag{4.15}$$

($a = \varrho e^{i\varphi}$) consists of a scale transformation by scale factor ϱ, rotation through angle φ, followed by translation by b. The transformation has a fixed point at ∞ and, if $a \neq 1$, a second fixed point at

$$z = \frac{b}{1 - a} \tag{4.16}$$

(ii) *Reciprocal transformation*

The *reciprocal transformation*, or *inversion* is defined by

$$w = 1/z \tag{4.17}$$

If $z = x + iy = re^{i\varphi}$, then $w = u + iv = \varrho e^{-i\varphi}$ with $\varrho = 1/r$; in terms of real variables,

$$u = \frac{x}{r^2}, \quad v = -\frac{y}{r^2} \quad \text{and} \quad x = \frac{u}{\varrho^2}, \quad y = -\frac{v}{\varrho^2} \tag{4.18}$$

The point $z = \infty$ is defined to be the image of 0 under the reciprocal transformation. The fixed points of the transformation are at $z = \pm 1$.

(iii) *Linear fractional transformation*

The general *linear fractional* (or *bilinear*) *transformation* is defined by

$$w = \frac{az+b}{cz+d} \tag{4.19}$$

with $\Delta \equiv ad - bc \neq 0$ (if $\Delta = 0$, the mapping becomes $w = a/c$, a constant). The inverse transformation

$$z = \frac{-dw+b}{cw-a} \tag{4.20}$$

is also a linear fractional transformation (hence the term bilinear). The transformation is thus a one-to-one mapping of the complex plane (including the point at ∞) into itself; the transformation has at most two fixed points, at

$$z = \frac{a-d \pm \sqrt{(a-d)^2 + 4bc}}{2c} \tag{4.21}$$

The point $z = \infty$ is mapped into $w = a/c$, while the point $z = -d/c$ is mapped into the point $w = \infty$.

The linear fractional transformation is the most general one-to-one mapping of the complex plane including the point at ∞ onto itself—as will be seen shortly, any more complicated mapping will not be one to one. With the mapping of the points z_1, z_2, z_3 into the points w_1, w_2, w_3 is associated the unique linear fractional transformation

$$\frac{(w-w_1)(w_2-w_3)}{(w-w_3)(w_2-w_1)} = \frac{(z-z_1)(z_2-z_3)}{(z-z_3)(z_2-z_1)} \tag{4.22}$$

Remark.   There is a one-to-one correspondence between linear fractional transformations and 2×2 matrices $\mathbf{A}$ with $\det \mathbf{A} = 1$—see Problem 2. $\square$

(iv) *Powers and roots*

Consider the transformation

$$w = z^2 \tag{4.23}$$

This transformation is conformal except at $z = 0, \infty$ (note that $w'(0) = 0$). However it is not one to one; as z ranges over the entire z-plane, the w-plane is covered twice, since $z = r\,e^{i\theta} \mapsto w = r^2\,e^{2i\theta}$ and a single circle around $z = 0$, corresponding to θ ranging from 0 to 2π, covers the corresponding circle in the w-plane twice. Hence each point in the w-plane is the image of two points in the z-plane.

The inverse transformation

$$z = \sqrt{w} \tag{4.24}$$

is analytic and conformal except for the points $w = 0, \infty$, which correspond to the points in the z-plane where the map (4.23) is not conformal. However, the map (4.24) is double valued: to each point in the w-plane, there are two corresponding points in the z-plane.

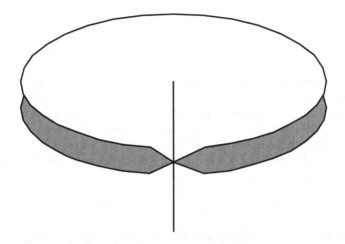

Figure 4.1: Mapping of the z-plane into the two-sheeted Riemann surface corresponding to a double covering of the w-plane. The unshaded upper half z-plane is mapped into the top (unshaded) copy of the w-plane, while the shaded lower half z-plane is mapped into the bottom (shaded) copy of the w-plane The straight line is the branch cut separating the two sheets of the Riemann surface.

The map can be made single valued if we view the image of the z-plane under the transformation (4.23) not as a single complex plane, but as a two-sheeted surface (the *Riemann surface* for the map (4.24)) whose sheets are connected by a line drawn from $w = 0$ to $w = \infty$, as shown in Fig. 4.1. Passage from one sheet to the other is by circling the point $w = 0$ once, corresponding to a path from a point z_0 in the z-plane to the point $-z_0$. A second circuit around $w = 0$ leads back to the original sheet, as it corresponds to a return from $-z_0$ to z_0 in the z-plane.

The line in the w-plane from 0 to ∞ separating the two sheets is a *branch cut*. It can be drawn along any path from 0 to ∞, although it is usually convenient to draw it along either the positive real axis or the negative real axis—in Fig. 4.1 it is drawn along the positive real axis. The points $w = 0$ and $w = \infty$ are *branch points* (in w) of the function $z = \sqrt{w}$.

Remark. In general, a *branch point* z_0 of an analytic function $f(z)$ is a point z_0 near which the function is analytic but not single valued, in the sense that if a circle of arbitrarily small radius is drawn around z_0, and the values of the function followed continuously around the circle, the value of the function does not return to its original value after one complete circle. This definition will be made more formal in Section 4.3. □

More generally, if n is any integer ≥ 2, the transformation

$$w = z^n \tag{4.25}$$

is conformal except at $z = 0, \infty$ [again $w'(0) = 0$]. However, there are n distinct points in the z-plane that map into a single point in the w-plane: if $w_0 = z_0^n$, then the points $z_k \equiv z_0 \exp(2\pi i k/n)$ also map into w_0 ($k = 1, \dots, n-1$). In order to make the inverse transformation

$$z = \sqrt[n]{w} \tag{4.26}$$

unique, the image of the z-plane under the transformation (4.25) must be constructed as an n-sheeted Riemann surface, on which passage from one sheet to the next is made traversing a closed path around the point $w = 0$. Encircling the point $w = 0$ a total of n times in the same direction returns to the original sheet. The sheets are again separated by a branch cut in the w-plane from 0 to ∞; as above, the actual location of the cut can be chosen at will.

It is also possible to consider fractional powers: if p and q are integers with no common factors, then the transformation

$$w = z^{p/q} \tag{4.27}$$

is still conformal except at $z = 0, \infty$. The transformation can be viewed as a one-to-one mapping from a q-sheeted Riemann surface with a branch cut from $z = 0$ to $z = \infty$ to a p-sheeted Riemann surface with a branch cut from $w = 0$ to $w = \infty$.

(v) *Exponential* and *logarithm*

The transformation

$$w = e^z \tag{4.28}$$

is conformal except at ∞. However, since $e^{2\pi i} = 1$, the points z and $z \pm 2n\pi i$ with $n = 1, 2, \ldots$ map to the same value of w. Thus to make the inverse transformation

$$z = \ln w \tag{4.29}$$

unique, the image of the z-plane under (4.28) must be represented as a Riemann surface with an infinite number of sheets, separated by a branch cut in the w-plane from 0 to ∞. On this Riemann surface, $w = \varrho e^{i\varphi}$, with $-\infty < \varphi < \infty$, and to each point on this surface corresponds a unique

$$z = \ln \varrho + i\varphi \tag{4.30}$$

If we take the branch cut in the w-plane along the negative real axis (this is a generally useful choice), then each strip

$$(2n - 1)\pi \leq \operatorname{Im} z < (2n + 1)\pi \tag{4.31}$$

($n = 0, \pm 1, \pm 2, \ldots$) in the z-plane is mapped onto one sheet of the w-plane. Circling the origin $w = 0$ once in a counterclockwise (clockwise) sense corresponds to increasing (decreasing) the value of ϕ by 2π.

The exponential function and the trigonometric functions are closely related in view of *deMoivre's formula*

$$e^{iz} = \cos z + i \sin z \tag{4.32}$$

This formula is valid throughout the complex plane, since both sides of the equation have the same power series expansion, with an infinite radius of convergence.

Many useful identities can be derived from Eq. (4.32). For example,

$$\cos z = \frac{e^{iz} + e^{-iz}}{2} = \cosh iz \tag{4.33}$$

and

$$\sin z = \frac{e^{iz} - e^{-iz}}{2i} = i \sinh iz \tag{4.34}$$

These equations serve to define the trigonometric functions throughout the complex plane. Also,

$$e^{2iz} = \frac{1 + i \tan z}{1 - i \tan z} \tag{4.35}$$

from which it follows that

$$\tan^{-1} z = \frac{1}{2i} \ln \left(\frac{1 + iz}{1 - iz} \right) \tag{4.36}$$

Derivation of these and further identities is left to the reader.

❑ **Example 4.1.** As a final example, note that the map

$$w = \sin z \tag{4.37}$$

is conformal except at $z = \pm(2n+1)\pi/2$ ($n = 0, 1, 2, \ldots$) and $z = \infty$, but not one to one since z, $z' \equiv \pi - z$, $z \pm 2n\pi$, and $z' \pm 2n\pi$ ($n = 1, 2, \ldots$) all map to the same value of w. To make the inverse transformation

$$z = \sin^{-1} w \tag{4.38}$$

one to one, it is necessary to represent the image of the z-plane under the mapping (4.37) as a Riemann surface with an infinite number of sheets connected by branch cuts joining the branch points $w = \pm 1$. If we take the branch cut to be

$$-\infty < w \le -1 \qquad \text{and} \qquad 1 \le w < \infty$$

then each strip

$$(2n - 1) \frac{\pi}{2} \le \mathrm{Re}\, e\, z < (2n + 1) \frac{\pi}{2} \tag{4.39}$$

($n = 0, \pm 1, \pm 2, \ldots$) is mapped onto one sheet of the w-surface. Note that it is allowed (and useful in this case) to take the branch cut from -1 to 1 through the point ∞ in w. ∎

4.2   Integration in the Complex Plane

4.2.1   Integration Along a Contour

Consider the integral

$$I \equiv \int_C f(z)\,dz \equiv P + iQ \tag{4.40}$$

of the function $f(z) \equiv u(x,y) + iv(x,y)$ over the curve C joining points z_1 and z_2 in the complex z-plane. If C can be represented parametrically as

$$C: \quad x = x(t),\ y = y(t) \quad (t_1 \le t \le t_2) \tag{4.41}$$

with $x(t)$ and $y(t)$ having piecewise continuous derivatives on $(t_1 \le t \le t_2)$, then C is a *contour*. The integral (4.40) is a *contour integral*, which can be expressed in terms of real single-variable integrals as

$$P = \int_{t_1}^{t_2} \{u[x(t), y(t)]x'(t) - v[x(t), y(t)]y'(t)\}\,dt \tag{4.42}$$

$$Q = \int_{t_1}^{t_2} \{v[x(t), y(t)]x'(t) + u[x(t), y(t)]y'(t)\}\,dt \tag{4.43}$$

using $dz = dx + i\,dy$, and expressing the integrand in terms of x, y.

A natural question to ask whether the integral (4.40) is independent of the actual contour joining the endpoints z_1 and z_2 of the contour.[1]

If we write the integrals P and Q in the form

$$P = \int (u\,dx - v\,dy), \quad Q = \int (v\,dx + u\,dy) \tag{4.44}$$

then this question is equivalent to the question of whether the integrands are exact differentials. Are there functions $\phi(x,y)$ and $\psi(x,y)$ such that

$$u(x,y)dx - v(x,y)dy = d\phi(x,y) \tag{4.45}$$

$$v(x,y)dx + u(x,y)dy = d\psi(x,y) \tag{4.46}$$

If such functions exist, then the integrals

$$P = \phi(x_2, y_2) - \phi(x_1, y_1), \quad Q = \psi(x_2, y_2) - \psi(x_1, y_1) \tag{4.47}$$

are independent of contour.[2]

---

[1] This question is also encountered in the discussion of conservative forces in mechanics: is the work done by a force independent of the path joining two endpoints? If so, the force can be derived from a potential energy function.
[2] The functions ϕ and ψ here are analogous to the potential energy in mechanics.

Equations (4.45) and (4.46) imply that u and v can be expressed as partial derivatives of ϕ and ψ according to

$$u(x,y) = \frac{\partial \phi}{\partial x} = \frac{\partial \psi}{\partial y} \tag{4.48}$$

$$v(x,y) = \frac{\partial \psi}{\partial x} = -\frac{\partial \phi}{\partial y} \tag{4.49}$$

If further $u(x,y)$ and $v(x,y)$ have continuous first partial derivatives in a simply connected region $\mathcal{R}$ containing the contour C, then the mixed second partial derivatives of ϕ and ψ can be taken in either order, so that

$$\frac{\partial^2 \phi}{\partial y \partial x} = \frac{\partial u}{\partial y} = -\frac{\partial v}{\partial x} = \frac{\partial^2 \phi}{\partial x \partial y} \tag{4.50}$$

$$\frac{\partial^2 \psi}{\partial y \partial x} = \frac{\partial v}{\partial y} = \frac{\partial u}{\partial x} = \frac{\partial^2 \psi}{\partial x \partial y} \tag{4.51}$$

in $\mathcal{R}$. Equations (4.50) and (4.51) are precisely the Cauchy–Riemann conditions (4.5) for the analyticity of $f(z)$ in $\mathcal{R}$.

4.2.2   Cauchy's Theorem

The connection between independence of path of the contour integral and analyticity of $f(z)$ is stated formally as

Cauchy's Theorem. Let $f(z)$ be analytic in the simply connected region $\mathcal{R}$, and let $f'(z)$ be continuous in $\mathcal{R}$. Then if z_1 and z_2 are any two points in $\mathcal{R}$, the contour integral

$$\int_{z_1}^{z_2} f(z)\, dz$$

is independent of the actual contour in $\mathcal{R}$ joining z_1 and z_2. Then also for any *closed* contour in $\mathcal{R}$, the contour integral

$$\oint_C f(z)\, dz = 0 \tag{4.52}$$

The assumption of continuity of $f'(z)$ is actually not needed, and its elimination is central to the theory of functions of a complex variable. However, this requires a lengthy technical proof, and we refer the interested reader to the standard mathematics texts for details.

The result (4.52) has important implications for the evaluation of integrals in the complex plane. It means that we can deform the contour of an integral whose integrand is analytic in any way we wish, so long as we keep the endpoints of the integral fixed and do not cross any singularities of the integrand. Furthermore, since an integral around the total boundary of a multiply connected region can be expressed as a sum of integrals around boundaries of simply connected regions, Eq. (4.52) holds as well when C is the total boundary of a multiply connected region in which the integrand is analytic. Thus when we deform a contour of integration to cross an isolated singularity of the integrand, we can express the resulting integral as the sum of the original integral and an integral around a tiny circle enclosing the singularity of the integrand.

4.2.3   Cauchy's Integral Formula

Cauchy's theorem (4.52) states that the integral of an analytic function around a closed contour vanishes if the function is analytic everywhere inside the contour. If the function has one isolated singularity inside the contour, it turns out that we can also evaluate the integral in terms of the behavior of the function near the singular point.

First note that if K_ϱ is a circle of radius ϱ about $z = 0$, then

$$\oint_{K_\varrho} z^n \, dz = i\varrho^{n+1} \int_0^{2\pi} e^{i(n+1)\varphi} \, d\varphi = \begin{cases} 2\pi i & n = -1 \\ 0 & \text{otherwise} \end{cases} \tag{4.53}$$

where it is to be understood that the circle K_ϱ is followed a counterclockwise sense. Hence if C is *any* closed contour that encloses z_0 in a counterclockwise sense, then

$$\oint_C (z - z_0)^n \, dz = \begin{cases} 2\pi i & n = -1 \\ 0 & \text{otherwise} \end{cases} \tag{4.54}$$

These results lead to the

Cauchy integral formula. Let $f(z)$ be analytic in the simply connected region $\mathcal{R}$ enclosed by the closed contour C. Then for z in the interior of $\mathcal{R}$, we have

$$\oint_C \frac{f(\xi)}{\xi - z} \, d\xi = 2\pi i f(z) \tag{4.55}$$

Proof. Let K_ϱ be a circle of radius ϱ about z lying entirely in $\mathcal{R}$. Then

$$\oint_C \frac{f(\xi)}{\xi - z} \, d\xi = f(z) \oint_{K_\varrho} \frac{1}{\xi - z} \, d\xi + \oint_{K_\varrho} \frac{f(\xi) - f(z)}{\xi - z} \, d\xi \tag{4.56}$$

and note that the second integral on the right-hand side vanishes, since the integrand is analytic on and within K_ϱ. ∎

The integral formula (4.55) can also be differentiated under the integral sign to give

$$f'(z) = \frac{1}{2\pi i} \oint_C \frac{f(\xi)}{(\xi - z)^2} \, d\xi \tag{4.57}$$

and further to give the nth derivative,

$$f^{(n)}(z) = \frac{n!}{2\pi i} \oint_C \frac{f(\xi)}{(\xi - z)^{n+1}} \, d\xi \tag{4.58}$$

→ **Exercise 4.1.** Let $f(z)$ be analytic within the circle $K_\rho : |z - z_0| = \rho$, and suppose $|f(z)| < M$ on K_ρ. Then

$$|f^{(n)}(z_0)| \leq \frac{n! \, M}{\rho^n}$$

$(n = 0, 1, 2, \ldots)$. ☐

It is also true that integration is the inverse of differentiation. If $f(z)$ is analytic in the simply connected region $\mathcal{R}$, then the integral

$$F(z) \equiv \int_{z_0}^{z} f(\xi)\, d\xi \tag{4.59}$$

is analytic in $\mathcal{R}$, and $F'(z) = f(z)$, just as in ordinary calculus. A further consequence is that if $f(z)$ is continuous in a simply connected region $\mathcal{R}$, and if $\oint_C f(\xi)\, d\xi = 0$ for every closed contour C in $\mathcal{R}$, then $f(z)$ is analytic in $\mathcal{R}$. In this case the integral in Eq. (4.59) is independent of path in $\mathcal{R}$, and thus defines an analytic function $F(z)$ whose derivative is $f(z)$.

Remark. An important consequence of the Cauchy integral theorem is that the values of an analytic function in a region of analyticity are completely determined by the values on the boundary of the region. Indeed, if $f(\xi)$ is continuous on the closed contour C, then

$$f(z) \equiv \frac{1}{2\pi i} \oint_C \frac{f(\xi)}{\xi - z}\, d\xi \tag{4.60}$$

defines a function analytic within the region bounded by C. Moreover, if a function is analytic in a region $\mathcal{R}$, then it has derivatives of all orders in $\mathcal{R}$, and the derivatives are analytic in $\mathcal{R}$. These are all obtained by differentiating Eq. (4.60) with respect to z. □

4.3 Analytic Functions

4.3.1 Analytic Continuation

We have introduced analytic functions as functions differentiable in a region of the complex plane, and as functions that define a conformal mapping from one complex plane (or Riemann surface) to another. However, the properties of contour integrals of analytic functions, and the power series representations derived from them, can be used to extend the domain of definition of an analytic function by a process known as *analytic continuation*. This leads to a global concept of an analytic function.

To begin, note that if $u_1(z)$, $u_2(z)$, ... are analytic in a region $\mathcal{R}$, and if the series

$$f(z) \equiv \sum_{n=1}^{\infty} u_n(z) \tag{4.61}$$

is uniformly convergent in $\mathcal{R}$, then $f(z)$ is analytic in $\mathcal{R}$. To show this, simply let C be any closed contour in $\mathcal{R}$. Then

$$\oint_C f(z)\, dz = \sum_{n=1}^{\infty} \oint_C u_n(z)\, dz = 0 \tag{4.62}$$

since the uniform convergence of the series permits the interchange of summation and integration.[3]

In particular, the power series

$$f(z) \equiv \sum_{n=0}^{\infty} a_n (z - z_0)^n \tag{4.63}$$

which is uniformly convergent within its circle of convergence is analytic within its circle of convergence and, as already shown in Chapter 1, has derivative given by

$$f'(z) \equiv \sum_{n=0}^{\infty} (n+1) a_{n+1} (z - z_0)^n \tag{4.64}$$

On the other hand, suppose $f(z)$ is analytic in the simply connected region $\mathcal{R}$. Let z_0 be an interior point of $\mathcal{R}$, and suppose the circle $K_\varrho : |z - z_0| = \varrho$ and its interior lie entirely in $\mathcal{R}$. Then for z inside K_ϱ, we can expand the denominator in the Cauchy formula to give

$$f(z) \equiv \frac{1}{2\pi i} \oint_{K_\varrho} \frac{f(\xi)}{\xi - z} d\xi \ = \ \frac{1}{2\pi i} \sum_{n=0}^{\infty} \left[\oint_{K_\varrho} \frac{f(\xi)}{(\xi - z_0)^{n+1}} d\xi \right] (z - z_0)^n$$

$$= \sum_{n=0}^{\infty} \frac{f^{(n)}(z_0)}{n!} (z - z_0)^n \tag{4.65}$$

Thus any function analytic in a region can be expanded in a Taylor series about any interior point z_0 of the region. The series converges inside any circle in which the function is analytic; hence the radius of convergence of a power series about z_0 is the distance to the nearest singular point of analytic function defined by the power series.

❏ **Example 4.2.** $f(z) = 1/(1 + z^2)$ is not analytic at $z = \pm i$. Thus the Taylor series expansion of $f(z)$ about $z = 0$ has radius of convergence $\varrho = 1$. (Verify this by constructing the series.) ∎

❏ **Example 4.3.** $f(z) = \ln(1 + z)$ is not analytic at $z = -1$; it has a branch point there. Hence we know that the Taylor series expansion of $f(z)$ about $z = 0$ has radius of convergence $\varrho = 1$, even without computing explicitly the coefficients in the series. ∎

An analytic function is uniquely defined by its values in any finite region; even by its values on a bounded sequence of points within its domain of analyticity. This follows from the

---

[3] Uniform convergence means that given $\varepsilon > 0$, there is some integer N such that

$$\left| f(z) - \sum_{n=1}^{N} u_n(z) \right| < \varepsilon$$

everywhere on the contour C. Hence, if L is the total length of the contour, then

$$\left| \oint_C \left[f(z) - \sum_{n=1}^{N} u_n(z) \right] \right| < \varepsilon L$$

Theorem 4.1. Suppose $f(z)$ is analytic in the region $\mathcal{R}$, and $\{z_k\}$ is a sequence of points in $\mathcal{R}$ with limit point z_0 in $\mathcal{R}$, and suppose $f(z_k) = 0$ $(k = 1, 2, \ldots)$. Then $f(z) = 0$ everywhere.

Proof. Since $f(z)$ is analytic, the Taylor series (4.63) for $f(z)$ about z_0 has radius of convergence $r > 0$. Now suppose $a_0 = 0, a_1 = 0, \ldots, a_N = 0$ but $a_{N+1} \neq 0$. Then if $0 < \varrho < r$, the sequence $\{a_n \varrho^n\}$ is bounded, and a short calculation shows that there is some positive constant M such that

$$|f(z) - a_{N+1}(z - z_0)^{N+1}| > M|z - z_0|^{N+2} \tag{4.66}$$

for $|z - z_0| < \varrho$. Then also

$$|f(z)| > |a_{N+1} - M(z - z_0)||(z - z_0)|^{N+1} \tag{4.67}$$

(think of $f(z)$, $a_{N+1}(z-z_0)^{N+1}$ and $M(z-z_0)^{N+2}$ as three sides of a triangle in the complex plane). But the right-hand side of Eq. (4.67) is positive for $|z - z_0|$ small enough, so that $f(z) \neq 0$ in some neighborhood of z_0, contrary to assumption that $f(z_k) = 0$ $(k = 1, 2, \ldots)$. Hence every coefficient in the Taylor series must vanish, and $f(z) = 0$ everywhere.   ■

Now suppose $\mathcal{R}_1$ and $\mathcal{R}_2$ are overlapping regions, with $f_1(z)$ analytic in $\mathcal{R}_1$ and $f_2(z)$ analytic in $\mathcal{R}_2$, and suppose

$$f_1(z) = f_2(z) \tag{4.68}$$

in some neighborhood of a point z_0 in $\mathcal{R}_1 \cap \mathcal{R}_2$. Then the preceding argument shows that $f_1(z) - f_2(z) = 0$ everywhere in $\mathcal{R}_1 \cap \mathcal{R}_2$. Thus the function $f(z)$ defined by

$$f(z) \equiv f_1(z) \tag{4.69}$$

for z in $\mathcal{R}_1$, and by

$$f(z) \equiv f_2(z) \tag{4.70}$$

for z in $\mathcal{R}_2$, is analytic in $\mathcal{R}_1 \cup \mathcal{R}_2$. $f(z)$ defines the (unique) *analytic continuation* of $f_1(z)$ from $\mathcal{R}_1$ to $\mathcal{R}_1 \cup \mathcal{R}_2$.

❑ **Example 4.4.**   The function

$$f(z) \equiv \frac{1}{1 + z} \tag{4.71}$$

is the analytic continuation of the geometric series $\sum_{n=0}^{\infty} (-1)^n z^n$ from $|z| < 1$ to the entire z-plane except for $z = -1$.   ∎

Remark.   The uniqueness is based on the result that an analytic function that vanishes in an interval (even on a bounded infinite set of points) vanishes everywhere. The corresponding statement for functions of a real variable is not true, as can be seen from the example

$$f(x) = \begin{cases} \exp(-1/x^2) & x > 0 \\ 0 & x \leq 0 \end{cases} \tag{4.72}$$

which, as a function of the real variable x, has continuous derivatives of any order (even at $x = 0$), vanishes on an interval, and yet does not vanish everywhere.   □

A formal method of defining analytic continuation uses power series. Suppose

$$f(z) = \sum_{n=0}^{\infty} a_n (z - z_0)^n \tag{4.73}$$

with radius of convergence $r > 0$. If $|z_1 - z_0| < r$, then $f(z)$ is analytic at z_1, the power series

$$f(z) = \sum_{n=0}^{\infty} b_n (z - z_1)^n \tag{4.74}$$

has radius of convergence $r_1 > 0$, and the new coefficients b_n can be computed from the a_n. If the circle $|z - z_1| < r_1$ extends beyond the circle $|z - z_0| < r$, then $f(z)$ has been analytically continued beyond the circle of convergence of the series in Eq. (4.73). Proceeding in this way, we can analytically continue $f(z)$ into a maximal region, the *domain of holomorphy* of $f(z)$, whose boundary, if any, is the *natural boundary* of $f(z)$. Each power series is a *functional element* of $f(z)$, and the complete collection of functional elements defines the *analytic function* $f(z)$. The values of $f(z)$ obtained by the power series method are the same as those obtained by any other method; the analytic continuation is unique.

4.3.2   Singularities of an Analytic Function

The analytic continuation process leads to a domain of holomorphy for an analytic function that may be a region in the z-plane, the entire z-plane, or a multisheeted Riemann surface, introduced in the discussion of conformal maps, with sheets separated by branch cuts joining pairs of branch points of the function. Within this domain of holomorphy there may be isolated singular points where the function is not analytic. The task now is to classify these singular points and characterize the behavior of an analytic function near such points.

Definition 4.2.   The analytic function $f(z)$ is *singular* at the point z_0 if it is not analytic there. The point z_0 is an *isolated singularity* of $f(z)$ if $f(z)$ is analytic at every point in some neighborhood of z_0 except at z_0 itself.   ∎

The isolated singularities of an analytic function fall into three classes: (i) poles, (ii) (isolated) essential singularities[4], and (iii) branch points. As we have seen, branch points come in pairs, and are associated with nontrivial Riemann surfaces, but each branch point is isolated in the examples we have seen. However, the general series expansions of functions about poles and essential singularities are not valid near a branch point; expansions about a branch point need special attention.

Definition 4.3.   The analytic function $f(z)$ has a *pole* at the isolated singularity z_0 if there is a positive integer n such that $(z - z_0)^n f(z)$ is analytic at z_0. The *order* of the pole is the smallest n for which this is the case. If $f(z)$ is analytic and single valued in a neighborhood of the singular point z_0, but there is no positive integer for which $(z - z_0)^n f(z)$ is analytic at z_0, then z_0 is an *essential singularity* of $f(z)$.   ∎

---

[4]If a $f(z)$ has a sequence of singular points with a limit point z_0, then z_0 is also described as an essential singularity of $f(z)$, but it is not isolated. In the present context, only isolated essential singularities are considered.

Definition 4.4. The singular point z_0 is a *branch point* of the analytic function $f(z)$ if in each sufficiently small neighborhood of z_0 there is a circle $K_\varrho : |z - z_0| = \varrho$ such that

(i) $f(z)$ is analytic on and within K_ϱ, except at z_0 itself, and

(ii) the functional element of $f(z)$ obtained by analytic continuation once around K_ϱ from any point z_1 on K_ϱ is distinct from the initial functional element. ∎

Remark. Condition (ii) essentially requires that $\oint_{K_\varrho} f'(z)\, dz \neq 0$ for all ϱ less than some $\rho_0 > 0$, since $f(z)$ does not return to the same value after one trip around the circle. □

The complex plane can be extended to include the point ∞ as the image of $z = 0$ under the reciprocal transformation. Thus $f(z)$ is analytic at ∞ if $g(\xi) \equiv f(1/\xi)$ is analytic at $\xi = 0$, and $f(z)$ has a pole of order n (or an essential singularity) at ∞ if $g(\xi)$ has a pole of order n (or an essential singularity) at $\xi = 0$.

❑ **Example 4.5.** $f(z) = (z - z_0)^{-n}$ has a pole of order n at z_0, Thus $f(z) = z^n$ has a pole of order n at ∞. ∎

❑ **Example 4.6.** $f(z) = e^z$ has an essential singularity at ∞, and $f(z) = \exp[1/(z - z_0)]$ has an essential singularity at z_0. ∎

❑ **Example 4.7.** The functions $f(z) = \ln(z - z_0)$ and $g_\alpha(z) = (z - z_0)^\alpha$, with α not being an integer, have a branch point at z_0. Each of these functions has a second branch point at ∞, and the functions can be made single valued in the complex plane with an arbitrary branch cut from z_0 to ∞. ∎

If $f(z)$ has a pole of order n at z_0, then the function $\phi(z) \equiv (z - z_0)^n f(z)$ is analytic at z_0 and the power series

$$\phi(z) = \sum_{k=0}^{\infty} a_k (z - z_0)^k \tag{4.75}$$

has radius of convergence $r > 0$. Then also

$$f(z) = \frac{a_0}{(z - z_0)^n} + \frac{a_1}{(z - z_0)^{n-1}} + \cdots + \frac{a_{n-1}}{z - z_0} + \sum_{k=0}^{\infty} a_{k+n}(z - z_0)^k \tag{4.76}$$

and the series is absolutely convergent for $0 < |z - z_0| < r$. There is a similar expansion near an essential singularity, to be derived below, that includes an infinite number of negative powers of $z - z_0$.

The behavior of an analytic function near a singularity depends on the type of singularity. If $f(z)$ is (single valued), bounded and analytic in the region $0 < |z - z_0| < \varrho$, then $f(z)$ is analytic at z_0, so that if z_0 is an isolated singularity of $f(z)$, then $f(z)$ must be unbounded in any neighborhood of z_0. If z_0 is a pole of order n as in Eq. (4.76), then we can let $b_n \equiv 1/a_0 \neq 0$, and expand

$$\frac{1}{f(z)} = \sum_{k=n}^{\infty} b_k (z - z_0)^k \tag{4.77}$$

with positive radius of convergence, so that $1/f(z)$ is analytic and bounded in some neighborhood of z_0. On the other hand, if $f(z)$ has an essential singularity at z_0, then $|f(z)|$ assumes arbitrarily small values in any neighborhood of z_0, for otherwise $1/f(z)$ would be bounded in the neighborhood, and hence analytic at z_0. Since an essential singularity of $f(z)$ is also an essential singularity of $f(z) - \alpha$ for any complex α, this means that $f(z)$ assumes values arbitrarily close to any complex number α in any neighborhood of an isolated essential singularity.[5]

A branch point is a singularity that limits the circle of convergence of a power series expansion of an analytic function. However the function need not be unbounded in any neighborhood of a branch point (consider, for example, $\sqrt{z}$ near $z = 0$). Branch points of an analytic function occur in pairs. If the analytic function $f(z)$ has one branch point (at z_1, say), it necessarily has a second branch point z_2 (perhaps at ∞). To see this, let K_n be the circle $|z - z_1| = n$ ($n = 1, 2, \ldots$). If $f(z)$ has no branch point other than z_1 within K_n for all n, then then the point $z = \infty$, which is the image of $w = 0$ under the reciprocal transformation $w = 1/z$ introduced earlier, can be classified as a branch point of $f(z)$.

A *branch cut* for the analytic function $f(z)$ is a line joining a pair of branch points of $f(z)$, such that (i) there is a region $\mathcal{R}$ surrounding the cut in which $f(z)$ is analytic, except possibly on the cut itself, and (ii) if z_1 and z_2 are two points in $\mathcal{R}$, the functional element of $f(z)$ obtained by analytic continuation from z_1 to z_2 is the same for any path in $\mathcal{R}$ that does not cross the cut. While the branch points of a function are well defined, the location of the cut joining a pair of branch points is not unique; it may be chosen for convenience.

4.3.3   Global Properties of Analytic Functions

From the classification of isolated singularities, we can proceed to study the global properties of analytic functions.

Definition 4.5. $f(z)$ is an *entire function* if it is analytic for all finite z.

❑ **Example 4.8.**   Any polynomial $p(z)$ is an entire function.   ∎

❑ **Example 4.9.**   The function e^z is an entire function.   ∎

The power series expansion of an entire function about any point in the finite z-plane has infinite radius of convergence. Hence an entire function is either a polynomial of some degree n, in which case it has a pole of order n at ∞, or it is defined by an infinite power series with coefficients a_n that rapidly tend to zero for large n, in which case it has an essential singularity at ∞. Thus a nonconstant analytic function must be unbounded somewhere. This can be expressed as a formal theorem:

Theorem 4.2. If the entire function $f(z)$ is bounded for all finite z, then $f(z)$ is a constant.

Proof. If $f(z)$ is bounded for all finite z, it is bounded in a neighborhood of ∞, and hence analytic there, so its power series expansion about $z = 0$ can have only a constant term.   ∎

---

[5]The precise statement is that in any neighborhood of an essential singularity, an analytic function $f(z)$ assumes all complex values with at most one exception (*Picard's theorem*). A careful discussion is given in the book by Copson.

Not only the singular points, but the points where an analytic function vanishes are of interest. We have the formal definition:

Definition 4.6. $f(z)$ has a *zero* of *order* n at z_0 if $1/f(z)$ has a pole of order n at z_0. If $f(z)$ is a polynomial, then z_0 is a *root* (of *multiplicity* n) of the polynomial.

The properties of analytic functions lead to the most elegant proof of the algebraic fact that a polynomial of degree n has n roots, when multiple roots are counted with their multiplicity. This results follows from the

Fundamental theorem of algebra. Every nonconstant polynomial has at least one root.

Proof. If $p(z)$ is a nonconstant polynomial, then $p(z) \to \infty$ as $z \to \infty$, whence $1/p(z) \to 0$. If $1/p(z)$ is also bounded in every finite region, it must be a constant. Since we assumed that $p(z)$ is a nonconstant polynomial, $1/p(z)$ must be unbounded in the neighborhood of some finite point z_0, which is then a root of $p(z)$. ∎

It follows that if $p(z)$ is a polynomial of degree n, there exist unique points $z_1, \ldots, z_n$, not necessarily distinct, and a unique constant α such that

$$p(z) = \alpha(z - z_1) \cdots (z - z_n) \tag{4.78}$$

For if z_1 is a root of $p(z)$, then $p_1(z) \equiv p(z)/(z - z_1)$ is a polynomial of degree $n - 1$, which, if $n > 1$, has at least one root z_2. Then further $p_2(z) \equiv p(z)/(z - z_1)(z - z_2)$ is a polynomial of degree $n - 2$. Proceeding in this way, we reach a constant polynomial $p(z)/(z - z_1) \cdots (z - z_n)$ (whose value we denote by α).

Remark. Thus every polynomial of degree n has exactly n roots, counted according to multiplicity, and we can formally carry out the factorization in Eq. (4.78), even if we cannot compute the roots explicitly. □

Definition 4.7. $f(z)$ is a *rational function* if $f(z) = p(z)/q(z)$ where $p(z)$ and $q(z)$ are polynomials with no common root (hence no common factor).

The rational function $r(z) = p(z)/q(z)$ is analytic in the finite z-plane except for poles at the roots of $q(z)$. If $q(z)$ has a root of multiplicity n at z_0, then $r(z)$ has a pole of order n at z_0. If $p(z)$ has degree p, and $q(z)$ has degree q, then

(i) if $q > p$, then $r(z)$ has a zero of order $q - p$ at ∞,
(ii) if $q = p$, then $r(z)$ is analytic at ∞, with $r(\infty) \neq 0$,
(iii) if $q < p$, then $r(z)$ has a pole of order $p - q$ at ∞.

These properties are more or less evident by inspection. Less obvious, but nonetheless true, is that a function analytic everywhere including ∞, except for a finite number of poles, is a rational function. For if $f(z)$ is analytic in the finite plane except for poles at $z_1, \ldots, z_m$ of order $n_1, \ldots, n_m$, then $(z - z_1)^{n_1} \cdots (z - z_m)^{n_m} f(z)$ is an entire function. If this function has at most a pole at ∞, then it is a polynomial, so that $f(z)$ is a rational function.

Definition 4.8. A function $f(z)$ analytic in a region $\mathcal{R}$ except for a finite number of poles is *meromorphic* in $\mathcal{R}$.

The preceding discussion shows that a function meromorphic in the entire complex plane including the point at ∞ is a rational function.

4.3.4   Laurent Series

The discussion following Eq. (4.76) suggests that there might be a power series expansion of an analytic function valid even in the neighborhood of an isolated singularity z_0, if negative powers of $z - z_0$ are included. Such an expansion can be derived for a function $f(z)$ analytic in an annular region

$$\mathcal{R}: \ \varrho_1 \leq |z - z_0| \leq \varrho_2$$

Let K_1 be the circle $|z - z_0| = \varrho_1$, and let K_2 be the circle $|z - z_0| = \varrho_2$. Then for z in $\mathcal{R}$, the Cauchy integral formula (4.55) gives

$$f(z) = \frac{1}{2\pi i} \left\{ \oint_{K_2} \frac{f(\xi)}{\xi - z} \, d\xi - \oint_{K_1} \frac{f(\xi)}{\xi - z} \, d\xi \right\} \tag{4.79}$$

(both integrals are to be taken counterclockwise). Expanding the denominators in these integrals leads to the expansion

$$f(z) = \sum_{n=0}^{\infty} \left[a_n (z - z_0)^n + b_n (z - z_0)^{-n-1} \right] \tag{4.80}$$

where

$$a_n = \frac{1}{2\pi i} \oint_{K_2} \frac{f(\xi)}{(\xi - z_0)^{n+1}} \, d\xi \tag{4.81}$$

and

$$b_n = \frac{1}{2\pi i} \oint_{K_1} f(\xi)(\xi - z_0)^n \, d\xi \tag{4.82}$$

Now the series

$$\sum_{n=0}^{\infty} a_n (z - z_0)^n$$

is absolutely convergent for $|z - z_0| < \varrho_2$, uniformly within any circle $|z - z_0| = r_2 < \varrho_2$, and the circle of convergence of the series can be extended up to the singularity of $f(z)$ outside K_2 closest to z_0. Also, the series

$$\sum_{n=0}^{\infty} b_n (z - z_0)^{-n-1}$$

is absolutely convergent for $|z - z_0| > \varrho_1$, uniformly within any circle $|z - z_0| = r_1 > \varrho_1$, and the circle of convergence of the series can be extended up to the singularity of $f(z)$ inside K_1 closest to z_0. The complete series (4.80) is thus absolutely convergent for $\varrho_1 < |z - z_0| < \varrho_2$, uniformly within any ring $r_1 \leq |z - z_0| \leq r_2$ (with $\varrho_1 < r_1 < r_2 < \varrho_2$).

The series (4.80) is a *Laurent series* for $f(z)$ about z_0. It is unique for fixed ϱ_1, ϱ_2. If $f(z)$ is analytic within the circle $K : |z - z_0| = \varrho$, except possibly at z_0 itself, then the inner radius ϱ_1 in the expansion (4.80) can be shrunk to zero, and the function has the unique Laurent series expansion

$$f(z) = \sum_{n=-\infty}^{\infty} a_n(z - z_0)^n \tag{4.83}$$

about z_0, where

$$a_n = \frac{1}{2\pi i} \oint_K \frac{f(\xi)}{(\xi - z)^{n+1}} \, d\xi \tag{4.84}$$

The series (4.83) is absolutely convergent for $0 < |z - z_0| < \varrho$ and the circle of convergence can be extended to the singularity of $f(z)$ closest to z_0. Here

$$R(z) \equiv \sum_{n=0}^{\infty} a_n(z - z_0)^n \tag{4.85}$$

is the *regular part* of $f(z)$ at z_0, and

$$S(z) \equiv \sum_{n=1}^{\infty} \frac{a_{-n}}{(z - z_0)^n} \tag{4.86}$$

is the *singular part* of $f(z)$ at z_0. $S(z) = 0$ if (and only if) $f(z)$ is analytic at z_0. If $f(z)$ has a pole at z_0, then $S(z)$ is a nonzero finite sum, while if $f(z)$ has an essential singularity at z_0, then $S(z)$ is an infinite series convergent for all $|z - z_0| > 0$.

Definition 4.9. The coefficient a_{-1} in Eq. (4.83) is the *residue* of $f(z)$ at z_0.
This coefficient has a special role in evaluating integrals of the function $f(z)$, as will be seen in the next section.

❑ **Example 4.10.**   The *logarithmic derivative* of the analytic function $f(z)$ is defined by

$$\frac{d}{dz} [\ln f(z)] = \frac{f'(z)}{f(z)} \tag{4.87}$$

If $f(z)$ has a zero of order n, then near z_0 we have

$$f(z) \simeq a(z - z_0)^n \quad \text{and} \quad f'(z) \simeq na(z - z_0)^{n-1} \tag{4.88}$$

so that the logarithmic derivative has a simple pole with residue n at z_0. Also, if $f(z)$ has a pole of order n at z_0, the logarithmic derivative has a simple pole with residue $-n$. This result will be used to derive the useful formula, Eq. (4.105).   ∎

4.3.5   Infinite Product Representations

There are useful representations of analytic functions in the form of series and products that are not simply power series expansions of the forms (4.63) and (4.83). In particular, there are representations that explicitly display the zeros and poles of a function by expressing the function as a product of its zeros times entire functions chosen to make the product converge, or as a sum over the singular parts of a function at its poles, modified by suitable polynomials to insure convergence of the sum.

❏ **Example 4.11.**   The function $(\sin \pi z)/\pi$ is a function with simple zeros at each of the integers. If we are careful to combine the zeros at $z = \pm n$, we can write

$$\sin \pi z = \pi z \, h(z) \prod_{n=1}^{\infty} \left(1 - \frac{z^2}{n^2} \right) \tag{4.89}$$

The infinite product is absolutely convergent for all finite z, according to the criteria developed in Section 1.2, so the function $h(z)$ on the right-hand side must be an entire function with no zeros. It takes a further calculation to show that in fact $h(z) = 1$.   ∎

❏ **Example 4.12.**   A closely related function is $\pi \cot \pi z$, which is meromorphic in the z-plane with an essential singularity at ∞. This function has simple poles at each of the integers, with residue = 1, suggesting the expansion

$$\pi \cot \pi z = \frac{1}{z} + \sum_{n=1}^{\infty} \left(\frac{1}{z - n} + \frac{1}{z + n} \right) = \frac{1}{z} + \sum_{n=1}^{\infty} \frac{2z}{z^2 - n^2} \tag{4.90}$$

This series is absolutely convergent in the finite z-plane except on the real axis at the integers, although here, too, a further calculation is needed to show that a possible entire function that might have been added to the right-hand side is not present.   ∎

First let $H(z)$ be an entire function with no zeros in the finite plane. Then there is an entire function $h(z)$ such that

$$H(z) = e^{h(z)} \tag{4.91}$$

since if we define $h(z) \equiv \ln H(z)$ by taking the principal branch of the logarithm, there can be no singularities of $h(z)$ in the finite plane. An entire function $F(z)$ with a finite number of zeros at $z_1, \ldots, z_m$ with multiplicities $q_1, \ldots, q_m$ can evidently be written in the form

$$F(z) = H(z) \prod_{k=1}^{m} (z - z_k)^{q_k} \tag{4.92}$$

where $H(z)$ is an entire function with no zeros in the finite plane. If $F(z)$ is an entire function with an infinite number of zeros at $z_1, z_2, \ldots$ with corresponding multiplicities $q_1, q_2, \ldots$, then this representation must be modified, since the infinite product in Eq. (4.92) does not converge as written.

The *Weierstrass factorization theorem* asserts that a general representation of an entire function with zeros at $z_1, z_2, \ldots$ with corresponding multiplicities $q_1, q_2, \ldots$ is given by

$$F(z) = H(z) \prod_{k=1}^{\infty} \left[\left(1 - \frac{z}{z_k} \right) e^{\gamma_k(z)} \right]^{q_k} \tag{4.93}$$

where the functions $\gamma_k(z)$ are polynomials that can be chosen to make the product converge if it is actually an infinite product, and $H(z)$ is an entire function with no zeros in the finite plane that has the representation (4.91).

To derive Eq. (4.93), we construct polynomials $\gamma_k(z)$ that insure the convergence of the infinite product. First, note that the product

$$\prod_{k=1}^{\infty} \left(1 - \frac{z}{z_k} \right)^{q_k} \tag{4.94}$$

is absolutely convergent if and only if the series

$$\sum_{k=1}^{\infty} q_k \left| \frac{z}{z_k} \right| \tag{4.95}$$

is convergent, as explained in Section 1.2. There is no guarantee that this will be the case, even though the sequence $\{z_k\} \to \infty$ for an entire function (it is left as an exercise to show this). However, it is always possible to find a sequence of integers $s_1, s_2, \ldots$ such that the series

$$S \equiv \sum_{k=1}^{\infty} q_k \left| \frac{z}{z_k} \right|^{s_k} \tag{4.96}$$

is absolutely convergent for every finite z. This is true because for any finite z, there is an integer N such that $|z/z_k| < 1/2$ for every $k > N$, and if we choose $s_k = k + q_k$, for example, we will have

$$q_k \left| \frac{z}{z_k} \right|^{s_k} < \frac{q_k}{2^{q_k}} \left(\frac{1}{2} \right)^k < \left(\frac{1}{2} \right)^k \tag{4.97}$$

and the series S will converge by comparison with the geometric series. Once a suitable sequence $s_1, s_2, \ldots$ has been found, it is sufficient to choose the polynomial $\gamma_k(z)$ to be

$$\gamma_k(z) = \sum_{m=1}^{s_k - 1} \frac{1}{m} \left(\frac{z}{z_k} \right)^m \tag{4.98}$$

The reader is invited to construct a careful proof of this statement.

There is also a standard representation for a function that is meromorphic in the finite plane. Suppose $f(z)$ has poles of order $p_1, \ldots, p_m$ at the points $z_1, \ldots, z_m$, and suppose the singular part (4.86) of $f(z)$ at z_k is given by

$$S_k(z) = \sum_{n-1}^{p_k} \frac{a_{k,n}}{(z - z_k)^n} \tag{4.99}$$

Then $f(z)$ can be expressed as

$$f(z) = \sum_{k=1}^{m} S_k(z) + Q(z) \tag{4.100}$$

where $Q(z)$ is an entire function. If $f(z)$ has a pole at ∞, then $Q(z)$ is a polynomial, and the expansion corresponds to the partial fraction expansion of the polynomial in the denominator of $f(z)$ when it is expressed as a ratio of two polynomials.

If $f(z)$ has an infinite number of poles, at a sequence $z_1, z_2, \ldots$ of points that converges to ∞, then the question of convergence of the sum over the singular parts of the function at the poles must be addressed. The result is the *Mittag–Leffler theorem*, which states that every function $f(z)$ meromorphic in the finite plane, with poles at $z_1, z_2, \ldots$ of order $p_1, p_2, \ldots$, can be expanded in the form

$$f(z) = \sum_{k=1}^{\infty} [S_k(z) - \sigma_k(z)] + E(z) \tag{4.101}$$

where $E(z)$ is an entire function, and the functions $\sigma_k(z)$ are polynomials that can be chosen to make the sum converge if it is actually an infinite series. The proof of this theorem is similar to the proof of the Weierstrass factorization theorem, in that polynomials can be constructed to approximate the singular parts $S_k(z)$ as closely as required when $|z/z_k| \ll 1$.

Important examples of expansions of the types (4.93) and (4.101) appear in the discussion of the Γ-function in Appendix A.

4.4   Calculus of Residues: Applications

4.4.1   Cauchy Residue Theorem

Suppose $\mathcal{R}$ is a simply connected region bounded by the closed contour C, and that $f(z)$ is an analytic function. From Cauchy's theorem (Eq. (4.52)), we know that if $f(z)$ has no singular points in $\mathcal{R}$, then

$$\oint_C f(z)\,dz = 0$$

If $f(z)$ is analytic in $\mathcal{R}$ except for an isolated singularity at z_0, then we can insert the Laurent series expansion (4.83) into the contour integral to obtain

$$\oint_C f(z)\,dz = 2\pi i a_{-1} \tag{4.102}$$

where a_{-1} is the coefficient of $(z - z_0)^{-1}$ in the series (4.83). As noted earlier, this coefficient is the *residue* of $f(z)$ at z_0,

$$a_{-1} \equiv \operatorname{Res}_{z=z_0} f(z) \tag{4.103}$$

If $f(z)$ has several singularities within $\mathcal{R}$, then each singularity contributes to the contour integral, and we have the

Cauchy residue theorem. Suppose $f(z)$ is analytic in the simply connected region $\mathcal{R}$ bounded by the closed curve C, except for isolated singularities at $z_1, \ldots, z_N$. Then

$$\oint_C f(z)\, dz = 2\pi i \sum_{k=1}^{N} \operatorname{Res}_{z=z_k} f(z) \tag{4.104}$$

This theorem leads to both useful formal results and to practical methods for evaluating definite integrals.

The logarithmic derivative $f'(z)/f(z)$ of the analytic function $f(z)$ is defined by Eq. (4.87). Recall that if $f(z)$ has a zero of order n at z_0, the logarithmic derivative has a simple pole with residue n, while if $f(z)$ has a pole of order n at z_0, the logarithmic derivative has a simple pole with residue $-n$. It follows that if $f(z)$ is analytic in the region $\mathcal{R}$ bounded by the closed curve C, except for poles of order $p_1, \ldots, p_m$ at $z_1, \ldots, z_m$ respectively, and if $1/f(z)$ is analytic in $\mathcal{R}$ except for poles of order $q_1, \ldots, q_n$ at $\xi_1, \ldots, \xi_n$ respectively, corresponding to zeros of $f(z)$, then

$$\frac{1}{2\pi i} \oint_C \frac{f'(z)}{f(z)}\, dz = \sum_{j=1}^{n} q_j - \sum_{k=1}^{m} p_k \tag{4.105}$$

Thus the integral around a closed contour of the logarithmic derivative of an analytic function counts the number of zeros minus the number of poles of the function within the contour. One application of this in quantum mechanics is *Levinson's theorem* for scattering phase shifts.

4.4.2   Evaluation of Real Integrals

Consider now the integral

$$I_1 \equiv \int_{-\infty}^{\infty} f(x)\, dx \tag{4.106}$$

which we suppose is convergent when the integration is taken along the real axis. If $f(x)$ can be analytically continued into the upper half of the complex z-plane, and if $f(z)$ has a finite number of isolated singularities in the upper half-plane, then we can close the contour by a large semicircle in the upper half-plane, as shown in Fig. 4.2. If $zf(z) \to 0$ as $z \to \infty$ in the upper half-plane, then large semicircle does not contribute to the integral and the residue theorem gives

$$I_1 = 2\pi i \sum_{k} \operatorname{Res}_{z=z_k} f(z) \tag{4.107}$$

where the summation is over the singularities $z_1, z_2, \ldots$ of $f(z)$ in the upper half-plane. This method also works if the function can be analytically continued into the lower half-plane, with $zf(z) \to 0$ as $z \to \infty$ in the lower half-plane, so long as a minus sign is included to take into

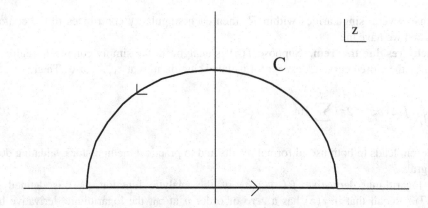

Figure 4.2: Contour for the evaluation of the integral I_1.

account the fact that the contour will now run clockwise around the singularities in the lower half-plane.

❑ **Example 4.13.** With $\alpha > 0$, $a > 0$, the integral

$$F(\alpha) \equiv \int_{-\infty}^{\infty} \frac{e^{i\alpha x}}{x^2 + a^2}\, dx \tag{4.108}$$

can be evaluated by closing the contour in the upper half-plane (Fig. 4.2), since $e^{i\alpha z} \to 0$ for $z \to \infty$ in the upper half-plane when $\alpha > 0$; hence the integrand satisfies the requirement that $zf(z) \to 0$ as $z \to \infty$ in the upper half-plane. The integrand has poles at $z = \pm ia$, but only the pole at $z = +ia$ is enclosed by the contour, so we end up with the result

$$F(\alpha) = \frac{\pi}{a} e^{-\alpha a} \tag{4.109}$$

This is one of the many examples of Fourier integrals (Chapter 6) that can be evaluated using contour integration methods.   ∎

Consider next the integral

$$I_2 \equiv \int_0^{2\pi} f(\cos\theta, \sin\theta)\, d\theta \tag{4.110}$$

With $z = e^{i\theta}$, the integral becomes

$$I_2 = \oint_K f\left[\frac{1}{2}\left(z + \frac{1}{z}\right), \frac{1}{2i}\left(z - \frac{1}{z}\right)\right] \frac{dz}{iz} \tag{4.111}$$

where K is the unit circle in the z-plane. If the integrand is analytic (in z) on K, with a finite number of isolated singularities within K, then the integral can be evaluated directly by the residue theorem.

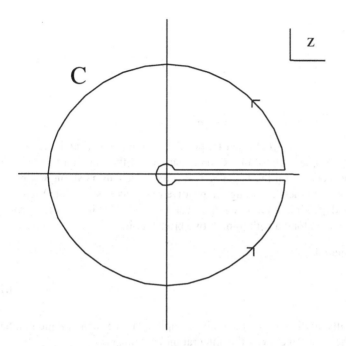

Figure 4.3: The standard keyhole contour.

❑ **Example 4.14.**   With $|b| < |a|$, the integral

$$F(a,b) \equiv \int_0^{2\pi} \frac{1}{a - b\cos\theta}\, d\theta = 2i \oint_K \frac{1}{bz^2 - 2az + b}\, dz = \frac{2\pi}{\sqrt{a^2 - b^2}} \qquad (4.112)$$

since one root of the denominator lies inside the unit circle, and one root lies outside (show this), and the integral picks up the residue at the pole inside.   ∎

A third type of integral has the generic form

$$I_3 \equiv \int_0^\infty x^\alpha f(x)\, dx \qquad (4.113)$$

If α is not an integer, then z^α has branch points at 0 and ∞, which we can connect by a branch cut along the positive real axis. Now suppose $f(x)$ can be analytically continued into the entire complex z-plane with a finite number of isolated singularities, and that $z^{\alpha+1}f(z) \to 0$ as $z \to \infty$, so that we can neglect the contribution from the large circle at ∞. Then the integral I_3 is related to an integral around the *standard keyhole contour* C shown in Fig. 4.3 by

$$(1 - e^{2\pi i\alpha})\, I_3 = \oint_C z^\alpha f(z)\, dz \qquad (4.114)$$

The contour integral can be evaluated by the residue theorem.

❑ **Example 4.15.** With $0 < |a| < 1$, we have the integral

$$I_a(z) \equiv \int_0^\infty \frac{1}{x^a(x-z)}\,dx = \left(\frac{1}{1-e^{-2\pi i a}}\right)\oint_C \frac{1}{\xi^a(\xi-z)}\,d\xi$$

(4.115)

$$= \left(\frac{1}{1-e^{-2\pi i a}}\right)\frac{2\pi i}{z^a} = \frac{\pi}{\sin \pi a}(-z)^{-a}$$

Here we can use the fact that $I_a(z)$ must be positive when z is real and negative to be certain we are on the proper sheet of the Riemann surface of the function $(-z)^{-a}$. Starting $(-z)^{-a}$ real and positive on the negative real z-axis, we can move anywhere in the z-plane with an implicit branch cut along the positive real z-axis, since $I_a(z)$ will have a different value there depending on how we approach the axis. This integral is a prototype for integrating a function along a path joining two branch points.   ∎

A closely related integral is

$$I_4 \equiv \int_0^\infty f(x)\,dx$$

(4.116)

If $f(x)$ can be analytically continued into the entire complex z-plane, with a finite number of isolated singularities, then we can convert this integral into an integral

$$I_4 = \frac{i}{2\pi}\oint_C (\ln z)f(z)\,dz$$

(4.117)

around the standard keyhole contour. For if we choose the branch cut of $\ln z$ along the positive real axis from 0 to ∞, then

$$\ln xe^{2\pi i} = \ln x + 2\pi i$$

(4.118)

and if $z(\ln z)f(z) \to 0$ as $z \to 0$ and as $z \to \infty$, we can again neglect the contribution from the large circle at ∞.

❑ **Example 4.16.** The integral

$$\int_0^\infty \frac{1}{1+x^3}\,dx = \frac{i}{2\pi}\oint_C \frac{\ln z}{1+z^3}\,dz$$

(4.119)

$$= \frac{i\pi}{3}\left\{\frac{5}{(\omega-\omega^*)(\omega^*+1)} - \frac{3}{(\omega+1)(\omega^*+1)} - \frac{1}{(\omega-\omega^*)(\omega+1)}\right\} = \frac{\pi}{3\sqrt{3}}$$

where

$$\omega = \frac{1}{2}(1+i\sqrt{3}) = e^{\frac{i\pi}{3}}\quad \text{and}\quad \omega^* = \frac{1}{2}(1-i\sqrt{3}) = e^{-\frac{i\pi}{3}}$$

(4.120)

are the complex cube roots of -1. The contour integral has picked up the contributions from all three of the cube roots.   ∎

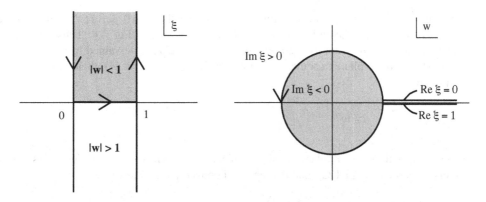

Figure 4.4: Mapping between a strip of the ξ-plane and the w-plane.

4.5 Periodic Functions; Fourier Series

4.5.1 Periodic Functions

Periodic functions are clearly very important in physics; here we look at some properties of periodic functions in the complex plane. We begin with the

Definition 4.10. A function $f(z)$ is *periodic* (with *period* a) if

$$f(z + a) = f(z) \tag{4.121}$$

for all z. If $f(z)$ has period a, then it also has periods $2a, 3a, \ldots$, so that we define the *fundamental period* (or *primitive period*) of $f(z)$ to be a period α such that α/n is not a period of $f(z)$ for any $n = 2, 3, \ldots$.[6]

Functions with fundamental period α are the exponential functions $\exp(\pm 2\pi i z/\alpha)$ or the corresponding trigonometric functions $\sin(2\pi z/\alpha)$ and $\cos(2\pi z/\alpha)$. However, any function $f(z)$ represented by a convergent series of the form

$$f(z) = \sum_{n=-\infty}^{\infty} c_n \, e^{2n\pi i z/\alpha} \tag{4.122}$$

with $c_1 \neq 0$ or $c_{-1} \neq 0$ will also have fundamental period α.

→ **Exercise 4.2.** What relations exist between the coefficients c_n in the expansion (4.122) if the function $f(z)$ is real for real z? (Assume α is also real.) ◻

The interesting question here is whether the converse holds true: Can any function $f(z)$ that has period α be represented by a series of the form (4.122)? To answer this question, introduce first the new variable $\xi \equiv z/\alpha$ so that the period of the function in the ξ-plane is 1. Then define the variable

$$w \equiv e^{2\pi i \xi} \tag{4.123}$$

---

[6]This definition allows $\pm\alpha$ to be fundamental periods. In general, we will choose a fundamental period α with $\operatorname{Im}\alpha > 0$, or if $\operatorname{Im}\alpha = 0$, then $\operatorname{Re}\alpha > 0$.

The function $f(z)$ is completely specified by its values in the strip $0 \le \mathrm{Re}\,\xi \le 1$ in the ξ-plane, which is mapped into the entire w-plane by (4.123) as shown in Fig. 4.4. Thus periodicity of $f(z)$ in z or ξ is equivalent to a requirement that $f(z)$ be a single-valued function of w. If f is analytic in the w-plane in some annular region $\mathcal{R} : w_1 < |w| < w_2$, then we can expand f in a Laurent series

$$f = \sum_{n=-\infty}^{\infty} c_n\, w^n \tag{4.124}$$

convergent within the annular region $\mathcal{R}$. When we return to the variable z, this series is precisely of the form (4.122), and the region of convergence is a strip of the form

$$\mathcal{S} : -\ln w_2 < 2\pi\, \mathrm{Im}\left(\frac{z}{\alpha}\right) < -\ln w_1$$

The coefficients c_n in the Laurent expansion (4.124) can be expressed as contour integrals

$$c_n = \frac{1}{2\pi i} \oint_C \frac{f(w)}{w^{n+1}}\, dw \tag{4.125}$$

where C is a circle of radius w_0 inside $\mathcal{R}$. This circle corresponds to the line

$$L : \{\xi = x + i\eta \mid 0 \le x \le 1, w_0 = e^{-2\pi\eta}\}$$

in the ξ-plane, and the integral formula (4.125) can also be written as

$$c_n = \frac{1}{2\pi w_0^n} \int_0^1 f(x)\, e^{-2\pi i n x}\, dx \tag{4.126}$$

When $f(z)$ is an analytic function of z in a strip of the form $\mathcal{S}$, then the expansion (4.122) is convergent inside the strip, uniformly in any closed region lying entirely within the strip. The expansion (4.122) is known as a (complex) *Fourier series*. The series has the alternative expression in terms of trigonometric functions as

$$f(z) = \sum_{n=0}^{\infty} a_n \cos\left(\frac{2n\pi z}{\alpha}\right) + \sum_{n=1}^{\infty} b_n \sin\left(\frac{2n\pi z}{\alpha}\right) \tag{4.127}$$

in which the coefficients a_n and b_n are real if $f(z)$ is real for real z. The coefficients c_n, or a_n and b_n, are the *Fourier coefficients* of $f(z)$.

→ **Exercise 4.3.**   Express the function $\tan \pi\xi$ as a single-valued function of the variable $w = e^{2\pi i\xi}$. Then find Fourier series expansions of $\tan \pi\xi$ and $\cot \pi\xi$ valid for $\mathrm{Im}\,\xi > 0$.   ☐

Periodic functions on the real z-axis that cannot be analytically continued into the complex plane also have Fourier series expansions of the forms (4.122) and (4.127), with coefficients given by Eq. (4.126) with $w_0 = 1$. However, the convergence properties are not the same, since the series cannot converge away from the real axis (the strip of convergence has zero width). These expansions will be examined further in Chapter 6 where Fourier series are revisited in the context of linear vector space theory.

A function $f(z)$ for which

$$f(z + a) = \mu f(z) \tag{4.128}$$

for all z (μ is a constant) is *quasi-periodic*, or *almost periodic*, with period a. A quasi-periodic function can be transformed into a periodic function with the same period: if $f(z)$ satisfies (4.128) with $\mu = e^{ika}$, then the function

$$\phi(z) \equiv e^{-ikz} f(z) \tag{4.129}$$

is strictly periodic with period a.

Remark. Almost periodic functions appear as solutions of linear differential equations with periodic coefficients that arise in the analysis of wave motion in periodic structures such as crystal lattices. These equations have solutions of the type (4.128). That they can be expressed in the form of Eq. (4.129) is a result known in the physics literature as *Bloch's theorem*. □

4.5.2 Doubly Periodic Functions

If α is a fundamental period of $f(z)$, then the periods $n\alpha$ ($n = \pm 1, \pm 2, \ldots$) lie on a line in the z-plane. If $f(z)$ has no periods other than these, the $f(z)$ is *simply periodic*. However, there are functions with additional periods, and a second fundamental period β such that every period Ω of $f(z)$ has the form

$$\Omega = m\alpha + n\beta \tag{4.130}$$

with m, n integers. Such functions are *doubly periodic*, and have some interesting properties that we will now explore.[7]

Suppose $f(z)$ is a doubly periodic function with fundamental periods α and β. We can choose α to be the period of $f(z)$ with the smallest magnitude, and again introduce the variable $\xi \equiv z/\alpha$, so that the fundamental periods of f in the ξ-plane are 1 and $\tau \equiv \beta/\alpha$. Then the values of f in the ξ-plane are determined by its values in the *fundamental parallelogram* (or *primitive cell*) shown by the shaded area in Fig. 4.5, and denoted by $\mathcal{P}$. If f is meromorphic in $\mathcal{P}$, then f is an *elliptic function*, which we will assume to be the case from here on.

Remark. The primitive cell $\mathcal{P}$, with opposite sides identified, is equivalent to the torus introduced in Chapter 3—see Fig. 3.4. Thus doubly periodic functions are analytic functions on a torus with a *complex structure*. □

---

[7] A natural question to ask is whether a function can have further periods that cannot be expressed in terms of two fundamental periods in the form (4.130). It is not possible for a nonconstant analytic function, since the existence of a third fundamental period γ would mean that any complex number of the form $m\alpha + n\beta + p\gamma$ with m, n, p integers would also be a period. But we can find numbers of this form arbitrarily close to any number in the complex plane (show this), so the analytic function must be constant.

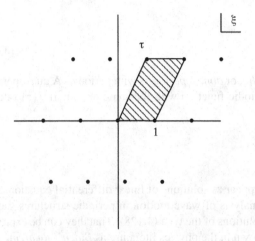

Figure 4.5: Fundamental parallelogram $\mathcal{P}$ (shaded) for a doubly periodic function with periods 1 and τ. The dots in the figure indicate the different periods of the function in the ξ-plane of the form $m + n\tau$.

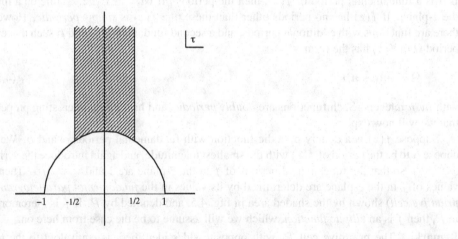

Figure 4.6: Region in the τ-plane needed to specify the periods of a doubly periodic function.

The second fundamental period of $f(z)$ can be chosen so that the ratio τ of fundamental periods lies in the shaded region of the τ plane shown in Fig. 4.6.[8] The origin can be chosen so that no singularities of f lie on the boundary of the primitive cell $\mathcal{P}$. The function f must then have poles within $\mathcal{P}$, else it would be a bounded entire function, and thus a constant. The *order* of an elliptic function $f(z)$ is the sum of the orders of the poles of $f(z)$ in $\mathcal{P}$.

---

[8] If τ is a fundamental period, so are $-\tau$ and $\tau \pm 1$. Hence the second fundamental period can be shifted to lie in the strip $-\frac{1}{2} < \mathrm{Re}\,\tau \le \frac{1}{2}$ with $\mathrm{Im}\,\tau > 0$. Also, $|\tau| \ge 1$ since we have rescaled z by the fundamental period of smallest magnitude.

If K is the boundary of $\mathcal{P}$, then the periodicity of $f(z)$ implies

$$\oint_K f(\xi)\, d\xi = 0 \tag{4.131}$$

since the contributions to the integral from opposite sides of $\mathcal{P}$ cancel. Thus the sum of the residues of f at the poles in $\mathcal{P}$ must be zero. Hence f cannot have just one simple pole inside $\mathcal{P}$; the minimum order of an elliptic function is 2, corresponding to a function that has either a double pole with zero residue, or a pair of simple poles with equal and opposite residues.

The order n of an elliptic function f is also equal to the number of zeros of f inside $\mathcal{P}$ (counted with appropriate multiplicity), since double periodicity also implies

$$\frac{1}{2\pi i} \oint_K \frac{f'(\xi)}{f(\xi)}\, d\xi = 0 \tag{4.132}$$

as in Eq. (4.131). We know from Eq. (4.105) that this integral is equal to the number of zeros minus the number of poles of f inside $\mathcal{P}$. Then f assumes *any* complex value c exactly n times inside $\mathcal{P}^9$, since if f is an elliptic function of order n, so is $f - c$.

Some elliptic functions that occur in physics are introduced in Appendix A of Chapter 5.

A   Gamma Function; Beta Function

A.1   Gamma Function

An important special function that has already made a brief appearance in Chapter 1 is the Γ-function, defined for Re $z > 0$ by the integral

$$\Gamma(z) \equiv \int_0^\infty t^{z-1} e^{-t}\, dt \tag{4.A1}$$

When z is a positive integer $n+1$, the integral can be evaluated using integration by parts, for example, to give

$$\Gamma(n+1) = \int_0^\infty t^n e^{-t}\, dt = n! \tag{4.A2}$$

($n = 0, 1, 2, \ldots$). Also, when z is a half-integer, we can let $t = u^2$ to obtain

$$\Gamma(n + \tfrac{1}{2}) = \int_0^\infty t^{n-\frac{1}{2}} e^{-t}\, dt = 2\int_0^\infty u^{2n} e^{-u^2}\, du = \frac{(2n)\,!}{2^{2n}\, n\,!} \sqrt{\pi} \tag{4.A3}$$

($n = 0, 1, 2, \ldots$).

The Γ-function defined by Eq. (4.A1) is the analytic continuation of the factorial function defined for integers into the half-plane Re $z > 0$. Note that it is analytic, since the derivative

$$\Gamma'(z) = \int_0^\infty (\ln t) t^{z-1} e^{-t}\, dt \tag{4.A4}$$

---

[9] The reader is invited to provide a modification of this statement for a value c that occurs on the boundary of $\mathcal{P}$.

is well defined for Re $z > 0$. Integration by parts in Eq. (4.A1) gives the recursion formula

$$z\Gamma(z) = \Gamma(z+1) \tag{4.A5}$$

and then

$$z(z+1)\cdots(z+n)\,\Gamma(z) = \Gamma(z+n+1) \tag{4.A6}$$

$(n = 1, 2, \ldots)$ for Re $z > 0$. But the right-hand side is analytic for Re $z > -n - 1$, so this equation provides the analytic continuation of $\Gamma(z)$ to the half-plane Re $z > -n - 1$. It follows that $\Gamma(z)$ is analytic everywhere in the finite z-plane except for simple poles at $z = 0, -1, -2, \ldots$, with residue at the pole at $z = -n$ given by $(-1)^n/n!$. These analyticity properties also follow from the expansion

$$
\begin{aligned}
\Gamma(z) &= \int_0^1 t^{z-1} e^{-t}\,dt + \int_1^\infty t^{z-1} e^{-t}\,dt \\
&= \sum_{n=0}^\infty \frac{(-1)^n}{n!} \frac{1}{z+n} + \int_1^\infty t^{z-1} e^{-t}\,dt \tag{4.A7}
\end{aligned}
$$

where we note that the sum is absolutely convergent everywhere except at the poles, and the integral is an entire function with an essential singularity at ∞.

Consider now the product $\Gamma(z)\Gamma(1-z)$, which has poles at each integer, with the residue of the pole at $z = n$ given simply by $(-1)^n$. It is thus plausible that

$$\Gamma(z)\Gamma(1-z) = \frac{\pi}{\sin \pi z}. \tag{4.A8}$$

To show that this is actually the case, suppose $0 < \mathrm{Re}\, z < 1$. Then

$$
\begin{aligned}
\Gamma(z)\Gamma(1-z) &= \int_0^\infty t^{z-1} e^{-t}\,dt \int_0^\infty u^{-z} e^{-u}\,du \\
&= 4 \int_0^\infty x^{2z-1} e^{-x^2}\,dx \int_0^\infty y^{1-2z} e^{-y^2}\,dy \\
&= 4 \int_0^\infty r\, e^{-r^2}\,dr \int_0^{\frac{\pi}{2}} (\cot \theta)^{2z-1}\,d\theta \\
&= 2 \int_0^\infty \frac{s^{2z-1}}{1+s^2}\,ds = \frac{\pi}{\sin \pi z} \tag{4.A9}
\end{aligned}
$$

where the last integral is evaluated using the standard keyhole contour introduced in the preceding section. The result (4.A8), derived for $0 < \mathrm{Re}\, z < 1$, is then true everywhere by analytic continuation.

An infinite product representation of the Γ-function is derived from the expression

$$
\begin{aligned}
\Gamma(z) &= \lim_{n\to\infty} \int_0^n \left(1 - \frac{t}{n}\right)^n t^{z-1}\,dt = \lim_{n\to\infty} \left\{ n^z \int_0^1 (1-u)^n u^{z-1}\,du \right\} \\
&= \lim_{n\to\infty} \left\{ n^z \frac{n!}{z(z+1)\cdots(z+n)} \right\} \tag{4.A10}
\end{aligned}
$$

valid for Re $z > 0$. It follows that

$$\frac{1}{\Gamma(z)} = z \lim_{n\to\infty} \left\{ n^{-z} \prod_{m=1}^{n} \left(1 + \frac{z}{m}\right) \right\}$$

$$= z \lim_{n\to\infty} \left\{ \exp\left[\left(\sum_{m=1}^{n} \frac{1}{m} - \ln n\right) z\right] \prod_{m=1}^{n} \left[\left(1 + \frac{z}{m}\right) e^{-\frac{z}{m}}\right] \right\}$$

$$= z\, e^{\gamma z} \prod_{m=1}^{\infty} \left[\left(1 + \frac{z}{m}\right) e^{-\frac{z}{m}}\right] \tag{4.A11}$$

where γ is the Euler–Mascheroni constant defined by

$$\gamma \equiv \lim_{n\to\infty} \left(\sum_{m=1}^{n} \frac{1}{m} - \ln n\right) \tag{4.A12}$$

(see Eq. (1.72) and Problem 1.12 for further details). The infinite product is convergent for all finite z due to the inclusion of the factor $\exp(-z/m)$. This is an example of the factor $\exp[\gamma_m(z)]$ introduced in Eq. (4.93).

→ **Exercise 4.A1.**   Using the infinite product representation (4.A11) or otherwise, derive the *duplication formula*

$$\Gamma(z)\Gamma(z + \tfrac{1}{2}) = \frac{\sqrt{\pi}\,\Gamma(2z)}{2^{2z-1}} \tag{4.A13}$$

This result and Eq. (4.A8) are quite useful.   □

The infinite product representation (4.A11) can be used to derive an integral representation for the logarithmic derivative of the Γ-function, which appears in some physics contexts. Taking the logarithm of both sides of Eq. (4.A11) gives

$$\ln \Gamma(z) = -\gamma z - \ln z + \sum_{m=1}^{\infty} \left[\frac{z}{m} - \ln\left(1 + \frac{z}{m}\right)\right] \tag{4.A14}$$

and then, by differentiation,

$$\frac{\Gamma'(z)}{\Gamma(z)} = \frac{d}{dz}\left[\ln\Gamma(z)\right] = -\gamma - \frac{1}{z} + z \sum_{m=1}^{\infty} \frac{1}{m(z+m)} \tag{4.A15}$$

Now recall the result

$$\gamma = \int_0^1 \left(\frac{1 - e^{-t}}{t}\right) dt - \int_1^\infty \frac{e^{-t}}{t}\, dt \tag{4.A16}$$

from Problem (1.12), and note that

$$\frac{1}{z + m} = \int_0^\infty e^{-(z+m)t}\, dt \tag{4.A17}$$

We then get the integral representation

$$\frac{d}{dz}\left[\ln\Gamma(z)\right] = \int_0^\infty \left(\frac{e^{-t}}{t} - \frac{e^{-zt}}{1 - e^{-t}}\right) dt \tag{4.A18}$$

valid for Re $z > 0$. Since

$$\int_0^\infty \frac{e^{-t} - e^{-zt}}{t}\, dt = \ln z \tag{4.A19}$$

we have the alternative representation (still for Re $z > 0$)

$$\frac{d}{dz}\left[\ln\Gamma(z)\right] = \ln z + \int_0^\infty \left(\frac{1}{t} - \frac{1}{1 - e^{-t}}\right) e^{-zt}\, dt \tag{4.A20}$$

The *Bernoulli numbers* B_{2n} are defined by the expansions

$$\frac{t}{1 - e^{-t}} = 1 + \tfrac{1}{2}t + \sum_{n=1}^{\infty} \frac{B_{2n}}{(2n)!}\, t^{2n} \tag{4.A21}$$

$$\frac{t}{e^t - 1} = 1 - \tfrac{1}{2}t + \sum_{n=1}^{\infty} \frac{B_{2n}}{(2n)!}\, t^{2n} \tag{4.A22}$$

The term in parentheses in the integrand in Eq. (4.A20) can be expanded to obtain the asymptotic expansion

$$\frac{d}{dz}\left[\ln\Gamma(z)\right] \sim \ln z - \frac{1}{2z} - \sum_{n=1}^{\infty} \frac{B_{2n}}{2n}\frac{1}{z^{2n}} \tag{4.A23}$$

for $z \to \infty$ in any sector $|\arg z| \le \tfrac{1}{2}\phi - \delta$ $(0 < \delta < \tfrac{\pi}{2})$. This expansion can be integrated term-by-term to give

$$\ln\Gamma(z) \sim (z - \tfrac{1}{2})\ln z - z + C + \sum_{n=0}^{\infty} \frac{B_{2n+2}}{(2n+1)(2n+2)}\frac{1}{z^{2n+1}} \tag{4.A24}$$

where the constant C must be determined separately. The calculation in Section 1.4 gives (see Eq. (1.111))

$$C = \tfrac{1}{2}\ln(2\pi) \tag{4.A25}$$

The corresponding asymptotic expansion of the Γ-function itself is then given by

$$\Gamma(z) \sim \sqrt{\frac{2\pi}{z}}\left(\frac{z}{e}\right)^z \exp\left\{\sum_{n=0}^{\infty} \frac{B_{2n+2}}{(2n+1)(2n+2)}\right\}$$

$$= \sqrt{\frac{2\pi}{z}}\left(\frac{z}{e}\right)^z \left\{1 + \frac{1}{12z} + \frac{1}{288z^2} + O\left(\frac{1}{z^3}\right)\right\} \tag{4.A26}$$

(*Stirling's series*). The expansion of $n!$ for large n obtained from this using Eq. (4.A2) is important in statistical physics.

A.2   Beta Function

Now consider the product

$$
\begin{aligned}
\Gamma(\alpha)\Gamma(\beta) &= \int_0^\infty x^{\alpha-1} e^{-x}\, dx \int_0^\infty y^{\beta-1} e^{-y}\, dy \\
&= \int_0^\infty \int_0^u x^{\alpha-1}(u-x)^{\beta-1} e^{-u}\, dx\, du \\
&= \int_0^\infty u^{\alpha+\beta-1} e^{-u}\, du \int_0^1 t^{\alpha-1}(1-t)^{\beta-1}\, dt \\
&= \Gamma(\alpha+\beta) \int_0^1 t^{\alpha-1}(1-t)^{\beta-1}\, dt
\end{aligned}
\tag{4.A27}
$$

where $u = x + y$ and $x = ut$.

The *beta function* $B(\alpha, \beta)$ is defined by

$$
B(\alpha,\beta) \equiv \int_0^1 t^{\alpha-1}(1-t)^{\beta-1}\, dt = \frac{\Gamma(\alpha)\Gamma(\beta)}{\Gamma(\alpha+\beta)}
\tag{4.A28}
$$

where the integral is defined for Re $\alpha > 0$, Re $\beta > 0$. The ratio of the Γ-functions then provides the analytic continuation of $B(\alpha, \beta)$ to all α, β. A second integral representation of the *beta function* is

$$
B(\alpha,\beta) = \int_0^\infty (1 - e^{-w})^{\beta-1} e^{-\alpha w}\, dw
\tag{4.A29}
$$

which follows from the substitution $t = e^{-w}$ in Eq. (4.A28).

For fixed β, $B(\alpha, \beta)$ is an analytic function of α with simple poles at $\alpha = 0, -1, -2, \ldots$ corresponding to the poles of $\Gamma(\alpha)$. The residue at the pole at $\alpha = -n$ is given by

$$
\begin{aligned}
\mathrm{Res}_{\alpha=-n} B(\alpha,\beta) &= \frac{(-1)^n}{n!} \frac{\Gamma(\beta)}{\Gamma(-n+\beta)} \\
&= \frac{(n-\beta)(n-\beta-1)\cdots(1-\beta)}{n!}
\end{aligned}
\tag{4.A30}
$$

($n = 0, 1, 2, \ldots$). Note that if $\beta = m$ is a positive integer, the apparent poles at $\alpha = -m, -m-1, \ldots$ are not actually present.

The B-function can be expanded as a sum over its poles according to

$$
\frac{\Gamma(\alpha)\Gamma(\beta)}{\Gamma(\alpha+\beta)} = \sum_{n=0}^\infty \frac{(n-\beta)(n-\beta-1)\cdots(1-\beta)}{n!} \frac{1}{\alpha+n}
\tag{4.A31}
$$

If Re $\beta > 0$, the series is absolutely convergent for all α except at the poles.

→ **Exercise 4.A2.**   Prove the convergence of the series (4.A31).   □

Bibliography and Notes

An introduction to the classical theory of functions of a complex variable is given in the book of Whittaker and Watson cited in Chapter 1. Another brief introduction is contained in the two paperback volumes

> Konrad Knopp, *Theory of Functions* (Parts I and II), Dover (1945, 1947).

These two volumes cover the basics of analytic functions, power series, contour integration, and Cauchy's theorems. The proofs of the Weierstrass factorization theorem and the Mittag-Leffler theorem sketched here are given in detail in the second volume, which also has an introduction to the theory of doubly periodic functions (elliptic functions).

A classic work that is still quite readable today is

> E. T. Copson, *Theory of Functions of a Complex Variable*, Clarendon Press, Oxford (1935).

After a general discussion of the properties of analytic functions, there is extensive treatment of many important special functions, including the Gamma function, hypergeometric, Legendre and Bessel functions, and the various elliptic functions. There is also a nice derivation of Picard's theorem.

Good modern textbooks at an introductory level are

> J. W. Brown and R, V. Churchill, *Complex Variables and Applications* (7th edition), McGraw-Hill (2004).

This standard is still a favorite after more than 50 years. It covers the basics and includes some useful discussion of applications of conformal mappings.

> E. B. Saff and A. D. Snider, *Fundamentals of Complex Analysis* (3rd edition), Prentice-Hall (2003).

This book also has a good treatment of conformal mappings, and also has several sections on numerical methods in various contexts.

> J. E. Marsden and M. E. Hoffman, *Basic Complex Analysis* (3rd edition), W. H. Freeman (1998).

In addition to a well-written introduction to the basics, this book has nice treatments of conformal mappings and of asymptotic methods, and a chapter on the Laplace transform.

A more advanced textbook is

> G. F. Carrier, M. Krook, and C. E. Pierson, *Functions of a Complex Variable: Theory and Technique*, Hod Books, Ithaca (1983).

This excellent book has a thorough introduction to the theory of functions of a complex variable and its applications to other problems. While there are few formal proofs, the mathematics is explained clearly and illustrated with well-chosen examples and exercises. There are lengthy chapters on conformal mappings, special functions, asymptotic methods, and integral transform methods, all of which get to the meat of the applications of these methods. There is also a useful chapter on the Wiener–Hopf method and singular integral equations.

Problems

1. For each of the mappings

   (i) $\quad w = \dfrac{z^2 + 1}{z^2 - 1}$

   (ii) $\quad w = \tan z$

   explain: For what values of z does the mapping fail to be conformal? What are the branch points of the map in the w-plane? What is the structure of the Riemann surface corresponding to the w-plane? Find appropriate branch cuts in the w-plane, and indicate what region of the z-plane is mapped onto each sheet of the w-plane. Sketch some details of the mapping.

2. Consider the linear fractional transformation

   $$z \mapsto w = \frac{az + b}{cz + d}$$

   with $\Delta \equiv ad - bc = 1$. With this transformation we can associate the 2×2 matrix

   $$\mathbf{A} \equiv \begin{pmatrix} a & b \\ c & d \end{pmatrix}$$

   (note that $\det \mathbf{A} = \Delta = 1$).

   (i) Show that the result of the linear fractional transformation with associated matrix $\mathbf{A}_1$ followed by the linear fractional transformation with associated matrix $\mathbf{A}_2$ is a linear fractional transformation whose associated matrix is the matrix product $\mathbf{A}_2 \mathbf{A}_1$.

   Remark. Thus the transformation group $SL(2)$ defined by 2×2 matrices $\mathbf{A}$ with $\det \mathbf{A} = 1$ has a *nonlinear realization* by linear fractional transformations of the complex plane. $\qquad\qquad\square$

   (ii) Show that if the matrix $\mathbf{A}$ is real, then the associated linear fractional transformation maps the upper half z-plane into the upper half w-plane.

3. Discuss the analytic properties of each of the functions below. In particular, locate and classify the singularities of each function, calculate the residue at each pole, join each pair of branch points with a suitable cut and calculate the discontinuity across the cut, and describe the Riemann surface on which each function is defined.

   (i) $\quad \dfrac{z}{(z^2 - a^2)^2}$

   (ii) $\quad \dfrac{\sin z}{z}$

   (iii) $\quad \dfrac{1}{\sqrt{z}} \ln \left(\dfrac{1 + \sqrt{z}}{1 - \sqrt{z}} \right)$

   (iv) $\quad \displaystyle\int_0^\infty \frac{e^{-t}}{t + z}\, dt$

4. Is it always true that

$$\ln z_1 z_2 = \ln z_1 + \ln z_2$$

when z_1 and z_2 are complex? Explain your answer with words and pictures.

5. Consider the function

$$f(z) = \frac{1}{a + bz + i\sqrt{z}}$$

where a, b are real constants. $f(z)$ is defined on the two-sheeted Riemann surface associated with $\sqrt{z}$. Choose the branch cut in the z-plane to run from 0 to ∞ along the positive real axis, and define the principal sheet to be where $\sqrt{z}$ is positive as the real axis is approached from the upper half z-plane (note that Im $\sqrt{z} > 0$ on this sheet). Find the pole(s) of $f(z)$ on the Riemann surface, noting carefully on which sheet each pole is located. Consider all possible sign combinations for a, b.

6. Derive the power series

$$\tan^{-1} z = \sum_{n=0}^{\infty} (-1)^n \frac{z^{2n+1}}{2n+1}$$

7. Use the arctangent series from the preceding problem to show that

$$\frac{\pi}{8} = \sum_{n=0}^{\infty} \frac{1}{(4n+1)(4n+3)}$$

8. Show that

$$\frac{1}{\sin^2 z} = \sum_{n=-\infty}^{\infty} \frac{1}{(z+n\pi)^2}$$

From this result, show that

$$\zeta(2) = \sum_{n=1}^{\infty} \frac{1}{n^2} = \frac{\pi^2}{6}$$

Then evaluate the series

$$\Sigma_4(z) \equiv \sum_{n=-\infty}^{\infty} \frac{1}{(z+n\pi)^4}$$

and the sum

$$\zeta(4) = \sum_{n=1}^{\infty} \frac{1}{n^4}$$

9. *Planck's radiation law.* In an ideal blackbody at (absolute) temperature T, the electromagnetic energy density between frequencies ν and $\nu + d\nu$ is given by

$$u_\nu d\nu = \left(\frac{8\pi h \nu^3}{c^3}\right) \frac{1}{\exp\left(\frac{h\nu}{kT}\right) - 1} d\nu$$

where h is Planck's original constant ($h = 2\pi\hbar$), k is the Boltzmann constant, and c is the speed of light.

(i) Show that the total energy density is proportional to the fourth power of temperature,

$$\mathcal{U}(T) \equiv \int_0^\infty u_\nu \, d\nu = aT^4$$

Remark. This is the *Stefan–Boltzmann law* for blackbody radiation; the power radiated from a blackbody is proportional to its energy density, and hence also to the fourth power of the temperature. □

(ii) Evaluate the Stefan–Boltzmann constant a in terms of h, k, and c.

10. Consider the expansion

$$\frac{1}{\sqrt{1 - 2hz + h^2}} = \sum_{n=0}^\infty c_n(z) \, h^n$$

For fixed $h > 0$, find the region in the z-plane for which this series is convergent.

11. Find the conditions under which the series

$$S(z) \equiv \sum_{n=1}^\infty \frac{z^n}{(1 - z^n)(1 - z^{n+1})}$$

is convergent and evaluate the sum.

12. If $F(z)$ is an entire function with zeros at $z_1, z_2, \ldots$, then the sequence $\{z_k\} \to \infty$.

13. Evaluate the integrals

(i) $$\int_0^\infty \frac{1}{1 + x^6} \, dx$$

(ii) $$\int_{-\infty}^\infty \frac{1}{(1 + x^2)(1 - 2\alpha x + x^2)} \, dx$$

with α being a fixed complex number.

14. Suppose m and n are positive integers with $m < n$. Evaluate the integral

$$I_{m,n} \equiv \int_0^\infty \frac{x^{2m}}{1 + x^{2n}} \, dx$$

15. Evaluate the integrals

(i) $$\int_0^\pi \frac{1}{1 + \sin^2 \theta} \, d\theta$$

(ii) $$\int_0^{2\pi} \frac{\sin^2 \theta}{a + b \cos \theta} \, d\theta$$

(iii) $$\int_{-\infty}^\infty \frac{\cos ax - \cos bx}{x^2} \, dx$$

(iv) $$\int_0^\infty \frac{\cos ax}{1 + x^4} \, dx$$

with a, b real and positive.

16. Evaluate the integral

$$I_n(a) \equiv \int_0^\infty \frac{x^{a-1}}{1 + x^{2n}} \, dx$$

with n being a positive integer, and $0 < a < 2n$.

17. Evaluate the integral

$$\int_0^\infty \frac{\ln x}{1 + x^2} \, dx$$

18. Evaluate the integral

$$I(a) \equiv \int_0^\infty \frac{dx}{\cosh ax}$$

19. Consider the function $h(z)$ defined by

$$h(z) \equiv \frac{2}{\sqrt{\pi}} \int_0^\infty \frac{x^2 e^{-x^2}}{1 - z e^{-x^2}} \, dx$$

(i) Evaluate $h(0)$.

(ii) Find the power series expansion of $h(z)$ about $z = 0$. For what region in the z-plane is the series convergent?

(iii) For what region in the z-plane is $h(z)$ defined by the integral? Locate the singularities of $h(z)$. Find suitable branch cuts connecting branch points of $h(z)$, and describe the Riemann surface of the function.

20. (i) Compute the volume $\mathcal{V}(\mathbf{S}^n)$ of the sphere $\mathbf{S}^n$. First, consider the n-dimensional integral

$$I_n \equiv \int_{-\infty}^{\infty} \cdots \int_{-\infty}^{\infty} e^{-r^2} \, dx_1, \ldots, dx_n$$

with $r^2 = x_1^2 + \cdots + x_n^2$ (r is the radial coordinate in n-dimensional spherical coordinates). In Cartesian coordinates, the integral can be expressed as a product of n identical one-dimensional integrals. In spherical coordinates, it is given by

$$I_n = \mathcal{V}(\mathbf{S}^{n-1}) \int_0^{\infty} r^{n-1} e^{-r^2} \, dr$$

which can be evaluated in terms of a Γ-function. Then evaluate $\mathcal{V}(\mathbf{S}^{n-1})$.

(ii) Evaluate the volume $\mathcal{V}(\mathbf{B}^n)$ of an n-dimensional ball $\mathbf{B}^n$ of radius R.

(iii) Compare your results with the standard results for $n = 1, 2, 3$.

Remark. Once these results are expressed in terms of Γ-functions, there is no need for the dimension n to be an integer. Extending these results to noninteger n is useful in renormalization group analyses both in quantum field theory and in statistical mechanics of many-body systems. $\qquad\square$

21. Show that $1/\Gamma(z)$ can be written as a contour integral according to

$$\frac{1}{\Gamma(z)} = \frac{1}{2\pi i} \int_C (-t)^{-z} e^{-t} \, dt$$

where C is the standard keyhole contour shown in Fig. 4.3.

22. Show that

$$\int_0^1 \frac{x^{a-1} (1-x)^{b-1}}{(x+p)^{a+b}} \, dx = \frac{\Gamma(a)\,\Gamma(b)}{\Gamma(a+b)} \frac{1}{(1+p)^a \, p^b}$$

for $\mathrm{Re}\, a > 0, \mathrm{Re}\, b > 0, p > 0$.

23. Show that

$$I_n \equiv \int_0^{\frac{\pi}{2}} \sin^{2n} \theta \, d\theta = 2^{2n-1} \frac{[\Gamma(n+\frac{1}{2})]^2}{\Gamma(2n+1)}$$

for $n = 0, 1, 2, \ldots$. Evaluate this for $n = 0$ to show that

$$\Gamma\left(\tfrac{1}{2}\right) = \sqrt{\pi}$$

Finally, show that

$$I_n = \frac{\pi}{2^{2n+1}} \frac{(2n)!}{(n!)^2}$$

24. Recall the *Riemann ζ-function* defined in Chapter 1 for Re $s > 1$ by

$$\zeta(s) \equiv \sum_{n=1}^{\infty} \frac{1}{n^s}$$

(i) Show that

$$\zeta(s) = \frac{1}{\Gamma(s)} \int_0^{\infty} \frac{x^{s-1}}{e^x - 1}\, dx$$

for Re $s > 1$.

(ii) Show that

$$\zeta(s) = \frac{\Gamma(1-s)}{2\pi i} \int_C \frac{(-t)^{s-1}}{e^t - 1}\, dt$$

where C is the standard keyhole contour shown in Fig. 4.3. Then show that $\zeta(s)$ is analytic in the finite s-plane except for a simple pole at $s = 1$ with residue 1.

(iii) Evaluate $\zeta(-p)$ $(p = 0, 1, 2, \ldots)$ in terms of Bernoulli numbers (Eqs. (4.A21) and (4.A22)).

25. Find the sum of the series

$$\sum_{n=0}^{\infty} e^{-nb} \cos 2\pi n z$$

$(b > 0)$. For what values of z does the series converge?

26. Suppose the elliptic function $f(z)$ of order n has zeros at $a_1, \ldots, a_n$ and poles at $b_1, \ldots, b_n$ in a primitive cell $\mathcal{P}$. Show that

$$\sum_{k=1}^{n} a_k - \sum_{k=1}^{n} b_k = p\alpha + q\beta$$

where α, β are the fundamental periods of $f(z)$, and p, q are integers.
Hint. Evaluate the integral

$$\frac{1}{2\pi i} \oint_K z\, \frac{f'(z)}{f(z)}\, dz$$

around the boundary K of $\mathcal{P}$.

5 Differential Equations: Analytical Methods

In Chapter 3, we introduced the concept of vector fields and their lines of flow (integral curves) as solutions of a system of first-order differential equations. It is relatively straightforward to generate a solution to such a system, numerically if necessary, starting from an initial point in the n-dimensional manifold on which the equations are defined. This solution will be unique, provided of course that the initial point is not a singular point of the vector field defined by the equations. However, there are also questions of long-term stability of solutions—do solutions that start from nearby points remain close to each other? and if so, for how long?—that are especially important both in principle and in practice, and many of these questions are still open to active research. They are only mentioned briefly here to remind the reader that there is more to using mathematics than providing input to a computer.

Moreover, not all systems of differential equations in physics are posed as initial value problems. A vibrating string, for example, is described by a linear second-order differential equation with boundary conditions at the endpoints of the string. Thus we need to consider methods to find general solutions to a differential equation, and then deal with the questions of existence and uniqueness of solutions satisfying various possible constraints. For linear systems, the theory of linear vector spaces and linear operators in Chapters 2, 6, and 7 is useful, but analytic techniques are still needed to construct explicit solutions.

We begin with the standard form of a system of n first-order differential equations

$$\frac{d}{dz} u_k(z) = h_k(u_1, \ldots, u_n; z)$$

($k = 1, \ldots, n$), where $u_1, \ldots, u_n$ are the functions to be determined, and z is the independent variable; here $u_1, \ldots, u_n$ and z may complex. Extending the range of the variables to the complex plane allows us to make use of the theory of functions of a complex variable elaborated in Chapter 4. We note that a general nth-order differential equation for a single function u can be reduced to a system of n first-order equations by introducing as independent variables the function u and its first $n - 1$ derivatives $u', \ldots, u^{(n-1)}$.

The simplest differential equation is the linear first-order equation

$$\frac{d}{dz} u(z) + p(z)u(z) = f(z)$$

We construct the general solution to the homogeneous equation [$f(z) = 0$] with initial condition $u(z_0) = u_0$, and show how the analytic properties of the solution are related to the analytic properties of the coefficient $p(z)$. We then show how to find the solution to the inhomogeneous equation.

Introduction to Mathematical Physics. Michael T. Vaughn
Copyright © 2007 WILEY-VCH Verlag GmbH & Co. KGaA, Weinheim
ISBN: 978-3-527-40627-2

We also look at some special nonlinear first-order equations. The Bernoulli equation is reduced to a linear equation by a change of variables. The first-order Ricatti equation is related to a linear second-order equation. We show how to solve a first-order equation that is an exact differential. Of course there is no universal method to reduce the general first-order equation to an exact differential!

We present the nth-order linear differential equation as a linear operator equation

$$L[u] \equiv u^{(n)}(z) + p_1(z)u^{(n-1)}(z) + \cdots + p_{n-1}(z)u'(z) + p_n(z)u(z) = f(z)$$

on the linear vector space C^n of n-times differentiable functions of the variable z. The linear equation $L[u] = 0$ with constant coefficients is related to the equation of a linear dynamical system introduced in Section 2.5; its solution is a linear combination of exponential functions of the form $\exp(\alpha z)$ with constants α expressed as the roots of a polynomial of degree n. We also find solutions when the roots of polynomial are degenerate.

The general solution of the homogeneous equation $L[u] = 0$ is obtained formally as a linear combination of n fundamental solutions that define an n-dimensional vector space. If the coefficients $p_1(z), \ldots, p_n(z)$ are analytic functions of the complex variable z, then the solutions to $L[u] = 0$ are analytic except perhaps where one or more of the coefficients is singular. The behavior of solutions near a singular point is of special interest, for the singular points often correspond to the boundary of a physical region in the variable z. A *regular singular point* is one at which the singularity of the general solution is at most a pole or a branch point; the general solution has an essential singularity at an *irregular singular point*. We note that the general solution of the inhomogeneous equation $L[u] = f$ is given as a particular solution of $L[u] = f$ plus any solution of the homogeneous equation.

Second-order linear equations are especially important in mathematical physics, since they arise in the analysis of partial differential equations involving the Laplacian operator when these equations are reduced to ordinary differential equations by separating variables, as will appear in Chapter 8. Here we show how to extract the leading behavior of solutions near singular points. For equations with one or two regular singular points, we can obtain a general solution in terms of elementary functions.

The second-order linear equation with three regular singular points is the *hypergeometric equation*, and its solutions are hypergeometric functions. We describe the Legendre equation and some of its solutions as a model of this equation, and devote Appendix A to a detailed analysis of the general hypergeometric equation and the hypergeometric functions.

When two of the regular singular points are merged in a special way to create an irregular singular point, the hypergeometric equation becomes the *confluent hypergeometric equation*. Bessel's equation is the version of this equation most often met, and we describe the important properties of several types of Bessel functions. We also provide an extensive analysis of the general confluent hypergeometric equation and the confluent hypergeometric equations that satisfy it in Appendix B.

In Appendix C, we analyze a nonlinear differential equation that arises in the classical description of the motion of a pendulum, or of an anharmonic (nonlinear) oscillator The solution to this equation is expressed in terms of elliptic functions, which are examples of the doubly periodic functions introduced in Chapter 4.

5.1 Systems of Differential Equations

5.1.1 General Systems of First-Order Equations

Differential equations with one independent variable are *ordinary differential equations*; those with more than one independent variable are *partial differential equations*. Chapter 8 is about partial differential equations; in this chapter we deal with ordinary differential equations, in particular, with systems of first-order differential equations of the form

$$\frac{d}{dz} u_k(z) = h_k(u_1, \ldots, u_n; z) \tag{5.1}$$

$(k = 1, \ldots, n)$, in which z is the independent variable, and $u_1, \ldots, u_n$ are the functions to be determined. The number n of independent equations is the *order n* of the system.

Remark. The general nth order ordinary differential equation has the form

$$F(v, v', \ldots, v^{(n)}; z) = 0 \tag{5.2}$$

where $v = v(z)$ is the function to be determined, and $v', \ldots, v^{(n)}$ denote the derivatives of v with respect to z. We assume that $\partial F / \partial v^{(n)}$ not identically zero, so that F actually depends on $v^{(n)}$, and that Eq. (5.2) can be solved, perhaps not uniquely, for $v^{(n)}$ so that we have

$$v^{(n)} = f(v, v', \ldots, v^{(n-1)}; z) \tag{5.3}$$

This nth order equation can be expressed as a system of n first-order equations, so that there is no loss of generality in starting with a system (5.1) of first-order equations. To see this, simply define $u_1 = v, u_2 = v', \ldots, u_n = v^{(n-1)}$. Then Eq. (5.3) is equivalent to the system

$$
\begin{aligned}
u'_n &= f(u_1, \ldots, u_n; z) \\
u'_{n-1} &= u_n \\
&\ \ \vdots \\
u'_1 &= u_2
\end{aligned}
\tag{5.4}
$$

of n first-order equations of the form (5.1) for the $u_1, \ldots, u_n$. □

In Chapter 3, the system (5.1) defined a vector field, and its solutions were the integral curves of the vector field when the independent variable z was a real parameter and the functions $h_k(u_1, \ldots, u_n; z)$ were independent of z. In many systems, the independent variable has the physical meaning of time, and Eqs. (5.1) are equations of motion. However, if the functions $h_1, \ldots, h_n$ are defined and well behaved for complex z, it is natural look for ways to make use of the theory of functions of a complex variable introduced in Chapter 4, whether or not complex values of z have any obvious physical meaning.

The system (5.1) does not by itself determine a set of functions $u_1, \ldots, u_n$. In order to obtain a definite solution, it is necessary to specify a set of conditions, such as the values of the $u_1, \ldots, u_n$ at some fixed value z_0 of the variable z. If we are given the initial conditions

$$u_1(z_0) = u_{10}, \ldots, u_n(z_0) = u_{n0} \tag{5.5}$$

then we can imagine generating a solution to Eq. (5.1) by an iterative procedure. Start from the initial conditions and take a small step δz. Then advance from z_0 to $z_0 + \delta z$ using the differential equations, so that

$$u_k(z_0 + \delta z) = u_{k0} + h_k(u_{10}, \dots, u_{n0}, z_0)\,\delta z \tag{5.6}$$

($k = 1, \dots, n$). A sequence of steps from z_0 to $z_1 \equiv z_0 + \delta z$ to $z_2 \equiv z_1 + \delta z$ and so on, generates an approximate solution along an interval of the real axis, or even a curve in the complex plane. We can imagine that the accuracy of the approximation could be improved by taking successively smaller steps. This procedure, which can be implemented numerically with varying degrees of sophistication, has an internal consistency check: if we advance from z_0 to z_1 and then return from z_1 to z_0, perhaps using different steps, then we should reproduce the initial conditions. Similarly, if we move around a closed path in the complex plane and return to z_0, then we should return to the initial conditions, unless by chance we have encircled a branch point of the solution.

A more formal method of generating a solution to the system of equations (5.1) is to evaluate higher derivatives at z_0 according to

$$u_k'' = \frac{du_k'}{dz} = \sum_{m=1}^{n} \frac{\partial h_k}{\partial u_m}\frac{du_m}{dz} + \frac{\partial h_k}{\partial z} = \sum_{m=1}^{n} h_m \frac{\partial h_k}{\partial u_m} + \frac{\partial h_k}{\partial z} \tag{5.7}$$

($k = 1, \dots, n$), and so forth, assuming the relevant partial derivatives exist. If they exist, the higher derivatives can all be evaluated at z_0 in terms of the initial conditions (5.5), leading to a formal power series expansion of the solutions $u_1, \dots, u_n$ about z_0. If these series have a positive radius of convergence, then a complete solution can be constructed in principle by analytic continuation.

These ideas lead to many important questions in the theory of differential equations. Among these are:

(i) Do the solutions obtained by the iterative method converge to a solution in the limit $\delta z \to 0$? Do the formal power series solutions actually converge? If they do, what is their radius of convergence?

(i) What about the existence and uniqueness of solutions that satisfy other types of conditions. For example, does a second-order equation have a solution with specified values at two distinct points? If so, is this solution unique?

(ii) What is the behavior of solutions near singular points of the equations, where one of the functions $h_k(u_1, u_2, \dots, u_n; z)$ in Eq. (5.1), or the function $f(v, v', \dots, v^{(n-1)}; z)$ in Eq. (5.3), becomes singular? Are there other singularities in the solutions that cannot be identified as singularities of the equations? The only singularities of solutions to linear equations are singularities of the coefficients, but the solutions of nonlinear equations often have singular points that depend on the initial conditions, or boundary conditions, imposed on the solutions.

(iii) Can we understand qualitative properties of solutions without necessarily constructing them explicitly. This involves the use of various schemes for constructing approximate solutions in a local region, as well as tools for characterizing global properties of solutions.

(iv) What approximate schemes are known for computing numerical solutions? Since few differential equations allow complete analytic solutions, it is important to have reliable methods for determining solutions numerically.

Various aspects of these issues are discussed in this and subsequent chapters, with the notable exception of numerical methods, which are mentioned only briefly. With the widespread availability of powerful computers, numerical methods are certainly important, but computational physics is a subject that is complementary to the mathematics discussed in this book.

5.1.2 Special Systems of Equations

Certain special classes of differential equations have been studied extensively. A system described by Eq. (5.1) is *linear* if each of the functions $h_k(u_1, \ldots, u_n; z)$ is linear in the variables $u_1, \ldots, u_n$, though not necessarily in z. Similarly, Eq. (5.3) is linear if the function $f(v, v', \ldots, v^{(n-1)}; z)$ is linear in the variables $v, v', \ldots, v^{(n-1)}$. Linear equations are especially important in physics; many fundamental equations of physics are linear (Maxwell's equations, the wave equation, the diffusion equation and the Schrödinger equation, for example). They also describe small oscillations of a system about an equilibrium configuration, and the analysis of the stability of equilibria depends heavily on the properties of these linear equations.

An *autonomous* system of equations is one in which the functions $h_k(u_1, \ldots, u_n; z)$ in Eq. (5.1) are independent of z. Such a system is unchanged by a translation of the variable z; the solutions can be characterized by curves in the space of the variables $u_1, \ldots, u_n$ on which the location of the point $z = 0$ is arbitrary. Equations (3.47) that define a vector field are autonomous, and we have seen how the solutions define a family of integral curves of the vector field.

Remark. An autonomous system of order n can generally be reduced to a system of order $n - 1$ by choosing one variable $u_n \equiv \tau$, say, and rewriting the remaining equations in (5.1) in the form

$$h_n(u_1, u_2, \ldots; \tau) \frac{du_k}{d\tau} = h_k(u_1, u_2, \ldots; \tau) \tag{5.8}$$

for $k = 1, \ldots, n - 1$. Conversely, a nonautonomous system of order n can always be transformed into an autonomous system of order $n + 1$ by introducing a new dependent variable $u_{n+1} \equiv z$. These transformations may, or may not, be useful. $\square$

A system of differential equations of the form (5.1) is *scale invariant* if it is invariant under the scale transformation $z \to az$ (a is a constant). To be precise, this means that

$$\frac{du}{dz} = h(u; z) = h(u; az) \tag{5.9}$$

If we define $g(u; z) = zh(u; z)$, then we have $g(u; z) = g(u; az)$, so that $g(u; z)$ is actually independent of z. If we then introduce the variable $\tau = \ln z$, we have

$$z \frac{du}{dz} = \frac{du}{d\tau} = g(u) \tag{5.10}$$

which is an autonomous equation in the variable τ. This argument evidently applies as well to a system of the form (5.1).

→ **Exercise 5.1.** A differential equation is *scale covariant*, or *isobaric*, if it is invariant under the scale transformation $z \to az$ together with $u \to a^p u$ for some p. Show that if we let

$$u(z) \equiv z^P w(z)$$

then the resulting equation for w will be scale invariant.   □

   While the properties of autonomy and scale invariance allow the reduction of a differential equation to one of lower order, there is no guarantee that this is a simplification. For example, an nth order linear equation is reduced to a nonlinear equation of order $n - 1$, which is seldom helpful. But there is such a variety of differential equations to deal with that there are no universal methods; all potentially useful tools are of interest.

5.2   First-Order Differential Equations

5.2.1   Linear First-Order Equations

The simplest differential equation is the linear homogeneous first-order equation. This has the standard form

$$u'(z) + p(z)u(z) = 0 \tag{5.11}$$

If we write this as

$$\frac{du}{u} = -p(z)\,dz \tag{5.12}$$

then we find the standard general solution

$$u(z) = u(z_0) \exp\left\{ -\int_{z_0}^{z} p(\xi)\,d\xi \right\} \tag{5.13}$$

if $p(z)$ is analytic at z_0. The solution thus obtained is uniquely determined by the value $u(z_0)$, and is analytic wherever $p(z)$ is analytic.

   We can also see that if $p(z)$ has a simple pole at z_1 with residue α, then

$$u(z) \sim (z - z_1)^{-\alpha} \tag{5.14}$$

as $z \to z_1$. This follows from the observation that

$$\int_{z_0}^{z} p(\xi)\,d\xi = \int_{z_0}^{z} \frac{\alpha}{\xi - z_1}\,d\xi + C = \alpha \ln\left(\frac{z - z_1}{z_0 - z_1} \right) + C \tag{5.15}$$

where C is a constant of integration. Thus for $z \simeq z_1$, we have

$$u(z) \simeq Au(z_0) \exp\left[-\alpha \ln\left(\frac{z - z_1}{z_0 - z_1} \right) \right] = Au(z_0) \left(\frac{z_0 - z_1}{z - z_1} \right)^{\alpha} \tag{5.16}$$

for $z \to z_1$, with $A = \exp(-C)$ another constant. Note that if α is a negative integer, then the solution $u(z)$ will be analytic at z_1 even though the equation is singular; a singularity of the equation does not always give rise to a singularity of the solution. If $p(z)$ has a pole of higher order or an essential singularity at z_1, then the integral of $p(\xi)$ will have a pole or essential singularity, and hence the solution will have an essential singularity there. Thus the behavior of the solution near a singularity of $p(z)$ is governed by the properties of the singularity.

To solve the corresponding inhomogeneous equation

$$u'(z) + p(z)u(z) = f(z) \tag{5.17}$$

we can make the standard substitution

$$u(z) = w(z)h(z) \tag{5.18}$$

where

$$h(z) = \exp\left\{-\int_{z_0}^z p(\xi)\,d\xi\right\} \tag{5.19}$$

is a solution of the homogeneous equation if $p(z)$ is analytic at z_0. If $f(z) = 0$, then $w(z)$ is constant, but in general $w(z)$ satisfies the simple equation

$$h(z)\,w'(z) = f(z) \tag{5.20}$$

that has the general solution

$$w(z) = u(z_0) + \int_{z_0}^z \frac{f(\xi)}{h(\xi)}\,d\xi \tag{5.21}$$

if $f(z)$ is analytic at z_0. The solution is uniquely determined by $u(z_0)$.

→ **Exercise 5.2.**   Find the solution $u(z)$ of the differential equation

$$\frac{du(z)}{dz} + \alpha u(z) = A\cos\omega z$$

satisfying the initial condition $u(0) = 1$. Here A, α, and ω are real constants.   □

A nonlinear equation closely related to the linear equation is the *Bernoulli equation*

$$u'(z) + \rho(z)u(z) = f(z)[u(z)]^\alpha \tag{5.22}$$

with $\alpha \neq 1$ a constant. If we let

$$w(z) = [u(z)]^{1-\alpha} \tag{5.23}$$

then $w(z)$ satisfies the linear inhomogeneous first-order equation

$$w'(z) + (1-\alpha)\rho(z)w(z) = (1-\alpha)f(z) \tag{5.24}$$

We can solve this equation using the standard methods just described.

5.2.2   Ricatti Equation

The *Ricatti equation* is

$$u'(z) = q_0(z) + q_1(z)u(z) + q_2(z)[u(z)]^2 \tag{5.25}$$

There is no completely general solution to this equation, but if we know a particular solution $u_0(z)$, then we can find the general solution in the form

$$u(z) = u_0(z) - \frac{c}{v(z)} \tag{5.26}$$

where c is a constant. Here $v(z)$ satisfies the linear inhomogeneous first-order equation

$$v'(z) + [q_1(z) + 2q_2(z)u_0(z)]v(z) = cq_2(z) \tag{5.27}$$

that can be solved by the standard method. It is also true that if $u(z)$ is a solution of Eq. (5.25), then $w(z) \equiv -1/u(z)$ is a solution of the Ricatti equation

$$w'(z) = q_2(z) - q_1(z)w(z) + q_0(z)[w(z)]^2 \tag{5.28}$$

with coefficients $q_0(z)$ and $q_2(z)$ interchanged.

❑ **Example 5.1.**   Consider the equation

$$u'(z) = 1 - [u(z)]^2 \tag{5.29}$$

One solution of this equation is

$$u_0(z) = \tanh z \tag{5.30}$$

To find the general solution, let

$$u(z) = \tanh z - \frac{1}{v(z)} \tag{5.31}$$

Then $v(z)$ must satisfy the linear equation

$$v'(z) = 2 \tanh z \, v(z) - 1 \tag{5.32}$$

To solve Eq. (5.32), let

$$v(z) = w(z) \exp\left\{2 \int_{z_0}^{z} \tanh \xi \, d\xi\right\} = w(z) \cosh^2 z \tag{5.33}$$

[recall $\int \tanh z \, dz = \ln \cosh z$]. Then $w(z)$ must satisfy

$$w'(z) = -\exp\left\{2 \int_{z_0}^{z} \tanh \xi \, d\xi\right\} = -\frac{1}{\cosh^2 z} \tag{5.34}$$

This equation has the solution

$$w(z) = A - \tanh z \tag{5.35}$$

where A is a constant of integration, and working backwards to the solution of Eq. (5.29), we find the general solution

$$u(z) = \tanh z - \frac{1}{\cosh z(A \cosh z - \sinh z)} = \frac{A \sinh z - \cosh z}{A \cosh z - \sinh z} \tag{5.36}$$

Note that $u(0) = -1/A$. ∎

→ **Exercise 5.3.** Solve the equation

$$u'(z) = 1 + [u(z)]^2 \tag{5.37}$$

by finding suitable changes of variables. □

→ **Exercise 5.4.** Find the general solution of the differential equation

$$(1 - z^2)u'(z) - zu(z) = z[u(z)]^2$$

by one method or another. □

The Ricatti equation (5.25) is nonlinear, but it can be converted to a second-order linear equation by the substitution

$$u(z) = -\frac{1}{q_2(z)} \frac{y'(z)}{y(z)} \tag{5.38}$$

A short calculation then shows that $y(z)$ satisfies the linear second-order equation

$$y''(z) - \left[q_1(z) + \frac{q_2'(z)}{q_2(z)} \right] y'(z) + q_0(z)q_2(z)y(z) = 0 \tag{5.39}$$

Conversely, the Ricatti equation can be used to factorize a second-order linear differential equation into a product of two first-order linear equations, as shown in the following exercise:

→ **Exercise 5.5.** Show that the second-order linear differential operator

$$L \equiv \frac{d^2}{dz^2} + p(z) \frac{d}{dz} + q(z)$$

can be factored into the form

$$L = \left[\frac{d}{dz} + s(z) \right] \left[\frac{d}{dz} + t(z) \right]$$

if $s(z) + t(z) = p(z)$ and $t(z)$ is a solution of the Ricatti equation

$$t'(z) = [t(z)]^2 - p(z)t(z) + q(z)$$

Then find the general solution to the factorized equation

$$L[u(z)] = f(z)$$

with $u(z_0) = u_0$, expressed in terms of certain definite integrals. □

5.2.3   Exact Differentials

The general first-order equation can be written in the form

$$g(u, z)du + h(u, z)dz = 0 \tag{5.40}$$

If there is a function $F(u, z)$ such that

$$g(u, z) = \frac{\partial F}{\partial u} \quad \text{and} \quad h(u, z) = \frac{\partial F}{\partial z} \tag{5.41}$$

then Eq. (5.40) is *exact*; it has the form

$$dF = \frac{\partial F}{\partial u} du + \frac{\partial F}{\partial z} dz = 0 \tag{5.42}$$

Then the surface of constant $F(u, z)$ defines a solution to Eq. (5.40). Evidently, Eq. (5.40) is exact if and only if

$$\frac{\partial}{\partial z} g(u, z) = \frac{\partial}{\partial u} h(u, z) \tag{5.43}$$

A first-order equation that is not exact can be converted into an exact equation if there is a factor $\lambda(u, z)$ such that the equation

$$\lambda(u, z)g(u, z)\, du + \lambda(u, z)h(u, z)\, dz = 0 \tag{5.44}$$

is exact; such a factor $\lambda(u, z)$ is an *integrating factor* for Eq. (5.40). While the existence of such an integrating factor can be proved under fairly broad conditions on the functions $g(u, z)$ and $h(u, z)$, there are few general techniques for explicitly constructing the integrating factor; intuition and experience are the principal guides.

Remark.   The above discussion can be amplified using the language of differential forms. In a manifold $\mathcal{M}$ with u and z as coordinates, we define the form

$$\sigma = g(u, z)du + h(u, z)dz \tag{5.45}$$

Equation (5.40) defines a slice of the cotangent bundle $\mathcal{T}^*(\mathcal{M})$. If we introduce the vector field

$$v = h(u, z)\frac{\partial}{\partial u} - g(u, z)\frac{\partial}{\partial z} \tag{5.46}$$

then Eq. (5.40) requires

$$\langle \sigma, v \rangle = 0 \tag{5.47}$$

so the solution curves for Eq. (5.40) are orthogonal to the integral curves of v. Equation (5.43) is equivalent to $d\sigma = 0$; if it is satisfied, then there is (at least locally) a function F such that $\sigma = dF$, corresponding to Eq. (5.41).   □

5.3   Linear Differential Equations

5.3.1 nth Order Linear Equations

Many physical systems show linear behavior; that is, the response of the system to an input is directly proportional to the magnitude of the input, and the response to a sum of inputs is the sum of the responses to the individual inputs. This behavior is a direct consequence of the fact that the laws governing such systems are themselves linear. Since these laws are often expressed in the form of differential equations, we are led to study the general linear nth order differential equation, which has the standard form

$$L[u] \equiv u^{(n)}(z) + p_1(z)u^{(n-1)}(z) + \cdots + p_{n-1}(z)u'(z) + p_n(z)u(z) = f(z) \quad (5.48)$$

with corresponding homogeneous equation

$$L[u] = 0 \tag{5.49}$$

The notation $L[u]$ is meant to suggest that we think of L as a *differential operator*

$$L \equiv \frac{d^n}{dz^n} + p_1(z) \frac{d^{n-1}}{dz^{n-1}} + \cdots + p_{n-1}(z) \frac{d}{dz} + p_n(z) \tag{5.50}$$

that acts linearly on functions $u(z)$, so that

$$L[c_1 u_1(z) + c_2 u_2(z)] = c_1 L[u_1(z)] + c_2 L[u_2(z)] \tag{5.51}$$

as with the linear operators introduced in Chapter 2. Note that the functions $u(z)$ themselves satisfy the axioms of a linear vector space; we will look at this more closely in Chapter 6.

The linearity of L expressed in Eq. (5.51) implies that if $u_1(z)$ and $u_2(z)$ are solutions of

$$L[u_1] = f_1(z) \quad \text{and} \quad L[u_2] = f_2(z) \tag{5.52}$$

then $u(z) \equiv c_1 u_1(z) + c_2 u_2(z)$ is a solution of

$$L[u] = c_1 f_1(z) + c_2 f_2(z) \tag{5.53}$$

In particular, if $v(z)$ is any solution of the inhomogeneous equation

$$L[v] = f(z) \tag{5.54}$$

and $u(z)$ is a solution of the homogeneous equation

$$L[u] = 0 \tag{5.55}$$

then $u(z) + v(z)$ is also a solution of the inhomogeneous equation (5.54). Indeed, the general solution of Eq. (5.54) has this form, the sum of a a particular solution $v(z)$ of the inhomogeneous equation and the general solution $u(z)$ of the homogeneous equation (5.55).

5.3.2   Power Series Solutions

If the variable z in Eq. (5.50) is meaningful as a complex variable, and if the coefficients $p_1(z), \ldots, p_n(z)$ are analytic functions of z, then we can make extensive use of the theory of analytic functions developed in Chapter 4. In this case, a point z_0 is a *regular point* of Eq. (5.49) if $p_1(z), \ldots, p_n(z)$ are all analytic at z_0, otherwise a *singular point*. Solutions to the homogeneous equation (5.49) can be constructed as Taylor series about any regular point z_0, with coefficients depending on the n arbitrary constants $u(z_0), u'(z_0), \ldots, u^{(n-1)}(z_0)$, and the Taylor series has a positive radius of convergence. It is plausible (and true) that the solution will be analytic in any region where $p_1(z), \ldots, p_n(z)$ are all analytic.

❑ **Example 5.2.**   Consider the second-order equation

$$u''(z) + zu(z) = 0 \tag{5.56}$$

whose coefficients are analytic everywhere in the finite z-plane. If we try to construct a series solution

$$u(z) = 1 + \sum_{n=1}^{\infty} a_n z^n \tag{5.57}$$

then we find that a_1 is arbitrary, since the first derivative $u'(0)$ is arbitrary, $a_2 = 0$, and the higher coefficients are determined from the recursion relation

$$a_{n+3} = -\frac{a_n}{(n+2)(n+3)} \tag{5.58}$$

Thus the solution to Eq. (5.56) satisfying

$$u(0) = 1 \text{ and } u'(0) = a \tag{5.59}$$

has the expansion

$$u(z) = 1 + \sum_{n=1}^{\infty} (-1)^n \frac{(3n-2)(3n-5)\ldots 1}{(3n)!} z^{3n}$$

$$+ az \left[1 + \sum_{n=1}^{\infty} (-1)^n \frac{(3n-1)(3n-4)\ldots 2}{(3n+1)!} z^{3n} \right] \tag{5.60}$$

Note that the series in Eq. (5.60) each have infinite radius of convergence. Hence the general solution to Eq. (5.56) is an entire function, which is not surprising since the coefficients in the equation are entire functions. ∎

➔ **Exercise 5.6.**   Find the basic solutions of the differential equation

$$u''(z) + z^2 u(z) = 0$$

at $z = 0$. Express these solutions as power series in z. What are the singularities of these solutions? ☐

5.3.3   Linear Independence; General Solution

To further explore the consequences of the linearity of equations (5.48) and (5.49), as expressed in Eqs. (5.51)–(5.53), we first introduce the concept of linear independence for functions, which is the same as the standard vector space definition in Chapter 2.

Definition 5.1. The functions $u_1(z), u_2(z), \ldots, u_n(z)$ are *linearly dependent* if there exist constants $c_1, c_2, \ldots, c_n$ (not all zero) such that

$$c_1 u_1(z) + c_2 u_2(z) + \cdots + c_n u_n(z) = 0 \tag{5.61}$$

Otherwise the functions are *linearly independent*. The left-hand side of Eq. (5.61) is a *linear combination* of the functions $u_1(z), u_2(z), \ldots, u_n(z)$ regarded as vectors.

One test for linear independence of the functions $u_1(z), u_2(z), \ldots, u_n(z)$ is to consider the *Wronskian* determinant defined by

$$W(u_1, u_2, \ldots, u_n; z) = \begin{vmatrix} u_1(z) & u_2(z) & \cdots & u_n(z) \\ u_1'(z) & u_2'(z) & \cdots & u_n'(z) \\ \vdots & \vdots & & \vdots \\ u_1^{(n-1)}(z) & u_2^{(n-1)}(z) & \cdots & u_n^{(n-1)}(z) \end{vmatrix} \tag{5.62}$$

If $u_1(z), u_2(z), \ldots, u_n(z)$ are linearly dependent, then $W(u_1, u_2, \ldots, u_n; z)$ vanishes everywhere. If the Wronskian does not vanish everywhere, then $u_1(z), u_2(z), \ldots, u_n(z)$ are linearly independent.

Remark. Note, however, that the vanishing of the Wronskian at a single point z_0 does not by itself imply that the functions are linearly dependent, for $W(u; z_0) = 0$ simply means that there are constants $a_1, a_2, \ldots, a_n$, not all zero, such that

$$a_1 u_1^{(k)}(z_0) + a_2 u_2^{(k)}(z_0) + \cdots + a_n u_n^{(k)}(z_0) = 0 \tag{5.63}$$

($k = 0, 1, \ldots, n - 1$). Then the linear combination

$$h(z) \equiv a_1 u_1(z) + a_2 u_2(z) + \cdots + a_n u_n(z) \tag{5.64}$$

is a function that vanishes together with its first $n - 1$ derivatives at z_0. This alone does not guarantee that $h(z) = 0$ everywhere, but if $h(z)$ also satisfies an nth order linear differential equation such as Eq. (5.48), then the nth and higher derivatives of $h(z)$ also vanish, and $h(z)$ does vanish everywhere, and $u_1(z), u_2(z), \ldots, u_n(z)$ are linearly dependent. □

If the functions $u_1(z), u_2(z), \ldots, u_n(z)$ are solutions to the linear equations (5.48) or (5.49) of order n, then the Wronskian satisfies the linear equation

$$\frac{d}{dz} W(u_1, u_2, \ldots, u_n; z) + p_1(z) W(u_1, u_2, \ldots, u_n; z) = 0 \tag{5.65}$$

➜ **Exercise 5.7.** Derive Eq. (5.65) from the definition (5.62).   □

Equation (5.65) has the standard solution

$$W(u_1, u_2, \ldots, u_n; z) = W(u_1, u_2, \ldots, u_n; z_0) \exp\left\{ -\int_{z_0}^{z} p_1(z)\, dz \right\} \qquad (5.66)$$

Hence if the Wronskian of a set of n solutions to Eqs. (5.48) vanishes at a single point, it vanishes everywhere, and the solutions are linearly dependent. Conversely, if the Wronskian is nonzero at a point where $p_1(z)$ is analytic, then it can vanish only at a singularity of $p_1(z)$ or at ∞. The solution $u(z)$ of Eq. (5.49) that satisfies

$$u(z_0) = \xi_1 \qquad u'(z_0) = \xi_2 , \ldots, \; u^{(n-1)}(z_0) = \xi_n \qquad (5.67)$$

at a regular point z_0 can be uniquely expressed as

$$u(z) = \sum_{k=1}^{n} c_k u_k(z) \qquad (5.68)$$

with coefficients $c_1, c_2, \ldots, c_n$ are determined from the equations

$$c_1 u_1(z_0) + \cdots + c_n u_n(z_0) = \xi_1$$

$$c_1 u_1'(z_0) + \cdots + c_n u_n'(z_0) = \xi_2$$

$$\vdots \qquad\qquad (5.69)$$

$$c_1 u_1^{(n-1)}(z_0) + \cdots + c_n u_n^{(n-1)}(z_0) = \xi_n$$

These equations have a unique solution since the determinant of the coefficients on the left-hand side is just the Wronskian $W(u_1, u_2, \ldots, u_n; z_0)$, which is nonzero since the solutions $u_1(z), u_2(z), \ldots, u_n(z)$ are linearly independent.

Thus the general solution of Eq. (5.49) defines an n-dimensional linear vector space. The initial conditions (5.67) choose a particular vector from this solution space. We can introduce a *fundamental set* of solutions $u_1(z; z_0), \ldots, u_n(z; z_0)$ at a regular point z_0 of Eq. (5.49) by the conditions

$$u_k^{(m-1)}(z_0; z_0) = \delta^{km} \qquad (5.70)$$

$(k, m = 1, \ldots, n)$. These n solutions are linearly independent, since the Wronskian

$$W(u_1, u_2, \ldots, u_n; z_0) = 1 \qquad (5.71)$$

The general solution of Eq. (5.49) can be expressed as a linear combination of the fundamental solutions according to

$$u(z) = \sum_{k=1}^{n} u^{(k-1)}(z_0) u_k(z; z_0) \qquad (5.72)$$

so that the functions defined by Eq. (5.70) form a basis of the n-dimensional solution space.

5.3.4 Linear Equation with Constant Coefficients

Of special importance is the linear nth order equation

$$L[u] \equiv u^{(n)}(z) + \alpha_1 u^{(n-1)}(z) + \cdots + \alpha_n u(z) = 0 \tag{5.73}$$

with constant coefficients $\alpha_1, \ldots, \alpha_n$. This equation can be solved algebraically, since

$$L[e^{\lambda z}] = p(\lambda)e^{\lambda z} \tag{5.74}$$

where $p(\lambda)$ is the nth degree polynomial

$$p(\lambda) = \lambda^n + \alpha_1 \lambda^{n-1} + \cdots + \alpha_n \tag{5.75}$$

Hence $u(z) = e^{\lambda z}$ is a solution of Eq. (5.73) if and only if λ is a root of $p(\lambda)$.

If the roots $\lambda_1, \ldots, \lambda_n$ of $p(\lambda)$ are distinct, then the general solution of Eq. (5.73) is

$$u(z) = c_1 e^{\lambda_1 z} + \cdots + c_n e^{\lambda_n z} \tag{5.76}$$

with coefficients $c_1, \ldots, c_n$ determined by the values of $u(z)$ and its first $n - 1$ derivatives at some point z_0. If the roots are not distinct, however, then further linearly independent solutions must be found. These appear immediately if we note that

$$L[z^m e^{\lambda z}] = \frac{\partial^k}{\partial \lambda^k}\left[p(\lambda)e^{\lambda z}\right] = \sum_{k=0}^{m} \binom{m}{k} p^{(k)}(\lambda) z^{m-k} e^{\lambda z} \tag{5.77}$$

Hence $u(z) = z^m e^{\lambda z}$ is a solution of Eq. (5.73) if and only if

$$p(\lambda) = 0 \qquad p'(\lambda) = 0, \ \ldots, \ p^{(m)}(\lambda) = 0 \tag{5.78}$$

that is, if λ is a root of $p(\lambda)$ of multiplicity $\geq m + 1$. If this is the case, then $u(z) = \rho(z) e^{\lambda z}$ is also a solution for any polynomial $\rho(z)$ of degree $\leq m + 1$.

The final result is that if the polynomial $p(\lambda)$ has the distinct roots $\lambda_1, \ldots, \lambda_q$ with multiplicities $m_1, \ldots, m_q$, respectively, then the general solution of Eq. (5.73) is

$$u(z) = \rho_1(z)e^{\lambda_1 z} + \cdots + \rho_q(z)e^{\lambda_n z} \tag{5.79}$$

where $\rho_1(z), \ldots, \rho_q(z)$ are arbitrary polynomials of degrees $d_1 \leq m_1 - 1, \ldots, d_q \leq m_q - 1$, respectively.

A more general system of linear first-order equations with constant coefficients is

$$u'_k(z) = \sum_{\ell} A_{k\ell} u_\ell(z) \tag{5.80}$$

($k = 1, \ldots, n$), with $\mathbf{A} = (A_{k\ell})$ a matrix of constants. As explained in Section 2.5, this system can also be reduced to a set of algebraic equations. Comparison of this system to Eq. (5.73) is left to Problem 6.

5.4   Linear Second-Order Equations

5.4.1   Classification of Singular Points

Linear second-order differential equations often appear in physics, both in systems of coupled oscillators and as a result of separation of variables in partial differential equations (see Chapter 8) such as Laplace's equation or the wave equation. Thus we go into great detail in examining these equations and some of their solutions. The linear homogeneous second-order differential equation has the standard form

$$L[u] \equiv u''(z) + p(z)u'(z) + q(z)u(z) = 0 \tag{5.81}$$

also known as the *Sturm–Liouville equation*. Just as for the general linear equation, z_0 is a *regular* (or *ordinary*) *point* of $L[u] = 0$ if $p(z)$ and $q(z)$ are analytic at z_0, otherwise a singular point. The singular point z_0 is a *regular singular point* if $(z - z_0)\,p(z)$ and $(z - z_0)^2 q(z)$ are analytic at z_0, otherwise an *irregular singular point*. As we shall see, the solutions near a regular singular point z_0 behave no worse than a power of $(z - z_0)$, perhaps multiplied by $\ln(z - z_0)$. At an irregular singular point, the general solution will have an essential singularity.

Remark.   To characterize the nature of the point at ∞, let $\xi \equiv 1/z$. Then Eq. (5.81) becomes

$$\xi^4 \frac{d^2}{d\xi^2}\left[u\left(\frac{1}{\xi}\right)\right] + \left\{2\xi^3 - \xi^2 p\left(\frac{1}{\xi}\right)\right\} \frac{d}{d\xi}\left[u\left(\frac{1}{\xi}\right)\right] + q\left(\frac{1}{\xi}\right) u\left(\frac{1}{\xi}\right) = 0 \tag{5.82}$$

Thus ∞ is a regular point if $[z^2 p(z) - 2z]$ and $z^4 q(z)$ are analytic at ∞, a regular singular point if $z\,p(z)$ and $z^2 q(z)$ are analytic at ∞, and an irregular singular point otherwise. $\square$

The standard method for solving linear equations with analytic functions as coefficients is to look for power series solutions.[1] If z_0 is a regular point of Eq. (5.81), then the general solution is analytic at z_0. It can be expanded in a power series about z_0, and the series will converge inside the largest circle about z_0 containing no singular points of the equation. The solution and its first derivative at z_0 must be determined from initial conditions. Singular points require more discussion, which we now provide.

5.4.2   Exponents at a Regular Singular Point

If z_0 is a regular singular point, then let

$$P(z) \equiv (z - z_0)p(z) = \sum_{n=0}^{\infty} p_n(z - z_0)^n \tag{5.83}$$

$$Q(z) \equiv (z - z_0)^2 q(z) = \sum_{n=0}^{\infty} q_n(z - z_0)^n \tag{5.84}$$

Because z_0 is a regular singular point, both $P(z)$ and $Q(z)$ are analytic at z_0; hence both series have positive radius of convergence.

---

[1] The following analysis is often called the *method of Frobenius*.

Now try a solution of the form

$$u(z) = (z - z_0)^\alpha \left[1 + \sum_{n=1}^{\infty} a_n (z - z_0)^n \right] \tag{5.85}$$

Equations (5.83)–(5.85) can be inserted into Eq. (5.81) and coefficients of each power of $(z - z_0)$ set equal. For the lowest power, this gives the *indicial equation*

$$\alpha (\alpha - 1) + \alpha p_0 + q_0 = 0 \tag{5.86}$$

that has roots

$$\alpha = \tfrac{1}{2} \left[1 - p_0 \pm \sqrt{(1 - p_0)^2 - 4q_0} \right] \equiv \alpha_\pm \tag{5.87}$$

The roots $\alpha_\pm$ are the *exponents* of the singularity at z_0; note also that

$$p_0 = 1 - \alpha_+ - \alpha_- \qquad q_0 = \alpha_+ \alpha_- \tag{5.88}$$

The coefficients in the power series can then be determined from the equations

$$[\alpha(\alpha + 1) + (\alpha + 1)p_0 + q_0] \, a_1 + \alpha p_1 + q_1 = 0$$

$$\vdots \tag{5.89}$$

$$[(\alpha + n)(\alpha + n - 1) + (\alpha + n) p_0 + q_0] \, a_n + [(\alpha + n - 1) p_1 + q_1] \, a_{n-1}$$

$$+ \cdots + [(\alpha + 1) p_{n-1} + q_{n-1}] \, a_1 + \alpha p_n + q_n = 0$$

If $\alpha_+ - \alpha_- \neq 0, 1, 2, \ldots$, we have two linearly independent solutions to Eq. (5.81),

$$u_\pm(z) = (z - z_0)^{\alpha_\pm} \left[1 + \sum_{n=1}^{\infty} a_n^{(\pm)} (z - z_0)^n \right] \tag{5.90}$$

The general solution is an arbitrary linear combination of $u_+(z)$ and $u_-(z)$.

If $\alpha_+ = \alpha_-$, there is only one power series solution. If

$$\alpha_+ = \alpha_- + n \tag{5.91}$$

($n = 1, 2, \ldots$), then the solution $u_-(z)$ is not uniquely determined, since the coefficient of $a_n^{(-)}$ in Eq. (5.89) is

$$(\alpha_- + n)(\alpha_- + n - 1) + (\alpha_- + n)(1 - \alpha_+ - \alpha_-) + \alpha_+ \alpha_- = 0 \tag{5.92}$$

In either case, one solution of Eq. (5.81) is given by

$$u_+(z) = (z - z_0)^{\alpha_+} f_+(z) \tag{5.93}$$

with $f_+(z)$ analytic at z_0 and $f_+(z_0) = 1$ by convention.

To find a second solution, let

$$u(z) \equiv w(z) \, u_+(z) \tag{5.94}$$

This substitution is generally useful when one solution to a differential equation is known, but it is especially needed here since the power series method fails. $w(z)$ must satisfy the equation

$$u_+(z) \, w''(z) + \left[2u'_+(z) + p(z) \, u_+(z) \right] w'(z) = 0 \tag{5.95}$$

This can be solved in the usual way to give

$$w'(z) = \frac{C}{[u_+(z)]^2} \exp\left\{ -\int^z p(\xi) \, d\xi \right\} \tag{5.96}$$

with C a constant. Since

$$p(z) = \frac{1 - \alpha_+ - \alpha_-}{z - z_0} + r(z) = \frac{n + 1 - 2\alpha_+}{z - z_0} + r(z) \tag{5.97}$$

with $r(z)$ analytic at z_0, we can use Eq. (5.93) to write

$$w'(z) = \frac{K}{(z - z_0)^{n+1}} \frac{1}{[f_+(z)]^2} \exp\left\{ -\int_{z_0}^z r(\xi) \, d\xi \right\} \equiv K \frac{g(z)}{(z - z_0)^{n+1}} \tag{5.98}$$

where K is another constant, and

$$g(z) = \frac{1}{[f_+(z)]^2} \exp\left\{ -\int_{z_0}^z r(\xi) \, d\xi \right\} \tag{5.99}$$

is analytic at z_0 with $g(z_0) = 1$. The function $w(z)$ is then given by

$$w(z) = K_1 + K \int^z \frac{g(\xi)}{(\xi - z_0)^{n+1}} \, d\xi \tag{5.100}$$

where K_1 is yet another constant, which corresponds to a multiple of the first solution $u_+(z)$ in the solution $u(z)$ and can be ignored here. Expanding $g(\xi)$ about z_0 and integrating term-by-term leads to

$$w(z) = \frac{K}{(z - z_0)^n} \left\{ \sum_{\substack{m=0 \\ m \neq n}}^{\infty} \frac{g^{(n)}(z_0)}{n!} \ln(z - z_0) + \frac{g^{(m)}(z_0)}{m - n} \frac{(z - z_0)^m}{m!} \right\} \tag{5.101}$$

Then a second linearly independent solution to Eq. (5.81) is given by

$$u_-(z) \equiv (z - z_0)^{\alpha_-} f_-(z) + \frac{g^{(n)}(z_0)}{n!} u_+(z) \ln(z - z_0) \tag{5.102}$$

where

$$f_-(z) = f_+(z) \sum_{\substack{m=0 \\ m \neq n}}^{\infty} \frac{g^{(m)}(z_0)}{m - n} \frac{(z - z_0)^m}{m!} \tag{5.103}$$

is analytic at z_0. Thus when $\alpha_+ - \alpha_- = n = 0, 1, 2, \ldots$, the general solution has a logarithmic branch point at z_0 unless $g^{(n)}(z_0) = 0$. Hence, an integer exponent difference at a regular singular point signals the possible presence of a logarithmic singularity in addition to the usual power law behavior.

5.4.3 One Regular Singular Point

Linear second-order equations with one, two or three singular points have been thoroughly studied. The second-order equation whose only singular point is a regular singular point at z_0 is uniquely given by

$$u''(z) + \frac{2}{z - z_0} u'(z) = 0 \qquad (5.104)$$

The exponents of the singularity are $\alpha = 0, -1$ and the general solution of the equation is

$$u(z) = a + \frac{b}{z - z_0} \qquad (5.105)$$

with a and b arbitrary constants.

Remark. Note that if $\alpha \neq -1$, the equation

$$u''(z) + \frac{1 - \alpha}{z - z_0} u'(z) = 0 \qquad (5.106)$$

has a regular singular point at ∞ in addition to the regular singular point at z_0. However, the general solution

$$u(z) = a + \frac{b}{(z - z_0)^\alpha} \qquad (5.107)$$

is analytic at ∞ if $\alpha = 2, 3, \ldots$. Thus in exceptional cases, a regular singular point of an equation may not be reflected as a singularity of the general solution. $\square$

➜ **Exercise 5.8.** Find the general solution to Eq. (5.106) if $\alpha = 0$. $\square$

5.4.4 Two Regular Singular Points

The general equation with two regular singular points at z_1 and z_2 is given by

$$u''(z) + \left[\frac{1 - \alpha - \beta}{z - z_1} + \frac{1 + \alpha + \beta}{z - z_2} \right] u'(z) \left[\frac{\alpha\beta(z_1 - z_2)^2}{(z - z_1)^2(z - z_2)^2} \right] u(z) = 0 \quad (5.108)$$

with exponents (α, β) at z_1 and $(-\alpha, -\beta)$ at z_2. The exponents at z_1 and z_2 must be equal and opposite in order to have ∞ as a regular point. A standard form of the equation, with the singular points at 0 and ∞, is reached by the transformation

$$\xi = \frac{z - z_1}{z - z_2} \qquad (5.109)$$

that maps z_1 to 0, z_2 to ∞, and transforms Eq. (5.108) into

$$u''(\xi) + \left(\frac{1 - \alpha - \beta}{\xi} \right) u'(\xi) + \frac{\alpha\beta}{\xi^2} u(\xi) = 0 \qquad (5.110)$$

The general solution of this equation is

$$u(\xi) = a\xi^\alpha + b\xi^\beta \tag{5.111}$$

if $\alpha \neq \beta$; if $\alpha = \beta$, the general solution is

$$u(\xi) = a\xi^\alpha(1 + c\ln\xi) \tag{5.112}$$

The two regular singular points in Eq. (5.108) can be merged into one irregular singular point. Let

$$z_2 = z_1 + \eta \qquad \alpha = \frac{k}{\eta} = -\beta \tag{5.113}$$

and take the limit $\eta \to 0$ (this is a *confluence* of the singular points). Then Eq. (5.108) becomes

$$u''(z) + \frac{2}{z - z_1}\, u'(z) - \frac{k^2}{(z - z_1)^4}\, u(z) = 0 \tag{5.114}$$

which has an irregular singular point at z_1 and no other singular points.

The singular point can be moved to ∞ by the transformation

$$\xi = \frac{1}{z - z_1} \tag{5.115}$$

whence the equation becomes

$$u''(\xi) - k^2\, u(\xi) = 0 \tag{5.116}$$

with general solution

$$u(\xi) = c_+\, e^{k\xi} + c_-\, e^{-k\xi} \tag{5.117}$$

where $c_\pm$ are arbitrary constants.

Remark. The general solution to the original equation is evidently given by

$$u(z) = c_+\, e^{k/(z - z_1)} + c_-\, e^{-k/(z - z_1)} \tag{5.118}$$

but this is less obvious than the solution to the transformed equation.   □

→ **Exercise 5.9.** Derive Eq. (5.116) from Eq. (5.114).   □

The equation with three regular singular points can be transformed to the *hypergeometric equation*; a confluence of two of the singular points leads to the *confluent hypergeometric equation*. The general equations are studied in Appendices A and B. In the next two sections, we look at Legendre's equation as an example of the hypergeometric equation, and Bessel's equation as an example of the confluent hypergeometric equation.

5.5   Legendre's Equation

5.5.1   Legendre Polynomials

An important differential equation in physics is *Legendre's equation*

$$(z^2 - 1)u''(z) + 2zu'(z) - \lambda(\lambda + 1)u(z) = 0 \tag{5.119}$$

which has the standard form of Eq. (5.81), with

$$p(z) = \frac{2z}{z^2 - 1} \quad \text{and} \quad q(z) = -\frac{\lambda(\lambda + 1)}{z^2 - 1} \tag{5.120}$$

where λ is a parameter that can be complex in general, although we will find that it is required to be an integer in many physical contexts. The singular points of this equation are at $z = \pm 1$ and $z = \infty$; each of the singular points is regular. The indicial equations at $z = \pm 1$ have double roots at $\alpha = 0$, so we can expect to find solutions $u_\pm(z)$ that are analytic at $z = \pm 1$, and second solutions $v_\pm(z)$ with logarithmic singularities at $z = \pm 1$.

➜ **Exercise 5.10.**   Find the exponents of the singular point of Eq. (5.119) at ∞. $\square$

We look for solutions as power series around $z = 0$. Since $z = 0$ is a regular point of Eq. (5.119). we expect to find a general solution of the form

$$u(z) = \sum_{k=0}^{\infty} a_k z^k \tag{5.121}$$

with a_0, a_1 arbitrary and the remaining coefficients determined by the differential equation. Inserting this series into Eq. (5.119) and setting the coefficient of each term in the power series to zero, we have the recursion relation

$$(k + 1)(k + 2)a_{k+2} = [k(k + 1) - \lambda(\lambda + 1)]a_k \tag{5.122}$$

If we express this recursion relation as a ratio

$$\frac{a_{k+2}}{a_k} = \frac{k(k + 1) - \lambda(\lambda + 1)}{(k + 1)(k + 2)} = \frac{(k - \lambda)(k + \lambda + 1)}{(k + 1)(k + 2)} \tag{5.123}$$

then we can see that the ratio $|a_{k+2}/a_k|$ tends to 1 for large k. This means that the radius of convergence of the power series is equal to 1 in general, as expected from a solution that is analytic except for possible singularities at $z = \pm 1, \infty$.

However, these singularities will not be present if the infinite series actually terminates, which will be the case if λ is an integer.[2] If $\lambda = n \geq 0$, the recursion relation gives $a_{n+2} = 0$, and there is a polynomial solution $P_n(z)$ of degree n. If we choose the initial condition $P_n(1) = 1$, then $P_n(z)$ is the *Legendre polynomial* of degree n. Note also that only even or odd powers of z will appear in $P_n(z)$, depending on whether n is even or odd; thus we have

$$P_n(-z) = (-1)^n P_n(z) \tag{5.124}$$

---

[2]Note that if $\lambda = n$, we can take $n \geq 0$, since λ and $-\lambda - 1$ give rise to the same set of solutions to Eq. (5.119).

The first few Legendre polynomials can be found explicitly; we have

$$P_0(z) = 1 \qquad P_1(z) = 1 \qquad P_2(z) = \tfrac{1}{2}(3z^2 - 1) \tag{5.125}$$

➔ **Exercise 5.11.** Find explicit forms for $P_3(z)$ and $P_4(z)$.   □

Another way to write Legendre's equation is

$$L[u] = \left\{ \frac{d}{dz}(z^2 - 1)\frac{d}{dz} \right\} u(z) = \lambda(\lambda + 1)u(z) \tag{5.126}$$

This looks like an eigenvalue equation, and in Chapter 7, we will see that it is when we look at eigenvalue problems for linear differential operators from a vector space viewpoint. Here we derive some properties of the $P_n(z)$ that follow from a representation of the solution known as *Rodrigues' formula*,

$$P_n(z) = C_n \frac{d^n}{dz^n}(z^2 - 1)^n \tag{5.127}$$

First we note that $P_n(z)$ defined by Eq. (5.127) is a polynomial of degree n. Then let

$$u_n(z) = \left(z^2 - 1\right)^n \tag{5.128}$$

and note that

$$(z^2 - 1)u_n'(z) = 2nz u_n(z) \tag{5.129}$$

If we differentiate this equation $n + 1$ times, we have

$$(z^2 - 1)u_n^{(n+2)}(z) + 2(n + 1)z u_n^{(n+1)}(z) + n(n + 1)u_n^{(n)}(z) =$$
$$\tag{5.130}$$
$$= 2nz u_n^{(n+1)}(z) + 2n(n + 1)u_n^{(n)}(z)$$

which leads to Legendre's equation (5.119), since Eq. (5.130) is equivalent to

$$\frac{d}{dz}\left[(z^2 - 1)u_n^{(n+1)}(z) \right] = (z^2 - 1)u_n^{(n+2)}(z) + 2z u_n^{(n+1)}(z) = n(n+1)u_n^{(n)}(z) \tag{5.131}$$

which shows that $u_n^{(n)}(z)$ satisfies Legendre's equation.

➔ **Exercise 5.12.** The constant C_n in Eq. (5.127) is determined by the condition $P_n(1) = 1$. Note that

$$(z^2 - 1)^n = (z - 1)^n(z + 1)^n$$

and evaluate the derivative

$$\frac{d^n}{dz^n}(z^2 - 1)^n$$

for $z = 1$. Then find the constant C_n.   □

Many other useful properties of Legendre polynomials can be derived from Rodrigues' formula. For example, Cauchy's theorem allows us to write

$$P_n(z) = \frac{1}{2\pi i} \oint_C \frac{P_n(\xi)}{\xi - z} \, d\xi \tag{5.132}$$

where C is any closed contour encircling the point t once counterclockwise. Then inserting the Rodrigues formula and integrating by parts n times gives *Schläfli's integral representation*

$$P_n(z) = \frac{1}{2^{n+1}\pi i} \oint_C \frac{(\xi^2 - 1)^n}{(\xi - z)^{n+1}} \, d\xi \tag{5.133}$$

for the Legendre polynomials.

➜ **Exercise 5.13.** In the Schläfli representation (5.133) for the $P_n(z)$, suppose that z is real with $-1 \leq z \leq 1$, and let the contour C be a circle of radius $\sqrt{1 - z^2}$ about z. Show that Eq. (5.133) then leads to

$$P_n(z) = \frac{1}{2\pi} \int_0^{2\pi} \left(z + i\sqrt{1 - z^2} \sin\theta\right)^n d\theta \tag{5.134}$$

Equation (5.134) is *Laplace's integral formula* for the Legendre polynomials. □

Consider next the series

$$S(t, z) \equiv \sum_{n=0}^{\infty} t^n P_n(z) \tag{5.135}$$

If we use the Schläfli integral for the $P_n(z)$ and sum the geometric series, we obtain

$$\sum_{n=0}^{\infty} t^n P_n(z) = \frac{1}{2\pi i} \oint_C \sum_{n=0}^{\infty} \left(\frac{t}{2}\right)^n \frac{(\xi^2 - 1)^n}{(\xi - z)^{n+1}} \, d\xi \tag{5.136}$$

$$= \frac{1}{\pi i} \oint_C \frac{1}{2\xi - 2z + t(1 - \xi^2)} \, d\xi$$

The contour C must be chosen so that the geometric series inside the integral is absolutely and uniformly convergent on C, and such that only one of the two poles of the integrand of the last integral inside the contour. This can be done for $-1 \leq z \leq 1$ and t in a region inside the unit circle but off the real axis (the reader is invited to work out the details). The contour for the second integral can then be deformed in any way so long as that one pole of the integrand remains inside the contour. Evaluating the residue at the pole then gives the result

$$S(t, z) = \sum_{n=0}^{\infty} t^n P_n(z) = \frac{1}{\sqrt{1 - 2zt + t^2}} \tag{5.137}$$

The function $S(z, t)$ is a *generating function* for the Legendre polynomials. Note that if $-1 \leq t \leq 1$, the series converges for $|z| < 1$, but there is a larger region of convergence in the complex z-plane (see Problem 8). The series (5.137) is used in Chapter 6 to derive orthogonality relations for the Legendre polynomials. It also appears in the multipole expansion in electrostatics (see Chapter 8) and elsewhere.

→ **Exercise 5.14.** Use either the Rodrigues formula (5.127) or the generating function (5.137) to derive the recursion formulas

$$(2n+1)zP_n(z) = (n+1)P_{n+1}(z) + nP_n(z)$$

and

$$(z^2 - 1)P_n'(z) = nzP_n(z) - nP_{n-1}(z)$$

for the Legendre polynomials.   □

Legendre's equation (5.119) has a polynomial solution if the parameter λ is an integer. But in any event, it has a solution $P_\lambda(z)$ that is analytic at $z = 1$, and scaled so that $P_\lambda(1) = 1$. $P_\lambda(z)$ is the *Legendre function of the first kind*. In order to study this solution, we change the variable to

$$t = \tfrac{1}{2}(1 - z)$$

Then the interval $-1 \le z \le 1$ is mapped to the interval $0 \le t \le 1$, and $z = 1$ in mapped to $t = 0$. In terms of the variable t, Legendre's equation is

$$t(t-1)u''(t) + (2t-1)u'(t) - \lambda(\lambda+1)u(t) = 0 \tag{5.138}$$

with regular singular points at $t = 0, 1, \infty$. The solution that is analytic at $t = 0$ with $u(0) = 1$ has the power series expansion

$$u(t) = 1 + \sum_{k=1}^{\infty} c_k t^k \tag{5.139}$$

The differential equation then requires

$$(k_1)^2 c_{k+1} = [k(k+1) - \lambda(\lambda+1)]c_k = (k-\lambda)(k+\lambda+1)c_k \tag{5.140}$$

and thus

$$c_k = \frac{\Gamma(k+\lambda+1)\Gamma(k-\lambda)}{\Gamma(\lambda+1)\Gamma(-\lambda)} \cdot \frac{1}{(k!)^2} \tag{5.141}$$

This is a hypergeometric series, as introduced in Appendix A, with parameters $a = -\lambda$, $b = \lambda + 1$ and $c = 1$, and thus the Legendre function of the first kind can be expressed as a hypergeometric function,

$$P_\lambda(z) = F(-\lambda, \lambda+1 | \tfrac{1}{2}(1-z)) \tag{5.142}$$

→ **Exercise 5.15.** Show that the recursion formulas of Exercise 5.14 are also valid for the Legendre functions $P_n(z)$ even if n (= λ) is not an integer.   □

5.5.2 Legendre Functions of the Second Kind

We have seen that if the parameter λ in Legendre's equation is an integer, there is a polynomial solution. Any independent solution must have a logarithmic singularity at $z = \pm 1$, since the indicial equations at $z = \pm 1$ have double roots at $\alpha = 0$, as already noted. If we are interested in solutions for $|z| > 1$, which are occasionally relevant, we can let $\xi = 1/z$. In terms of the variable ξ, Legendre's equation is

$$\xi^2 \frac{d}{d\xi}\left\{ (\xi^2 - 1)\frac{d}{d\xi} \right\} u(\xi) + \lambda(\lambda + 1)u(\xi) = 0 \tag{5.143}$$

or, in a form from which the singularities can be seen directly,

$$\xi^2(\xi^2 - 1)u''(\xi) + 2\xi^3 u'(\xi) + \lambda(\lambda + 1)u(\xi) = 0 \tag{5.144}$$

This equation has regular singular points at $\xi = 0, \pm 1$. The exponents at $\xi = 0$ are given by

$$\alpha = \lambda + 1, -\lambda \tag{5.145}$$

A solution with exponent $\lambda + 1$ at $\xi = 0$ is proportional to the *Legendre function of second kind*, denoted by $Q_\lambda(z)$.

There are several ways to look at the properties of the $Q_\lambda(z)$. The most straightforward is to start from Eq. (5.144) and define

$$u(\xi) = \xi^{\lambda+1}v(\xi) \tag{5.146}$$

Then $v(\xi)$ satisfies the differential equation

$$\xi(\xi^2 - 1)v''(\xi) + 2[(\lambda + 1)(\xi^2 - 1) + \xi^2]v'(\xi) + (\lambda + 1)(\lambda + 2)\xi v(\xi) = 0 \tag{5.147}$$

This equation has a solution $v(\xi)$ with $v(0) = 1$ and a power series expansion

$$v(\xi) = 1 + \sum_{k=1}^{\infty} b_k \xi^k \tag{5.148}$$

with coefficients b_k determined by the differential equation. Inserting this power series into Eq. (5.147) and equating coefficients of ξ^{k+1} gives the recursion formula

$$(k + 2)(2\lambda + k + 3)b_{k+2} = (\lambda + k + 1)(\lambda + k + 2)b_k \tag{5.149}$$

after noting that

$$(\lambda + 1)(\lambda + 2) + 2k(\lambda + 2) + k(k - 1) = (\lambda + k + 1)(\lambda + k + 2) \tag{5.150}$$

From the recursion formula, we can see that $b_1 = 0$, since $b_{-1} = 0$, Hence $b_{2k+1} = 0$ ($k = 1, 2, \ldots$), and the solution $v(\xi)$ is a power series in ξ^2.

➙ **Exercise 5.16.** Use the recursion formula to express the coefficients b_{2k} ($k = 1, 2, \ldots$) in terms of Γ-functions and thus find an explicit series representation of $v(\xi)$. What is the radius of convergence of this series? □

→ **Exercise 5.17.**   Change the variable in Eq. (5.147) to $\eta = \xi^2$ and determine the resulting differential equation for $w(\eta) = v(\xi)$. Compare this with the hypergeometric equation in Appendix A and then express $w(\eta)$ as a hypergeometric function.   □

Another approach is to start with the Schläfli representation (5.133), and consider a contour integral of the form

$$I_\lambda(z) = K_\lambda \int_C \frac{(1 - \xi^2)^\lambda}{(z - \xi)^{\lambda+1}} \, d\xi \tag{5.151}$$

over a contour C in the complex ξ-plane. If λ is not an integer, then the integrand has branch points at $\xi = \pm 1$ that were not present for $\lambda = n$, but that is not a problem so long as we pay attention to the branch cuts. The integral $I_\lambda(z)$ can lead to a solution of Legendre's equation with an appropriate choice of contour C, since we have, after some algebra,

$$\left\{ \frac{d}{dz}(z^2 - 1)\frac{d}{dz} - \lambda(\lambda + 1) \right\} \frac{(1 - \xi^2)^\lambda}{(z - \xi)^{\lambda+1}} = -\frac{d}{d\xi}\left\{ \frac{(1 - \xi^2)^{\lambda+1}}{(z - \xi)^{\lambda+2}} \right\} \tag{5.152}$$

We can then choose the contour C to simply be the line $-1 \le \xi \le 1$; then we have

$$Q_\lambda(z) = \frac{1}{2^{\lambda+1}} \int_{-1}^1 \frac{(1 - \xi^2)^\lambda}{(z - \xi)^{\lambda+1}} \, d\xi \tag{5.153}$$

Here the choice of branch cuts of the integrand is such that $(1 - \xi^2)^\lambda$ is real and positive on the interval $-1 \le \xi \le 1$, as is $(z - \xi)^{\lambda+1}$ when z is real and $z > 1$. The choice of constant factor is standard.

→ **Exercise 5.18.**   Use Eq. (5.153) to show that

$$Q_\lambda(z) \to \frac{[\Gamma(\lambda + 1)]^2}{2\Gamma(2\lambda + 2)} \left(\frac{2}{z} \right)^{\lambda+1}$$

for $z \to \infty$.   □

When $\lambda = n$ is a positive integer, then Eq. (5.153) can be used to show that

$$Q_n(z) = \frac{1}{2} \int_{-1}^1 \frac{P_n(t)}{z - t} \, dt \tag{5.154}$$

(see Problem 10). It then follows that

$$Q_n(z) = \frac{1}{2} P_n(z) \ln \frac{z + 1}{z - 1} - q_{n-1}(z) \tag{5.155}$$

where $q_{n-1}(z)$ is a polynomial of degree $n - 1$ given by

$$q_n(z) = \frac{1}{2} \int_{-1}^1 \frac{P_n(z) - P_n(t)}{z - t} \, dt \tag{5.156}$$

[Show that $q_n(z)$ actually is a polynomial of degree $n - 1$]. We have the explicit forms

$$Q_0(z) = \frac{1}{2} \ln \frac{z + 1}{z - 1} \qquad Q_1(z) = \frac{z}{2} \ln \frac{z + 1}{z - 1} - 1 \tag{5.157}$$

→ **Exercise 5.19.**   Find explicit forms for $Q_2(z)$ and $Q_3(z)$.   □

5.6   Bessel's Equation

5.6.1   Bessel Functions

Another frequently encountered differential equation is *Bessel's equation*

$$\frac{d^2u}{dz^2} + \frac{1}{z}\frac{du}{dz} + \left(1 - \frac{\lambda^2}{z^2}\right)u = 0 \tag{5.158}$$

The parameter λ is often an integer in practice, but it need not be. Bessel's equation often appears when the Laplacian operator is expressed in cylindrical coordinates; hence the solutions are sometimes called *cylinder functions*.

Bessel's equation has a regular singular point at $z = 0$ with exponents $\pm\lambda$, and an irregular singular point at ∞. The solution with exponent λ can be expressed as a power series

$$u(z) = z^\lambda \left(1 + \sum_{k=1}^\infty a_k z^k\right) \tag{5.159}$$

The differential equation then requires

$$[(k+\lambda+2)^2 - \lambda^2]a_{k+2} = (k+2)(k+2\lambda+2)a_{k+2} = -a_k \tag{5.160}$$

Only even powers of z will appear in the infinite series, and we have

$$a_{2m} = -\frac{1}{4m(m+\lambda)}\,a_{2m-2} \tag{5.161}$$

The function $J_\lambda(z)$ defined by the series

$$J_\lambda(z) \equiv \left(\frac{z}{2}\right)^\lambda \sum_{m=0}^\infty (-1)^m \frac{1}{m!\,\Gamma(m+\lambda+1)}\left(\frac{z}{2}\right)^{2m} \tag{5.162}$$

is a solution of Eq. (5.158) unless λ is a negative integer; it is the *Bessel function* of *order* λ.

➜ **Exercise 5.20.**   Show that the Bessel function $J_\lambda(z)$ can be related to a confluent hypergeometric function by

$$J_\lambda(z) = \frac{1}{\Gamma(\lambda+1)}\left(\frac{z}{2}\right)^\lambda e^{-iz}F(\lambda + \tfrac{1}{2}|2\lambda+1|2iz)$$

Thus many further properties of Bessel functions can be obtained from the general properties of confluent hypergeometric functions derived in Appendix B.   □

If λ is not an integer, then $J_{-\lambda}(z)$ is a second independent solution. of Bessel's equation. However, if λ is an integer $n \geq 0$, we have

$$J_{-n}(z) = (-1)^n J_n(z) \tag{5.163}$$

since $1/\Gamma(m+\lambda+1)$ vanishes when $m + \lambda$ is a negative integer.

A second independent solution of Bessel's equation for all λ is defined by

$$N_\lambda(z) = \frac{J_\lambda(z) \cos \pi\lambda - J_{-\lambda}(z)}{\sin \pi\lambda} \tag{5.164}$$

$N_\lambda(z)$ is a *Bessel function of second kind*, or *Neumann function*.

→ **Exercise 5.21.** Show that the Wronskian

$$W(N_\lambda, J_\lambda) = \frac{A}{z}$$

and evaluate the constant A. One way to do this is to consider the behavior for $z \to 0$. This shows that $J_\lambda(z)$ and $N_\lambda(z)$ are independent solutions for all λ.   □

The Bessel functions satisfy the recursion relations

$$J_{\lambda-1} + J_{\lambda+1} = \frac{2\lambda}{z} J_\lambda(z)$$

$$\tag{5.165}$$

$$J_{\lambda-1} - J_{\lambda+1} = 2J'_\lambda(z)$$

that follow from direct manipulation of the power series (5.162). The Neumann functions satisfy the same recursion relations, as do the Hankel functions to be introduced soon.

A generating function for the Bessel functions of integer order is

$$F(t, z) = e^{z(t-\frac{1}{t})} = \sum_{n=-\infty}^{\infty} t^n J_n(z) \tag{5.166}$$

From the generating function, it follows that $J_n(z)$ can be expressed as a contour integral

$$J_n(z) = \frac{1}{2\pi i} \oint_C \exp\left[\frac{z}{2}\left(t - \frac{1}{t}\right)\right] \frac{dt}{t^{n+1}} \tag{5.167}$$

where the contour C in the t-plane encircles $t = 0$ once in a counterclockwise direction. If we choose the contour to be the unit circle in the t-plane and let $t = \exp(i\theta)$, then this integral becomes

$$J_n(z) = \frac{1}{2\pi} \int_0^{2\pi} e^{iz\sin\theta} e^{-in\theta} \, d\theta \tag{5.168}$$

and then, since $J_0(z)$ is real,

$$J_n(z) = \frac{1}{\pi} \int_0^\pi \cos(n\theta - z\sin\theta) \, d\theta \tag{5.169}$$

→ **Exercise 5.22.**   Show that the coefficient of t^0 in Eq. (5.166) is $J_0(z)$ by computing this coefficient as a power series in z. Then show that the other terms satisfy the recursion relations (5.165) by looking at the partial derivatives of $F(t, z)$.   □

Remark.   The choice of contour in Eq. (5.167) leads to an integral representation for the Bessel function $J_n(z)$. Other choices of contour can lead to different solutions and different integral representations. See Problem 13 for one example.   □

5.6.2 Hankel Functions

The Bessel functions $J_\lambda(z)$ and $N_\lambda(z)$ are defined mainly by their properties near $z = 0$. Functions with well-defined behavior for $z \to \infty$ are the *Hankel functions*, or *Bessel functions of the third kind*, defined by

$$H_\lambda^{(1)}(z) = J_\lambda(z) + iN_\lambda(z) = \frac{i}{\sin \pi\lambda} \left\{ e^{-i\pi\lambda} J_\lambda(z) - J_{-\lambda}(z) \right\}$$

$$H_\lambda^{(2)}(z) = J_\lambda(z) - iN_\lambda(z) = \frac{-i}{\sin \pi\lambda} \left\{ e^{i\pi\lambda} J_\lambda(z) + J_{-\lambda}(z) \right\} \tag{5.170}$$

In terms of the Whittaker functions defined by the integral representations (5.B47) and (5.B48), the Hankel functions can be expressed as

$$H_\lambda^{(\alpha)}(z) = \frac{2}{\sqrt{\pi}} (2z)^\lambda U_\alpha(\lambda + \tfrac{1}{2} | 2\lambda + 1 | 2iz) \tag{5.171}$$

$(\alpha = 1, 2)$. Including the extra factors in the integral representations leads to the integral formulas for the Hankel functions

$$H_\lambda^\pm(z) = \frac{1}{\Gamma(\lambda + \tfrac{1}{2})} \sqrt{\frac{2}{\pi z}} \, e^{\pm iz} e^{\mp \frac{1}{2}\pi i(\lambda + \frac{1}{2})} I_\lambda(\pm 2iz) \tag{5.172}$$

where $H_\lambda^+ = H_\lambda^{(1)}$ and $H_\lambda^- = H_\lambda^{(2)}$. Here $I_\lambda(\xi)$ is the integral

$$I_\lambda(\xi) = \int_0^\infty e^{-u} u^{\lambda - \frac{1}{2}} \left(1 - \frac{u}{\xi} \right)^{\lambda - \frac{1}{2}} du$$

is an integral that was introduced in Section 1.4 as a prototype for one method of generating asymptotic series. Here it is clear that keeping only the leading term gives the asymptotic behavior

$$H_\lambda^\pm(z) \sim \sqrt{\frac{2}{\pi z}} \, e^{\pm iz} e^{\mp \frac{1}{2}\pi i(\lambda + \frac{1}{2})} \tag{5.173}$$

for $z \to \infty$ in a suitable sector of the z-plane.

The asymptotic expansions of the Bessel and Neumann functions for $z \to \infty$ then have leading terms

$$J_\lambda(z) \sim \sqrt{\frac{2}{\pi z}} \, \cos \left(z - \tfrac{1}{2}\pi\lambda - \tfrac{1}{4}\pi \right)$$

$$\tag{5.174}$$

$$N_\lambda(z) \sim \sqrt{\frac{2}{\pi z}} \, \sin \left(z - \tfrac{1}{2}\pi\lambda - \tfrac{1}{4}\pi \right)$$

Thus the Bessel and Neumann functions behave somewhat like the usual trigonometric sine and cosine functions, while the Hankel functions behave like complex exponentials.

5.6.3   Spherical Bessel Functions

The power series for the Bessel function $J_{1/2}(z)$ is

$$J_{\frac{1}{2}}(z) = \sqrt{\frac{z}{2}} \sum_{m=0}^{\infty} \frac{(-1)^m}{m!\, \Gamma(m + \frac{3}{2})} \left(\frac{z}{2}\right)^{2m} \tag{5.175}$$

From the duplication formula for the Γ-function (Exercise 4.A1), we have

$$m!\, \Gamma(m + \tfrac{3}{2}) = \frac{\sqrt{\pi}\, \Gamma(2m + 2)}{2^{2m+1}} \tag{5.176}$$

and then

$$J_{\frac{1}{2}}(z) = \sqrt{\frac{2}{\pi z}} \sum_{m=0}^{\infty} \frac{(-1)^m}{(2m+1)!} z^{2m+1} = \sqrt{\frac{2}{\pi z}}\, \sin z \tag{5.177}$$

The recursion formulas (5.165) then lead to

$$J_{-\frac{1}{2}}(z) = J_{\frac{1}{2}}'(z) + \frac{1}{2z} J_{\frac{1}{2}}(z) = \sqrt{\frac{2}{\pi z}}\, \cos z = -N_{\frac{1}{2}}(z) \tag{5.178}$$

Thus the Bessel functions of order $\frac{1}{2}$ can be expressed in terms of elementary functions. In fact, it follows from the recursion relations that any of the Bessel functions of order $n + \frac{1}{2}$ (with n integer) can be expressed in terms of trigonometric functions and odd powers of $\sqrt{z}$.

→ **Exercise 5.23.**   Find explicit formulas for $J_{\pm 3/2}(z)$, $N_{\pm 3/2}(z)$, and $H^{(1,2)}_{\pm 3/2}(z)$.   □

Spherical Bessel functions are defined in terms of the (cylindrical) Bessel functions by

$$j_n(z) = \sqrt{\frac{\pi}{2z}}\, J_{n+\frac{1}{2}}(z) \qquad n_n(z) = \sqrt{\frac{\pi}{2z}}\, N_{n+\frac{1}{2}}(z) \tag{5.179}$$

with corresponding definitions of spherical Hankel functions. These functions appear as solutions of the differential equations resulting from expressing the Laplacian in spherical coordinates. The spherical Bessel functions satisfy the differential equation

$$\frac{d^2 u}{dz^2} + \frac{2}{z}\frac{du}{dz} + \left[1 - \frac{n(n+1)}{z^2}\right] u = 0 \tag{5.180}$$

The $j_n(z)$ are entire functions, while the $n_n(z)$ and the Hankel functions $h_n^{(1)}(z)$, $h_n^{(2)}(z)$ are entire functions except for a pole of order n at $z = 0$. In particular, these functions have no branch points and they are single-valued in the complex z-plane.

→ **Exercise 5.24.**   Find explicit formulas for the spherical Bessel functions $j_1(z)$, $n_1(z)$, $h_0^{(1,2)}(z)$, and $h_1^{(1,2)}(z)$ in terms of elementary functions.   □

A   Hypergeometric Equation

A.1   Reduction to Standard Form

The general linear second-order differential equation whose only singular points are three regular singular points at z_1, z_2, and z_3 can be written in the standard form

$$u''(z) + p(z)\, u'(z) + q(z)u(z) = 0 \tag{5.A1}$$

with

$$p(z) = \frac{1 - \alpha_1 - \beta_1}{z - z_1} + \frac{1 - \alpha_2 - \beta_2}{z - z_2} + \frac{1 - \alpha_3 - \beta_3}{z - z_3} \tag{5.A2}$$

and

$$q(z) = \frac{1}{(z - z_1)(z - z_2)(z - z_3)} \left\{ \frac{\alpha_1\beta_1(z_1 - z_2)(z_1 - z_3)}{z - z_1} \right.$$

$$\left. + \frac{\alpha_2\beta_2(z_2 - z_1)(z_2 - z_3)}{z - z_2} + \frac{\alpha_3\beta_3(z_3 - z_1)(z_3 - z_2)}{z - z_3} \right\} \tag{5.A3}$$

Expressions (5.A2) and (5.A3) are designed to make it clear that (α_1, β_1), (α_2, β_2), and (α_3, β_3) are the exponent pairs at z_1, z_2, and z_3, respectively. For ∞ to be a regular point, it is necessary that

$$\sum_{n=1}^{3} (\alpha_n + \beta_n) = 1 \tag{5.A4}$$

In this form, Eq. (5.A1) is known as the *Papperitz equation*; its general solution is denoted by the *Riemann P-symbol*

$$u(z) \equiv P \left\{ \begin{array}{cccc} z_1 & z_2 & z_3 & \\ \alpha_1 & \alpha_2 & \alpha_3 & z \\ \beta_1 & \beta_2 & \beta_3 & \end{array} \right\} \tag{5.A5}$$

A standard form of the equation, with the three regular singular points at $0, 1, \infty$, is reached by the linear fractional transformation

$$\xi = \left(\frac{z - z_1}{z - z_3} \right) \left(\frac{z_2 - z_3}{z_2 - z_1} \right) \tag{5.A6}$$

(see Eqs. (4.19) and (4.22)). This puts the equation in the form

$$\xi(\xi - 1)\, u''(\xi) + \left[(1 - \alpha_1 - \beta_1)(\xi - 1) + (1 - \alpha_2 - \beta_2)\xi \right] u'(\xi)$$

$$+ \left(\alpha_3\beta_3 - \frac{\alpha_1\beta_1}{\xi} - \frac{\alpha_2\beta_2}{1 - \xi} \right) u(\xi) = 0 \tag{5.A7}$$

The general solution of Eq. (5.A7) can be expressed as

$$u(\xi) = P \left\{ \begin{array}{ccc} 0 & 1 & \infty \\ \alpha_1 & \alpha_2 & \alpha_3 \\ \beta_1 & \beta_2 & \beta_3 \end{array} \; \xi \right\} \tag{5.A8}$$

In this form, (α_1, β_1), (α_2, β_2), and (α_3, β_3) are the exponents of the regular singular points at $0, 1, \infty$, respectively. Now two of these exponents can be made to vanish by a suitable factorization of the unknown function. If we let

$$u(\xi) = \xi^{\alpha_1}(\xi - 1)^{\alpha_2} w(\xi) \tag{5.A9}$$

then $w(\xi)$ must satisfy

$$\xi(\xi - 1)\, w''(\xi) + [(1 + \alpha_1 - \beta_1)(\xi - 1) + (1 + \alpha_2 - \beta_2)\xi]\, w'(\xi)$$

$$+ (\alpha_1 + \alpha_2 + \alpha_3)(\alpha_1 + \alpha_2 + \beta_3)\, w(\xi) = 0 \tag{5.A10}$$

There are only three independent parameters in this equation, since two of the exponents have been forced to vanish by Eq. (5.A9), and there is still one relation imposed by Eq. (5.A4). A standard choice of parameters is given by

$$a = \alpha_1 + \alpha_2 + \alpha_3 \qquad b = \alpha_1 + \alpha_2 + \beta_3 \qquad c = 1 + \alpha_1 - \beta_1 \tag{5.A11}$$

Then Eq. (5.A4) requires

$$1 + \alpha_2 - \beta_2 = a + b - c + 1 \tag{5.A12}$$

and Eq. (5.A10) can be written as

$$\xi(\xi - 1)\, w''(\xi) + [(a + b + 1)\xi - c]\, w'(\xi) + ab\, w(\xi) = 0 \tag{5.A13}$$

This is the standard form of the *hypergeometric equation*.

➡ **Exercise 5.A1.**   Show that a formal general solution to Eq. (5.A13) is given by the P-symbol

$$w(\xi) = P \left\{ \begin{array}{ccc} 0 & 1 & \infty \\ 0 & 0 & a \\ 1 - c & c - a - b & b \end{array} \; \xi \right\}$$

from which it follows that the exponents of the singularity at $\xi = 0$ are 0 and $1 - c$, so that one solution of Eq. (5.A13) is analytic at $\xi = 0$.   ◻

A.2   Power Series Solutions

The particular solution of Eq. (5.A13) that is analytic at $\xi = 0$ and has $w(0) = 1$ is defined by the power series

$$F(a, b|c|\xi) = 1 + \frac{ab}{c}\xi + \frac{a(a + 1)b(b + 1)}{c(c + 1)}\xi^2 + \cdots$$

$$= \frac{\Gamma(c)}{\Gamma(a)\Gamma(b)} \sum_{n=0}^{\infty} \frac{\Gamma(a + n)\Gamma(b + n)}{\Gamma(c + n)\, n!}\xi^n \tag{5.A14}$$

unless $1 - c$ is a positive integer, a case to be dealt with later. The series on the right-hand side of Eq. (5.A14) is the *hypergeometric series*, and its analytic continuation is the *hypergeometric function* $F(a, b|c|\xi)$. The series has radius of convergence $r = 1$ in general, and the hypergeometric function has singularities, which turn out to be branch points in general, at 1 and ∞, unless a or b is a negative integer, in which case the series (5.A14) has only a finite number of terms, and the hypergeometric function is simply a polynomial.

Remark.   The polynomial solutions are especially important in physics, since they have no singularities in the finite ξ-plane, and the physical context in which the hypergeometric equation appears often demands solutions that are nonsingular. The Legendre polynomials described in Section 5.5, as well as other polynomials to be considered in Chapter 6, are solutions to special forms of the hypergeometric equation.   □

To find a second solution of the hypergeometric equation (5.A13) near $\xi = 0$, let

$$w(\xi) = \xi^{1-c} v(\xi) \tag{5.A15}$$

Then $v(\xi)$ satisfies another hypergeometric equation

$$\xi(\xi - 1)\, v''(\xi) + [(a + b - 2c + 3)\xi + c - 2]\, v'(\xi) + (a - c + 1)(b - c + 1)\, v(\xi) = 0 \tag{5.A16}$$

that has a solution analytic and nonzero at $\xi = 0$ given by the hypergeometric function $F(a - c + 1, b - c + 1|2 - c|\xi)$, unless $c - 1$ is a positive integer.

Thus, if c is not an integer, the general solution to the hypergeometric equation (5.A13) is

$$w(\xi) = A\, F(a, b|c|\xi) + B\xi^{1-c} F(a - c + 1, b - c + 1|2 - c|\xi) \tag{5.A17}$$

where A and B are arbitrary constants. If c is an integer, a second solution must be found by the method described in the previous section, or by analytic continuation in c in one or another of the representations derived below.

To study the behavior of the solution (5.A17) near $\xi = 1$, let $\eta \equiv 1 - \xi$ and let

$$v(\eta) \equiv w(1 - \eta) \tag{5.A18}$$

Then $v(\eta)$ satisfies the hypergeometric equation

$$\eta(\eta - 1)\, v''(\eta) + [(a + b + 1)\eta - (a + b - c + 1)]\, v'(\eta) + ab\, v(\eta) = 0 \tag{5.A19}$$

with the general solution

$$v(\eta) = C\, F(a, b|a + b - c + 1|\eta) + D\eta^{c-a-b} F(c - a, c - b|c - a - b + 1|\eta) \tag{5.A20}$$

(unless $c - a - b$ is an integer), where C and D are arbitrary constants. Thus the general solution of the hypergeometric equation (5.A13) can also be written as

$$\begin{aligned} w(\xi) &= C\, F(a, b|a + b - c + 1|1 - \xi) \\ &\quad + D(1 - \xi)^{c-a-b} F(c - a, c - b|c - a - b + 1|1 - \xi) \end{aligned} \tag{5.A21}$$

A.3   Integral Representations

Now recall the binomial expansion

$$(1 - u\xi)^{-a} = \sum_{m=0}^{\infty} \frac{\Gamma(a+m)}{\Gamma(a)} \frac{(u\xi)^m}{m!} \tag{5.A22}$$

and the integral (Eq. (4.A28))

$$\int_0^1 u^{b+m-1}(1-u)^{c-b-1}\, du = \frac{\Gamma(b+m)\Gamma(c-b)}{\Gamma(c+m)} \tag{5.A23}$$

($m = 0, 1, 2, \ldots$). These two can be combined to give the integral representation

$$F(a,b|c|\xi) = \frac{\Gamma(c)}{\Gamma(b)\Gamma(c-b)} \int_0^1 (1-u\xi)^{-a} u^{b-1}(1-u)^{c-b-1}\, du \tag{5.A24}$$

($\operatorname{Re} c > \operatorname{Re} b > 0$). Equation (5.A24) provides the analytic continuation of $F(a,b|c|\xi)$ to the complex ξ-plane with a branch cut along the positive real axis from 1 to ∞. With a new integration variable $t = 1/u$, this has the alternate form

$$F(a,b|c|\xi) = \frac{\Gamma(c)}{\Gamma(b)\Gamma(c-b)} \int_1^{\infty} (t-\xi)^{-a} t^{a-c}(t-1)^{c-b-1}\, dt \tag{5.A25}$$

Now suppose $0 < \operatorname{Re} b < \operatorname{Re} c < \operatorname{Re} a + 1$, and $0 < \operatorname{Re} a < 1$, and consider the integral

$$I \equiv \frac{1}{2i} \int_C (\xi-t)^{-a} t^{a-c}(1-t)^{c-b-1}\, dt \tag{5.A26}$$

where $0 < \xi < 1$ and C is the contour shown in Fig. 5.1. If we choose

   (i) the branch cut of t^{a-c} along the negative real axis from $-\infty$ to 0,

   (ii) the branch cut of $(\xi - t)^{-a}$ along the positive real axis from ξ to ∞, and

   (iii) the branch cut of $(1-t)^{c-b-1}$ along the positive real axis from 1 to ∞,

then the integral I can be written as

$$I = \sin \pi(a-c) \int_{-\infty}^0 (\xi-t)^{-a}(-t)^{a-c}(1-t)^{c-b-1}\, dt$$

$$+ \sin \pi a \int_{\xi}^1 (t-\xi)^{-a} t^{a-c}(1-t)^{c-b-1}\, dt \tag{5.A27}$$

$$+ \sin \pi(a+b-c+1) \int_1^{\infty} (t-\xi)^{-a} t^{a-c}(t-1)^{c-b-1}\, dt = 0$$

With $t = 1 - u$, the first term in I can be expressed as a hypergeometric function,

$$\int_{-\infty}^0 (\xi-t)^{-a}(-t)^{a-c}(1-t)^{c-b-1}\, dt =$$

$$= \int_1^{\infty} (u-1+\xi)^{-a} u^{c-b-1}(u-1)^{a-c}\, du \tag{5.A28}$$

$$= \frac{\Gamma(b)\Gamma(a-c+1)}{\Gamma(a+b-c+1)} F(a,b|a+b-c+1|1-\xi)$$

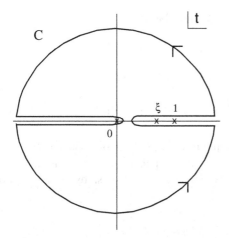

Figure 5.1: The contour C for the evaluation of the integral I in Eq. (5.A26).

Also, with $t - 1 - (1 - \xi)/w$, the second term can be expressed as

$$\int_{\xi}^{1} (t - \xi)^{-a} t^{a-c} (1 - t)^{c-b-1}\, dt =$$

$$= (1 - \xi)^{c-a-b} \int_{1}^{\infty} (w - 1 + \xi)^{a-c} w^{b-1} (w - 1)^{-a}\, dw \qquad (5.A29)$$

$$= \frac{\Gamma(c - b)\Gamma(1 - a)}{\Gamma(c - a - b + 1)} (1 - \xi)^{c-a-b} F(c - a, c - b | c - a - b + 1 | 1 - \xi)$$

The third term in I is directly related to the hypergeometric function in Eq. (5.A25).
From Eqs. (5.A27)–(5.A29) and the result

$$\sin \pi z = \frac{\pi}{\Gamma(z)\Gamma(1 - z)} \qquad (5.A30)$$

(see Eq. (4.A8)), it finally follows that

$$F(a, b | c | \xi) = \frac{\Gamma(c)\Gamma(c - a - b)}{\Gamma(c - a)\Gamma(c - b)} F(a, b | a + b - c + 1 | 1 - \xi)$$

$$\qquad (5.A31)$$

$$+ \frac{\Gamma(a + b - c)\Gamma(c)}{\Gamma(a)\Gamma(b)} (1 - \xi)^{c-a-b} F(c - a, c - b | c - a - b + 1 | 1 - \xi)$$

This formula provides the connection between the solutions of the hypergeometric equation
near $\xi = 0$ and the solutions near $\xi = 1$. The connection formula can be analytically continued
to all ξ, and to all values of a, b, c for which the functions are defined.

B   Confluent Hypergeometric Equation

B.1   Reduction to Standard Form

Another common linear second-order equation has the general form

$$u''(z) + \left(\frac{1 - \alpha - \beta}{z}\right) u'(z) + \left(\frac{\alpha\beta}{z^2} + \frac{2\kappa}{z} - \gamma^2\right) u(z) = 0 \qquad (5.\text{B}32)$$

where the choice of constants is motivated by the form of the solutions obtained below. This equation has a regular singular point at $z = 0$, with exponents α and β, and an irregular singular point at ∞. We can make one of the exponents at $z = 0$ vanish by the substitution

$$u(z) = z^\alpha v(z) \qquad (5.\text{B}33)$$

that leads to the equation

$$v''(z) + \left(\frac{1 + \alpha - \beta}{z}\right) v'(z) + \left(\frac{2\kappa}{z} - \gamma^2\right) v(z) = 0 \qquad (5.\text{B}34)$$

whose exponents at $z = 0$ are 0 and $\beta - \alpha$.

The asymptotic form of this equation for $z \to \infty$ is

$$v''(z) - \gamma^2 v(z) = 0 \qquad (5.\text{B}35)$$

This equation has solutions $\exp(\pm\gamma z)$, which suggests the substitution

$$v(z) = e^{-\gamma z} w(z) \qquad (5.\text{B}36)$$

Then $w(z)$ satisfies the equation

$$w''(z) + \left[\frac{1 + \alpha - \beta}{z} - 2\gamma\right] w'(z) + \left[\frac{2\kappa - \gamma(1 + \alpha - \beta)}{z}\right] w(z) = 0 \qquad (5.\text{B}37)$$

Now let $\xi = 2\gamma z$, $2\gamma a = \gamma(1 + \alpha - \beta) - 2\kappa$ and $c = 1 + \alpha - \beta$. Then Eq. (5.B37) becomes

$$\xi w''(\xi) + (c - \xi) w'(\xi) - a w(\xi) = 0 \qquad (5.\text{B}38)$$

This is the standard form of the *confluent hypergeometric equation*. It has a regular singular point at $\xi = 0$ with exponents 0 and $1 - c$, and an irregular singular point at ∞.

The solution $F(a|c|\xi)$ of Eq. (5.B38) that is analytic at $\xi = 0$ with $F(0) = 1$ is defined by the power series

$$F(a|c|\xi) = 1 + \frac{a}{c}\xi + \frac{a(a+1)}{c(c+1)}\xi^2 + \cdots = \frac{\Gamma(c)}{\Gamma(a)} \sum_{n=0}^{\infty} \frac{\Gamma(a+n)}{\Gamma(c+n)} \frac{\xi^n}{n!} \qquad (5.\text{B}39)$$

(again unless $1 - c$ is a positive integer, which will be considered later). $F(a|c|\xi)$ is the *confluent hypergeometric series*; it defines an entire function. It is a polynomial if a is zero or a negative integer; otherwise, it has an essential singularity at ∞.

Remark. Here again the polynomial solutions are very important in physics. The *Laguerre polynomials* and *Hermite polynomials* that appear in the quantum mechanics of the hydrogen atom and the simple harmonic oscillator can be expressed in terms of confluent hypergeometric functions. (see Section 6.5). Bessel functions (see Section 5.6) are not polynomials, but they also appear in solutions of the wave equation and other equations in physics. The various forms of *Bessel functions* are closely related to confluent hypergeometric functions (see also Exercise 5.20, for example).   □

To obtain a second solution of Eq. (5.B38) near $\xi = 0$, let

$$w(\xi) = \xi^{1-c}v(\xi) \tag{5.B40}$$

Then $v(\xi)$ satisfies a confluent hypergeometric equation

$$\xi v''(\xi) + (2 - c - \xi)v'(\xi) - (a - c + 1)v(\xi) = 0 \tag{5.B41}$$

This has a solution that is analytic and nonzero at $\xi = 0$ given by the confluent hypergeometric function $F(a - c + 1|2 - c|\xi)$, unless $c - 1$ is a positive integer.

Thus the general solution to the confluent hypergeometric equation (5.B38) is given by

$$w(\xi) = AF(a|c|\xi) + B\xi^{1-c}F(a - c + 1|2 - c|\xi) \tag{5.B42}$$

if c is not an integer, where A and B are arbitrary constants. If c is an integer, a second solution must be found by other methods, such as those discussed above.

Remark. Equation (5.B38) can be obtained from the hypergeometric equation

$$z(z - 1)w''(z) + [(a + b + 1)z) - c]w'(z) + abw(z) = 0 \tag{5.A13}$$

by letting $\xi \equiv bz$ and passing to the limit $b \to \infty$. Then the regular singular points at $\xi = b$ ($z = 1$) and ∞ in the hypergeometric equation merge to become the irregular singular point at ∞ in the confluent hypergeometric equation. Then also

$$F(a|c|\xi) = \lim_{b \to \infty} F(a, b|c|\xi/b) \tag{5.B43}$$

so the solution (5.B42) is a limit of the solution (5.A17) to the hypergeometric equation (5.A13) by the same confluence.   □

B.2   Integral Representations

$F(a|c|\xi)$ also has the integral representation

$$F(a|c|\xi) = \frac{\Gamma(c)}{\Gamma(a)\Gamma(c - a)} \int_0^1 e^{\xi t} t^{a-1} (1 - t)^{c-a-1} \, dt \tag{5.B44}$$

if $\mathrm{Re}\, c > \mathrm{Re}\, a > 0$. This can be verified by expanding the exponential $e^{\xi t}$ inside the integral and integrating term-by-term.

From this integral representation we have

$$F(a|c|\xi)\frac{\Gamma(c)}{\Gamma(a)\Gamma(c-a)}\left\{\int_{-\infty}^{1} e^{\xi t}t^{a-1}(1-t)^{c-a-1}\,dt\right.$$

$$\left.+\int_{0}^{-\infty} e^{\xi t}t^{a-1}(1-t)^{c-a-1}\,dt\right\} \tag{5.B45}$$

Now let $u = \xi(1-t)$ in the first integral, and $u = -\xi t$ in the second. Then we have

$$F(a|c|\xi) = \frac{\Gamma(c)}{\Gamma(a)\Gamma(c-a)}\left\{\xi^{a-c}e^{\xi}\int_{0}^{\infty} e^{-u}u^{c-a-1}\left(1-\frac{u}{\xi}\right)^{a-1}\,du\right.$$

$$\left.+(-\xi)^{-a}\int_{0}^{\infty} e^{-u}u^{a-1}\left(1+\frac{u}{\xi}\right)^{c-a-1}\,du\right\} \tag{5.B46}$$

The representation (5.B46) can be used to derive an asymptotic series for $F(a|c|\xi)$ by expanding the integrand in a power series in u/ξ, as explained in Section 1.4. In the resulting expansion, the first integral dominates for $\mathrm{Re}\,\xi > 0$, the second for $\mathrm{Re}\,\xi < 0$.

Another pair of linearly independent solutions of the confluent hypergeometric equation are the functions $U_1(a|c|\xi)$ and $U_2(a|c|\xi)$ defined by

$$U_1(a|c|\xi) \equiv \frac{1}{\Gamma(c-a)}\xi^{a-c}e^{\xi}\int_{0}^{\infty} e^{-u}u^{c-a-1}\left(1-\frac{u}{\xi}\right)^{a-1}\,du \tag{5.B47}$$

for $\mathrm{Re}\,c > \mathrm{Re}\,a > 0, 0 < \arg\xi < 2\pi$, and

$$U_2(a|c|\xi) \equiv \frac{1}{\Gamma(a)}(-\xi)^{-a}\int_{0}^{\infty} e^{-u}u^{a-1}\left(1+\frac{u}{\xi}\right)^{c-a-1}\,du \tag{5.B48}$$

for $\mathrm{Re}\,a > 0, -\pi < \arg\xi < \pi$. These functions are *confluent hypergeometric functions of the third kind*, or *Whittaker functions*. They are distinguished because they capture the two distinct forms of asymptotic behavior of solutions of the confluent hypergeometric equation as $z \to \infty$. It follows from Eq. (5.B46) that

$$F(a|c|\xi) = \frac{\Gamma(c)}{\Gamma(a)}U_1(a|c|\xi) + \frac{\Gamma(c)}{\Gamma(c-a)}U_2(a|c|\xi) \tag{5.B49}$$

for $0 < \arg\xi < \pi$.

Comparison of the asymptotic behavior of the integral representations also leads

$$U_1(a|c|\xi) = \frac{\Gamma(1-c)}{\Gamma(1-a)}e^{i\pi(a-c)}F(a|c|\xi)-\frac{\Gamma(c-1)}{\Gamma(c-a)}e^{i\pi a}\xi^{1-c}F(a-c+1|2-c|\xi) \tag{5.B50}$$

for $0 < \arg\xi < 2\pi$, and

$$U_2(a|c|\xi) = \frac{\Gamma(1-c)}{\Gamma(a-c+1)}e^{i\pi a}F(a|c|\xi)+\frac{\Gamma(c-1)}{\Gamma(a)}e^{i\pi a}\xi^{1-c}F(a-c+1|2-c|\xi) \tag{5.B51}$$

for $-\pi < \arg\xi < \pi$, which express the Whittaker functions in terms of the solutions defined near $\xi = 0$.

C   Elliptic Integrals and Elliptic Functions

A nonlinear differential equation that arises in mechanics is

$$\left(\frac{du}{dz}\right)^2 = (1 - u^2)(1 - k^2 u^2) \tag{5.C52}$$

where k^2 is a parameter that for now we will assume to satisfy $0 \le k^2 < 1$. Note that if $k^2 > 1$, we can introduce rescaled variables $w \equiv ku$, $\xi \equiv kz$. Then w satisfies

$$\left(\frac{dw}{d\xi}\right)^2 = (1 - w^2)(1 - \alpha^2 w^2) \tag{5.C53}$$

with $\alpha^2 \equiv 1/k^2 < 1$. It is left to the problems to show how this equation arises in the problem of a simple pendulum, and the problem of a nonlinear oscillator.

A solution $u(z)$ of this equation with $u(0) = 0$ is obtained from the integral

$$z = \int_0^u \frac{1}{\sqrt{(1 - t^2)(1 - k^2 t^2)}} \, dt \tag{5.C54}$$

with $t - \sin\phi$, this solution can also be written as

$$z = \int_0^\theta \frac{1}{\sqrt{1 - k^2 \sin^2\phi}} \, d\phi \tag{5.C55}$$

with $u = \sin\theta$. As u increases from 0 to 1 (or θ increases from 0 to $\pi/2$), z increases from 0 to $K(k)$, where

$$K(k) \equiv \int_0^1 \frac{1}{\sqrt{(1 - t^2)(1 - k^2 t^2)}} \, dt = \int_0^{\frac{\pi}{2}} \frac{1}{\sqrt{1 - k^2 \sin^2\phi}} \, d\phi \tag{5.C56}$$

is the *complete elliptic integral of the first kind*, of *modulus* k.

➜ **Exercise 5.C2.**   Show that the complete elliptic integral $K(k)$ defined by Eq. (5.C56) can be expressed as a hypergeometric function according to

$$K(k) = \tfrac{1}{2}\pi F(\tfrac{1}{2}, \tfrac{1}{2}|1|k^2) \tag{5.C57}$$

Hint. A brute force method is to expand the integrand in a power series and integrate term by term. It is more elegant to show that $K(k)$ satisfies a hypergeometric equation.   □

The function inverse to the function $z(u)$ defined by Eq. (5.C54) is

$$u(z) \equiv \operatorname{sn}(z, k) = \sin\theta \tag{5.C58}$$

This defines the *(Jacobi) elliptic function* $\operatorname{sn}(z, k)$ as a function that increases from 0 to 1 for $0 \le z \le K(k)$. The modulus k is not always written explicitly when it is fixed, so it is also common to write simply $u = \operatorname{sn} z$.

Now Eq. (5.C54) is derived from Eq. (5.C52) by choosing

$$\frac{du}{dz} = +\sqrt{(1 - u^2)(1 - k^2 u^2)} \tag{5.C59}$$

but as z increases beyond $K(k)$, the sign of the derivative must change, since it follows from a direct calculation that

$$\frac{d^2 u}{dz^2} = -4u \left[1 - \tfrac{1}{2}(1 + k^2)u^2\right] \tag{5.C60}$$

Thus $u''(z) < 0$ when $u = 1$ (we have assumed $0 \leq k^2 < 1$). In the dynamical systems described by this equation, the points $u = \pm 1$ correspond to the turning points of the system. Thus when we continue the integral in Eq. (5.C54) in the complex t-plane, we wrap around the lower side of the branch cut of $\sqrt{1 - t^2}$, which we can take to run from -1 to 1 as shown in Fig. 5.2. A complete clockwise circuit of the contour C around the branch cut from -1 to 1 corresponds to an increase of z by $4K(k)$, while $u = u(z)$ has returned to its original location on the Riemann surface of the integrand. Thus

$$\mathrm{sn}(z + 4K, k) = \mathrm{sn}(z, k) \tag{5.C61}$$

so that $\mathrm{sn}(z, k)$ is periodic with period $4K(k)$.

From Fig. 5.2, it is clear that there are other closed contours on the Riemann surface of the integrand in Eq. (5.C54) that cannot be deformed into C without crossing one or more branch points of the integrand[3] (these branch points are at $t = \pm 1$ and at $t = \pm \alpha \equiv \pm 1/k$; the integrand is analytic at ∞). Note that these branch points are of the square root type, so the integrand changes sign on encircling one of the branch points. Hence if we make a circuit around a closed contour L that encircles any *two* of the branch points, the integral will return to its original value. If we let

$$P \equiv \oint_L \frac{1}{\sqrt{(1 - t^2)(1 - k^2 t^2)}} \, dt \neq 0 \tag{5.C62}$$

then P will be a period for the function $u = u(z)$.

A closed contour that cannot be deformed into the contour C is the contour C' shown in Fig. 5.2, which encircles the branch points of the integrand at $t = +1$ and $t = +\alpha$. The corresponding period is given by

$$P = 2i \int_1^\alpha \frac{1}{\sqrt{(t^2 - 1)(1 - k^2 t^2)}} \, dt = 2i \int_0^1 \frac{1}{\sqrt{(1 - \xi^2)(1 - k'^2 \xi^2)}} \, d\xi \tag{5.C63}$$

$$= 2iK(k') \equiv 2iK'(k)$$

Here $k'^2 \equiv 1 - k^2$ (k' is the *complementary modulus* of the elliptic integral, and the second integral is obtained by the change of variable

$$\xi^2 \equiv \frac{1 - k^2 t^2}{1 - k^2} \tag{5.C64}$$

---

[3]When we speak of deformations of contours here and below, we understand that the deformations are forbidden to cross singularities of the integrand unless explicitly stated otherwise.

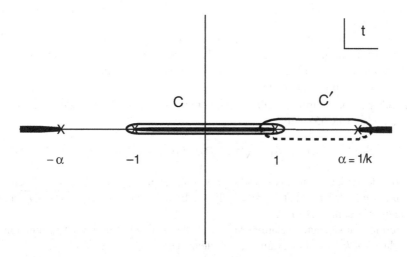

Figure 5.2: The contour C for continuation of the integral in Eq. (5.C54) beyond $u = 1$; a complete circuit of C returns the integrand to its original value. Shown also is a second closed contour C' around which the integrand returns to its original value. The solid portion of C' lies on the first (visible) sheet of the Riemann surface of the integrand; the dashed portion lies on the second sheet reached by passing through any of the cuts.

Thus $\mathrm{sn}(z, k)$ is a doubly periodic function of z with periods

$$\Omega \equiv 4K(k) \qquad \Omega' \equiv 2iK(k') \tag{5.C65}$$

Note that any contour on the Riemann surface shown in Fig. 5.2 must encircle an even number of branch points in order to return to the starting value of the integrand (multiple circuits are possible, but each circuit must be included in the count); only such contours are closed *on the Riemann surface*. Furthermore, the integral around a circle centered at $t = 0$, with radius $r > \alpha$, vanishes since the integrand is analytic at ∞ (thus the radius of the circle can be made arbitrarily large without changing the value of the integral, and the integral evidently vanishes for $r \to \infty$).

Any contour that encircles an even number of branch points of the integrand can be deformed into a contour that makes m circuits of C and m' circuits of C', plus a contour around which the integral vanishes, such as one or more circuits of circle of radius $r > \alpha$; here m and m' are integers that may be positive (for clockwise circuits), negative (for counterclockwise circuits) or zero. It requires a bit of sketching with pencil and paper to verify this statement, but it is true. The period associated with such a contour is given by

$$P = m\Omega + m'\Omega' \tag{5.C66}$$

as might be expected from the general argument given in Chapter 4 that there can be no more than two fundamental periods for a nonconstant analytic function. Now Ω and Ω' are in fact fundamental periods of $\mathrm{sn}(z, k)$, since there are no smaller contours than C and C' that might subdivide these periods.

Remark. The two-sheeted Riemann surface illustrated in Fig. 4.1 is isomorphic to a torus T^2, since the elliptic integral provides a diffeomorphism from the Riemann surface onto the unit cell of the lattice shown in Fig. 4.5 (the elliptic function sn z provides the inverse mapping). This unit cell is evidently equivalent to the rectangle in Fig. 3.4, which we have seen in Chapter 3 to be equivalent to a torus. □

There are other Jacobi elliptic functions related to the integrals (5.C54) and (5.C55). With z and θ related by Eq. (5.C55), define

$$\mathrm{cn}(z, k) \equiv \cos\theta \qquad \mathrm{dn}(z, k) \equiv \sqrt{1 - k^2 \sin^2\theta} \tag{5.C67}$$

where again the dependence on k is not always written explicitly. Jacobi also defined $\theta \equiv$ am z, the *amplitude* of z. Note that for real θ, dn z is bounded between $k' = \sqrt{1 - k^2}$ and 1, so that it can never become negative.

These functions satisfy many relations similar to those for the trigonometric functions to which they reduce for $k \to 0$. For example, it is clear that

$$\mathrm{sn}^2 z + \mathrm{cn}^2 z = 1 \tag{5.C68}$$

$$k^2 \mathrm{sn}^2 z + \mathrm{dn}^2 z = 1 \tag{5.C69}$$

The derivatives of the functions are easily evaluated, since Eq. (5.C55) implies

$$\frac{dz}{d\theta} = \frac{1}{\sqrt{1 - k^2 \sin^2\theta}} = \frac{1}{\mathrm{dn}\, z} \tag{5.C70}$$

Then we have

$$\frac{d}{dz}\mathrm{sn}\, z = \cos\theta\frac{d\theta}{dz} = \mathrm{cn}\, z\, \mathrm{dn}\, z \tag{5.C71}$$

and similarly,

$$\frac{d}{dz}\mathrm{cn}\, z = -\mathrm{sn}\, z\, \mathrm{dn}\, z \qquad \frac{d}{dz}\mathrm{dn}\, z = -k^2 \mathrm{sn}\, z\, \mathrm{cn}\, z \tag{5.C72}$$

It follows directly from the definitions that

$$\mathrm{sn}(-z) = -\mathrm{sn}\, z \qquad \mathrm{cn}(-z) = \mathrm{cn}\, z \qquad \mathrm{dn}(-z) = \mathrm{dn}\, z \tag{5.C73}$$

Also, since $z \to z + 2K$ corresponds to $\theta \to \theta + \pi$, we have

$$\mathrm{sn}(z + 2K) = -\mathrm{sn}\, z \qquad \mathrm{cn}(z + 2K) = -\mathrm{cn}\, z \qquad \mathrm{dn}(z + 2K) = \mathrm{dn}\, z \tag{5.C74}$$

Thus cn z has a real period $4K$, while dn z has a real period $2K$.

To see what happens when z increases by $2iK'$, recall that this corresponds to one circuit of the contour C' in Fig. 5.2. As we go around this circuit, cn $z = \sqrt{1 - t^2}$ and dn $z = \sqrt{1 - k^2 t^2}$ change sign, while sn z returns to its initial value; thus we have

$$\mathrm{sn}(z + 2iK') = \mathrm{sn}\, z \qquad \mathrm{cn}(z + 2iK') = -\mathrm{cn}\, z \qquad \mathrm{dn}(z + 2iK') = -\mathrm{dn}\, z \tag{5.C75}$$

Equations (5.C74) and (5.C75) show that cn z has a complex fundamental period $2K + 2iK'$, while Eq. (5.C75) shows that dn z has a second fundamental period $4iK'$.

To locate the poles of sn z, we return to the integral (5.C54) and let $u \to \infty$. Then

$$z = \int_0^1 \frac{1}{\sqrt{(1 - t^2)(1 - k^2 t^2)}}\, dt + i \int_1^{1/k} \frac{1}{\sqrt{(t^2 - 1)(1 - k^2 t^2)}}\, dt$$

$$\pm \int_{1/k}^{\infty} \frac{1}{\sqrt{(t^2 - 1)(k^2 t^2 - 1)}}\, dt \tag{5.C76}$$

The sign ambiguity in z reflects the fact that there are two distinct routes from 1 to ∞, one passing above the cut in Fig. 5.2, the other passing below. The first integral is equal to $K(k)$, the second integral is equal to $K'(k)$, while the third integral is also equal to $K(k)$, which is most easily shown by changing variables to $\xi \equiv 1/kt$. Thus the values of z for which $u = \operatorname{sn} z \to \infty$ are given by

$$z = iK' \qquad 2K + iK' \tag{5.C77}$$

together with the points obtained by translation along the lattice generated by the fundamental periods $4K$ and $2iK'$. These points correspond to contours in the t-plane that make extra circuits of the branch points shown in Fig. 5.2.

To show that these points are simple poles of $u = \operatorname{sn} z$, note that

$$z - iK' = + \int_u^{\infty} \frac{1}{\sqrt{(t^2 - 1)(k^2 t^2 - 1)}}\, dt \tag{5.C78}$$

$$z - 2K - iK' = - \int_u^{\infty} \frac{1}{\sqrt{(t^2 - 1)(k^2 t^2 - 1)}}\, dt \tag{5.C79}$$

Since the integral on the right-hand side is approximated by $1/ku$ for $u \to \infty$, we have

$$\lim_{z \to iK'} (z - iK')\, \operatorname{sn} z = \frac{1}{k} \tag{5.C80}$$

$$\lim_{z \to 2K + iK'} (z - 2K - iK')\, \operatorname{sn} z = -\frac{1}{k} \tag{5.C81}$$

Thus the singularities of sn z in its fundamental parallelogram are two simple poles, with equal and opposite residues. From the relations (5.C68) and (5.C69), it follows that cn z and dn z have poles at the same points.

➜ **Exercise 5.C3.**   Compute the residues of cn z and dn z at their poles at $z = iK'$ and at $z = 2K - iK'$.   □

➜ **Exercise 5.C4.**   Show that

$$E(e) = \tfrac{1}{2}\pi F(\tfrac{1}{2}, -\tfrac{1}{2}|1|e^2)$$

by one means or another.   □

Bibliography and Notes

There are many introductory books on differential equations. One highly recommended book is

    Martin Braun, *Differential Equations and Their Applications* (4th edition), Springer (1993).

This is a substantial book that covers many interesting examples and applications at a reasonably introductory level. Another excellent introduction is

    George F. Simmons, *Differential Equations with Applications and Historical Notes* (2nd edition), McGraw-Hill (1991).

In addition to covering standard topics, this book has digressions into various interesting applications, as well as many biographical and historical notes that enrich the book beyond the level of an ordinary textbook.

    An advanced book that covers the analytical treatment of differential equations in great depth is

    Carl M. Bender and Steven A. Orszag, *Advanced Mathematical Methods for Scientists and Engineers*, McGraw-Hill (1978).

This book treats a wide range of analytic approximation methods, and has an excellent discussion of asymptotic methods.

    The treatment of linear equations and the special functions associated with the linear second-order equation is classical; more detail can be found in the work of Whittaker and Watson cited in Chapter 1. Detailed discussions of the second-order differential equations arising from the partial differential equations of physics, and for understanding the properties of these solutions, can be found in the classic work of Morse and Feshbach cited in Chapter 8.

    Elliptic functions and elliptic integrals are treated in the book by Copson cited in Chapter 4, as well as in the book by Whittaker and Watson. There is also a book

    Harry E. Rauch and Aaron Lebowitz, *Elliptic Functions, Theta Functions and Riemann Surfaces*, Williams and Wilkins (1973).

that describes the Riemann surfaces associated with the elliptic functions in great detail. A modern treatment of elliptic functions can be found in

    K. Chandrasekharan, *Elliptic Functions*, Springe (1985).

This book gives a through survey of the varieties of elliptic functions and the related theta-functions and modular functions from a contemporary point of view.

    A nice treatment of elliptic functions in the context of soluble two-dimensional models in statistical mechanics is given by

    Rodney J. Baxter, *Exactly Solved Models in Statistical Mechanics*, Academic Press (1982).

It should not be surprising that doubly periodic functions are exceptionally useful in describing physics on a two-dimensional lattice.

Problems

1. Find the general solution of the differential equation

$$u(z)u''(z) - [u'(z)]^2 = 6z\,[u(z)]^2$$

2. Consider the differential equation

$$(1-z)\,u''(z) + z\,u'(z) - u(z) = (1-z)^2$$

Evidently $u(z) = z$ is a solution of the related homogeneous equation. Find the general solution of the inhomogeneous equation.

3. An nth order linear differential operator of the form (5.50) is *factorizable* if it can be written in the form

$$L = \left[\frac{d}{dz} + s_1(z)\right] \cdots \left[\frac{d}{dz} + s_n(z)\right]$$

$$= \prod_{k=1}^{n} \left[\frac{d}{dz} + s_k(z)\right]$$

with known functions $s_1(z), \ldots, s_n(z)$.

Show that if L is factorizable, then the general solution of the equation $L[u] = f$ can be constructed by solving the sequence of linear inhomogeneous first-order equations

$$\left[\frac{d}{dz} + s_k(z)\right] u_k(z) = u_{k-1}(z)$$

$(k = 1, \ldots, n)$ with $u_0(z) = f(z)$.

4. The time evolution of a radioactive decay chain (or sequence of irreversible chemical reactions) $1 \to 2 \to \ldots \to n$ is described by the set of coupled differential equations

$$u_1'(t) = -\lambda_1\,u_1(t)$$

$$u_2'(t) = -\lambda_2\,u_2(t) + \lambda_1\,u_1(t)$$

$$u_3'(t) = -\lambda_3\,u_3(t) + \lambda_2\,u_2(t)$$

$$\vdots$$

$$u_n'(t) = \lambda_{n-1}\,u_{n-1}(t)$$

is the population of species k at time t, and λ_k is the rate per unit time for the process $k \to k+1$.

(i) Find explicit solutions for $u_1(t)$, $u_2(t)$ and $u_3(t)$ in terms of the initial values $u_1(0)$, $u_2(0)$, and $u_3(0)$.

(ii) At what time will $u_2(t)$ reach its maximum value?

(iii) Find a set of basic solutions $\{u_k^{(m)}(t)\}$ that satisfy

$$u_k^{(m)}(0) = \delta_{km}$$

$(k, m = 1, \ldots, n)$, and interpret these solutions physically.

(iv) Express the solution $\{u_k(t)\}$ that satisfies

$$u_k(0) = c_k$$

$(k = 1, \ldots, n)$ in terms of these basic solutions.

5. In the preceding problem, suppose we start with a pure sample of species 1, so that

$$u_1(0) = N \qquad u_2(0) = 0 \qquad u_3(0) = 0$$

(i) How long will it take for $u_1(t)$ to fall to half its initial value?

(ii) At what time will $u_2(t)$ reach its maximum value? What is this maximum value?

(iii) If species 3 is the endpoint of the reaction (this is equivalent to setting $\lambda_3 = 0$), how long will it take for $u_3(t)$ to reach half its final value?

6. Consider the nth order linear differential equation with constant coefficients

$$L[u] \equiv u^{(n)}(z) + \alpha_1 u^{(n-1)}(z) + \cdots + \alpha_n u(z) = 0$$

with constant coefficients (Eq. (5.73)).

(i) Express this equation as a matrix equation

$$v' = \mathbf{A}v$$

with vector $v = (u, u', \ldots, u^{(n-1)})$.

(ii) Show that the eigenvalues of $\mathbf{A}$ are exactly the roots of the polynomial $p(\lambda)$ in Eq. (5.75).

(iii) Express the matrix $\mathbf{A} = \mathbf{D} + \mathbf{N}$, as defined in Eq. (2.151).

7. Consider the differential equation

$$\frac{d^2 u}{dz^2} + \left(\lambda - c\,\frac{e^{-\alpha z}}{z}\right) u(z) = 0$$

(i) Locate and classify the singular points of this equation.

(ii) Find a solution that is analytic at $z = 0$, neglecting terms that are $o(z^3)$.

(iii) Find a linearly independent solution near $z = 0$, neglecting terms that are $o(z^2)$.

8. (i) Find the radius of convergence of the series

$$S(t, z) = \sum_{n=0}^{\infty} t^n P_n(z) = \frac{1}{\sqrt{1 - 2zt + t^2}}$$

as a function of t, if z is real and $-1 \le z \le 1$.

(ii) Find the region of convergence in the complex z-plane if t is real with $|t| < 1$.

9. Find the coefficients $c_n(z)$ in the expansion

$$\frac{1 - t^2}{(1 - 2zt + t^2)^{3/2}} = \sum_{n=0}^{\infty} c_n(t) P_n(z)$$

where $|t| < 1$ and the $P_n(z)$ are the Legendre polynomials. For what values of z does this series converge?

10. Consider the representation (5.153)

$$Q_\lambda(z) = \frac{1}{2^{\lambda+1}} \int_{-1}^{1} \frac{(1 - \xi^2)^\lambda}{(z - \xi)^{\lambda+1}} \, d\xi$$

of the Legendre function of the second kind. Use this representation together with the Rodrigues formula to show that for $\lambda = n$ a positive integer,

$$Q_n(z) = \frac{1}{2} \int_{-1}^{1} \frac{P_n(t)}{z - t} \, dt$$

where $P_n(t)$ is the corresponding Legendre polynomial.

11. (i) Show that the Wronskian determinant (Eq. (5.62)) of $P_\lambda(z)$ and $Q_\lambda(z)$ defined by

$$W(P_\lambda, Q_\lambda) = P_\lambda Q'_\lambda - P'_\lambda Q_\lambda$$

satisfies the differential equation

$$\frac{1}{W} \frac{dW}{dz} = -\ln(z^2 - 1)$$

and thus

$$P_\lambda(z)Q'_\lambda(z) - P'_\lambda(z)Q_\lambda(z) = \frac{A}{z^2 - 1}$$

(ii) Evaluate the constant A. *Note.* One way to do this is to consider the asymptotic behavior of the equation for $z \to \infty$.

(iii) Show that if we define $Q_\lambda(z) = q_\lambda(z)P_\lambda(z)$, then $q_\lambda(z)$ satisfies

$$q'_\lambda(z)[P_\lambda(z)]^2 = \frac{A}{z^2 - 1}$$

(iv) Finally, show that

$$Q_\lambda(z) = AP_\lambda(z) \int_z^\infty \frac{1}{(t^2 - 1)[P_\lambda(t)]^2} \, dt$$

12. Show that the Legendre functions of the second kind satisfy the same recursion relations

$$(2\lambda + 1)zQ_\lambda(z) = (\lambda + 1)Q_{\lambda+1}(z) + \lambda Q_\lambda(z)$$

and

$$(z^2 - 1)Q'_\lambda(z) = \lambda z Q_\lambda(z) - \lambda Q_{\lambda-1}(z)$$

as the Legendre polynomials and functions of the first kind (see Exercise 5.14).

13. Show that the Bessel function $J_\lambda(z)$ has the integral representation

$$J_\lambda(z) = C_\lambda \left(\frac{z}{2}\right)^\lambda \int_{-1}^1 e^{izu}(1 - u^2)^{\lambda-\frac{1}{2}} \, du$$

$$= C_\lambda \left(\frac{z}{2}\right)^\lambda \int_0^\pi e^{iz\cos\theta}(\sin\theta)^{2\lambda} \, d\theta$$

and evaluate the constant C_λ.

14. Use the result of the previous problem to show that the spherical Bessel function

$$j_\lambda(z) = \sqrt{\frac{\pi}{2z}} \, J_{\lambda+\frac{1}{2}}(s)$$

has the integral representation

$$j_\lambda(z) = \sqrt{\pi}C_\lambda \left(\frac{z}{2}\right)^{\lambda+\frac{1}{2}} \int_0^\pi e^{iz\cos\theta}(\sin\theta)^{2\lambda+1} \, d\theta$$

Then show that for integer n,

$$j_n(z) = A_n \int_{-1}^1 e^{-iz\cos\theta} P_n(\cos\theta) \, d(\cos\theta)$$

and evaluate the constant A_n.

15. Show that the hypergeometric function satisfies the recursion formulas

(i) $aF(a+1, b|c|\xi) - bF(a, b+1|c|\xi) = (a - b) \, F(a, b|c|\xi)$

(ii) $F(a+1, b|c|\xi) - F(a, b+1|c|\xi) = \left(\dfrac{b - a}{c}\right) \xi \, F(a+1, b+1|c+1|\xi)$

(iii) $\dfrac{d}{d\xi} F(a, b|c|\xi) = \dfrac{ab}{c} F(a+1, b+1|c+1|\xi)$

16. Show that the general solution of the hypergeometric equation can be expressed as

$$w(\xi) = A_1 \xi^{-a} F(a, a - c + 1 | a - b + 1 | 1/\xi)$$

$$+ A_2 \, \xi^{-b} F(b - c + 1, b | b - a + 1 | 1/\xi)$$

and find the coefficients A_1 and A_2 when $w(\xi) = F(a, b | c | \xi)$.

17. Show that the confluent hypergeometric function satisfies the recursion formulas

(i) $aF(a + 1 | c + 1 | \xi) - cF(a | c | \xi) = (a - c)F(a | c + 1 | \xi)$

(ii) $c[F(a + 1 | c | \xi) - F(a | c | \xi)] = \xi F(a + 1 | c + 1 | \xi)$

(iii) $\dfrac{d}{d\xi} F(a | c | \xi) = \dfrac{a}{c} F(a + 1 | c + 1 | \xi)$

18. Show that the confluent hypergeometric function satisfies

$$F(c - a | c | \xi) = e^{\xi} F(a | c | - \xi)$$

using the integral representations given in Appendix B.

19. Conservation of energy for a simple pendulum (mass m, length L, angular displacement θ) has the form

$$\frac{1}{2} m L^2 \left(\frac{d\theta}{dt} \right)^2 + mgL(1 - \cos \theta) = E$$

(i) Introduce the variable $u \equiv \sin \frac{1}{2}\theta$ and rescale the time variable in order to reduce the conservation of energy equation to an equation of the form (5.C52).

(ii) Derive an expression for the period of the pendulum in terms of a complete elliptic integral.

20. Consider an anharmonic one-dimensional oscillator with potential energy

$$V(x) = \tfrac{1}{2} m \omega^2 x^2 + \tfrac{1}{2} m \sigma x^4$$

Sketch the potential energy function for each of the cases:

(i) $\sigma > 0$, $\omega^2 > 0$ (ii) $\sigma > 0$, $\omega^2 < 0$ (iii) $\sigma < 0$, $\omega^2 > 0$

In each case, introduce new variables so that the conservation of energy equation

$$\frac{m}{2} \left(\frac{dx}{dt} \right)^2 + V(x) = E$$

takes the form of Eq. (5.C52). Under what conditions can the variables in the resulting equation be chosen so that the parameter k^2 satisfies $0 < k^2 < 1$?

21. Show that the change of variable $x = u^2$ transforms Eq. (5.C52) into

$$\left(\frac{dx}{dz}\right)^2 = 4x(1-x)(1-k^2x)$$

Remark. Thus the quartic polynomial on the right-hand side of Eq. (5.C52) can be transformed into a cubic. This does not reduce the number of branch points of dx/dz, but simply relocates one of them to ∞. □

22. Show that the circumference of the ellipse

$$\frac{x^2}{a^2} + \frac{y^2}{b^2} = 1$$

is given by $4aE(e)$, where $e = \sqrt{a^2 - b^2}/a$ is the *eccentricity* of the ellipse and

$$E(e) \equiv \int_0^1 \sqrt{\frac{1 - e^2t^2}{1 - t^2}}\, dt$$

is the *complete elliptic integral* of the *second kind* (of *modulus e*).

6 Hilbert Spaces

Linear vector spaces are often useful in the description of states of physical systems and their time evolution, as already noted in Chapter 2. However, there are many systems whose state space is infinite dimensional, i.e., there are infinitely many degrees of freedom of the system. For example, a vibrating string allows harmonics with any integer multiple of the fundamental frequency of the string. More generally, any system described by a linear wave equation will have an infinite set of normal modes and corresponding frequencies. Thus we need to understand how the theory of infinite-dimensional linear vector spaces differs from the finite-dimensional theory described in Chapter 2.

The theory of linear operators and their spectra is also more complicated in infinite dimensions. Especially important is the existence of operators with spectra that are continuous, or unbounded, or even nonexistent. In Chapter 7, we will examine the theory of linear operators, especially differential and integral operators, in infinite-dimensional spaces.

In this chapter, we deal first with the subtle issues of convergence that arise in infinite dimensional spaces. We then proceed to discuss the expansion of functions in terms of various standard sets of basis functions—the Fourier series in terms of trigonometric functions, and expansions of functions in terms of certain sets of orthogonal polynomials that arise in the solution of second-order differential equations. The reader who is more interested in practical applications can proceed directly to Section 6.3 where Fourier series are introduced.

As a prototype for an infinite-dimensional vector space, we consider first the space $\ell^2(\mathbf{C})$ of sequences $x \equiv (\xi_1, \xi_2, \ldots)$ of complex numbers with

$$\sum_{k=1}^{\infty} |\xi_k|^2 < \infty$$

A scalar product of $x = (\xi_1, \xi_2, \ldots)$ and $y = (\eta_1, \eta_2, \ldots)$ given by

$$(x, y) \equiv \sum_{k=1}^{\infty} \xi_k{}^* \eta_k$$

satisfies the axioms appropriate to a unitary vector space, and we show that the series converges if x and y are in $\ell^2(\mathbf{C})$. The space $\ell^2(\mathbf{C})$ satisfies two new axioms that characterize a Hilbert space: the axiom of separability, which requires the existence of a countable basis, and the axiom of completeness, which requires that the limit of a Cauchy sequence to belong to the space. Thus the space $\ell^2(\mathbf{C})$ provides a model Hilbert space in the same way that n-tuples of real or complex numbers provided model finite-dimensional vector spaces.

Introduction to Mathematical Physics. Michael T. Vaughn
Copyright © 2007 WILEY-VCH Verlag GmbH & Co. KGaA, Weinheim
ISBN: 978-3-527-40627-2

Functions defined on a domain Ω in $\mathbf{R}^n$ or $\mathbf{C}^n$ also have a natural vector space structure, and a scalar product

$$(f, g) = \int_\Omega f^*(x)g(x)w(x)dx$$

can be introduced that satisfies the standard axioms. Here $w(x)$ is a nonnegative weight function that can be introduced into the scalar product; often $w(x) = 1$ everywhere on Ω, but there are many problems for which other weight functions are useful. To construct a Hilbert space, the space of continuous functions must be enlarged to include the limits of Cauchy sequences of continuous functions, just as the set of rational numbers needed to be enlarged to include the limits of Cauchy sequences of rationals. This expanded space, denoted by $L^2(\Omega)$, includes "functions" for which

$$\|f\|^2 \equiv \int_\Omega |f(x)|^2 \, w(x) \, d\Omega$$

is finite. Here the integral understood as a Lebesgue integral, which is a generalization of the Riemann integral of elementary calculus. We discuss briefly the Lebesgue integral after a short introduction to the concept of measure, on which Lebesgue integration is based. We also note that the elements of $L^2(\Omega)$ need not be functions in the classical sense, since they need not be continuous, or even well defined at every point of Ω.

Convergence of sequences of vectors in Hilbert space is subtle: there are no less than four types of convergence: uniform convergence and pointwise convergence as defined in classical analysis, weak convergence as defined in Chapter 1 and again here in a vector space context, and strong convergence in Hilbert space, also known as convergence in the mean (weak and strong convergence are equivalent in a finite-dimensional space).

Every continuous function can be expressed as the limit of a uniformly convergent sequence of polynomials (Weierstrass approximation theorem). On the other hand, a Cauchy sequence of polynomials need not have a pointwise limit. While these subtleties are not often crucial in practice, it is important to be aware of their existence in order to avoid drawing false conclusions on the basis of experience with finite-dimensional vector spaces.

There are various sets of basis functions in a function space that are relevant to physical applications. The classical example is *Fourier series* expansion of a periodic function in terms of complex exponential functions (or in terms of the related trigonometric functions). If $f(t)$ is a well-behaved function with period T, then it can be expressed as a Fourier series in terms of complex exponential functions as

$$f(t) = \sqrt{\frac{1}{T}} \sum_{n=-\infty}^{\infty} c_n \exp\left(\frac{2\pi int}{T}\right)$$

with complex Fourier coefficients c_n given by

$$c_n = \sqrt{\frac{1}{T}} \int_0^T \exp\left(-\frac{2\pi int}{T}\right) f(t) \, dt$$

Note that $c_{-n} = c_n^*$ if $f(t)$ is real. Each Fourier series represents a function with period T as a superposition of functions corresponding to pure oscillators with frequencies $f_n = nf_0$, where

$f_0 = 1/T$ is the fundamental frequency corresponding to period T; the Fourier coefficients c_n and c_{-n} are amplitudes of the frequency component f_n in the function $f(t)$. One peculiarity we note is the Gibbs phenomenon in the Fourier series expansion of a step function; this is a phenomenon that generally occurs in the approximation of discontinuous functions by series of continuous functions.

Other important sets of bases are certain families of orthogonal polynomials that appear as solutions to second-order linear differential equations. In particular, there are several families of orthogonal polynomials related to the hypergeometric equation and the confluent hypergeometric equation discussed in Chapter 5. Some important properties of the Legendre polynomials are derived in detail, and the corresponding properties of Gegenbauer, Jacobi, Laguerre, and Hermite polynomials are summarized in Appendix A.

A nonperiodic function $f(t)$ that vanishes for $t \to \pm\infty$ can be expressed as a superposition of functions with definite (angular) frequency ω by the *Fourier integral*

$$f(t) = \frac{1}{2\pi} \int_{-\infty}^{\infty} c(\omega) e^{-i\omega t} \, d\omega$$

which is obtained from the Fourier series by carefully passing to the limit $T \to \infty$. The function $c(\omega)$ is obtained from $f(t)$ by a similar integral

$$c(\omega) = \int_{-\infty}^{\infty} e^{i\omega t} f(t) \, dt$$

The functions $f(t)$ and $c(\omega)$ are *Fourier transforms* of each other. If $f(t)$ describes the evolution of a system as a function of time t, then $c(\omega)$ describes the frequency spectrum ($\omega = 2\pi f$) associated with the system. If t is actually a spatial coordinate x, then the corresponding Fourier transform variable is often denoted by k (the *wave number*, or *propagation vector*).

Another integral transform, useful for functions $f(t)$ defined for $0 \le t < \infty$, is the *Laplace transform*

$$\mathcal{L}f(p) \equiv \int_{0}^{\infty} f(t) e^{-pt} \, dt$$

which is obtained from the Fourier transform by rotation of the integration contour in the complex t-plane. These transforms, as well as the related Mellin transform introduced as a problem, have various applications. Here the transforms are used to evaluate integrals, and to find solutions to certain second-order linear differential equations.

There are many physical problems in which it is important to analyze the behavior of a system on various scales. For example, renormalization group and block spin methods in quantum field theory and statistical mechanics explicitly consider the behavior of physical quantities on a wide range of scales. This leads to the concept of *multiresolution analysis* using a set of basis functions designed to explore the properties of a function at successively higher levels of resolution. This concept is illustrated here with the Haar functions. The new basis functions introduced in passing from one resolution to a finer resolution are *wavelets*. Some methods of constructing wavelet bases are described in Section 6.6.

6.1   Infinite-Dimensional Vector Spaces

6.1.1   Hilbert Space Axioms

The basic linear vector space axioms admit infinite-dimensional vector spaces, and further axioms are needed to define more clearly the possibilities. For example, consider the linear vector space $\mathbf{C}^\infty$ whose elements are sequences of complex numbers of the form $x = (\xi_1, \xi_2, \ldots)$, $y = (\eta_1, \eta_2, \ldots)$, $\ldots$, with addition and multiplication by scalars defined in the obvious way. This vector space satisfies the basic axioms, and it has a basis

$$\phi_1 = (1, 0, 0, \ldots)$$
$$\phi_2 = (0, 1, 0, \ldots) \tag{6.1}$$
$$\phi_3 = (0, 0, 1, \ldots)$$
$$\vdots$$

that is *countable*, i.e., in one-to-one correspondence with the positive integers. This space is too large for most purposes; if we introduce a scalar product by

$$(x, y) \equiv \sum_{k=1}^{\infty} \xi_k^* \eta_k \tag{6.2}$$

then the infinite sum on the right-hand side does not converge in general. However, we can restrict the space to include only those vectors $x = (\xi_1, \xi_2, \ldots)$ for which

$$\|x\|^2 \equiv \sum_{k=1}^{\infty} |\xi_k|^2 < \infty \tag{6.3}$$

that is, vectors of finite length $\|x\|$ as defined by Eq. (6.3). This defines a space $\ell^2(\mathbf{C})$, which is evidently a vector space; the addition of two vectors of finite length leads to a vector of finite length since

$$|\xi_k + \eta_k|^2 \leq 2(|\xi_k|^2 + |\eta_k|^2) \tag{6.4}$$

It is also a unitary vector space; if $\|x\|$ and $\|y\|$ are finite, then the scalar product defined by Eq. (6.2) is an absolutely convergent series since

$$|\xi_k \eta_k| \leq \tfrac{1}{2}(|\xi_k|^2 + |\eta_k|^2) \tag{6.5}$$

The two important properties of $\ell^2(\mathbf{C})$ that require axioms beyond those for a finite-dimensional unitary vector space are:

1. $\ell^2(\mathbf{C})$ is *separable*. This means that $\ell^2(\mathbf{C})$ has a countable basis.[1] One such basis in $\ell^2(\mathbf{C})$ has been defined by Eq. (6.1).

---

[1]Mathematicians generally define separability in terms of the existence of a countable set of elements that is everywhere dense (see Definition (1.9)). In the present context, that is equivalent to the existence of a countable basis.

2. $\ell^2(\mathbf{C})$ is *complete*, or *closed*. This means that every Cauchy sequence defined in $\ell^2(\mathbf{C})$ has a limit in $\ell^2(\mathbf{C})$.

The proof of the latter statement is similar to the corresponding proof in the finite-dimensional case. If $\{x_1 = (\xi_{11}, \xi_{12}, \ldots), x_2 = (\xi_{21}, \xi_{12}, \ldots), \ldots\}$ is a sequence of vectors in $\ell^2(\mathbf{C})$, then $x_1, x_2, \ldots$ is a Cauchy sequence if and only if $\{\xi_{1k}, \xi_{2k}, \ldots\}$ is a Cauchy sequence for every $k = 1, 2, \ldots$, since

$$|\xi_{mk} - \xi_{nk}|^2 \le \|x_m - x_n\|^2 \le \sum_{k=1}^{\infty} |\xi_{mk} - \xi_{nk}|^2 \tag{6.6}$$

Thus the sequence $x_1, x_2, \ldots$ converges to a limit x if and only if the sequence $\{\xi_{1k}, \xi_{2k}, \ldots\}$ converges to a limit, call it ξ_k, for every $k = 1, 2, \ldots$; in the case of convergence, we have

$$x = (\xi_1, \xi_2, \ldots) \tag{6.7}$$

Remark. The crucial point here is that $\ell^2(\mathbf{C})$ has been constructed so that the limit (6.7) is in the space. The space $\pi^2(\mathbf{C})$ of vectors $x = (\xi_1, \xi_2, \ldots)$ with only a finite number of nonzero components satisfies all the axioms except completeness, since the limit vector in (6.7) need not have a finite number of nonzero components (this shows the independence of the axiom of completeness). In fact, the space $\ell^2(\mathbf{C})$ is obtained from $\pi^2(\mathbf{C})$ precisely by including the limit points of all Cauchy sequences in $\pi^2(\mathbf{C})$, in much the same way as the real numbers were obtained from the rational numbers by including the limit points of all Cauchy sequences of rationals. Thus we can say that $\ell^2(\mathbf{C})$ is the *completion*, or *closure*, of $\pi^2(\mathbf{C})$. □

An infinite-dimensional unitary vector space that is separable and complete is a *Hilbert space*.[2] The example $\ell^2(\mathbf{C})$ provides a model Hilbert space. Any complex Hilbert space is isomorphic and isometric to $\ell^2(\mathbf{C})$ in the same way that any n-dimensional complex unitary vector space is isomorphic and isometric to $\mathbf{C}^n$. Similarly, any real Hilbert space is isomorphic and isometric to the space $\ell^2(\mathbf{R})$ of sequences of real numbers with finite length as defined by (6.3).

A finite-dimensional linear manifold in a Hilbert space is always closed, but an infinite-dimensional linear manifold may or may not be. If $\phi_1, \phi_2, \ldots$ is an orthonormal system in the Hilbert space $\mathcal{H}$, then $\mathcal{M}(\phi_1, \phi_2, \ldots)$ denotes the linear manifold consisting of linear combinations of $\phi_1, \phi_2, \ldots$ with a finite number of vanishing terms. Including the limit points of all Cauchy sequences of vectors in $\mathcal{M}(\phi_1, \phi_2, \ldots)$ gives a *subspace* $\mathcal{M}[\phi_1, \phi_2, \ldots]$, the *closure* of $\mathcal{M}(\phi_1, \phi_2, \ldots)$.

As in the finite-dimensional case, the orthonormal system $\phi_1, \phi_2, \ldots$ is *complete* if and only if the only vector orthogonal to all the ϕ_k is the zero vector θ. Equivalent conditions are contained in the

Theorem 6.1. The orthonormal system $\phi_1, \phi_2, \ldots$ in the Hilbert space $\mathcal{H}$ is complete if and only if

---

[2] A finite-dimensional unitary space is sometimes called a Hilbert space as well, since the axioms of separability and completeness are satisfied automatically (i.e., as a consequence of the other axioms) for such spaces. Here we reserve the term Hilbert space for infinite-dimensional spaces.

(i) for every vector x in $\mathcal{H}$, we have

$$x = \sum_{k=1}^{\infty} (\phi_k, x)\, \phi_k \tag{6.8}$$

(ii) for every vector x in $\mathcal{H}$, we have

$$\|x\|^2 = \sum_{k=1}^{\infty} |(\phi_k, x)|^2 \tag{6.9}$$

(iii) for every pair of vectors x and y in $\mathcal{H}$, we have

$$(x, y) = \sum_{k=1}^{\infty} (x, \phi_k)(\phi_k, y) \tag{6.10}$$

(iv) $\mathcal{H} = M[\phi_1, \phi_2, \ldots]$, that is, $\mathcal{H}$ is a closed linear manifold spanned by the orthonormal system $\phi_1, \phi_2, \ldots$.

Any one of these conditions is necessary and sufficient for all of them. It is left as an exercise for the reader to show this.

Note, however, that while the expansion (6.8) in terms of a complete orthonormal system is always possible, there are nonorthogonal bases for which the only vector orthogonal to every element of the basis is the zero vector, but for which there are vectors that do not admit an expansion as a linear combination of the basis vectors.

❑ **Example 6.1.** Let $\phi_1, \phi_2, \ldots$ be a complete orthonormal system in the Hilbert space $\mathcal{H}$, and define vectors $\psi_1, \psi_2, \ldots$ by

$$\psi_n = \sqrt{\tfrac{1}{2}}\, (\phi_n - \phi_{n+1}) \tag{6.11}$$

Then the $\psi_1, \psi_2, \ldots$ are linearly independent, and $(\psi_n, x) = 0$ for all n if and only if $x = \theta$. However, the $\psi_1, \psi_2, \ldots$ are not orthogonal, and not every vector in $\mathcal{H}$ has a convergent expansion in terms of the $\psi_1, \psi_2, \ldots$. For example, we have the formal expansion

$$\phi_1 \sim \sqrt{2}\,(\psi_1 + \psi_2 + \psi_3 + \cdots) \sim \sqrt{2} \sum_{k=1}^{\infty} \psi_k \tag{6.12}$$

But

$$\left\| \phi_1 - \sqrt{2} \sum_{k=1}^{N} \psi_k \right\| = \|\phi_{N+1}\| = 1 \tag{6.13}$$

Hence the expansion (6.12) does not converge strongly to ϕ_1, although it does converge weakly. This example shows that it is necessary to use extra caution with nonorthogonal bases in Hilbert space. ∎

6.1.2 Convergence in Hilbert space

The concept of weak convergence, introduced in the discussion of sequences of functions in Chapter 1, is also relevant to sequences in Hilbert space. Here we have the

Definition 6.1. If the sequence $\{x_k\}$ in a unitary vector space $\mathcal{V}$ is such that the sequence of numbers $\{(x_k, y)\}$ converges to (x, y) for every y in $\mathcal{V}$, then the sequence $\{x_k\}$ is *weakly convergent* to x, denoted by

$$\{x_k\} \rightharpoonup x \tag{6.14}$$

(look closely at the half arrow here). ∎

In a finite-dimensional space, weak convergence is equivalent to ordinary (strong) convergence, since if $\phi_1, \dots, \phi_n$ is a complete orthonormal system, and if

$$\{(x_k, \phi_p)\} \to \xi_p \tag{6.15}$$

for $p = 1, \dots, n$, then for any $\varepsilon > 0$, there is an N such that

$$|(x_k, \phi_p) - \xi_p| < \varepsilon \tag{6.16}$$

for all $p = 1, \dots, n$ whenever $k > N$. Then also

$$\left\| x_k - \sum_{p=1}^{n} \xi_p \phi_p \right\|^2 < n\varepsilon^2 \tag{6.17}$$

so that $\{x_k\} \to \sum_{p=1}^{n} \xi_p \phi_p$ in the usual (strong) sense. In an infinite-dimensional space, we can no longer be sure that an N exists for which the inequality (6.16) is satisfied for all p, and there are sequences that converge weakly, but not strongly.

❑ **Example 6.2.** Let $\phi_1, \phi_2, \dots$ be an infinite orthonormal system in a Hilbert space $\mathcal{H}$. Then $\{(\phi_n, x)\} \to 0$ for any x in $\mathcal{H}$, since

$$\|x\|^2 \le \sum_{n=1}^{\infty} |(\phi_n, x)|^2 \tag{6.18}$$

is finite (the sequence of terms of a convergent series must converge to zero). Thus the sequence $\{\phi_n\}$ converges weakly to the zero vector θ. On the other hand, $\|\phi_n\| = 1$ for all n, so the $\{\phi_n\}$ cannot converge strongly to zero. ∎

→ **Exercise 6.1.** Let $\mathcal{M}$ be an infinite-dimensional linear manifold in the Hilbert space $\mathcal{H}$. Then

(i) $\mathcal{M}^\perp$ is closed,
(ii) $\mathcal{M} = (\mathcal{M}^\perp)^\perp$ if and only if $\mathcal{M}$ is closed, and
(iii) $\mathcal{H} = \mathcal{M} \oplus \mathcal{M}^\perp$. □

→ **Exercise 6.2.** Show that if $\{x_n\} \rightharpoonup x$ and $\{y_n\} \to y$, then $\{(x_n, y_n)\} \to (x, y)$. Is it true that if $\{x_n\} \rightharpoonup x$ and $\{y_n\} \rightharpoonup y$, then $\{(x_n, y_n)\} \to (x, y)$?   □

An important result is that a weakly convergent sequence must be bounded.[3]

Theorem 6.2. If the sequence $\{x_n\}$ of vectors in the Hilbert space $\mathcal{H}$ is weakly convergent to x, then it is bounded (there is an $M > 0$ such that $\|x_n - x\| \leq M$ for all $n = 1, 2, \ldots$).

The proof is long and somewhat technical, and we will not present it here. The previous example might make it plausible. The converse of this theorem is an analog of the Bolzano–Weierstrass theorem for real or complex numbers (see Exercise 1.4).

Theorem 6.3. Any infinite bounded set in the Hilbert space $\mathcal{H}$ contains a weakly convergent subsequence.

Proof. Let $\{\phi_1, \phi_2, \ldots\}$ be a complete orthonormal system in $\mathcal{H}$, and let $\{x_n\}$ be a bounded infinite set in $\mathcal{H}$. Then there is a subsequence $\{x_{1n}\}$ such that the sequence $\{\xi_{1n} \equiv (\phi_1, x_{1n})\}$ of numbers is convergent to a limit, call it ξ_1. This follows from the plain version of the Bolzano–Weierstrass theorem in Exercise 1.1.4. The subsequence $\{x_{1n}\}$ is bounded; hence we can choose from it a subsequence $\{x_{2n}\}$ such that the sequence of numbers $\{\xi_{2n} \equiv (\phi_2, x_{2n})\}$ is convergent to a limit, call it ξ_2. Proceeding in this way, we obtain a nested set of subsequences $\{x_{kn}\}$ such that the sequences of numbers $\{\xi_{kn} \equiv (\phi_k, x_{kn})\}$ converge to limits ξ_k, for each $k = 1, 2, \ldots$. Let

$$x_* = \sum_{k=1}^{\infty} \xi_k \phi_k \tag{6.19}$$

Then the sequence $\{y_n\}$, with $y_n \equiv x_{nn}$, converges (weakly) to x_*.   ■

6.2   Function Spaces; Measure Theory

6.2.1   Polynomial Approximation; Weierstrass Approximation Theorem

Consider spaces of (complex-valued) functions defined on some domain Ω, which may be an interval of the real axis, a region in the complex plane, or some higher dimensional domain. There is a natural vector space structure for such functions, and properties such as continuity, differentiability, and integrability are preserved by vector addition and multiplication by scalars. A scalar product can be introduced on the function space by

$$(f, g) = \int_{\Omega} f^*(x) w(x) g(x) d\Omega \tag{6.20}$$

where $d\Omega$ is a suitably defined volume element on Ω, and $w(x)$ is a *weight function* satisfying $w(x) > 0$ on Ω except possibly at the boundary of Ω (often $w(x) = 1$ everywhere, but other weight functions are sometimes convenient). This scalar product allows us to introduce norm and distance between functions in the usual way.[4]

---

[3] This is the analog of the uniform boundedness principle, or Banach–Steinhaus theorem, in real analysis.

[4] As noted above, norm and length can be defined with no associated scalar product. Function spaces with such norms are *Banach spaces*. These are also important in mathematics; further discussion may be found in books cited at the end of the chapter.

Complications arise with sequences of functions in a function space; the limiting processes involved in the definitions of continuity and differentiability need not be interchangeable with taking the limit of a sequence (see Section 1.3). Furthermore, there are several distinct concepts of convergence of sequences of functions. In addition to the classical pointwise and uniform convergence, and the weak convergence introduced in Section 1.3, there are the concepts of strong and weak convergence in Hilbert space as defined above. Thus we examine the conditions under which sequences of functions converge in the various senses, and the properties of the limit functions.

One important result is that a continuous function can be approximated on a finite interval by a uniformly convergent sequence of polynomials. This is stated formally as the

Theorem 6.4. (Weierstrass Approximation theorem). If the function $f(t)$ is continuous on the closed interval $a \le t \le b$, then there is a sequence $p_n(t)$ of polynomials that converges uniformly to $f(t)$ for $a \le t \le b$.

Remark.   If the function $f(t)$ is k times continuously differentiable on the interval, then the sequence of polynomials can be constructed so that the sequence of derivatives $p_n^{(k)}(t)$ also converges uniformly to the derivative $f^{(k)}(t)$ on the interval.   □

We will not give a complete proof, but we can construct a sequence of polynomials that is uniformly convergent. Let

$$J_n = \int_{-1}^{1} (1 - t^2)^n \, dt = 2 \int_{0}^{1} (1 - t^2)^n \, dt > \frac{2}{n + 1} \tag{6.21}$$

where the last inequality follows from noting that $1 - t^2 > 1 - t$ of $0 \le t < 1$, and define

$$\Pi_n(t|f) \equiv \frac{1}{J_n} \int_{0}^{1} f(u) \left[1 - (u - t)^2 \right]^n \, du \tag{6.22}$$

Then the sequence $\{\Pi_n(t|f)\}$ convergent uniformly to $f(t)$ for $a \le t \le b$.

Remark.   It is important to realize that the existence of a uniformly convergent sequence of polynomials is *not* equivalent to a power series expansion. For example, the function $f(t) = \sqrt{t}$ is continuous on $0 \le t \le 1$, and hence can be approximated uniformly by a sequence of polynomials in t. Yet we know that it has no convergent power series expansion around $t = 0$, since it has a branch point singularity at $t = 0$ in the complex t-plane. The point is that the coefficient of a fixed power of t in the approximating polynomial can depend on n.   □

Remark.   This theorem underlies approximate methods for computing transcendental (non-polynomial) functions. Since computers are only able to compute polynomials directly, any computation of a transcendental function is some kind of polynomial approximation. The theorem assures us that we can choose fixed coefficients to approximate the function over a definite interval to a specified degree of accuracy. It remains, of course, to determine the most efficient polynomial for any particular function.   □

6.2.2   Convergence in the Mean

A sequence $\{f_n\}$ of functions converges to a limit f in the (strong) Hilbert space sense if $\{\|f_n - f\|\} \to 0$. With Hilbert space norm from the scalar product (6.20),

$$\int_\Omega |f_n(x) - f(x)|^2 \, w(x) \, d\Omega \to 0 \tag{6.23}$$

for $n \to \infty$. In this case, we also say that $\{f_n\}$ *converges in the mean* to f, since the mean square deviation of $\{f_n\}$ from the limit function tends to zero. Convergence in the mean does not imply uniform convergence, or even pointwise convergence, although it does imply weak convergence in the sense that

$$\left\{ \int_\Omega g^*(x) \, w(x) \, [f_n(x) - f(x)] \, d\Omega \right\} \to 0 \tag{6.24}$$

for every integrable function $g(x)$. Uniform convergence implies convergence in the mean, but there are sequences of functions that converge to a limit at every point of an interval, but do not converge in the mean, or even weakly in the Hilbert space sense.

❑ **Example 6.3.**   Consider the sequence of functions $\{f_n(t)\}$ defined on $[-1, 1]$ by

$$f_{n+1}(t) \equiv 2 \frac{(n+1)}{J_{n+1}} t^2 (1 - t^2)^n \tag{6.25}$$

with J_n defined in Eq. (6.21). Now $\{f_n(t)\} \to 0$ for every t, but if $g(t)$ is continuous on $[-1, 1]$, then

$$\int_{-1}^1 g(t) f_{n+1}(t) \, dt = \frac{1}{J_{n+1}} \int_{-1}^1 \frac{d}{dt} [tg(t)] \, (1 - t^2)^{n+1} \, dt \tag{6.26}$$

The sequence of integrals on the right-hand side converges to $g(0)$ for every continuous function $g(t)$, so the sequence $\{f_n(t)\} \not\to 0$. The sequence converges weakly as a sequence of linear functionals; we have

$$\{f_n(t)\} \to t\delta'(t) \tag{6.27}$$

where $\delta'(t)$ is the derivative of the Dirac δ-function introduced in Chapter 1, but this limit is not in the Hilbert space of functions on $[-1, 1]$. Note that each of the $\{f_n(t)\}$ defined by Eq. (6.25) has continuous derivatives of all orders on $[-1, 1]$, but the lack of uniform convergence allows the weak limit to be nonzero.   ∎

A variety of function spaces on a domain Ω are of interest:
C: $C(\Omega)$ contains functions continuous on Ω.
C^k: $C^k(\Omega)$ contains functions continuous together with their first k derivatives on Ω.
D: $D(\Omega)$ contains functions piecewise continuous on Ω.
D^k: $D^k(\Omega)$ contains functions whose kth derivative is piecewise continuous on Ω.[5]

---

[5] Note that any function in $D^k(\Omega)$ is also in $C^{k-1}(\Omega)$, since the integral of the piecewise continuous kth derivative is continuous in Ω.

Each of these function spaces is a linear vector space that can be made unitary with a scalar product of the form (6.20). It then satisfies all the Hilbert space axioms except completeness. The limit of a Cauchy sequence of functions in one of these spaces need not be a function in the space, and not every function $f(x)$ for which

$$\|f\|^2 \equiv \int_\Omega |f(x)|^2 \, w(x) \, d\Omega \tag{6.28}$$

is defined belongs to one of these classes on Ω. To satisfy the axiom of completeness, we need to introduce a new kind of integral, the *Lebesgue integral*, which is a generalization, based on the concept of *measure*, of the usual (Riemann) integral. In the next section, we give a brief description of measure, and then explain how the Lebesgue integral is defined. To obtain a complete function space, we need to understand the integral (6.28) as a Lebesgue integral, and include all functions f on the domain Ω for which the integral is finite.

6.2.3   Measure Theory

The *measure* of a set S of real numbers, complex numbers, or, for that matter, a set of vectors in $\mathbf{R}^n$ or $\mathbf{C}^n$, is a non-negative number associated with the set that corresponds roughly to the length, or area, or volume of the set. The measure $\mu(S)$ must have the property that if S_1 and S_2 are disjoint (nonoverlapping) sets, then

$$\mu(S_1 \cup S_2) = \mu(S_1) + \mu(S_2) \tag{6.29}$$

Thus the measure is *additive*. A set is of *measure zero* if it can be contained within a set of arbitrarily small measure.

Measure can be defined on the real axis in a natural way by defining the measure of an interval (open or closed) to be the length of the interval. Additivity then requires that the measure of a collection of disjoint intervals is the sum of the length of the intervals.

A countable set of real numbers has measure zero. If the elements of the set are $x_1, x_2, \ldots$, then enclose x_1 inside an interval of measure $\varepsilon/2$, x_2 inside an interval of measure $\varepsilon/4$, and so on, with x_n enclosed inside an interval of length $\varepsilon/2^n$. The total length of the enclosing intervals is then ε, which can be made arbitrarily small. Hence the set has measure zero. An example of an uncountable set of measure zero is the Cantor set in the following exercise:

→ **Exercise 6.3.**   Consider the closed interval $0 \le t \le 1$. Remove the open interval

$$\tfrac{1}{3} < t < \tfrac{2}{3}$$

(the "middle open third") from this interval. From each of the remaining intervals, remove the open middle third (thus excluding $\frac{1}{9} < t < \frac{2}{9}$ and $\frac{7}{9} < t < \frac{8}{9}$). Proceed in this manner *ad infinitum*, and let S denotes the set of numbers that remain.

(i) Show that S is a set of measure zero. (*Hint.* What is the total measure of the pieces that have been removed from $[0,1]$?)

(ii) Show that S is uncountable (note that S does *not* consist solely of rational numbers of the form $m/3^n$ with m, n integer). ☐

Remark.   The set S is the *Cantor set*. It is an uncountable set of measure zero. The Cantor set and its generalizations underlie much modern work on fractal geometry. ☐

A *step function* $\sigma(x)$ is a function that assumes a finite set of values $\sigma_1, \ldots, \sigma_n$ on sets $\mathcal{S}_1, \ldots, \mathcal{S}_n$ covering a domain Ω. Integration is defined for a step function by

$$\int_\Omega \sigma(x)\,dx = \sum_{k=1}^n \sigma_k\,\mu(\mathcal{S}_k) \tag{6.30}$$

Note that if the sets $\mathcal{S}_1, \ldots, \mathcal{S}_n$ consist of discrete subintervals of an interval of the real axis, then this definition coincides with the usual (Riemann) integral.

Remark.   With this definition of the integral, the value of a function on a set of measure zero is irrelevant, so long as it is finite. Two functions are be identified if they differ only on a set of measure zero, since they have the same integral over any finite interval. In general, any property such as equality, or continuity, etc., that holds everywhere except on a set of measure zero is said to hold *almost everywhere*. This identification of functions that are equal almost everywhere is an extension of the classical notion of a function needed to define the completion of function spaces. $\square$

A function $F(x)$ is *measurable* on a domain Ω if there is a sequence $\sigma_1, \sigma_2, \ldots$ of step functions that converges to $F(x)$ almost everywhere on Ω. If $F(x)$ is measurable on Ω, and if $\sigma_1, \sigma_2, \ldots$ is a sequence of step functions that converges to $F(x)$ almost everywhere on Ω, then form the sequence $I_1, I_2, \ldots$ defined by

$$I_n \equiv \int_\Omega \sigma_n(x)\,dx \tag{6.31}$$

If this sequence converges to a finite limit I, then $F(x)$ is (*Lebesgue*) *integrable*, or *summable*, on Ω, and we define the *Lebesgue integral* of $F(x)$ over Ω by

$$\int_\Omega F(x)\,dx \equiv I = \lim_{n\to\infty} I_n \tag{6.32}$$

If the ordinary (Riemann) integral of $F(x)$ over Ω exists, then so does Lebesgue integral, and it is equal to the Riemann integral. However, the Lebesgue integral may exist when the Riemann integral does not.

The function space $L^2(\Omega)$ is the space of complex-valued functions $f(x)$ on the domain Ω for which $f(x)$ is measurable on Ω, and the integral (6.28) exists as a Lebesgue integral. It follows from the basic properties of Lebesgue integrals that $L^2(\Omega)$ is a normed linear vector space, and it is unitary with scalar product defined by Eq. (6.20). The crucial property of $L^2(\Omega)$ is that it is a Hilbert space; every Cauchy sequence in $L^2(\Omega)$ converges to a limit in $L^2(\Omega)$. This is a consequence of the

Theorem 6.5. (Riesz–Fischer theorem). If $f_1, f_2, \ldots$ is a Cauchy sequence of functions in $L^2(\Omega)$, then there is a function f in $L^2(\Omega)$ such that

$$\lim_{n\to\infty} \|f - f_n\| = 0 \tag{6.33}$$

The proof is too complicated to give here; details can be found in books on real analysis. The main point of the theorem is that there actually exists a function space that satisfies the completeness axiom, unlike the more familiar function spaces of the type C^k or D^k,

6.3   Fourier Series

6.3.1   Periodic Functions and Trigonometric Polynomials

The classical expansion of functions along an orthogonal basis is the Fourier series expansion in terms of either trigonometric functions or the corresponding complex exponential functions. Fourier series were introduced here in Chapter 4 as an expansion of periodic analytic functions. Now we look at expansions in terms of various Hilbert space complete orthonormal systems formed from the trigonometric functions.

If $f(t)$ is a periodic function of the real variable t (often time) with fundamental period T, the *fundamental frequency* ν and *angular frequency* ω are defined by[6]

$$\nu \equiv \frac{\omega}{2\pi} \equiv \frac{1}{T} \tag{6.34}$$

Such periodic functions have a natural linear vector space structure, since a linear combination of functions with period T has the same period T. A natural closure of this vector space is the Hilbert space $L^2(0, T)$ with scalar product defined by

$$(f, g) = \int_0^T f^*(t)g(t)dt \tag{6.35}$$

The trigonometric functions $\sin \omega t$, $\cos \omega t$, or the corresponding complex exponential functions $\exp(\pm i\omega t)$, have a special place among functions with period T: the trigonometric functions describe simple harmonic motion with frequency $\nu = \omega/2\pi$, and the complex exponential functions describe motion with constant angular velocity ω around a circle in the complex plane (the sign defines the sense of rotation). To include all periodic functions with the same period, it is necessary to include also functions whose frequency is an integer multiple of the fundamental frequency (these frequencies $\nu_n \equiv n\nu, n = 2, 3, \ldots$ are *harmonics* of the fundamental frequency). Thus we have a set of complex exponential functions

$$\phi_n(t) \equiv \sqrt{\frac{1}{T}} e^{in\omega t} \tag{6.36}$$

$(n = 0, \pm, 1 \pm 2, \ldots)$, and a set of trigonometric functions

$$C_0(t) \equiv \sqrt{\frac{1}{T}} \qquad C_n(t) \equiv \sqrt{\frac{2}{T}} \cos n\omega t \qquad S_n(t) \equiv \sqrt{\frac{2}{T}} \sin n\omega t \tag{6.37}$$

$(n = 1, 2, \ldots)$. That each of the sets $\{\phi_n\}$ and $\{C_0, \{C_n, S_n\}\}$ is an orthonormal system follows from some elementary integration. The Weierstrass approximation theorem can be adapted to show that any continuous periodic function can be uniformly approximated by a sequence of finite linear combinations of these functions ("trigonometric polynomials"), so that each of the sets is actually a complete orthonormal system.

---

[6]Note that ω is sometimes called the frequency, although it isn't. Just remember that $\omega = 2\pi/T$.

6.3.2   Classical Fourier Series

A standard Fourier series expansion is obtained by introducing the variable

$$x \equiv \frac{2\pi t}{T} = \omega t \tag{6.38}$$

so that the period of the function in the variable x is 2π, and choosing the primary domain of the function to be the interval $-\pi \leq x \leq \pi$. Then any function $f(x)$ in $L^2(-\pi, \pi)$ with period 2π has the Fourier series expansion

$$f(x) = \sum_{n=-\infty}^{\infty} c_n \phi_n(x) = \frac{1}{\sqrt{2\pi}} \sum_{n=-\infty}^{\infty} c_n e^{inx} \tag{6.39}$$

with Fourier coefficients c_n given by

$$c_n = (\phi_n, f) = \frac{1}{\sqrt{2\pi}} \int_{-\pi}^{\pi} e^{-inx} f(x)\, dx \tag{6.40}$$

Note that $c_{-n} = c_n^*$ if $f(x)$ is real.

The corresponding real form of the Fourier series is

$$f(x) = \frac{a_0}{\sqrt{2\pi}} + \frac{1}{\sqrt{\pi}} \sum_{n=1}^{\infty} [a_n \cos nx + b_n \sin nx] \tag{6.41}$$

with Fourier coefficients a_n and b_n given by

$$a_0 = c_0 = \frac{1}{\sqrt{2\pi}} \int_{-\pi}^{\pi} f(x)\, dx \tag{6.42}$$

$$a_n = (C_n, f) = \frac{1}{\sqrt{\pi}} \int_{-\pi}^{\pi} \cos nx\, f(x)\, dx \tag{6.43}$$

$$b_n = (S_n, f) = \frac{1}{\sqrt{\pi}} \int_{-\pi}^{\pi} \sin nx\, f(x)\, dx \tag{6.44}$$

$(n = 1, 2, \ldots)$. These expansions correspond to the series (4.122) and (4.127), although here they are defined not just for analytic functions, but for all functions in $L^2(-\pi, \pi)$. Note that the factors of $1/\sqrt{2\pi}$ were absorbed in the Fourier coefficients in Chapter 4; here they are placed symmetrically between the series (6.39) (or (6.41)) and the coefficients in Eq. (6.40) (or Eqs. (6.43)–(6.44)). There is no universal convention for defining these factors; whatever is most convenient for the problem at hand will do. In the vector space context, we want to work with orthonormal vectors as defined by Eqs. (6.36) or (6.37); in the context of analytic function expansions, there is no reason to insert factors of $1/\sqrt{2\pi}$.

The vector space norm of the function $f(x)$ is given by

$$\|f\|^2 = \int_{-\pi}^{\pi} |f(x)|^2\, dx = \sum_{n=-\infty}^{\infty} |c_n|^2 = |a_0|^2 + \sum_{n=1}^{\infty} \left(|a_n|^2 + |b_n|^2 \right) \tag{6.45}$$

This result is known as *Parseval's theorem*. It requires the Fourier coefficients a_n, b_n, and c_n to vanish for large n, and fast enough that the series converge.

→ **Exercise 6.4.**   Show that if

$$f(x) = \frac{1}{\sqrt{2\pi}} \sum_{n=-\infty}^{\infty} c_n e^{inx} \quad \text{and} \quad g(x) = \frac{1}{\sqrt{2\pi}} \sum_{n=-\infty}^{\infty} d_n e^{inx}$$

are any two functions in $L^2(-\pi, \pi)$, then

$$(f, g) \equiv \int_{-\pi}^{\pi} f^*(x) g(x)\, dx = \sum_{n=-\infty}^{\infty} c_n^* d_n$$

This is the standard form for the scalar product in terms of vector components.   □

→ **Exercise 6.5.**   Find the Fourier series expansions of each of the following functions defined by the formulas for $-\pi < x < \pi$, and elsewhere by $f(x + 2\pi) = f(x)$.

(i) $f(x) = x$ (ii) $f(x) = x^2$ (iii) $f(x) = \sin|x|$ □

6.3.3   Convergence of Fourier Series

The Fourier series converges in the Hilbert space sense for functions in $L^2(-\pi, \pi)$, while it converges absolutely and uniformly if the function $f(x)$ is analytic in some strip $-a < \mathrm{Im}\, x < a$ about the real axis in the complex x-plane. To see what happens in the intermediate cases, consider the partial sums of the series (6.39) given by

$$f_N(x) \equiv \sum_{n=-N}^{N} c_n \phi_n(x) = \frac{1}{2\pi} \sum_{n=-N}^{N} \int_{-\pi}^{\pi} e^{in(x-y)} f(y)\, dy \tag{6.46}$$

Now

$$\sum_{n=-N}^{N} e^{inu} = \frac{\sin(N + \frac{1}{2})u}{\sin \frac{1}{2} u} \equiv D_N(u) \tag{6.47}$$

(the series is a a geometric series). The function $D_N(u)$ is the *Dirichlet kernel*. It is an even function of u and

$$\int_{-\pi}^{\pi} D_N(u)\, du = 2\pi \tag{6.48}$$

which follows from integrating the defining series (6.47) term by term. Thus we have

$$f_N(x) = \frac{1}{2\pi} \int_{-\pi}^{\pi} D_N(x - y) f(y)\, dy = \frac{1}{2\pi} \int_{-\pi-x}^{\pi-x} D_N(u) f(x + u)\, du \tag{6.49}$$

Now suppose the function f is piecewise continuous, so that the one-sided limits

$$f_\pm(x_0) \equiv \lim_{\varepsilon \to 0^+} f(x_0 \pm \varepsilon) \tag{6.50}$$

exist at every point x_0. Then we can write

$$f_n(x_0) - \tfrac{1}{2}[f_+(x_0) + f_-(x_0)] = \frac{1}{2\pi} \int_{-\pi}^{0} D_N(u)[f(x_0 + u) - f_-(x_0)]\, du$$

$$+ \frac{1}{2\pi} \int_{0}^{\pi} D_N(u)[f(x_0 + u) - f_+(x_0)]\, du$$

(6.51)

If the right side vanishes for large N, then the Fourier series will converge to the average of the left- and right-hand limits of $f(x)$ at x_0, and to the value $f(x_0)$ if the function is continuous.

The two integrals on the right-hand side of Eq. (6.51) can be written as

$$\Delta_{\pm}^{N} = \pm \frac{1}{\pi} \int_{0}^{\pm\frac{1}{2}\pi} \left\{ \frac{f(x_0 \pm 2v) - f_{\pm}(x_0)}{\sin v} \right\} \sin(2N+1)v\, dv$$

(6.52)

(here $v = \tfrac{1}{2}u$). These integrals are Fourier coefficients for the functions in braces, and hence vanish for large N if the functions are in a Hilbert space such as $L^2(0, \pm\tfrac{1}{2}\pi)$. This will certainly be the case if $f(x)$ has one-sided derivative at x_0 and may be the case even if it does not. Thus there are reasonably broad conditions that are sufficient for the pointwise convergence of the Fourier series. The convergence properties can be illustrated further with a pair of examples.

❑ **Example 6.4.** Consider the function $f(x)$ defined for $-\pi \le x \le \pi$ by

$$f(x) = |x|$$

(6.53)

Then the Fourier coefficients of $f(x)$ are given by

$$\sqrt{2\pi}\, c_0 = 2 \int_{0}^{\pi} x\, dx = \pi^2$$

(6.54)

and

$$\sqrt{2\pi}\, c_n = \int_{0}^{\pi} e^{-inx} x\, dx - \int_{-\pi}^{0} e^{-inx} x\, dx = 2 \int_{0}^{\pi} x \cos nx\, dx$$

(6.55)

$$= \begin{cases} -4/n^2 & n \text{ odd} \\ 0 & n \text{ even} \end{cases}$$

which leads to the Fourier series

$$f(x) = \frac{\pi}{2} - \frac{4}{\pi} \sum_{n=0}^{\infty} \frac{\cos(2n+1)x}{(2n+1)^2}$$

(6.56)

The series is absolutely and uniformly convergent on the entire real axis. However, it diverges if $\mathrm{Im}\, x \ne 0$, since the series then contains an exponentially increasing part. This should not be surprising, since the function $f(x)$ is not analytic. ∎

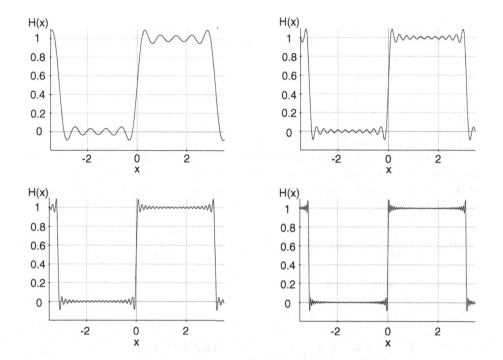

Figure 6.1: Approximations $H_N(x)$ to the function $H(x)$ by the truncated Fourier series (6.61) for $N = 4$ (upper left), $N = 10$ (upper right), $N = 20$ (lower left), and $N = 40$ (lower right).

❑ **Example 6.5.**   Consider the (Heaviside) step function $H(x)$ defined for $-\pi \le x \le \pi$ by

$$H(x) = \begin{cases} 1 & 0 < x < \pi \\ 0 & -\pi < x < 0 \end{cases} \tag{6.57}$$

and elsewhere by $f(x + 2\pi) = f(x)$. The Fourier coefficients of $H(x)$ are given by

$$\sqrt{2\pi}\, c_0 = \int_0^\pi dx = \pi \tag{6.58}$$

and for $n > 0$ by

$$\sqrt{2\pi}\, c_n = \int_0^\pi e^{-inx}\, dx = \frac{1}{in}\left(1 - e^{in\pi}\right) = \begin{cases} 2/in & n \text{ odd} \\ 0 & n \text{ even} \end{cases} \tag{6.59}$$

which leads to the Fourier series

$$H(x) = \frac{1}{2} + \frac{2}{\pi}\sum_{n=0}^\infty \frac{\sin(2n+1)x}{(2n+1)} \tag{6.60}$$

The Fourier series (6.60) is convergent except at the points $0, \pm\pi, \pm 2\pi, \ldots$ of discontinuity of $H(x)$; at these points, the series has the value $\frac{1}{2}$. The partial sums $H_N(x)$ of the series, defined by

$$H_N(x) = \frac{1}{2} + \frac{2}{\pi} \sum_{n=0}^{N} \frac{\sin(2n+1)x}{(2n+1)} \tag{6.61}$$

are shown in Fig. 6.1 for $N = 4, 10, 20, 40$.

One interesting feature that can be seen in these graphs is that the partial sums overshoot the function near the discontinuity at $x = 0$. This overshoot does not disappear in the limit as $N \to \infty$; it simply moves closer to the discontinuity. To see this, note that Eq. (6.49) gives here

$$H_N(x) = \frac{1}{2\pi} \int_0^\pi D_{2N+1}(x-y)\, dy = \frac{1}{2\pi} \int_{-x}^{\pi-x} \frac{\sin(2N+\frac{3}{2})u}{\sin\frac{1}{2}u}\, du \tag{6.62}$$

and the discontinuity function

$$\Delta_N(x) = H_N(x) - H_N(-x) = \tag{6.63}$$

$$= \frac{1}{2\pi} \int_{-x}^{x} \frac{\sin(2N+\frac{3}{2})u}{\sin\frac{1}{2}u}\, du + \frac{1}{2\pi} \int_{\pi-x}^{\pi+x} \frac{\sin(2N+\frac{3}{2})u}{\sin\frac{1}{2}u}\, du$$

A short calculation shows that the first maximum of $\Delta_N(x)$ for $x > 0$ occurs for

$$x = \frac{\pi}{2(N+1)} \equiv x_N \tag{6.64}$$

The value of $\Delta_N(x)$ at this maximum is given by

$$\Delta_N(x_N) = \frac{1}{\pi} \int_0^{x_N} \frac{\sin(2N+\frac{3}{2})u}{\sin\frac{1}{2}u}\, du + \frac{1}{2\pi} \int_{\pi-x_N}^{\pi+x_N} \frac{\sin(2N+\frac{3}{2})u}{\sin\frac{1}{2}u}\, du \tag{6.65}$$

$$\simeq \frac{2}{\pi} \int_0^\pi \frac{\sin\xi}{\xi}\, d\xi + \cdots$$

where the neglected terms are $O(1/N)$ for $N \to \infty$. This last result follows from substituting $\xi = (2N + \frac{3}{2})u$ and approximating $\sin\frac{1}{2}u \simeq \frac{1}{2}u$ in the first integral on the right-hand side of Eq. (6.63). The final integral has the numerical value $1.179\ldots$. Thus the maximum of the partial sum of the series overshoots the function by about 18% of the magnitude of the discontinuity on each side.

This effect is known as the *Gibbs phenomenon*; it is a consequence of the attempt to approximate a discontinuous function by a sequence of smooth functions. It is a general property of Fourier series, and other series as well, though the magnitude of the overshoot may depend on the particular set of smooth functions used.   ∎

These examples illustrate the characteristic convergence properties of the Fourier series expansion of a function in L^2. While the series is guaranteed to converge in the Hilbert space sense, the expansion of a discontinuous function converges slowly, especially near points of discontinuity. The smoother the function, the more rapidly the Fourier coefficients a_n, b_n (or c_n) vanish for large n. The analytic properties of the series are best seen in the complex exponential form (6.39) of the series. Write

$$f(x) = f_0 + \sum_{n=1}^{\infty} f_{-n} e^{-inx} + \sum_{n=1}^{\infty} f_n e^{inx} \equiv f_0 + f_-(x) + f_+(x) \tag{6.66}$$

($f_n = c_n/\sqrt{2\pi}$). Then it is clear that if the series converges in the Hilbert space for real x, the function $f_+(x)$ [$f_-(x)$] is analytic in the half-plane $\operatorname{Im} x > 0$ [$\operatorname{Im} x < 0$], since the imaginary part of x provides an exponentially decreasing factor that enhances the convergence of the series. If the function $f(x)$ is actually analytic in a strip including the real axis, then the coefficients $c_{\pm n}$ must vanish faster than any power of n for large n. However, we have seen that Fourier series converge on the real axis alone for a much broader class of functions, and there is no guarantee even of pointwise convergence on the real axis.

6.3.4 Fourier Cosine Series; Fourier Sine Series

There are other important trigonometric series. Consider, for example, the set of functions $\{\psi_n(x)\}$ defined by

$$\psi_0 \equiv \sqrt{\frac{1}{\pi}} \qquad \psi_n(x) \equiv \sqrt{\frac{2}{\pi}} \cos nx \tag{6.67}$$

($n = 1, 2, \ldots$). This set is a complete orthonormal system on the interval $0 \leq x \leq \pi$, and any function $f(x)$ in $L^2(0, \pi)$ can be expanded as

$$f(x) = \sum_{n=0}^{\infty} a_n \psi_n(x) \tag{6.68}$$

with

$$a_0 = (\psi_0, f) = \sqrt{\frac{1}{\pi}} \int_0^{\pi} f(x)\, dx \tag{6.69}$$

$$a_n = (\psi_n, f) = \sqrt{\frac{2}{\pi}} \int_0^{\pi} \cos nx\, f(x)\, dx \tag{6.70}$$

The series (6.68) is a *Fourier cosine series*; it can be obtained from the classical Fourier series (6.41) by defining $f(-x) \equiv f(x)$ ($0 \leq x \leq \pi$), extending $f(x)$ to be an even function of x on the interval $-\pi \leq x \leq \pi$. Then the only nonvanishing Fourier coefficients in the classical series (6.41) are those of the $\cos nx$ terms ($n = 0, 1, 2, \ldots$).

Another complete orthonormal system on the interval $0 \leq x \leq \pi$ is the set $\{\phi_n(x)\}$ defined by

$$\phi_n(x) \equiv \sqrt{\frac{2}{\pi}} \sin nx \tag{6.71}$$

$(n = 1, 2, \ldots)$. Any function $f(x)$ in $L^2(0, \pi)$ can be extended to an odd function of x on the interval $-\pi \leq x \leq \pi$ by defining $f(-x) \equiv -f(x)$ $(0 \leq x \leq \pi)$. Then the only nonvanishing Fourier coefficients in the classical series (6.41) are those of the $\sin nx$ terms $(n = 1, 2, \ldots)$, and $f(x)$ can be expanded as a *Fourier sine series*

$$f(x) = \sum_{n=1}^{\infty} b_n \phi_n(x) \tag{6.72}$$

with

$$b_n = (\phi_n, f) = \sqrt{\frac{2}{\pi}} \int_0^{\pi} \sin nx f(x)\, dx \tag{6.73}$$

$(n = 1, 2, \ldots)$.

A function $f(x)$ in $L^2(0, \pi)$ has both a Fourier sine series and a Fourier cosine series. These series are unrelated in general, since one is the classical Fourier series of the odd extension of $f(x)$ to the interval $-\pi \leq x \leq \pi$, while the other is the classical Fourier series of the even extension.

❑ **Example 6.6.**   Consider the function $f(x) = 1$ on $0 \leq x \leq \pi$. The function has a one term Fourier cosine series $f(x) = 1$, while the Fourier sine series is given by

$$f(x) = \frac{4}{\pi} \sum_{n=0}^{\infty} \frac{\sin(2n+1)x}{(2n+1)} \tag{6.74}$$

since the odd extension of $f(x)$ is simply given by

$$f_{\text{odd}}(x) = H(x) - H(-x) \tag{6.75}$$

where $H(x)$ is the step function introduced above in Eq. (6.199).   ∎

→ **Exercise 6.6.**   Find the Fourier sine series and cosine series expansions of the functions defined by

(i) $f(x) = x$

(ii) $f(x) = x(\pi - x)$

(iii) $f(x) = e^{-\alpha x}$

(iv) $f(x) = \begin{cases} 1 & 0 \leq x < \frac{\pi}{2} \\ -1 & \frac{\pi}{2} \leq x < \pi \end{cases}$ \tag{6.76}

for $0 < x < \pi$.   ❑

6.4   Fourier Integral; Integral Transforms

6.4.1   Fourier Transform

The Fourier series expansion of a function with period T expresses the function in terms of oscillators with frequencies ν_n that are integer multiples of the fundamental frequency $\nu = 1/T$. The longer the period, the smaller the interval between the frequencies, and it is plausible that in the limit $T \to \infty$, the spectrum of frequencies becomes continuous. To see how this happens, write the Fourier series expansion of $f(t)$ in the form

$$f(t) = \frac{1}{T} \sum_{n=-\infty}^{\infty} c_n \exp\left(-\frac{2\pi int}{T}\right) \tag{6.77}$$

(here the minus sign in the exponential is a convention), with

$$c_n = \int_{-\frac{1}{2}T}^{\frac{1}{2}T} \exp\left(\frac{2\pi int}{T}\right) f(t)\, dt \tag{6.78}$$

If we let $\omega_n \equiv 2\pi n/T$ and $\Delta\omega \equiv 2\pi/T$, then the series (6.77) becomes

$$f(t) = \frac{1}{2\pi} \sum_{n=-\infty}^{\infty} c(\omega_n) \exp\left(-i\omega_n t\right) \Delta\omega \tag{6.79}$$

In the limit $T \to \infty$, the series becomes the integral

$$f(t) = \frac{1}{2\pi} \int_{-\infty}^{\infty} c(\omega) e^{-i\omega t}\, d\omega \tag{6.80}$$

with

$$c(\omega) = \int_{-\infty}^{\infty} e^{i\omega t} f(t)\, dt \tag{6.81}$$

Equation (6.80) defines $f(t)$ as a *Fourier integral*, the continuous version of the Fourier series (6.77). The function $c(\omega)$ is the *Fourier transform* of $f(t)$, and $f(t)$ is the (inverse) Fourier transform of $c(\omega)$.[7] Note that if $f(t)$ is real, then

$$c^*(\omega) = c(-\omega) \tag{6.82}$$

The Fourier integrals (6.80) and (6.81) exist under various conditions. If

$$\|f\|^2 = \int_{-\infty}^{\infty} |f(t)|^2\, dt < \infty \tag{6.83}$$

---

[7]There is no universal notation here: the Fourier transform $c(\omega)$ of $f(t)$ defined by the integral (6.81) is denoted variously by $\tilde{f}(\omega)$, $\hat{f}(\omega)$, $F(\omega)$, and $\mathcal{F}[f]$. Moreover, there are various arrangements of the factor of 2π between the Fourier transform and its inverse. Finally, the signs in the exponentials can be reversed, which is equivalent to the substitution $c(\omega) \to c(-\omega)$.

so that $f(t)$ is in $L^2(-\infty, \infty)$, then the integrals (6.80) and (6.81) exist as vector space limits. Then also

$$\|c\|^2 = \int_{-\infty}^{\infty} |c(\omega)|^2 \, d\omega = 2\pi \|f\|^2 \tag{6.84}$$

a result known as *Plancherel's formula*, the integral version of Parseval's formula (6.45). On the other hand, if the integral (6.81) is absolutely convergent, so that

$$\int_{-\infty}^{\infty} |f(t)| \, dt < \infty \tag{6.85}$$

then the integral (6.81) defines $c(\omega)$ as a continuous function of ω.

Remark. The condition (6.85) is distinct from (6.83). For example, the function

$$f(t) \equiv \frac{|t|}{1 + t^2} \tag{6.86}$$

is in $L^2(-\infty, \infty)$, but the Fourier integral is not absolutely convergent. On the other hand, the function

$$f(t) \equiv \frac{e^{-\alpha|t|}}{\sqrt{|t|}} \tag{6.87}$$

has an absolutely convergent Fourier integral if $\operatorname{Re}\alpha > 0$. However, $f(t)$ is not in $L^2(-\infty, \infty)$ due to the singularity of $|f(t)|^2$ at $t = 0$, and neither is its Fourier transform $c(\omega)$, whose evaluation is left as a problem.   □

The Fourier transform is defined by (6.81) for real ω. However, the definition provides a natural analytic continuation to complex ω, since we can write

$$f(t) = f_+(t) + f_-(t) \tag{6.88}$$

with

$$f_+(t) = \begin{cases} f(t) & t > 0 \\ 0 & t < 0 \end{cases} \quad \text{and} \quad f_-(t) = \begin{cases} 0 & t > 0 \\ f(t) & t < 0 \end{cases} \tag{6.89}$$

and corresponding Fourier transforms

$$c_+(\omega) = \int_0^{\infty} e^{i\omega t} f(t) \, dt \quad \text{and} \quad c_-(\omega) = \int_{-\infty}^0 e^{i\omega t} f(t) \, dt \tag{6.90}$$

If the Fourier transform of $f(t)$ exists for real ω, then $c_+(\omega)$ is analytic in the upper half ω-plane $\operatorname{Im}\omega > 0$, since $\exp(i\omega t)$ is then exponentially damped for $t \to +\infty$. Similarly, $c_-(\omega)$ is analytic in the lower half ω-plane $\operatorname{Im}\omega < 0$. This analytic continuation to the complex ω-plane corresponds to the continuation to the complex x-plane in Eq. (6.66). The analytic properties of the Fourier transform can often be used to evaluate the Fourier integral (6.80) by the contour integration methods of Chapter 4.

❑ **Example 6.7.**   Consider the function $f(t)$ defined by

$$f(t) = \begin{cases} e^{-\alpha t} & t > 0 \\ 0 & t < 0 \end{cases} \tag{6.91}$$

(Re $\alpha > 0$). The Fourier transform of $f(t)$ is given by

$$c(\omega) = \int_0^\infty e^{i\omega t} e^{-\alpha t}\, dt = \frac{1}{\alpha - i\omega} \tag{6.92}$$

which has a pole at $\omega = -i\alpha$. The corresponding Fourier integral

$$\int_{-\infty}^\infty c(\omega) \exp(-i\omega t)\, d\omega \tag{6.93}$$

can be evaluated as a contour integral; see Eq. (4.108) for a similar integral.   ∎

One especially useful property of the Fourier transform is that differentiation with respect to t becomes multiplication by $-i\omega$ in the Fourier transform: if $f(t)$ has Fourier transform $c(\omega)$ defined by (6.81), and if the derivative $f'(t)$ has a Fourier transform, the Fourier transform of $f'(t)$ is given by $-i\omega c(\omega)$. This result can be used to transform differential equations in which derivatives appear with constant coefficients.

❑ **Example 6.8.**   Consider a forced, damped harmonic oscillator for which the equation of motion is

$$\frac{d^2 x}{dt^2} + 2\gamma \frac{dx}{dt} + \omega_0^2 x = f(t) \tag{6.94}$$

with $\gamma > 0$ to ensure damping (and ω_0 real), and suppose that the forcing term $f(t)$ can be expressed as a Fourier integral according to Eq. (6.80). If we also write

$$x(t) = \frac{1}{2\pi} \int_{-\infty}^\infty a(\omega) e^{-i\omega t}\, d\omega \tag{6.95}$$

then the equation of motion is reduced to the algebraic equation

$$\left(-\omega^2 - 2i\gamma\omega + \omega_0^2\right) a(\omega) = c(\omega) \tag{6.96}$$

Then we have the solution

$$a(\omega) = \frac{c(\omega)}{\omega_0^2 - \omega^2 - 2i\gamma\omega} \tag{6.97}$$

from which $x(t)$ is obtained by evaluating the Fourier integral. Note that the zeroes of the denominator at

$$\omega = -i\gamma \pm \sqrt{\omega_0^2 - \gamma^2} \equiv \omega_\pm \tag{6.98}$$

are both in the lower half ω-plane when $\gamma > 0$. Thus if the forcing term vanishes for $t < 0$, so that $c(\omega)$ is analytic in the upper half ω-plane, then $a(\omega)$ is also analytic in the upper half-plane, and the solution $x(t)$ also vanishes for $t < 0$, as expected from causality. Thus it satisfies the initial conditions $x(0) = 0$ and $x'(0) = 0$ if $f(t) = 0$ for $t < 0$, although it has no apparent dependence on the initial conditions.

To find a solution for $t > 0$ that satisfies the general initial conditions

$$x(0) = x_0 \qquad x'(0) = v_0 \tag{6.99}$$

it is necessary to add to $x(t)$ a solution

$$A_+ e^{-i\omega_+ t} + A_- e^{-i\omega_- t} \tag{6.100}$$

of the homogeneous equation, with

$$A_+ + A_- = x_0 \tag{6.101}$$

and

$$\omega_+ A_+ - \omega_- A_- = i v_0 \tag{6.102}$$

in order to satisfy the initial conditions. ∎

6.4.2 Convolution Theorem; Correlation Functions

The *convolution* of the functions $f(t)$ and $g(t)$, denoted by $f * g$, is defined by

$$(f * g)(t) \equiv \int_{-\infty}^{\infty} f(t - u) g(u) \, du \tag{6.103}$$

If $f(t)$ and $g(t)$ are expressed as Fourier integrals according to

$$f(t) = \frac{1}{2\pi} \int_{-\infty}^{\infty} c(\omega) e^{-i\omega t} \, d\omega \qquad g(t) = \frac{1}{2\pi} \int_{-\infty}^{\infty} d(\omega) e^{-i\omega t} \, d\omega \tag{6.104}$$

then we can write

$$(f * g)(t) = \frac{1}{2\pi} \int_{-\infty}^{\infty} \int_{-\infty}^{\infty} c(\omega) e^{-i\omega(t - u)} g(u) \, du d\omega \tag{6.105}$$

But the Fourier inversion formula (6.81) gives

$$\int_{-\infty}^{\infty} e^{i\omega u} g(u) \, du = d(\omega) \tag{6.106}$$

so that the convolution integral is given by

$$(f * g)(t) = \frac{1}{2\pi} \int_{-\infty}^{\infty} c(\omega) \, d(\omega) e^{-i\omega t} \, d\omega \tag{6.107}$$

Thus the Fourier transform of the convolution $f * g$ is given by the ordinary product of the Fourier transforms of f and g and conversely, the Fourier transform of the ordinary product of f and g is the convolution $f * g$.

❑ **Example 6.9.**   The solution (6.97) to Eq. (6.94) in the preceding example has the form

$$a(\omega) = k(\omega)c(\omega) \tag{6.108}$$

where $c(\omega)$ is the Fourier transform of the forcing term, and

$$k(\omega) = \frac{1}{\omega_0^2 - \omega^2 - 2i\gamma\omega} = -\frac{1}{(\omega - \omega_+)(\omega - \omega_-)} \tag{6.109}$$

where $\omega_\pm$ are defined by Eq. (6.98). Thus the solution $x(t)$ can be written as a convolution

$$x(t) = \int_{-\infty}^{\infty} K(t - u) f(u) \, du \tag{6.110}$$

with

$$K(\tau) = \frac{1}{2\pi} \int_{-\infty}^{\infty} k(\omega)e^{-i\omega\tau} \, d\tau = \begin{cases} i(e^{-i\omega_+\tau} - e^{-i\omega_-\tau})/\Delta & (\tau > 0) \\ 0 & (\tau < 0) \end{cases} \tag{6.111}$$

(here $\Delta = \omega_+ - \omega_- = 2\sqrt{\omega_0^2 - \gamma^2}$). The result (6.111) is derived by closing the contour in the upper half ω-plane for $\tau < 0$ and in the lower half ω-plane for $\tau > 0$ (note that the two poles of $k(\omega)$ lie in the lower half-plane). The function $K(\tau)$ is the *Green function* (or *response function*) for the forced oscillator. Green functions will be discussed in greater detail in Chapters 7 and 8.   ∎

Closely related to the convolution $f * g$ is the *correlation function*[8]

$$\mathcal{C}_{f,g}(T) \equiv \int_{-\infty}^{\infty} f^*(t) \, g(t + T) \, dt \tag{6.112}$$

The correlation function can also be expressed as a Fourier integral according to

$$\mathcal{C}_{f,g}(T) = \frac{1}{2\pi} \int_{-\infty}^{\infty} c^*(\omega) \, d(\omega)e^{-i\omega T} \, d\omega \tag{6.113}$$

following the analysis given above. A special case of interest is the correlation function of f with itself (the *autocorrelation function* of f), defined by

$$\mathcal{A}_f(T) \equiv \int_{-\infty}^{\infty} f^*(t) \, f(t + T) \, dt = \frac{1}{2\pi} \int_{-\infty}^{\infty} |c(\omega)|^2 \, e^{-i\omega T} \, d\omega \tag{6.114}$$

Thus $|c(\omega)|^2$ is the Fourier transform of the autocorrelation function; if $|c(\omega)|^2$ has a peak at $\omega = \omega_0$, say, then the autocorrelation function will have an important component with period $2\pi/\omega_0$. $|c(\omega)|^2$ is also known as the *spectral density* of $f(t)$.

---

[8]In some contexts, the correlation function is normalized by dividing by $\|f\|\|g\|$.

6.4.3   Laplace Transform

Suppose now that $f(t)$ is a function that vanishes for $t < 0$, and consider the integral

$$c(\omega) = \int_0^\infty f(t)e^{i\omega t}\, dt \qquad (6.115)$$

If the standard Fourier integral (6.81) converges for real ω, then the integral (6.115) converges for $\mathrm{Im}\,\omega > 0$ as already noted, but the class of functions for which this integral converges for some complex ω is larger, since it includes functions $f(t)$ that grow no faster than exponentially as $t \to \infty$. In particular, if there are (real) constants M and a such that $|f(t)| \le Me^{at}$ for all $t \ge 0$, then the integral (6.115) converges and defines an analytic function of ω in the half-plane $\mathrm{Im}\,\omega > a$. If we define $\omega \equiv ip$, then we can define

$$\mathcal{L}f(p) \equiv c(ip) = \int_0^\infty f(t)e^{-pt}\, dt \qquad (6.116)$$

$\mathcal{L}f(p)$ is the *Laplace transform* of $f(t)$; it is analytic in the half-plane $\mathrm{Re}\,p > a$. The Laplace transform be inverted using the standard Fourier integral formula (6.80) and changing the integration variable from ω to p; this gives

$$f(t) = \frac{1}{2\pi i}\int_{b-i\infty}^{b+i\infty} \mathcal{L}f(p)e^{pt}\, dp \qquad (6.117)$$

where the integral is taken along a line $\mathrm{Re}\,p = b(< a)$ parallel to the imaginary p-axis.

The Laplace transform can be useful for solving differential equations since the derivative of a function has a fairly simple Laplace transform; we have

$$\frac{d}{dp}(\mathcal{L}f)(p) = -\int_0^\infty tf(t)e^{-pt}\, dt \qquad (6.118)$$

and

$$\int_0^\infty \frac{df(t)}{dt}e^{-pt}\, dt = p(\mathcal{L}f)(p) - f(0) \qquad (6.119)$$

❑ **Example 6.10.**   Recall the confluent hypergeometric equation (5.B38)

$$\xi f''(\xi) + (c - \xi)f'(\xi) - af(\xi) = 0 \qquad (6.120)$$

If

$$f(\xi) = \int_0^\infty h(t)e^{-\xi t}\, dt \qquad (6.121)$$

then $h(t)$ must satisfy the first-order equation

$$t(1 + t)h'(t) + [1 - a + (2 - c)t]h(t) = 0 \qquad (6.122)$$

This equation has the solution

$$h(t) = At^{a-1}(1+t)^{c-a-1} \tag{6.123}$$

(A is an arbitrary constant), which leads to

$$f(\xi) = A \int_0^\infty t^{a-1}(1+t)^{c-a-1}e^{-\xi t}\,dt \tag{6.124}$$

$$= \frac{A}{\xi^a}\int_0^\infty u^{a-1}(1+\frac{u}{\xi})^{c-a-1}e^{-u}\,du$$

Note that this solution depends only on one arbitrary constant A. This is due to the assumption (6.121) that $f(\xi)$ is actually a Laplace transform of some function $h(t)$, which implies that $f(\xi) \to 0$ for $\xi \to \infty$. Hence the solutions of Eq. (6.120) that do not vanish for $\xi \to \infty$ are lost. Note also that it is necessary that Re $a > 0$ for the integral (6.121) to exist; if this is not the case, then there are no solutions of Eq. (6.120) that vanish for $\xi \to \infty$. Finally, note that the solution (6.124) is just the Whittaker function of the second kind defined by Eq. (5.B51). ∎

In this example, a solution to a second-order differential equation was found as the Laplace transform of a function that satisfies a first-order equation. Problem 14 is an example of how the Laplace transform can be used to reduce directly a second-order equation to a first-order equation.

Fourier and Laplace transform methods are quite similar, since the two transforms are related in the complex plane (only the contours of integration are different). The classes of functions on which the two transforms are defined overlap, but are not identical. There are related transforms that deal with other classes of functions; one such transform is the Mellin transform introduced in Problem 15.

6.4.4 Multidimensional Fourier Transform

Multidimensional Fourier transforms are obtained directly by repeating the one-dimensional Fourier transform. A function f of the n-dimensional vector $\vec{x} = (x_1, \ldots, x_n)$ can be expressed as an n-dimensional Fourier integral

$$f(\vec{x}) = \frac{1}{(2\pi)^{\frac{1}{2}n}} \int e^{i\vec{k}\cdot\vec{x}}\phi(\vec{k})\,d^n k \tag{6.125}$$

with the Fourier transform $\phi(\vec{k})$ given by

$$\phi(\vec{k}) = \frac{1}{(2\pi)^{\frac{1}{2}n}} \int e^{-i\vec{k}\cdot\vec{x}}f(\vec{x})\,d^n x \tag{6.126}$$

Here $\vec{k} = (k_1, \ldots, k_n)$, $d^n k = dk_1 \ldots dk_n$, and $\vec{k}\cdot\vec{x} = k_1 x_1 + \cdots + k_n x_n$ is the usual scalar product. The convergence properties are again varied: $f(\vec{x})$ is in $L^2(\mathbf{R^n})$ if and only if $\phi(\vec{k})$ is in $L^2(\mathbf{R^n})$, but the Fourier integral (6.125) is also well defined for other $\phi(\vec{k})$. As in one dimension, there are variations in allocating the factors of 2π and the choice of $\pm i$ in the exponential.

6.4.5   Fourier Transform in Quantum Mechanics

The Fourier transform plays a special role in quantum mechanics, in which a particle is described by a probability amplitude (wave function) $\psi(\vec{x}, t)$ that depends on position $\vec{x}$ and time t. The corresponding variable in the spatial Fourier transform $\phi(\vec{k}, t)$ is the wave vector $\vec{k}$; the momentum $\vec{p}$ of the particle is related to $\vec{k}$ by

$$\vec{p} = \hbar \vec{k} \tag{6.127}$$

where $\hbar$ is Planck's constant. Thus the probability amplitudes $\psi(\vec{x}, t)$ for position and $\phi(\vec{k}, t)$ for momentum are Fourier transforms of each other. Furthermore, wave functions with definite frequency,

$$\psi(\vec{x}, t) = \psi_0(x)e^{-i\omega t} \tag{6.128}$$

correspond to particle states of definite energy $E = \hbar\omega$.

For a nonrelativistic particle of mass m, the free-particle wave function $\psi(\vec{x}, t)$ satisfies the Schrödinger equation

$$i\hbar \frac{\partial \psi}{\partial t} = -\frac{\hbar^2}{2m} \nabla^2 \psi \tag{6.129}$$

This equation has plane wave solutions

$$\psi_{\vec{k}}(\vec{x}, t) = A e^{i\vec{k} \cdot \vec{x} - i\omega_k t} \tag{6.130}$$

[although these are not in $L^2(\mathbf{R}^3)$] where the Schrödinger equation requires the relation

$$E = \hbar\omega_k = \frac{\hbar^2 k^2}{2m} = \frac{p^2}{2m} \tag{6.131}$$

which is exactly the energy–momentum relation for a free particle.

A complete solution to the Schrödinger equation (6.129) starting from an initial wave function $\psi(\vec{x}, 0)$ at $t = 0$ [assumed to be in $L^2(\mathbf{R}^3)$], can be obtained using the Fourier transform. If the initial wave function is expressed as

$$\psi(\vec{x}, 0) = \int A(\vec{k}) e^{i\vec{k} \cdot \vec{x}} \, d^3k \tag{6.132}$$

then the wave function

$$\psi(\vec{x}, t) = \int A(\vec{k}) e^{i\vec{k} \cdot \vec{x} - i\omega_k t} \, d^3k \tag{6.133}$$

is a solution of (6.129) that satisfies the initial conditions; it is unique, since the Fourier integral (6.132) uniquely defines $A(\vec{k})$ through the inverse Fourier transform. Further properties of the Schrödinger equation will be discussed in Chapter 8.

6.5   Orthogonal Polynomials

6.5.1   Weight Functions and Orthogonal Polynomials

Consider an interval $[a, b]$ on the real axis and weight function $w(t)$ in the scalar product

$$(f, g) = \int_a^b f^*(t) w(t) g(t) \, dt \tag{6.134}$$

as introduced in Eq. (6.20). it is straightforward in principle to construct a sequence $p_0(t)$, $p_1(t)$, $p_2(t)$, ... of polynomials of degree n that form an orthonormal system.[9] The functions $1, t, t^2, \ldots$ are linearly independent, and orthogonal (even orthonormal) polynomials can be constructed by the Gram–Schmidt process; note that these polynomials have real coefficients.

The orthonormal system $\pi_0(t), \pi_1(t), \pi_2(t), \ldots$ formed by normalizing these polynomials is complete in $L^2(a, b)$. Any function $f(t)$ in $L^2(a, b)$ can be expanded as

$$f(t) = \sum_{n=0}^{\infty} c_n \pi_n(t) \tag{6.135}$$

where the series converges in the Hilbert space sense; the expansion coefficients are given by

$$c_n = (\pi_n, f) = \int_a^b \pi_n^*(t) w(t) f(t) \, dt \tag{6.136}$$

One general property of the orthogonal polynomials constructed in this way is that the polynomial $\pi_n(t)$ has n simple zeroes, all within the interval $[a, b]$, To see this, suppose that $\pi_n(t)$ changes sign at the points $\xi_1, \ldots, \xi_m$ within $[a, b]$ (note that $m \le n$, since $\pi_n(t)$ is of degree n). Then the function

$$\pi_n(t)(t - \xi_1) \cdots (t - \xi_m)$$

does not change sign in $[a, b]$, so that the integral

$$\int_a^b \pi_n(t) w(t)(t - \xi_1) \cdots (t - \xi_m) \, dt \ne 0 \tag{6.137}$$

But $\pi_n(t)$ is orthogonal to all polynomials of degree $\le n - 1$ by construction; hence $m = n$ and $\pi_n(t)$ has the form

$$\pi_n(t) = C_n(t - \xi_1) \cdots (t - \xi_m) \tag{6.138}$$

The families of polynomials introduced here all appear as solutions to the linear second-order differential equation (5.81) with special coefficients. It is plausible that in order to have a polynomial solutions, the differential equation can have at most two singular points in the finite t-plane/ This requires the second-order equation to have the form of either the hypergeometric equation (5.A13) or the confluent hypergeometric equation (5.B38), perhaps with singular points shifted by a linear change of variable. The polynomial solutions of these equations are obtained from the general forms $F(a, b|c|\xi)$ or $F(a|c|\xi)$ by setting the parameter $a = -n$ $(n = 0, 1, 2, \ldots)$.

---

[9] This is true even for an infinite interval if the weight function $w(t)$ decreases rapidly enough.

6.5.2   Legendre Polynomials and Associated Legendre Functions

We look first for orthogonal polynomials on the interval $[-1, 1]$. It turns out that the Legendre polynomials introduced in Section 5.5 are orthogonal on $[-1, 1]$, with weight function $w(t) = 1$. To show this, we can use Rodrigues' formula from Section 5.5,

$$P_n(t) = \frac{1}{2^n n!} \frac{d^n}{dt^n} (t^2 - 1)^n = C_n u_n^{(n)}(t) \tag{5.127}$$

We have now evaluated the constant C_n, and here again $u_n(t) = (t^2 - 1)^n$. If $m < n$, we can integrate by parts m times to evaluate the scalar product

$$(P_m, P_n) = \int_{-1}^{1} P_m(t) P_n(t)\, dt = C_m C_n \int_{-1}^{1} u_m^{(m)}(t) u_n^{(n)}(t)\, dt = \cdots$$

$$= (-1)^m C_m C_n (2m)! \int_{-1}^{1} u_n^{(n-m)}(t)\, dt \tag{6.139}$$

$$= (-1)^m C_m C_n (2m)!\, u_n^{(n-m-1)} \Big|_{-1}^{1} = 0$$

Thus the $\{P_n(t)\}$ are orthogonal on $[-1, 1]$. If $m = n$, the same procedure gives

$$(P_n, P_n) = \int_{-1}^{1} [P_n(t)]^2\, dt = C_n^2 (2n)! \int_{-1}^{1} (1 - t^2)^n\, dt$$

$$\tag{6.140}$$

$$= C_n^2 (2n)!\, 2^{2n+1} \frac{[\Gamma(n+1)]^2}{\Gamma(2n+2)} = \frac{2}{2n+1}$$

Thus a set $\{\pi_n(t)\}$ of orthonormal polynomials on $[-1, 1]$ is given in terms of the $\{P_n(t)\}$ by

$$\pi_n(t) = \sqrt{\frac{2n+1}{2}}\, P_n(t) \tag{6.141}$$

These results can also be obtained using the generating function (5.137) (see Problem 16).

➔ **Exercise 6.7.**   Show that applying the Gram–Schmidt process to the linearly independent monomials $1, t, t^2, t^3$ leads to the first four polynomials $\pi_0(t), \dots, \pi_3(t)$.   ☐

We note here the recursion formulas for the Legendre polynomials

$$(n+1) P_{n+1}(t) = (2n+1) t P_n(t) - n P_n(z) \tag{6.142}$$

$$(t^2 - 1) P_n'(t) = nt P_n(t) - n P_{n-1}(t) \tag{6.143}$$

from Exercise 5.14. These relations can be derived by differentiating the generating function

$$S(\xi, t) = \frac{1}{\sqrt{1 - 2\xi t + \xi^2}} = \sum_{n=0}^{\infty} \xi^n P_n(t) \tag{5.137}$$

with respect to ξ and t, respectively. Equation (6.142) is useful for evaluating $P_n(t)$ numerically if n is not too large.

Also important are the *associated Legendre functions* $P_n^a(t)$ defined for $-1 \le t \le 1$ by

$$P_n^a(t) = (-1)^a (1 - t^2)^{\frac{1}{2}a} \frac{d^a}{dt^a} P_n(t) = (-1)^a C_n (1 - t^2)^{\frac{1}{2}a} u_n^{(n+a)}(t) \qquad (6.144)$$

with $a = 0, 1, 2, \ldots$ integer and $n = a, a + 1, a + 2, \ldots$. For fixed integer a, the $P_n^a(t)$ are orthogonal on $[-1, 1]$. To show this, suppose that $n < m$. We then have the scalar product

$$(P_m^a, P_n^a) = \int_{-1}^{1} P_m^a(t) P_n^a(t)\, dt =$$

$$= C_m C_n \int_{-1}^{1} u_m^{(m+a)}(t)(1 - t^2)^a u_n^{(n+a)}(t)\, dt = \cdots \qquad (6.145)$$

$$= C_m C_n A_{(n,a)} \int_{-1}^{1} u_n^{(n-m)}(t)\, dt = C_m C_n A_{(n,a)} \, u_n^{(n-m-1)} \big|_{-1}^{1} = 0$$

after integrating by parts $n + a$ times; the constant $A_{(n,a)}$ is given by

$$A_{(n,a)} = (-1)^{n+a} \frac{d^{n+a}}{dt^{n+a}} \left\{ (1 - t^2)^a \, u_n^{(n+a)}(t) \right\} = (-1)^n \, (2n)! \, \frac{(n+a)!}{(n-a)!} \qquad (6.146)$$

Thus the $\{P_n^a(t)\}$ are orthogonal on $[-1, 1]$. To find the normalization, we can evaluate the integral for $m = n$. We have

$$(P_n^a, P_n^a) = \int_{-1}^{1} [P_n^a(t)]^2 \, dt = C_n^2 (-1)^n A_{(n,a)} \int_{-1}^{1} (1 - t^2)^n \, dt$$

$$= C_n^2 \, (2n)! \, \frac{(n+a)!}{(n-a)!} \, 2^{2n+1} \, \frac{[\Gamma(n+1)]^2}{\Gamma(2n+2)} = \frac{2}{2n+1} \, \frac{(n+a)!}{(n-a)!} \qquad (6.147)$$

Then for fixed $a = 0, 1, 2, \ldots$, the functions $\pi_n^a(t)$ defined by

$$\pi_n^a(t) = \sqrt{\frac{2n+1}{2}} \, \sqrt{\frac{(n-a)!}{(n+a)!}} \, P_n^a(t) \qquad (6.148)$$

$(n = a, a + 1, a + 2, \ldots)$ form a complete orthonormal system on $[-1, 1]$.

→ **Exercise 6.8.**   Show that the associated Legendre function $P_n^a(t)$ is a solution of the differential equation

$$(1 - t^2) u''(t) - 2t u'(t) + \left\{ n(n+1) - \frac{a^2}{1-t^2} \right\} u(t) = 0$$

Note the conventional sign change from the original Legendre equation.   □

Remark.   From the preceding discussion, it follows that the polynomials $F_n^a(t)$ defined by

$$F_n^a(t) = (-1)^a \frac{d^a}{dt^a} P_{n+a}(t) = (-1)^a C_n u_{n+a}^{(n+2a)}(t) \qquad (6.149)$$

form a set of orthogonal polynomials on $[-1, 1]$ with weight function $w(t) = (1 - t^2)^a$. It is left to the reader to express these polynomials in terms of the Gegenbauer polynomials described in Appendix A.   □

6.5.3 Spherical Harmonics

Legendre polynomials and the associated Legendre functions appear in a number of contexts, but they first appear to most physicists in the study of partial differential equations involving the Laplacian, such as Poisson's equation or the Schrödinger equation. These equations will be analyzed at great length in Chapter 8. Here we simply note that for many systems, it is convenient to use the spherical coordinates r, θ, ϕ introduced in Eq. (3.173). In these coordinates, the Laplacian has the form

$$\Delta = \frac{1}{r^2} \left\{ \frac{\partial}{\partial r} \left(r^2 \frac{\partial}{\partial r} \right) + \frac{1}{\sin \theta} \frac{\partial}{\partial \theta} \left(\sin \theta \frac{\partial}{\partial \theta} \right) + \frac{1}{\sin^2 \theta} \frac{\partial^2}{\partial \phi^2} \right\} \qquad (6.150)$$

as derived in Chapter 3 (see Eq. (3.193)). We look for solutions of the relevant partial differential equation that have the form

$$f(r, \theta, \phi) = R(r) Y(\theta, \phi) \qquad (6.151)$$

In most such cases, the angular function $Y(\theta, \phi)$ must be a solution of the partial differential equation

$$\left\{ \frac{1}{\sin \theta} \frac{\partial}{\partial \theta} \left(\sin \theta \frac{\partial}{\partial \theta} \right) + \frac{1}{\sin^2 \theta} \frac{\partial^2}{\partial \phi^2} \right\} Y(\theta, \phi) = \lambda Y(\theta, \phi) \qquad (6.152)$$

with λ such that the $Y(\theta, \phi)$ is

(i) single valued as a function of ϕ, and

(ii) nonsingular as a function of θ over the range $0 \leq \theta \leq \pi$ including the endpoints.

To satisfy the first requirement, we can expand $Y(\theta, \phi)$ as a Fourier series

$$Y(\theta, \phi) = \sum_{m=0}^{\infty} [A_m(\cos \theta) \, \cos(m\phi) + B_m(\cos \theta) \, \sin(m\phi)]$$

$$= \sum_{m=-\infty}^{\infty} C_m(\cos \theta) \, e^{im\phi} \qquad (6.153)$$

Then the coefficients $A_m(\cos \theta), B_m(\cos \theta), C_m(\cos \theta)$ must satisfy the differential equation

$$\left\{ \frac{1}{\sin \theta} \frac{d}{d\theta} \left(\sin \theta \frac{d}{d\theta} \right) - \frac{m^2}{1 - \cos^s \theta} \right\} X_m(\cos \theta) = \lambda X_m(\cos \theta) \qquad (6.154)$$

(here $X_m = A_m, B_m, C_m$). Now introduce the variable $t = \cos \theta$, and express Eq. (6.154) in terms of t as

$$\left\{ \frac{d}{dt} (1 - t^2) \frac{d}{dt} - \frac{m^2}{1 - t^2} \right\} X_m(t) = \lambda X_m(t) \qquad (6.155)$$

From Exercise 6.8, we see that this is the differential equation for the associated Legendre function $P_n^m(t)$, with $\lambda = -n(n+1) \, (n = m, m+1, \dots)$.

Thus we have solutions to Eq. (6.152) of the form

$$C_{nm}(\theta, \phi) = P_n^{|m|}(\cos\theta) \, e^{im\phi} \tag{6.156}$$

for $n = 0, 1, 2, \ldots$; $m = 0, \pm 1, \ldots, \pm n$. Using the normalization integral (6.147), we find functions

$$Y_{nm}(\theta, \phi) = \sqrt{\frac{2n+1}{4\pi} \frac{(n-|m|)!}{(n+|m|)!}} P_n^{|m|}(\cos\theta) \, e^{im\phi} \tag{6.157}$$

that form a complete orthonormal system on the surface of the sphere S^2. The $Y_{nm}(\theta, \phi)$ are *spherical harmonics*.

One important result is the *spherical harmonic addition theorem*. Suppose we have two unit vectors

$$\hat{n} = \hat{n}(\theta, \phi) \quad \text{and} \quad \hat{n}' = \hat{n}'(\theta', \phi')$$

and let Θ be the angle between $\hat{n}$ and $\hat{n}'$. Then we have

$$\hat{n} \cdot \hat{n}' = \cos\Theta = \cos\theta \cos\theta' + \sin\theta \sin\theta' \cos(\phi - \phi') \tag{6.158}$$

The spherical harmonic addition theorem states that

$$P_n(\cos\Theta) = \frac{4\pi}{2n+1} \sum_{m=-n}^{n} Y_{nm}^*(\theta', \phi') Y_{nm}(\theta, \phi) \tag{6.159}$$

To show this, note that $\hat{n} \cdot \hat{n}'$ is a scalar; hence $P_n(\cos\Theta)$ is as well. Since the Laplacian Δ is a scalar, so is the differential operator in Eq. (6.152). Hence

$$\left\{ \frac{1}{\sin\theta} \frac{\partial}{\partial\theta} \left(\sin\theta \frac{\partial}{\partial\theta} \right) + \frac{1}{\sin^2\theta} \frac{\partial^2}{\partial\phi^2} \right\} P_n(\cos\Theta) = -n(n+1) P_n(\cos\Theta) \tag{6.160}$$

since this differential equation is certainly true in a coordinate system with the Z-axis chosen along $\hat{n}'$, as $\cos\Theta = \cos\theta$ in such a coordinate system. But any solution of Eq. (6.160) must be a linear combination of spherical harmonics $Y_{nm}(\theta, \phi)$ ($m = n, n-1, \ldots, -n+1, -n$). Hence we can write

$$P_n(\cos\Theta) = \sum_{m=-n}^{n} c_{nm}(\theta', \phi') \, Y_{nm}(\theta, \phi) \tag{6.161}$$

Since $\cos\Theta$ depends on ϕ and ϕ' only in the combination $(\phi - \phi')$, the coefficient $c_{nm}(\theta', \phi')$ must be proportional to $Y_{nm}^*(\theta', \phi')$ (note that c_{nm} must satisfy Eq. (6.160) in the primed variables). The overall scale on the right-hand side of Eq. (6.159) is fixed by considering the case when both $\hat{n}$ and $\hat{n}'$ are in the Z-direction.

6.6   Haar Functions; Wavelets

Fourier series and Fourier integral methods permit the extraction of frequency information from a function of time. However, these methods have two important limitations when applied to real signals. First, they require a knowledge of the signal over the entire domain of definition, so that analysis of the signal must wait until the entire signal has been received (although some analysis can often begin while the signal is being recorded). Furthermore, the standard Fourier analysis converts a function of time into a function of frequency, and does not deal directly with the important problem of a signal at a standard frequency (the *carrier frequency*) upon which is superimposed an information bearing signal either in the form of *amplitude modulation* (as in the case of AM radio), or *frequency modulation* (FM radio, television). Decoding such a signal involves analysis both in time and in frequency.

There are many problems that involve the structure of a function at various scales, both in the use of renormalization group methods in quantum field theory and the statistical mechanics of phase transitions, and in macroscopic applications to signal processing and analysis. While Fourier series can be used to analyze the behavior of functions with a fixed period, they are not well suited for the rescaling of the time interval or spatial distances ("zooming in" and "zooming out") needed in these problems. Thus it is important to have sets of functions that have the flexibility to resolve multiple scales.

A simple set of such functions consists of the *Haar functions* $h_0(t)$, $h_{n,k}(t)$,

$$h_0(t) = \begin{cases} 1 & 0 < t \leq 1 \\ 0 & \text{otherwise} \end{cases} \tag{6.162}$$

(the *characteristic function* of the interval $[0, 1]$, also known as the "box" function), and

$$h_{n,k}(t) = \begin{cases} 1 & 2k - 2 < 2^n t \leq 2k - 1 \\ -1 & 2k - 1 < 2^n t \leq 2k \\ 0 & \text{otherwise} \end{cases} \tag{6.163}$$

($n = 1, 2, \ldots; k = 1, \ldots, n$). The first four Haar functions are shown in Fig. 6.2.

The Haar functions are orthogonal, since

$$\int_0^1 h_0(t) h_{n,k}(t)\, dt = \int_0^1 h_{n,k}(t)\, dt = 0 \tag{6.164}$$

and

$$\int_0^1 h_{m,q}(t) h_{n,k}(t)\, dt = \left(\tfrac{1}{2}\right)^{n-1} \delta_{mn} \delta_{kq} \tag{6.165}$$

Thus the functions $\{\chi_{n,k}(t)\}$ defined by

$$\chi_{n,k}(t) \equiv \sqrt{2^{n-1}} h_{n,k}(t) \tag{6.166}$$

form an orthonormal system on the interval $0 \leq t \leq 1$, and the orthonormal system formed by $h_0(t)$ and the $\{\chi_{n,k}(t)\}$ is complete since any integrable function can be approximated by

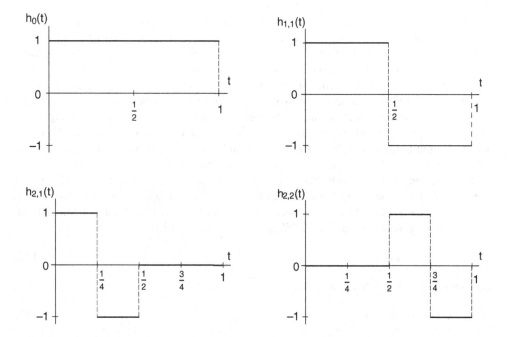

Figure 6.2: The first four Haar functions $h_0(t)$, $h_{1,1}(t)$, $h_{2,1}(t)$, and $h_{2,2}(t)$.

a sequence of step functions. Thus any function in $L^2(0, 1)$ can be expressed as a series

$$f(t) = c_0 + \sum_{n=1}^{\infty} \sum_{k=1}^{n} c_{n,k} h_{n,k}(t) \tag{6.167}$$

with

$$c_0 = \int_0^1 f(t)\, dt \quad \text{and} \quad c_{n,k} = 2^{n-1} \int_0^1 f(t) h_{n,k}(t)\, dt \tag{6.168}$$

Note that the Haar functions $\{h_{n,k}(t)\}$ are obtained from the basic function $h_{1,1}(t)$ with argument rescaled by powers of 2 and shifted by integers. Explicitly,

$$h_{n+1,k+1}(t) = h_{1,1}(2^n t - k) \tag{6.169}$$

$(n = 0, 1, 2, \ldots; k = 0, 1, \ldots, n - 1)$. The function $h_{1,1}(t)$ can in turn be expressed in terms of $h_0(t)$ as

$$h_{1,1}(t) = h_0(2t) - h_0(2t - 1) \tag{6.170}$$

so that the entire basis is derived from the single function $h_0(t)$ by dilation and translation of the argument.

Expansion (6.167) resolves a function in $L^2(0,1)$ into components $c_{n,k}$ with respect to the Haar basis. The index n defines a scale (2^{-n} in time or 2^n in frequency) on which the details of the function are resolved, and the index k defines the evolution of these details over the n steps from $t = 0$ to $t = 1$ at this scale. Thus the expansion can describe the time evolution of the function on a range of frequency scales, in contrast to the standard Fourier analysis, which describes either the time dependence of the function, or the frequency composition of the function, but not the time dependence of frequency components. The price of this expansion is that the frequency scales are multiples of the fundamental frequency times 2^n, rather than any integer multiple, and the time steps are submultiples of the fundamental period (hence dependent on the fundamental frequency chosen) by a power of 2, rather than continuous.

The preceding analysis can be extended to functions defined on the entire real axis, at least to those in $L^2(-\infty, \infty)$, with the added feature that the time scale can be expanded as well as contracted by powers of two. The Haar functions are used as an explicit example, but the analysis can be generalized to other families of functions. First let

$$\phi(t) \equiv h_0(t) \tag{6.171}$$

and consider the shifted functions

$$\phi_{0,k}(t) \equiv \phi(t - k) \tag{6.172}$$

($k = 0, \pm 1, \pm 2, \dots$). The $\{\phi_{0,k}\}$ form an orthonormal system that defines a subspace V_0 of $L^2(-\infty, \infty)$ consisting of elements of the form

$$a(t) = \sum_{k=-\infty}^{\infty} \alpha_k \phi_{0,k}(t) \tag{6.173}$$

with

$$\|a\|^2 = \int_{-\infty}^{\infty} |a(t)|^2 \, dt = \sum_{k=-\infty}^{\infty} |\alpha_k|^2 < \infty \tag{6.174}$$

V_0 is the (infinite-dimensional) space of functions that are piecewise constant on intervals of the form $k < t \le k + 1$. It is equivalent to the sequence space $\ell^2(\mathbf{R})$ or $\ell^2(\mathbf{C})$, depending on whether the α_k are real or complex.

For any function $f(t)$ in $L^2(-\infty, \infty)$, the component f_0 of f in V_0 is

$$f_0(t) = \sum_{k=-\infty}^{\infty} f_{0,k} \phi_{0,k}(t) \tag{6.175}$$

with coefficients

$$f_{0,k} = \int_{-\infty}^{\infty} \phi_{0,k}(t) f(t) \, dt = \int_{k}^{k+1} f(t) \, dt \tag{6.176}$$

($k = 0, \pm 1, \pm 2, \dots$).

f_0 is an approximation to f by step functions of unit length, and indeed the best approximation in the mean square sense with origin fixed (for a particular function, shifting the origin might improve the approximation, but that is beside the point here).

This approximation can be refined by reducing the scale of the interval. Thus consider the new basis functions

$$\phi_{1,k}(t) \equiv \sqrt{2}\phi(2t - k) = \sqrt{2}\phi_{0,k}(2t) \qquad (6.177)$$

($k = 0, \pm1, \pm2, \dots$), where the factor of $\sqrt{2}$ normalizes the $\phi_{1,k}$. The $\{\phi_{1,k}\}$ form an orthonormal system that defines a subspace $\mathcal{V}_1$ of $L^2(\infty, \infty)$ whose elements are functions that are piecewise constant on intervals of the form $k < 2t \leq k+1$. The space $\mathcal{V}_1$ includes $\mathcal{V}_0$ as a proper subspace (a function that is piecewise constant on intervals of length 1 is also piecewise constant on intervals of length $\frac{1}{2}$). Note that

$$\phi_{0,k}(t) = \sqrt{\tfrac{1}{2}}\left(\phi_{1,2k}(t) + \phi_{1,2k+1}(t)\right) \qquad (6.178)$$

gives an explicit expression for the basis vectors of $\mathcal{V}_0$ in terms of those of $\mathcal{V}_1$. The component f_1 of a function f in $\mathcal{V}_1$ is

$$f_1(t) = \sum_{k=-\infty}^{\infty} f_{1,k}\phi_{1,k}(t) \qquad (6.179)$$

with

$$f_{1,k} = \int_{-\infty}^{\infty} \phi_{1,k}(t)f(t)\,dt = \sqrt{2}\int_{k/2}^{(k+1)/2} f(t)\,dt$$

$$= \sqrt{\tfrac{1}{2}} \int_{k}^{k+1} f\left(\frac{u}{2}\right) du \qquad (6.180)$$

($k = 0, \pm1, \pm2, \dots$). The last integral shows how the detail of f on a finer scale is blown up in the construction of f_1.

Continuing this process leads to a sequence of nested subspaces

$$\cdots \subset \mathcal{V}_{-2} \subset \mathcal{V}_{-1} \subset \mathcal{V}_0 \subset \mathcal{V}_1 \subset \mathcal{V}_2 \subset \cdots \qquad (6.181)$$

(it is possible to go backwards from $\mathcal{V}_0$ to coarser scales by doubling the interval over which the functions are piecewise continuous), such that a complete orthonormal system on $\mathcal{V}_n$ is given by the set $\{\phi_{n,k}\}$ with

$$\phi_{n,k}(t) \equiv 2^{\frac{n}{2}} \phi(2^n t - k) \qquad (6.182)$$

($k = 0, \pm1, \pm2, \dots$) for each $n = 0, \pm1, \pm2, \dots$. A distinctive feature of this set is that all the functions are obtained from the single *scaling function* $\phi(t)$ by rescaling and shifting the argument. A nested sequence (6.181) of subspaces, with a complete orthonormal system on each subspace obtained from a single scaling function by a relation like Eq. (6.182), is called a *multiresolution analysis*.

The component f_n of a function f in $\mathcal{V}_n$ has the explicit representation

$$f_n(t) = \sum_{k=-\infty}^{\infty} f_{n,k} \phi_{n,k}(t) \tag{6.183}$$

with

$$f_{n,k} = \int_{-\infty}^{\infty} \phi_{n,k}(t) f(t)\, dt = 2^{\frac{n}{2}} \int_{k/2^n}^{(k+1)/2^n} f(t)\, dt \tag{6.184}$$

$$= \left(\frac{1}{2}\right)^{\frac{n}{2}} \int_{k}^{k+1} f\left(\frac{u}{2^n}\right) du$$

($k = 0, \pm 1, \pm 2, \dots$), again showing the resolution of the fine scale details of f. For any function f in $L^2(-\infty, \infty)$, the sequence of approximations $\{\dots, f_{-2}, f_{-1}, f_0, f_1, f_2, \dots\}$ converges to f as a vector space limit since, as already noted, every square integrable function can be approximated by a sequence of step functions.

At each stage in the sequence (6.181), a new space is added. If $\mathcal{W}_n$ is the orthogonal complement of $\mathcal{V}_n$ in $\mathcal{V}_{n+1}$, then

$$\mathcal{V}_{n+1} = \mathcal{V}_n \oplus \mathcal{W}_n \tag{6.185}$$

($n = 0, \pm 1, \pm 2, \dots$), and the convergence of the sequence $\{f_n\}$ is equivalent to

$$L^2(-\infty, \infty) = \oplus_{-\infty}^{\infty} \mathcal{W}_n \tag{6.186}$$

The space $\mathcal{W}_n$ is the *wavelet space* at level n; it is where new structure in a function is resolved that was not present in the component f_n. Equation (6.185) can be seen as an illustration of the split of a function in $\mathcal{V}_{n+1}$ into a low frequency part (in $\mathcal{V}_n$) and a high frequency part (in $\mathcal{W}_n$).

From the fundamental relation

$$\phi(t) = \phi(2t) + \phi(2t - 1) \tag{6.187}$$

(this can be seen by inspection) and Eq. (6.182) it follows that

$$\phi_{n,k}(t) = \sqrt{\tfrac{1}{2}} \left(\phi_{n+1,2k}(t) + \phi_{n+1,2k+1}(t) \right) \tag{6.188}$$

[see Eqs. (6.177) and (6.178)]. Then the set of vectors $\{\psi_{n,k}\}$ defined by

$$\psi_{n,k}(t) \equiv \sqrt{\tfrac{1}{2}} \left(\phi_{n+1,2k}(t) - \phi_{n+1,2k+1}(t) \right) \tag{6.189}$$

are orthogonal to the $\{\phi_{n,k}\}$ in $\mathcal{V}_{n+1}$. Hence the $\{\psi_{n,k}\}$ must be in $\mathcal{W}_n$. The $\{\psi_{n,k}\}$ are orthonormal by construction, and together with the $\{\phi_{n,k}\}$ are equivalent to the set $\{\phi_{n+1,k}\}$ which is a complete orthonormal system in $\mathcal{V}_{n+1} = \mathcal{V}_n \oplus \mathcal{W}_n$. Hence the $\{\psi_{n,k}\}$ form a complete orthonormal system in $\mathcal{W}_n$. The $\{\psi_{n,k}\}$ are *wavelets* at scale 2^n in frequency (or scale 2^{-n} in t); they are orthogonal to all functions in $\mathcal{V}_n$.

The $\{\psi_{n,k}(t)\}$ can be obtained from a single function $\psi(t)$, the *fundamental wavelet* (or *mother wavelet*), which is given here by

$$\psi(t) = \phi(2t) - \phi(2t - 1) \tag{6.190}$$

which is equivalent to Eq. (6.170) since $\psi(t) = h_{1,1}(t)$ [compare this with the corresponding relation (6.187) for $\phi(t)$]. Corresponding to Eq. (6.182) for the $\{\phi_{n,k}\}$, we have

$$\psi_{n,k}(t) \equiv 2^{\frac{n}{2}} \psi(2^n t - k) \tag{6.191}$$

The $\{\psi_{n,k}\}$ are related to the Haar functions $\{h_{n,k}\}$ by

$$\psi_{n,k}(t) \equiv 2^{\frac{n}{2}} h_{n+1,k+1}(t) \tag{6.192}$$

except that here n and k can range over all integers without restriction.

An important advance of the 1980s was the discovery that the multiresolution analysis described here with the Haar functions could also be carried out with other, smoother choices for the scaling function $\phi(t)$. The essential features of the multiresolution analysis are

(i) The space $L^2(-\infty, \infty)$ is decomposed into a nested sequence of subspaces $\{V_n\}$ as in (6.181), such that if f is in L^2, then the sequence $\{f_n\}$ (f_n is the component of f in V_n) converges to f. Also, if $f(t)$ is in V_n then $f(2t)$ is in V_{n+1} so that the space $\{V_{n+1}\}$ is an exact rescaling of the space $\{V_n\}$.

(ii) There is a *scaling function* $\phi(t)$ such that the translations of $\phi(t)$ by integers,

$$\phi_{0,k}(t) \equiv \phi(t - k) \tag{6.193}$$

$(k = 0, \pm 1, \pm 2, \dots)$, form a complete orthonormal system on V_0, and the dilations of $\phi(t)$ by 2^n, together with integer translations,

$$\phi_{n,k}(t) \equiv 2^{\frac{n}{2}} \phi(2^n t - k) \tag{6.194}$$

$(k = 0, \pm 1, \pm 2, \dots)$, form a complete orthonormal system on V_n for each integer n.

The nesting condition (6.181) implies that $\phi(t)$ is in V_1. Hence there must be an expansion

$$\phi(t) = \sum_{k=-\infty}^{\infty} c_k \phi(2t - k) \tag{6.195}$$

of $\phi(t)$ in terms of its rescaled and shifted versions. Equation (6.187) is a relation of this type for the "box" function $h_0(t)$. If we introduce the Fourier transform

$$\Phi(\omega) \equiv \int_{-\infty}^{\infty} e^{i\omega t} \phi(t)\, dt \tag{6.196}$$

then the relation (6.195) leads to

$$\Phi(\omega) = \sum_{k=-\infty}^{\infty} c_k \int_{-\infty}^{\infty} e^{i\omega t} \phi(2t - k)\, dt$$

$$\tag{6.197}$$

$$= \frac{1}{2} \sum_{k=-\infty}^{\infty} c_k e^{\frac{1}{2} i k \omega} \int_{-\infty}^{\infty} e^{\frac{1}{2} i \omega \tau} \phi(\tau)\, d\tau$$

Thus the Fourier transform is scaled according to

$$\Phi(\omega) = H(\tfrac{1}{2}\omega)\Phi(\tfrac{1}{2}\omega) \tag{6.198}$$

where

$$H(\omega) = \tfrac{1}{2} \sum_{k=-\infty}^{\infty} c_k e^{ik\omega} \tag{6.199}$$

Iterating this equation gives

$$\Phi(\omega) = \left\{ \prod_{k=1}^{\infty} H\left(\frac{\omega}{2^k}\right) \right\} \Phi(0) \tag{6.200}$$

and it will be seen shortly that $\Phi(0) = 1$. Thus the Fourier transform of the scaling function can be expressed formally as an infinite product. Proving convergence of the infinite product requires some heavy analysis in general, but it does converge for the examples described here.

❑ **Example 6.11.**   With $c_0 = c_1 = 1$ (and all other $c_n = 0$), we have the function

$$H(\omega) = \tfrac{1}{2}(1 + e^{i\omega}) = \frac{1 - e^{2i\omega}}{2(1 - e^{i\omega})} = e^{\frac{1}{2}i\omega} \cos \tfrac{1}{2}\omega \tag{6.201}$$

and

$$\Phi(\omega) = \lim_{m \to \infty} \frac{1 - e^{i\omega}}{2^m(1 - e^{i\omega/2^m})} = \frac{1 - e^{i\omega}}{-i\omega} = e^{\frac{1}{2}i\omega} \frac{\sin \tfrac{1}{2}\omega}{\tfrac{1}{2}\omega} \tag{6.202}$$

which is precisely the Fourier transform of $h_0(t)$.   ∎

➙ **Exercise 6.9.**   Show that the Fourier transform of the Haar function $h_0(t)$ defined by Eq. (6.162) is given by

$$\hat{h}_0(\omega) = \frac{1 - e^{i\omega}}{-i\omega} = e^{\frac{1}{2}i\omega} \frac{\sin \tfrac{1}{2}\omega}{\tfrac{1}{2}\omega}$$

as claimed in the preceding example.   ☐

If the functions $\{\phi_{0,k}(t)\}$ defined by Eq. (6.193) form an orthonormal system, then

$$I_k \equiv \int_{-\infty}^{\infty} \phi^*(t)\phi(t - k)\, dt = \frac{1}{2\pi} \int_{-\infty}^{\infty} |\Phi(\omega)|^2 e^{ik\omega}\, d\omega = \delta_{k0} \tag{6.203}$$

Now we can write

$$I_k = \frac{1}{2\pi} \sum_{m=-\infty}^{\infty} \int_0^{2\pi} |\Phi(\omega + 2m\pi)|^2 e^{ik\omega}\, d\omega \tag{6.204}$$

Then Eq. (6.203) implies

$$M_\phi(\omega) \equiv \sum_{m=-\infty}^{\infty} |\Phi(\omega + 2m\pi)|^2 = 1 \tag{6.205}$$

independent of ω, since Eq. (6.204) shows that the I_k are the Fourier coefficients of a 2π-periodic function $M_\phi(\omega)$. Note also that

$$\sum_{m=-\infty}^{\infty} |\Phi(2\omega + 2m\pi)|^2 = \sum_{m=-\infty}^{\infty} |H(\omega + m\pi)|^2 |\Phi(\omega + m\pi)|^2$$

$$= |H(\omega)|^2 + |H(\omega + \pi)|^2 \tag{6.206}$$

so that $|H(\omega)|^2 + |H(\omega + \pi)|^2 = 1$ in view of Eq. (6.205). Since $H(0) = 1$, this gives $H(\pi) = 0$, and thus

$$\Phi(2m\pi) = 0 \tag{6.207}$$

($m = \pm 1, \pm 2, \dots$) from Eq. (6.198). Equation (6.205) then gives $\Phi(0) = 1$, so the normalization of Φ follows from the normalization of ϕ.

On each space $\mathcal{W}_n$ in the split (6.185) a wavelet basis can be derived, as in the case of the Haar functions, from a fundamental wavelet $\psi(t)$ such that
   (i) the translations of $\psi(t)$ by integers,

$$\psi_{0,k}(t) \equiv \psi(t - k) \tag{6.208}$$

($k = 0, \pm 1, \pm 2, \dots$), form a complete orthonormal system on $\mathcal{W}_0$, and
   (ii) the dilations of $\psi(t)$ by 2^n, together with integer translations,

$$\psi_{n,k}(t) \equiv 2^{\frac{n}{2}} \psi(2^n t - k) \tag{6.209}$$

($k = 0, \pm 1, \pm 2, \dots$), form a complete orthonormal system on $\mathcal{W}_n$ for each integer n.

It follows from (i) and (ii) that the $\{\psi_{n,k}\}$ ($n, k = 0, \pm 1, \pm 2, \dots$) then form a complete orthonormal system (a *wavelet basis*) on $L^2(-\infty, \infty)$. We shall see that such a fundamental wavelet $\psi(t)$ can always be constructed from a multiresolution analysis with scaling function $\phi(t)$ whose integer translations $\{\phi(t - k)\}$ form an orthonormal system.

The fundamental wavelet $\psi(t)$ is in the space $\mathcal{V}_1$, as a consequence of Eq. (6.185). Hence it can be expanded as

$$\psi(t) = \sum_{k=-\infty}^{\infty} d_k \phi(2t - k) \tag{6.210}$$

of $\psi(t)$ in terms of the scaling function $\phi(t)$. Then the Fourier transform $\Psi(\omega)$ of the fundamental wavelet is given by

$$\Psi(\omega) \equiv \int_{-\infty}^{\infty} e^{i\omega t} \psi(t)\, dt = G(\tfrac{1}{2}\omega)\Phi(\tfrac{1}{2}\omega) \tag{6.211}$$

where

$$G(\omega) \equiv \frac{1}{2} \sum_{k=-\infty}^{\infty} d_k e^{ik\omega} \tag{6.212}$$

is a 2π-periodic function that corresponds to the function $H(\omega)$ introduced in the scaling relation for $\phi(t)$. If the functions $\{\psi_{0,k}(t)\}$ defined by Eq. (6.208) form an orthonormal system, then the function $G(\omega)$ must satisfy

$$|G(\omega)|^2 + |G(\omega + \pi)|^2 = 1 \tag{6.213}$$

Equation (6.213) is derived following the steps in Eqs. (6.203)–(6.206) above, with ψ in place of ϕ.

Orthogonality of the functions $\{\psi_{0,k}(t)\}$ and $\{\phi_{0,\ell}(t)\}$ leads to the condition

$$G^*(\omega)H(\omega) + G^*(\omega + \pi)H(\omega + \pi) = 0 \tag{6.214}$$

by a similar analysis. A "natural" choice of $G(\omega)$ that satisfies Eqs. (6.213) and (6.214) is

$$G(\omega) = e^{\pm i\omega} H^*(\omega + \pi) \tag{6.215}$$

This $G(\omega)$ can be multiplied by a phase factor $\exp(2im\omega)$ for any integer m; this corresponds simply to an integer translation of the wavelet basis. In terms of the coefficients c_n and d_n, the solution (6.215) gives

$$d_n = (-1)^{n+1} c_{-n\pm 1} \tag{6.216}$$

The fundamental wavelet $\psi(t)$ can then be derived from the scaling function $\phi(t)$ using Eq. (6.210), or $\Psi(\omega)$ can be determined from $\Phi(\omega)$ using Eq. (6.211).

❑ **Example 6.12.** For the Haar functions, with $H(\omega) = \frac{1}{2}(1 + e^{i\omega})$, we can choose

$$G(\omega) = \frac{1}{2}(1 - e^{i\omega}) = -e^{i\omega} H^*(\omega + \pi) \tag{6.217}$$

This gives the scaling function

$$\psi(t) = \phi(2t) - \phi(2t - 1) \tag{6.218}$$

which is just the Haar function $h_{1,1}(t)$. Note also that

$$\Psi(\omega) = i \frac{(1 - e^{\frac{1}{2}i\omega})^2}{\omega} = e^{\frac{1}{2}i\omega} \frac{\sin^2 \frac{1}{4}\omega}{\frac{1}{4}i\omega} \tag{6.219}$$

as expected from Eq. (6.211). ∎

Two types of scaling function have been studied extensively:

(i) *Spline functions.* A *spline function* of *order* n is a function that consists of polynomials of degree n on each of a finite number of intervals, joined together at points ("knots") with continuous derivatives up to order $n - 1$.

Wavelets starting from a scaling function $\psi(t)$ with Fourier transform

$$\Psi(\omega) = e^{i\frac{N}{2}\omega} \left(\frac{\sin \frac{1}{2}\omega}{\frac{1}{2}\omega} \right)^N \tag{6.220}$$

have been constructed by Battle and Lemarié. The function $\psi(t)$ obtained from $\Psi(\omega)$ is an Nth order spline function on the interval $0 < t < N$ (check this). The function

$$M_\psi(\omega) \equiv \sum_{m=-\infty}^{\infty} |\Psi(\omega + 2m\pi)|^2 \tag{6.221}$$

is not equal to one for $N > 1$, so the functions $\{\psi(t - k)\}$ are not orthonormal. An orthonormal system can be constructed from the function $\phi(t)$ whose Fourier transform is

$$\Phi(\omega) = \frac{\Psi(\omega)}{\sqrt{M_\psi(\omega)}} \tag{6.222}$$

$\phi(t)$ is no longer confined to the interval $0 < t < N$, although it decays exponentially for $t \to \pm\infty$. The linear spline function is studied further in the problems.

(ii) *Compactly supported wavelets*.[10] In 1988, Daubechies discovered new orthonormal bases of wavelets derived from a scaling function $\phi(t)$ that has compact support. Such wavelets can be useful for analyzing functions defined only on a finite interval, or on a circle. For these bases, the function $H(\omega)$ must be a trigonometric polynomial of the form

$$H(\omega) = \frac{1}{2} \sum_{k=0}^{N} c_k e^{ik\omega} \tag{6.223}$$

(with this definition, $\phi(t)$ vanishes outside the interval $0 \le t \le N$), with $H(0) = 1$ and such that Eq. (6.206) is satisfied. Since this implies $H(\pi) = 0$, we can write

$$H(\omega) = \left(\frac{1 + e^{i\omega}}{2} \right)^n K(\omega) \tag{6.224}$$

with $0 < n \le N$ and $K(\omega)$ a trigonometric polynomial of degree $N - n$. Then also

$$|H(\omega)|^2 = \left(\cos^2 \tfrac{1}{2}\omega \right)^n |K(\omega)|^2 = \left(\frac{1 + \cos\omega}{2} \right)^n |K(\omega)|^2 \tag{6.225}$$

and, since the c_k are real, we have

$$|K(\omega)|^2 = \sum_{k=0}^{N-n} a_k \cos k\omega \equiv P\left(\sin^2 \tfrac{1}{2}\omega \right) \tag{6.226}$$

---

[10]The *support* of a function is the smallest collection of closed intervals outside of which the function vanishes everywhere. Here *compact* means that the support is bounded. For example, the support of the Haar function $h_0(t)$ is the interval $0 \le t \le 1$, which is compact; the Haar function has compact support.

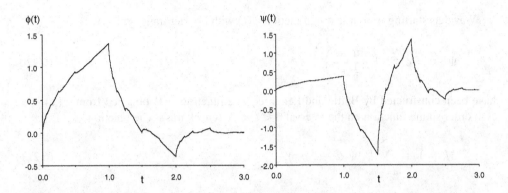

Figure 6.3: The scaling function $\phi(t)$ (left) and the fundamental wavelet $\psi(t)$ corresponding to the minimal polynomial with $n = 2$ (right).

with $P(y)$ a polynomial of degree $N - n$ in the variable $y \equiv \sin^2 \frac{1}{2}\omega = \frac{1}{2}(1 - \cos\omega)$.
Now

$$|H(\omega + \pi)|^2 = \left(\frac{1 - \cos\omega}{2}\right)^n \quad |K(\omega + \pi)|^2 = \left(\sin^2 \tfrac{1}{2}\omega\right)^n P\left(\cos^2 \tfrac{1}{2}\omega\right) \qquad (6.227)$$

so that Eq. (6.206) implies

$$(1 - y)^n P(y) + y^n P(1 - y) = 1 \qquad (6.228)$$

There is a (unique) polynomial $P_n(y)$ of degree $n - 1$

$$P_n(y) = \sum_{k=0}^{n-1} \binom{n + k - 1}{n - 1} y^k \qquad (6.229)$$

that satisfies Eq. (6.228). To this can be added a polynomial of the form $y^n R(y - \frac{1}{2})$, with $R(z)$ an odd polynomial in its argument, since such a polynomial adds nothing to the right-hand side of Eq. (6.228). Thus the most general solution to Eq. (6.228) has the form

$$P(y) = P_n(y) + y^n R(y - \tfrac{1}{2}) \qquad (6.230)$$

with $R(z)$ an odd polynomial in z (with real coefficients).

The polynomial $P(\sin^2 \frac{1}{2}\omega)$ gives $|K(\omega)|^2$ from Eq. (6.226). To extract $K(\omega)$, it is necessary to carry out some further algebra. A complete discussion may be found in the lectures of Daubechies cited at the end of the chapter. Here we just give two examples.

❑ **Example 6.13.** With $n = 1$, the minimal polynomial $P_1(y) = 1$ and the corresponding scale function is the Haar function $h_0(t)$, which has already been discussed at length. ∎

❑ **Example 6.14.** With $n = 2$, the minimal polynomial is

$$P_2(y) = 1 + 2y = 2 - \cos\omega = 2 - \tfrac{1}{2}\left(e^{i\omega} + e^{-i\omega}\right) \qquad (6.231)$$

To have

$$P_2(\sin^2 \tfrac{1}{2}\omega) = |a + be^{i\omega}|^2 \equiv |K(\omega)|^2 \qquad (6.232)$$

we need $a^2 + b^2 = 2$ and $ab = -\tfrac{1}{2}$ [note also that $a + b = 1$ since $P(0) = 1$], which leads to the solutions

$$a = \tfrac{1}{2}(1 \pm \sqrt{3}) \qquad b = \tfrac{1}{2}(1 \mp \sqrt{3}) \qquad (6.233)$$

Then

$$K(\omega) = \tfrac{1}{2}(1 \pm \sqrt{3}) + \tfrac{1}{2}(1 \mp \sqrt{3})e^{i\omega} \qquad (6.234)$$

and

$$H(\omega) = \tfrac{1}{8}\left\{1 \pm \sqrt{3} + (3 \pm \sqrt{3})e^{i\omega} + (3 \mp \sqrt{3})e^{2i\omega} + (1 \mp \sqrt{3})e^{3i\omega}\right\} \qquad (6.235)$$

The corresponding recursion coefficients are given by

$$c_0 = \tfrac{1}{4}(1 \pm \sqrt{3}) \quad c_1 = \tfrac{1}{4}(3 \pm \sqrt{3}) \quad c_2 = \tfrac{1}{4}(3 \mp \sqrt{3}) \quad c_3 = \tfrac{1}{4}(1 \mp \sqrt{3}) \qquad (6.236)$$

The scaling function $\phi(t)$ and the fundamental wavelet $\psi(t)$ associated with these coefficients are shown in Fig. 6.3. These functions are continuous, but not differentiable; smoother functions are obtained for larger intervals of support (see the Daubechies lectures for more details). Note that the two choices of sign in the roots do not correspond to truly distinct wavelet forms. Changing the sign of the roots is equivalent to the transformation

$$H(\omega) \rightarrow e^{3i\omega} H^*(\omega) \qquad (6.237)$$

that has the effect of sending $\phi(t)$ into its reflection about $t = \tfrac{3}{2}$.   ∎

A   Standard Families of Orthogonal Polynomials

The methods described here can be used to analyze other families of orthogonal polynomials. In the following pages, we summarize the properties of the families that satisfy either the hypergeometric equation or the confluent hypergeometric equation. These are the

- Gegenbauer polynomials $G_n^a(t)$: weight function $w(t) = (1 - t^2)^a$, and

- Jacobi polynomials $P_n^{(a,b)}(t)$: weight function $w(t) = (1 - t)^a(1 + t)^b$

on the interval $-1 \le t \le 1$, the

- (associated) Laguerre polynomials $L_n^a(t)$: weight function $w(t) = t^a e^{-t}$

on the interval $0 \le t < \infty$, and the

- Hermite polynomials $H_n(t)$: weight function: $w(t) = e^{-t^2}$

on the interval $-\infty \le t < \infty$. Derivation of the properties of these polynomials is left to the problems.

Remark.   The reader should be aware that, except for Legendre polynomials, there are various normalization and sign conventions for the polynomial families.   □

Gegenbauer polynomials $G_n^a(t)$: interval $-1 \le t \le 1$, weight function $w(t) = (1 - t^2)^a$

Rodrigues formula:

$$G_n^a(t) = \frac{(-1)^n}{2^n n!} (1 - t^2)^{-a} \frac{d^n}{dt^n} \left\{ (1 - t^2)^{n+a} \right\} \equiv C_n (1 - t^2)^{-a} u_n^{(n)}(t) \qquad (6.\text{A}1)$$

with $u_n(t) = (1 - t^2)^{n+a}$ and $C_n = (-1)^n / 2^n n!$ *orthogonality and normalization:*

$$\int_{-1}^{1} G_m^a(t) G_n^a(t) (1 - t^2)^a \, dt = (-1)^n C_n \int_{-1}^{1} u_n(t) \frac{d^n}{dt^n} G_m^a(t) \, dt$$

$$= \frac{2^{2a+1}}{2n + 2a + 1} \frac{[\Gamma(n + a + 1)]^2}{n! \, \Gamma(n + 2a + 1)} \delta_{mn} \qquad (6.\text{A}2)$$

differential equation:

$$(1 - t^2) G_n^{a}{}''(t) - 2(a + 1) t G_n^{a}{}'(t) + n(n + 2a + 1) G_n^a(t) = 0 \qquad (6.\text{A}3)$$

hypergeometric function:

$$G_n^a(t) = \frac{\Gamma(n + a + 1)}{n! \, \Gamma(a + 1)} F\left(-n, n + 2a + 1 | 1 + a | \tfrac{1}{2}(1 - t)\right) \qquad (6.\text{A}4)$$

This follows from the differential equation and the normalization

$$G_n^a(1) = \frac{\Gamma(n + a + 1)}{n! \, \Gamma(a + 1)} \qquad (6.\text{A}5)$$

generating function:

$$G(z, t) = \frac{1}{(1 - 2zt + z^2)^{a + \frac{1}{2}}} = \sum_{n=0}^{\infty} C_n^a(t) z^n \qquad (6.\text{A}6)$$

The $C_n^a(t)$ satisfy the differential equation for the Gegenbauer polynomials, since

$$(1 - t^2) \frac{\partial^2}{\partial t^2} G(z, t) - 2(a + 1) t \frac{\partial}{\partial t} G(z, t) =$$

$$= -z^2 \frac{\partial^2}{\partial z^2} G(z, t) - 2(a + 1) z \frac{\partial}{\partial z} G(z, t) \qquad (6.\text{A}7)$$

The normalization of the $C_n^a(t)$ is obtained from the binomial expansion for $t = 1$, so that the $C_n^a(t)$ are given in terms of the $G_n^a(t)$ by

$$C_n^a(t) = \frac{\Gamma(n + 2a + 1)\Gamma(a + 1)}{\Gamma(n + a + 1)\Gamma(2a + 1)} G_n^a(t) \qquad (6.\text{A}8)$$

Jacobi polynomials $P_n^{(a,b)}(t)$: $-1 \leq t \leq 1$, weight function $w(t) = (1-t)^a(1+t)^b$

Rodrigues formula:

$$P_n^{(a,b)}(t) = \frac{(-1)^n}{2^n n!} (1-t)^{-a}(1+t)^{-b} \frac{d^n}{dt^n} \left\{ (1-t)^{n+a}(1+t)^{n+b} \right\}$$

$$\equiv C_n (1-t)^{-a}(1+t)^{-b} u_n^{(n)}(t) \tag{6.A9}$$

with $u_n(t) = (1-t)^{n+a}(1+t)^{n+b}$ and $C_n = (-1)^n/2^n n!$.

orthogonality and normalization:

$$\int_{-1}^{1} P_m^{(a,b)}(t) P_n^{(a,b)}(t)(1-t)^a(1+t)^b \, dt =$$

$$= C_m C_n \int_{-1}^{1} u_m^{(m)}(t) u_n^{(n)}(t)(1-t)^{-a}(1+t)^{-b} \, dt \tag{6.A10}$$

$$= (-1)^n C_n \int_{-1}^{1} u_n(t) \frac{d^n}{dt^n} P_m^{(a,b)}(t) \, dt \tag{6.A11}$$

$$= \frac{2^{a+b+1}}{2n+a+b+1} \frac{\Gamma(n+a+1)\Gamma(n+b+1)}{n!\,\Gamma(n+a+b+1)} \delta_{mn}$$

differential equation:

$$(1-t^2)P_n^{(a,b)''}(t) + [b-a-(a+b+2)t]P_n^{(a,b)'}(t)$$

$$+ n(n+a+b+1)P_n^{(a,b)}(t) = 0 \tag{6.A12}$$

hypergeometric function:

$$P_n^{(a,b)}(t) = \frac{\Gamma(n+a+1)}{n!\,\Gamma(a+1)} F\left(-n, n+a+b+1 \middle| 1+a \middle| \tfrac{1}{2}(1-t)\right) \tag{6.A13}$$

This follows from the differential equation and the normalization

$$P_n^{(a,b)}(1) = \frac{\Gamma(n+a+1)}{n!\,\Gamma(a+1)} \tag{6.A14}$$

There is a generating function for the Jacobi polynomials, but it is not especially illuminating, so we omit it. Note that the Gegenbauer polynomials and the Legendre polynomials are special cases of the Jacobi polynomials,

- Gegenbauer polynomials: $G_n^a(t) = P_n^{(a,a)}(t)$.

- Legendre polynomials: $P_n(t) = G_n^0(t) = P_n^{(0,0)}(t)$.

Laguerre polynomials $L_n^a(t)$: interval $0 \leq t < \infty$, weight function $w(t) = t^a e^{-t}$

Rodrigues formula:

$$L_n^a(t) = \frac{1}{n!} t^{-a} e^t \frac{d^n}{dt^n} \left(t^{n+a} e^{-t} \right) \equiv C_n t^{-a} e^t \, u_n^{(n)}(t) \tag{6.A15}$$

with $u_n(t) = t^{n+a} e^{-t}$ and $C_n = 1/n!$.

orthogonality and normalization:

$$\int_0^\infty L_m^a(t) L_n^a(t) t^a e^{-t} \, dt = C_n \int_0^\infty L_m^a(t) u_n^{(n)}(t) \, dt =$$

$$= (-1)^n C_n \int_0^\infty u_n(t) \frac{d^n}{dt^n} L_m^a(t) \, dt = \frac{\Gamma(n+a+1)}{n!} \delta_{mn} \tag{6.A16}$$

differential equation:

$$t L_n^{a\,\prime\prime}(t) + (1 + a - t) L_n^{a\,\prime}(t) + n L_n^a(t) = 0 \tag{6.A17}$$

confluent hypergeometric function:

$$L_n^a(t) = \frac{\Gamma(n+a+1)}{n!\,\Gamma(a+1)} \, F(-n|1+a|t) \tag{6.A18}$$

This follows from the differential equation and the normalization at t=0.

generating function:

$$L(z,t) \equiv \frac{e^{-zt/(1-z)}}{(1-z)^{1+a}} = \sum_{n=0}^\infty L_n^a(t) z^n \tag{6.A19}$$

This follows here from the differential equation and the observation that

$$\left[t \frac{\partial^2}{\partial t^2} + (1 + a - t) \frac{\partial}{\partial t} + z \frac{\partial}{\partial z} \right] L(z,t) = 0 \tag{6.A20}$$

The Laguerre polynomials appear in solutions to the Schrödinger equation for two charged particles interacting through a Coulomb potential. See Example 8.5 where these solutions are derived.

Hermite polynomials $H_n(t)$: interval: $-\infty \leq t < \infty$, weight function: $w(t) = e^{-t^2}$

Rodrigues formula:

$$H_n(t) = (-1)^n e^{t^2} \frac{d^n}{dt^n}\left(e^{-t^2}\right) \equiv e^{t^2} u^{(n)}(t) \tag{6.A21}$$

with $u(t) = (-1)^n e^{-t^2}$.

orthogonality and normalization:

$$\int_{-\infty}^{\infty} H_m(t) H_n(t) e^{-t^2}\, dt = (-1)^n \int_{-\infty}^{\infty} e^{-t^2} \frac{d^n}{dt^n} H_n(t)\, dt$$
$$= 2^n n! \int_{-\infty}^{\infty} e^{-t^2}\, dt = 2^n n!\, \sqrt{\pi}\, \delta_{mn} \tag{6.A22}$$

differential equation:

$$H_n''(t) - 2t H_n'(t) + 2n H_n(t) = 0 \tag{6.A23}$$

confluent hypergeometric function:

$$H_{2n}(t) = (-1)^n \frac{(2n)!}{n!}\, F(-n\mid -\tfrac{1}{2}\mid t^2) \tag{6.A24}$$

$$H_{2n+1}(t) = (-1)^n \frac{(2n+1)!}{n!}\, t F(-n\mid \tfrac{1}{2}\mid t^2) \tag{6.A25}$$

generating function:

$$H(z,t) \equiv e^{-z^2+2zt} = \sum_{n=0}^{\infty} H_n(t) \frac{z^n}{n!} \tag{6.A26}$$

One derivation of this formula is to consider the sum

$$e^{-(z-t)^2} = \sum_{n=0}^{\infty}\left[\frac{d^n}{dz^n} e^{-(z-t)^2}\right]_{z=0} \frac{z^n}{n!} =$$
$$= \sum_{n=0}^{\infty} (-1)^n \left[\frac{d^n}{dt^n} e^{-t^2}\right] \frac{z^n}{n!} = e^{-t^2} \sum_{n=0}^{\infty} H_n(t) \frac{z^n}{n!} \tag{6.A27}$$

The Hermite polynomials and the related Hermite functions $h_n(t) = H_n(t) \exp(-\tfrac{1}{2}t^2)$ introduced in Problems 6.20 and 7.16 appear in wave functions for the quantum mechanical harmonic oscillator.

Bibliography and Notes

A sequel to the book by Halmos on finite-dimensional vector spaces cited in Chapter 2 is

Paul R. Halmos, *Introduction to Hilbert Space* (2nd edition), Chelsea (1957).

This is a brief and terse monograph, but it covers important material. To delve more deeply into Hilbert space, the dedicated student can also work through the excellent companion book

Paul R. Halmos, *A Hilbert Space Problem Book* (2nd edition), Springer (1982),

that has a rich collection of mathematical exercises and problems, with discussion, hints, solutions and references. A classic work on function spaces is

R. Courant and D. Hilbert, *Methods of Mathematical Physics* (vol. 1), Interscience (1953),

It contains a detailed treatment of the function spaces and orthonormal systems introduced here, as well as a more general discussion of Hilbert spaces and linear operators.

A more modern book that provides a good introduction to function spaces, and especially Fourier analysis, at a fairly elementary level is

Gerald B. Folland, *Fourier Analysis and Its Applications*, Wadsworth & Brooks/ Cole (1992).

The handbook by Abramowitz and Irene Stegun cited in Chapter 8 has a comprehensive list of formulas and graphs of special functions that arise in mathematical physics and elsewhere. It is an extremely useful reference.

Wavelet analysis has been an important topic of recent research in applied mathematics. An early survey of the uses of wavelet transforms by one of the prime movers is

Ingrid Daubechies, *Ten Lectures on Wavelets*, SIAM (1992).

A more recent survey, with history and applications, of wavelet analysis is

Stéphane Jaffard, Yves Meyer and Robert D. Ryan, *Wavelets: Tools for Science and Technology*, SIAM (2001),

Especially interesting is the broad range of applications discussed by the authors, although main emphasis of the book is on signal processing. A recent elementary introduction is

David F. Walnut, *Introduction to Wavelet Analysis*, Birkhäuser (2002).

Problems

1. Let $\phi_1, \phi_2, \ldots$ be a complete orthonormal system in a Hilbert space $\mathcal{H}$. Define vectors $\psi_1, \psi_2, \ldots$ by

    $$\psi_n = C_n \left(\sum_{k=1}^{n} \phi_k - n\phi_{n+1} \right)$$

    $(n = 1, 2, \ldots)$.

    (i) Show that $\psi_1, \psi_2, \ldots$ form an orthogonal system in $\mathcal{H}$.

    (ii) Find constants C_n that make $\psi_1, \psi_2, \ldots$ into an orthonormal system in $\mathcal{H}$.

    (iii) Show that if $(\psi_n, x) = 0$ for all $n = 1, 2, \ldots$, then $x = 0$.

    (iv) Verify that

    $$\|\phi_k\|^2 = 1 = \sum_{n=1}^{\infty} |(\psi_n, \phi_k)|^2$$

    $(k = 1, 2, \ldots)$.

2. If $\{x_n\} \rightharpoonup x$, then $\{x_n\} \to x$ if and only if $\{\|x_n\|\} \to \|x\|$.

3. Use Fourier series to evaluate the sums

    $$\zeta(2) = \sum_{n=1}^{\infty} \frac{1}{n^2} \quad \text{and} \quad \zeta(4) = \sum_{n=1}^{\infty} \frac{1}{n^4}$$

4. Find the Fourier series expansions of each of the following functions defined by the formulas for $-\pi < x < \pi$, and elsewhere by $f(x + 2\pi) = f(x)$.

    (i) $f(x) = \begin{cases} 1/2a & 0 < |x| < a \, (< \pi) \\ 0 & \text{otherwise} \end{cases}$

    (ii) $f(x) = \begin{cases} \cos ax & 0 < |x| < \pi/2a \, (< \pi) \\ 0 & \text{otherwise} \end{cases}$

    (iii) $f(x) = \cosh \alpha x$

    (iv) $f(x) = \sinh \alpha x$

5. Show that the functions $\{\phi_n(x)\}$ defined by

    $$\phi_n(x) \equiv \sqrt{\frac{2}{\pi}} \cos(n + \tfrac{1}{2})x$$

($n = 0, 1, 2, \ldots$) form a complete orthonormal system in $L^2(0, \pi)$. Show also that the functions $\{\psi_n(x)\}$ defined by

$$\psi_n(x) \equiv \sqrt{\frac{2}{\pi}} \, \sin(n + \tfrac{1}{2})x$$

($n = 0, 1, 2, \ldots$) form a complete orthonormal system in $L^2(0, \pi)$.

6. Show that the Fourier transform and its inverse are equivalent to the formal relation

$$\int_{-\infty}^{\infty} e^{ik(x-y)} \, dk = 2\pi \delta(x - y)$$

Give a corresponding formal expression for the discrete sum

$$\lim_{N \to \infty} D_N(x - y) = \sum_{k=-\infty}^{\infty} e^{ik(x-y)}$$

where $D_N(x-y)$ is the Dirichlet kernel defined by Eq. (6.47). Finally, derive the *Poisson summation formula* (here $c(\omega)$ is the Fourier transform of $f(t)$ defined by Eq. (6.81)).

$$\sum_{n=-\infty}^{\infty} f(n) = 2\pi \sum_{k=-\infty}^{\infty} c(2\pi k)$$

7. Find the Fourier transform $c(\omega)$ of the function $f(t)$ defined by

$$f(t) \equiv \frac{e^{-\alpha|t|}}{\sqrt{|t|}}$$

(Re $\alpha > 0$). Is $c(\omega)$ in $L^2(-\infty, \infty)$? Explain your answer.

8. Find the Fourier transform of the function $f(t)$ defined by

$$f(t) \equiv \frac{\sin \alpha t}{t}$$

with α real. Then use Plancherel's formula to evaluate the integral

$$\int_{-\infty}^{\infty} \left(\frac{\sin \alpha t}{t} \right)^2 dt$$

9. Consider the Gaussian pulse

$$f(t) = A e^{-i\omega_0 t} e^{-\frac{1}{2}\alpha(t-t_0)^2}$$

which corresponds to a signal of frequency ω_0 restricted to a time interval Δt of order $1/\sqrt{\alpha}$ about t_0 by the Gaussian modulating factor $\exp\{-\frac{1}{2}\alpha(t - t_0)^2\}$, which is sometimes called a *window function*.

(i) Find the Fourier transform of this pulse.

(ii) Show that the intensity $|\mathcal{F}f(\omega)|^2$ of the pulse as a function of frequency is peaked at $\omega = \omega_0$ with a finite width $\Delta\omega$ due to the Gaussian modulating factor. Give a precise definition to this width (there are several reasonable definitions), and compute it.

(iii) Use the same definition as in part (ii) to give a precise evaluation of the time duration Δt of the pulse, and show that

$$\Delta\omega \cdot \Delta t = \text{constant}$$

with constant of order 1 independent of the parameter α. (The precise constant will depend on the definitions of $\Delta\omega$ and Δt).

10. (i) Show that

$$\sum_{n=1}^{N} e^{\frac{2\pi ikn}{N}} e^{-\frac{2\pi iqn}{N}} = N\delta_{kq}$$

(ii) Show that a set of numbers $f_1, \ldots, f_N$ can be expressed in the form

$$f_n = \sum_{k=1}^{N} c_k e^{\frac{2\pi ikn}{N}} \tag{*}$$

with coefficients c_k given by

$$c_k = \frac{1}{N} \sum_{m=1}^{N} f_m e^{-\frac{2\pi ikm}{N}}$$

(iii) Suppose the $f_1, \ldots, f_N$ are related to a signal

$$f(t) = \sum_{k=-\infty}^{\infty} \xi_k e^{\frac{2\pi ikt}{T}}$$

by $f_n = f(nT/N)$. Express the coefficients c_k of the finite Fourier transform in terms of the coefficients ξ_k in the complete Fourier series.

(iv) Show that if the $f_1, \ldots, f_N$ are real, then $c_{N-k} = c_k^*$ ($k = 1, \ldots, N$).

Remark. The expansion (*) is known as the *finite Fourier transform*. If the sequence $f_1, \ldots, f_N$ results from the measurement of a signal of duration T at discrete intervals $\Delta t = T/N$, then (*) is an approximation to the Fourier integral transform, and c_k corresponds to the amplitude for frequency $\nu_k = k/T$ ($k = 1, \ldots, N$). The appearance of high frequency components of the complete signal in the c_k found in part (iii) is known as *aliasing*; it is one of the problems that must be addressed in reconstructing a signal from a discrete set of observations. The result in (iv) means that the frequencies actually sampled lie in the range $1/T \le \nu \le N/2T$ (explain this). $\square$

11. Show that

$$\sum_{n=1}^{N-1} \sin\frac{n\pi k}{N} \sin\frac{n\pi q}{N} = \frac{N}{2}\delta_{kq}$$

and thus that the set of numbers $\{f_1,\ldots,f_{N-1}\}$ can be expressed as

$$f_n = \sum_{k=1}^{N-1} b_k \sin\frac{k\pi n}{N}$$

with coefficients b_k given by

$$b_k = \frac{2}{N}\sum_{m=1}^{N-1} f_m \sin\frac{m\pi k}{N}$$

Remark. This expansion is known as the *finite Fourier sine transform.*   □

12. Find the Laplace transforms of the functions

    (i) $f(t) = \cos\alpha t$

    (ii) $f(t) = t^a L_n^a(t)$

where $L_n^a(t)$ is the associated Laguerre polynomial defined by Eq. (6.A15).

13. Show that if $f(t)$ is periodic with period τ, then the Laplace transform of $f(t)$ is given by

$$\mathcal{L}f(p) = \frac{F(p)}{1 - e^{-p\tau}}$$

where

$$F(p) = \int_0^\tau f(t)e^{-pt}\,dt$$

14. Consider the differential equation

$$tf''(t) + f'(t) + tf(t) = 0$$

(*Bessel's equation* of order zero), and let

$$u(p) \equiv \int_0^\infty f(t)e^{-pt}\,dt$$

be the Laplace transform of $f(t)$.

(i) Show that $u(p)$ satisfies the first-order differential equation

$$(p^2 + 1)u'(p) + pu(p) = 0$$

(ii) Find the solution of this equation that is the Laplace transform of the Bessel function $J_0(t)$. Recall that the Bessel function has the integral representation (see problem 5.13).

$$J_0(t) = \frac{1}{\pi} \int_0^\pi e^{it\cos\theta}\, d\theta$$

15. The *Mellin transform* of $f(t)$ is defined by

$$\mathcal{M}f(s) \equiv \int_0^\infty t^{s-1}f(t)\, dt$$

(i) Show that the Mellin transform of $f(t)$ exists for $s = i\xi$ (ξ real) if

$$\int_0^\infty |f(t)|^2 \frac{dt}{t} < \infty$$

Remark. This condition is sufficient but not necessary, as in the case of the Fourier transform. □

(ii) Show that if the preceding inequality is satisfied, then the inverse Mellin transform is

$$f(t) = \frac{1}{2\pi i} \int_{-i\infty}^{i\infty} \frac{\mathcal{M}f(s)}{t^s}\, ds$$

Also,

$$\int_{-i\infty}^{i\infty} |\mathcal{M}f(s)|^2\, ds = \int_0^\infty |f(t)|^2 \frac{dt}{t}$$

(iii) Use the preceding result to show that if $\sigma > 0$,

$$\int_0^\infty |\Gamma(\sigma + i\tau)|^2\, d\tau = \frac{\pi\Gamma(2\sigma)}{2^{2\sigma}}$$

16. Show that the generating function

$$S(t, z) = \frac{1}{\sqrt{1 - 2zt + t^2}} = \sum_{n=0}^\infty t^n P_n(z)$$

for the Legendre polynomials satisfies

$$\int_{-1}^1 S(u, z)S(v, z)\, dz = \frac{1}{\sqrt{uv}} \ln\left(\frac{1 + \sqrt{uv}}{1 - \sqrt{uv}}\right)$$

Use this result to derive the orthogonality (Eq. (6.139)) and normalization (Eq. (6.140)) of the $P_n(t)$.

17. Show that the Gegenbauer polynomials $C_n^a(t)$ defined by Eq. (6.A6) satisfy

$$(n+1)C_{n+1}^a(t) - (2n + 2a + 1)tC_n^a(t) + (n + 2a)C_{n-1}^a(t) = 0$$

and

$$\frac{d}{dt}C_n^a(t) = (2a+1)C_{n-1}^{a+1}(t)$$

18. Show that the Laguerre polynomials satisfy

$$tL_n^a(t) = (n+1)L_{n+1}^a(t) - (2n + a + 1)L_n^a(t) + (n+a)L_{n-1}^a(t)$$

and

$$\frac{d}{dt}L_n^a(t) = -L_{n-1}^{a+1}(t) = L_n^a(t) - L_n^{a+1}(t)$$

19. Show that the Hermite polynomials satisfy

$$H_n(-t) = (-1)^n H_n(t)$$

Then evaluate $H_n(0)$ and $H_n'(0)$ $(n = 0, 1, \ldots)$. Show further that

$$H^{n+1}(t) - 2tH_n(t) + 2nH_{n-1}(t) = 0$$

and

$$\frac{d}{dt}H_n(t) = 2nH_{n-1}(t)$$

20. Define *Hermite functions* $h_n(t)$ in terms of the Hermite polynomials $H_n(t)$ by

$$h_n(t) \equiv e^{-\frac{1}{2}t^2} H_n(t)$$

Show that

$$\int_{-\infty}^{\infty} h_n(t)e^{i\omega t}\, dt = i^n \sqrt{2\pi}\, h_n(\omega)$$

Remark. Thus $h_n(t)$ is its own Fourier transform apart from a constant factor.   □

21. The *Chebysheff polynomials* $T_n(t)$ are defined by

$$T_0(t) = 1 \qquad T_n(t) = \frac{1}{2^{n-1}}\cos(n\cos^{-1} t)$$

$(n = 1, 2, \ldots)$.

(i) Show that $T_n(t)$ is indeed a polynomial of degree n.

(ii) Show that the $\{T_n(t)\}$ are orthogonal polynomials on the interval $-1 \le t \le 1$ for suitably chosen weight function $w(t)$ [which you should find].

Find normalization constants C_n that make the $\{T_n(t)\}$ into a complete orthonormal system.

22. Consider the Bessel function $J_\lambda(x)$ defined by Eq. (5.162) as a solution of Bessel's equation (5.158)

$$\frac{d^2 u}{dx^2} + \frac{1}{x}\frac{du}{dx} + \left(1 - \frac{\lambda^2}{x^2}\right)u = 0$$

(i) Show that

$$\frac{d}{dx}\left\{x\left[J_\lambda(\alpha x)\frac{d}{dx}J_\lambda(\beta x) - J_\lambda(\beta x)\frac{d}{dx}J_\lambda(\alpha x)\right]\right\} = (\alpha^2 - \beta^2)x J_\lambda(\alpha x)J_\lambda(\beta x)$$

(ii) Then show that

$$\int_a^b x J_\lambda(\alpha x)J_\lambda(\beta x)\,dx = (\alpha^2 - \beta^2)x J_\lambda(\alpha x)J_\lambda(\beta x)\Big|_a^b$$

Remark. This shows that if α and β are chosen so that the endpoint terms vanish, then $J_\lambda(\alpha x)$ and $J_\lambda(\beta x)$ are orthogonal functions on the interval $[a, b]$ with weight function $w(x) = x$ (we assume $0 \le a < b$ here). □

23. Consider the spherical Bessel function $j_n(x)$ defined in Section 5.6.3 as a solution of the differential equation (5.180)

$$\frac{d^2 u}{dx^2} + \frac{2}{x}\frac{du}{dx} + \left[1 - \frac{n(n+1)}{x^2}\right]u = 0$$

(i) Show that

$$\frac{d}{dx}\left\{x^2\left[j_n(\alpha x)\frac{d}{dx}j_n(\beta x) - j_n(\beta x)\frac{d}{dx}j_n(\alpha x)\right]\right\} = (\alpha^2 - \beta^2)x^2 j_n(\alpha x)j_n(\beta x)$$

(ii) Then show that

$$\int_a^b x^2 j_n(\alpha x)j_n(\beta x)\,dx = (\alpha^2 - \beta^2)x^2 j_n(\alpha x)j_n(\beta x)\Big|_a^b$$

Remark. Again, if α and β are chosen so that the endpoint terms vanish, then $j_n(\alpha x)$ and $j_n(\beta x)$ are orthogonal functions on the interval $[a, b]$ with weight function $w(x) = x^2$ (again we assume $0 \le a < b$). □

24. Consider the "hat" function $g(t)$ defined by

$$g(t) = \begin{cases} t & 0 < t \le 1 \\ 2 - t & 1 < t < 2 \\ 0 & \text{otherwise} \end{cases}$$

(i) Show that $g(t)$ satisfies the relation

$$g(t) = \tfrac{1}{2}g(2t) + g(2t - 1) + \tfrac{1}{2}g(2t - 2)$$

(ii) Show that $g(t)$ is obtained from the Haar function $h_0(t)$ by convolution,

$$g(t) = \int_{-\infty}^{\infty} h_0(\tau)h_0(t - \tau)\, d\tau$$

(iii) Show that the Fourier transform of $g(t)$ is given by

$$\hat{g}(\omega) \equiv \int_{-\infty}^{\infty} e^{i\omega t} g(t)\, dt = e^{i\omega} \left(\frac{\sin \tfrac{1}{2}\omega}{\tfrac{1}{2}\omega} \right)^2$$

(iv) Evaluate the function $H(\omega)$ defined by Eq. (6.199), and verify Eq. (6.198).

(v) Evaluate the sum

$$M_g(\omega) \equiv \sum_{m=-\infty}^{\infty} |\hat{g}(\omega + 2m\pi)|^2$$

and then use Eq. (6.222) to construct the Fourier transform $\Phi(\omega)$ of a scaling function $\phi(t)$ for which the $\{\phi(t - k)\}$ form an orthonormal system.

(vi) From the analytic properties of the $\Phi(\omega)$ just constructed, estimate the rate of exponential decay of $\phi(t)$ for $t \to \pm\infty$.

25. For $n = 1$, the simplest nonminimal solution to Eq. (6.228) can be written as

$$P(y) = 1 + 2\nu(\nu + 2)y(y - \tfrac{1}{2})$$

(the parametrization in terms of ν is chosen for convenience in the solution). In terms of the Fourier transform variable ω, we have

$$P(\sin^2 \tfrac{1}{2}\omega) = 1 - \tfrac{1}{2}\nu(\nu + 2)\cos\omega(1 - \cos\omega)$$

Now find constants a, b, c such that

$$P(\sin^2 \tfrac{1}{2}\omega) = |a + be^{i\omega} + ce^{2i\omega}|^2 \equiv |K(\omega)|^2$$

where $K(\omega)$ has been introduced in Eq. (6.224) [note that $K(0) = 1$]. For what range of values of ν are these constants real? For what value(s) of ν does the function $H(\omega)$ reduce to the case of $n = 2$ with minimal polynomial? For what values of ν does the scaling function reduce to the "box" function?

7 Linear Operators on Hilbert Space

The theory of linear operators on a Hilbert space $\mathcal{H}$ requires deeper analysis than the study of finite-dimensional operators given in Chapter 2. The infinite dimensionality of $\mathcal{H}$ allows many new varieties of operator behavior, as illustrated by three examples.

First is a linear operator $\mathbf{A}$ with a complete orthonormal system of eigenvectors, whose eigenvalues are real but form an unbounded set. $\mathbf{A}$ cannot be defined on all of $\mathcal{H}$, but only on an open linear manifold which is, however, dense in $\mathcal{H}$. Thus care must be taken to characterize the domain of an operator on $\mathcal{H}$.

Second is the linear operator $\mathbf{X}$ that transforms a function $f(x)$ in $L_2(-1, 1)$ to the function $xf(x)$, which is also in $L_2(-1, 1)$. $\mathbf{X}$ is bounded, self-adjoint, and everywhere defined on $L_2(-1, 1)$, but it has no proper eigenvectors. We need to introduce the concept of a continuous spectrum to describe the spectrum of $\mathbf{X}$.

The third example is an operator $\mathbf{U}$ that shifts the elements of a complete orthonormal system so that the image of $\mathbf{U}$ is a proper (but still infinite-dimensional) subspace of $\mathcal{H}$. $\mathbf{U}$ preserves the length of vectors ($\mathbf{U}$ is *isometric*) but is not unitary, since $\mathbf{U}^\dagger$ is singular. $\mathbf{U}$ has no eigenvectors at all, while $\mathbf{U}^\dagger$ has an eigenvector for each complex number λ with $|\lambda| < 1$.

The general properties of linear operators introduced in Section 2.2 are reviewed with special attention to new features allowed by the infinite dimensionality of $\mathcal{H}$. Linear operators need not be bounded. Unbounded operators cannot be defined on the entire Hilbert space $\mathcal{H}$, but at best on a dense subset $\mathcal{D}$ of $\mathcal{H}$, the *domain* of the operator, which must be carefully specified. Bounded operators are continuous, which allows them to be defined on all of $\mathcal{H}$. Convergence of sequences of operators has three forms—uniform, strong, and weak—the latter two being related to strong and weak convergence of sequences of vectors.

Some linear operators on $\mathcal{H}$ behave more like their finite-dimensional counterparts. *Compact* operators transform every weakly convergent sequence of vectors into a strongly convergent sequence. *Hilbert–Schmidt* operators have a finite norm, and themselves form a Hilbert space. Many of the integral operators introduced later are of this type. Finally, there are operators of *finite rank*, whose range is finite dimensional.

The adjoint operator is defined by analogy with the finite-dimensional case, but questions of domain are important for unbounded operators, since there are operators that are formally Hermitian, but which have no self-adjoint extension. There are operators that are isometric but not unitary, for example, the shift operator $\mathbf{U}$ cited above. Further illustrations of potential pitfalls in infinite dimensions are given both in the text and in the problems.

The theory of the spectra of normal operators is outlined. Compact normal operators have a complete orthonormal system of eigenvectors belonging to discrete eigenvalues, and the eigenvalues form a sequence that converges to 0. A noncompact self-adjoint operator

Introduction to Mathematical Physics. Michael T. Vaughn
Copyright © 2007 WILEY-VCH Verlag GmbH & Co. KGaA, Weinheim
ISBN: 978-3-527-40627-2

A may have a continuous spectrum as well as a discrete spectrum. A formal construction admitting this possibility is obtained by introducing the concept of a *resolution of the identity*, a nondecreasing family $\mathbf{E}_\lambda$ of projection operators that interpolates between 0 for $\lambda \to -\infty$ and 1 for $\lambda \to \infty$. For every self-adjoint operator **A** there is a resolution of the identity in terms of which **A** can be expressed as

$$\mathbf{A} = \int \lambda \, d\mathbf{E}_\lambda$$

This generalizes the representation of a self-adjoint operator in terms of a complete orthonormal system of eigenvectors to include the continuous spectrum.

The definition of a linear differential operator as a Hilbert space operator must include associated boundary conditions. The formal adjoint of a differential operator is defined by integration by parts, which introduces boundary terms. In order to make an operator self-adjoint, a set of boundary conditions must be imposed that insure that these boundary terms vanish for any function satisfying the boundary conditions. There are various possible boundary conditions, each leading to a distinct self-adjoint extension of the formal differential operator.

These ideas are illustrated with the operator

$$\mathbf{P} = \frac{1}{i} \frac{d}{dx}$$

which is a momentum operator in quantum mechanics, and the second-order operator

$$\mathbf{L} = \frac{d}{dx} \left[p(x) \frac{d}{dx} \right] + s(x)$$

(the *Sturm–Liouville* operator), which often appears in physics after the reduction of second-order partial differential equations involving the Laplacian (see Chapter 8) to ordinary differential equations.

If Ω is a region in $\mathbf{R}^n$, an integral operator **K** on $L_2(\Omega)$ has the form

$$(\mathbf{K}f)(x) = \int_\Omega K(x, y) f(y) dy$$

If

$$\int_\Omega \int_\Omega |K(x, y)|^2 \, dx dy < \infty$$

then **K** is compact, and if **K** is self-adjoint, or even normal, it then must have a complete orthonormal system of eigenvectors belonging to discrete eigenvalues. Such an integral operator often appears as the inverse of a differential operator, in which context it is a *Green function*. If the Green function of a differential operator **L** has a discrete spectrum, so does the differential operator itself, although it is unbounded. This is most important in view of the other possibilities for the spectrum of a noncompact operator. If the Green function of **L** can be constructed explicitly, then it also provides the solution to inhomogeneous differential equation

$$\mathbf{L}u = f$$

7.1   Some Hilbert Space Subtleties

Many of the concepts that were introduced in the study of linear operators on finite-dimensional vector spaces are also relevant to linear operators on a Hilbert space $\mathcal{H}$.[1] However, there are new features in $\mathcal{H}$, as illustrated by the following three examples.

❑ **Example 7.1.** Let $\phi_1, \phi_2, \ldots$ be a complete orthonormal system on the Hilbert space $\mathcal{H}$, and define the linear operator $\mathbf{A}$ by

$$\mathbf{A}\phi_n = n\phi_n \tag{7.1}$$

($n = 1, 2, \ldots$). $\mathbf{A}$ has a complete orthonormal system of eigenvectors, and $\mathbf{A}$ is defined on any finite linear combination of the $\phi_1, \phi_2, \ldots$, a set that is everywhere dense in $\mathcal{H}$. However, $\mathbf{A}$ is not defined everywhere in $\mathcal{H}$. The limit as $N \to \infty$ of the vector

$$\phi_N \equiv \sum_{n=1}^{N} \frac{1}{n}\phi_n \tag{7.2}$$

is an element ϕ_∞ of $\mathcal{H}$, but the limit of the vector

$$\mathbf{A}\phi_N = \sum_{n=1}^{N} \phi_n \tag{7.3}$$

is not. Hence $\mathbf{A}\phi_\infty$ is not defined.

Moreover, if we try to define an inverse $\bar{\mathbf{A}}$ of $\mathbf{A}$ by

$$\bar{\mathbf{A}}\phi_n = \frac{1}{n}\phi_n \tag{7.4}$$

then we have $\mathbf{A}\bar{\mathbf{A}} = 1$ (there is no problem here since the range of $\bar{\mathbf{A}}$ is a subset of the domain of $\mathbf{A}$), but it is not exactly true that $\bar{\mathbf{A}}\mathbf{A} = 1$ since $\mathbf{A}$ is not defined on the entire Hilbert space. Thus it is necessary to pay careful attention to the domain of an operator.

In general, a linear operator $\mathbf{A}$ is defined on $\mathcal{H}$ by specifying a linear manifold $\mathcal{D}_\mathbf{A} \subset \mathcal{H}$ (the *domain* of $\mathbf{A}$) and the vectors $\mathbf{A}x$ at least for some basis of $\mathcal{D}_\mathbf{A}$. The vector $\mathbf{A}x$ is then defined for any finite linear combination of the basis vectors, but it may, or may not be possible to extend the domain of $\mathbf{A}$ to include the entire space $\mathcal{H}$, or even all the limit points of sequences in $\mathcal{D}_\mathbf{A}$. As we have just seen, the sequence $\{\phi_N\}$ is bounded (in fact, convergent), but for the linear operator $\mathbf{A}$ defined by Eq. (7.1), the sequence $\{\mathbf{A}\phi_N\}$ is unbounded, and hence has no limit in $\mathcal{H}$.

If a linear operator $\mathbf{A}'$ can be defined on a larger manifold $\mathcal{D}_{\mathbf{A}'} \supset \mathcal{D}_\mathbf{A}$, with $\mathbf{A}'x = \mathbf{A}x$ on $\mathcal{D}_\mathbf{A}$, then $\mathbf{A}'$ is an *extension* of $\mathbf{A}$ ($\mathbf{A}' \supset \mathbf{A}$ or $\mathbf{A} \subset \mathbf{A}'$). Thus, to be precise, we should write $\bar{\mathbf{A}}\mathbf{A} \subset 1$ for the operator $\mathbf{A}$ in the present example, but often we will just write $\bar{\mathbf{A}}\mathbf{A} = 1$ anyway.   ∎

---

[1] We remind the reader that by our convention, a Hilbert space is infinite dimensional unless otherwise stated.

Remark. The domain of the operator $\mathbf{A}$ defined by Eq. (7.1) can be characterized precisely. The vector

$$\phi = \sum_{n=1}^{\infty} c_n \phi_n \tag{7.5}$$

is in $\mathcal{D}_\mathbf{A}$ if and only if

$$\sum_{n=1}^{\infty} n^2 |c_n|^2 < \infty \tag{7.6}$$

$\mathcal{D}_\mathbf{A}$ is dense in $\mathcal{H}$, but does not include the vector ϕ_∞, for example.   □

❑ **Example 7.2.**   In the preceding example, the unboundedness of the operator $\mathbf{A}$ defined by Eq. (7.1) led to the lack of an extension of $\mathbf{A}$ to all of $\mathcal{H}$. In this example, we will see that even for a bounded operator, there can be something new in Hilbert space.

Consider the Hilbert space $L^2(-1, 1)$ of functions $f(x)$ that are square integrable on the interval $-1 \leq x \leq 1$, and define the linear operator $\mathbf{X}$ by

$$\mathbf{X}f(x) = xf(x) \tag{7.7}$$

$\mathbf{X}$ is self-adjoint, and it is bounded, since

$$\|\mathbf{X}f\|^2 = \int_{-1}^{1} |xf(x)|^2 \, dx < \int_{-1}^{1} |f(x)|^2 \, dx = \|f\|^2 \tag{7.8}$$

Moreover, it is defined everywhere on $L^2(-1, 1)$.

However, the eigenvalue equation

$$\mathbf{X}f(x) = \lambda f(x) \tag{7.9}$$

has no solution in $L^2(-1, 1)$, since the eigenvalue equation implies $f(x) = 0$ for $x \neq \lambda$, and this in turn means

$$\int_{-1}^{1} |f(x)|^2 \, dx = 0 \tag{7.10}$$

as a Lebesgue integral. There is a formal solution

$$f_\lambda(x) = \delta(x - \lambda) \tag{7.11}$$

for any value of λ between -1 and 1, but the Dirac δ-function is not in $L^2(-1, 1)$. Thus $\mathbf{X}$ is a bounded self-adjoint linear operator that does not have a complete orthonormal system of eigenvectors. The concept of discrete eigenvalues must be generalized to include a continuous spectrum associated with solutions to the eigenvalue equation that are almost, but not quite in the Hilbert space. This will be explained in detail in Section 7.3.   ∎

❑ **Example 7.3.** Let $\phi_1, \phi_2, \ldots$ be a complete orthonormal system on the Hilbert space $\mathcal{H}$, and define the linear operator $\mathbf{U}$ (the *shift operator*) by

$$\mathbf{U}\phi_n = \phi_{n+1} \tag{7.12}$$

Then

$$\mathbf{U}\left(\sum_{n=1}^{\infty} c_n \phi_n\right) = \sum_{n=1}^{\infty} c_n \phi_{n+1} \tag{7.13}$$

is defined for every vector $x = \sum c_n \phi_n$ in $\mathcal{H}$, and $\|\mathbf{U}x\| = \|x\|$, so that $\mathbf{U}^\dagger \mathbf{U} = 1$. Nevertheless, $\mathbf{U}$ is not unitary. We have

$$\mathbf{U}^\dagger \phi_{n+1} = \phi_n \tag{7.14}$$

$(n = 1, 2, \ldots)$, but

$$\mathbf{U}^\dagger \phi_1 = \theta \tag{7.15}$$

since $(\phi_1, \mathbf{U}x) = 0$ for every vector x. Thus

$$\mathbf{U}\mathbf{U}^\dagger = 1 - \mathbf{P}_1 \tag{7.16}$$

where $\mathbf{P}_1$ is the projection operator onto the linear manifold spanned by ϕ_1.

The spectrum of the operators $\mathbf{U}$ and $\mathbf{U}^\dagger$ is also interesting. $\mathbf{U}$ has no spectrum at all since there are no solutions to the eigenvalue equation

$$\mathbf{U}x = \lambda x \tag{7.17}$$

On the other hand, the eigenvalue equation

$$\mathbf{U}^\dagger x = \lambda x \tag{7.18}$$

requires the coefficients c_n in the expansion of x to satisfy the simple recursion relation

$$c_{n+1} = \lambda c_n \tag{7.19}$$

Thus $\mathbf{U}^\dagger$ has eigenvectors of the form

$$\phi_\lambda = N_\lambda \left(\phi_1 + \sum_{n=1}^{\infty} \lambda^n \phi_{n+1}\right) \tag{7.20}$$

that belong to $\mathcal{H}$ for any λ with $|\lambda| < 1$. Note that ϕ_λ is a unit vector if we chose the normalization constant N_λ so that

$$|N_\lambda|^2 = 1 - |\lambda|^2 \tag{7.21}$$

Eigenvectors belonging to different eigenvalues are not orthogonal, since

$$(\phi_\lambda, \phi_\mu) = \frac{N_\lambda^* N_\mu}{1 - \lambda^* \mu} \tag{7.22}$$

and there is no completeness relation for these eigenvectors.   ∎

7.2   General Properties of Linear Operators on Hilbert Space

7.2.1   Bounded, Continuous, and Closed Operators

A linear operator $\mathbf{A}$ on a Hilbert space $\mathcal{H}$ is defined on a domain $\mathcal{D}_{\mathbf{A}} \subset \mathcal{H}$. It is generally assumed that $\mathcal{D}_{\mathbf{A}}$ contains a basis of $\mathcal{H}$ (if there is a nonzero vector x in $\mathcal{D}_{\mathbf{A}}^{\perp}$, then $\mathbf{A}x$ can be defined arbitrarily). Then $\mathbf{A}$ is also defined on any finite linear combination of the basis vectors, and, in particular, on the complete orthonormal system formed from the basis by the Gram–Schmidt process. Hence we can assume that $\mathcal{D}_{\mathbf{A}}$ contains a complete orthonormal system of $\mathcal{H}$, and thus is dense in $\mathcal{H}$.

Definition 7.1.   The linear operator $\mathbf{A}$ is *bounded* if there is a constant $C > 0$ such that $\|\mathbf{A}x\| \leq C\|x\|$ for all x in $\mathcal{D}_{\mathbf{A}}$. The smallest C for which the inequality is true is the *bound* of $\mathbf{A}$ ($|\mathbf{A}|$).[2]   ∎

Definition 7.2. $\mathbf{A}$ is *continuous* at the point x_0 in $\mathcal{D}_{\mathbf{A}}$ if for every sequence $\{x_n\}$ of vectors in $\mathcal{D}_{\mathbf{A}}$ that converges to x_0, the sequence $\{\mathbf{A}x_n\}$ converges to $\mathbf{A}x_0$.   ∎

Since $\mathbf{A}$ is linear, it is actually true that $\mathbf{A}$ is continuous at every point in $\mathcal{D}_{\mathbf{A}}$ if it is continuous at $x = \theta$. For if $\{\mathbf{A}y_n\}$ converges to θ whenever $\{y_n\}$ converges to θ, then also $\{\mathbf{A}(x_n - x_0)\}$ converges to θ whenever $\{x_n - x_0\}$ converges to θ.

An important result is that if $\mathbf{A}$ is bounded, then $\mathbf{A}$ is continuous. For if $\mathbf{A}$ is bounded and $\{x_n\}$ converges to θ, then $\{\|\mathbf{A}x_n\|\}$ converges to 0, since $\|\mathbf{A}x_n\| \leq |\mathbf{A}| \, \|x_n\|$ and $\{\|x_n\|\}$ converges to 0. Conversely, if $\mathbf{A}$ is continuous (even at θ), then $\mathbf{A}$ is bounded. For suppose $\mathbf{A}$ is continuous at $x = \theta$, but unbounded. Then there is a sequence $\{y_n\}$ of unit vectors such that $\|\mathbf{A}y_n\| > n$ for all n. But if we let $x_n \equiv y_n/n$, then $\{x_n\}$ converges to θ, but $\|\mathbf{A}x_n\| > 1$ for all n, so $\{\mathbf{A}x_n\}$ cannot converge to θ, contradicting the assumption of continuity.

Definition 7.3. $\mathbf{A}$ is *closed* if for every sequence $\{x_n\}$ of vectors in $\mathcal{D}_{\mathbf{A}}$ that converges to x_0 for which the sequence $\{\mathbf{A}x_n\}$ is also convergent, the limit x_0 is in $\mathcal{D}_{\mathbf{A}}$, and the sequence $\{\mathbf{A}x_n\}$ converges to $\mathbf{A}x_0$.   ∎

If $\mathbf{A}$ is bounded, then $\mathbf{A}$ has a closed extension (denoted by $[\mathbf{A}]$) whose domain is $\mathcal{H}$. For if x is a vector, then there is a Cauchy sequence $\{x_n\}$ of vectors in $\mathcal{D}_{\mathbf{A}}$ that converges to x. But if $\mathbf{A}$ is bounded, then the sequence $\{\mathbf{A}x_n\}$ is also convergent, and we can define the limit to be $\mathbf{A}x$. An unbounded operator $\mathbf{A}$ may, or may not, have a closed extension. The operator $\mathbf{A}$ introduced in Example 7.1 is closed when its domain is extended to include all vectors

$$\phi = \sum_{n=1}^{\infty} a_n \phi_n \tag{7.23}$$

for which $\sum_{n=1}^{\infty} n^2 \, |a_n|^2 < \infty$.

---

[2]The bound $|\mathbf{A}|$ defined here is sometimes called the *norm* (or *operator norm*) of $\mathbf{A}$. The norm introduced below is then called the *Hilbert–Schmidt norm* to distinguish it from the operator norm.

However, consider the linear operator $\mathbf{B}$ defined by $\mathbf{B}\phi_n = \phi_1$ ($n = 1, 2, \ldots$), so that[3]

$$\mathbf{B}\left(\sum_{k=1}^{n} a_k \phi_k\right) = \left(\sum_{k=1}^{n} a_k\right)\phi_1 \tag{7.24}$$

If we let

$$x_n \equiv \frac{1}{n} \sum_{k=1}^{n} \phi_k \tag{7.25}$$

then $\{x_n\} \to \theta$ (note that $\|x_n\| = 1/n$). However,

$$\mathbf{B}x_n = \phi_1 \tag{7.26}$$

for all n, so $\{\mathbf{B}x_n\}$ is convergent, not to $\mathbf{B}\theta$ ($= \theta$), but to ϕ_1. Hence $\mathbf{B}$ is not closed.

Remark.   Note also that $\mathbf{B}$ is unbounded, even though $\|\mathbf{B}\phi_n\| = \|\phi_1\| = 1$ for every element of the complete orthonormal system $\{\phi_n\}$. Hence it is not enough that $\mathbf{B}$ is bounded on a complete orthonormal system for $\mathbf{B}$ to be bounded.   □

7.2.2   Inverse Operator

The inverse of a linear operator on a Hilbert space must be carefully defined. Recall that the linear operator $\mathbf{A}$ is *nonsingular* if $\mathbf{A}x = \theta$ only if $x = \theta$. Here we add the new

Definition 7.4.   The linear operator $\mathbf{A}$ is *invertible* if (and only if) it is bounded *and* there is a bounded linear operator $\mathbf{B}$ such that

$$\mathbf{BA} = \mathbf{AB} = 1 \tag{7.27}$$

If such a $\mathbf{B}$ exists, then $\mathbf{B} = \mathbf{A}^{-1}$ is the *inverse* of $\mathbf{A}$.   ∎

An invertible operator is nonsingular, but the converse need not be true. The operator $\mathbf{U}$ in Example 3 is nonsingular, bounded, and $\mathbf{U}^\dagger\mathbf{U} = 1$, but $\mathbf{U}$ is not invertible, since there is no operator $\mathbf{V}$ such that $\mathbf{UV} = 1$. For the operator $\mathbf{A}$ in Example 1, we have $\mathbf{A}\bar{\mathbf{A}} = 1$, and $\bar{\mathbf{A}}\mathbf{A} \subset 1$ (it can be extended to 1 on all of $\mathcal{H}$, but that is not the point here), although it is common to see $\bar{\mathbf{A}}$ treated as $\mathbf{A}^{-1}$ (and we will also do so when no harm is done). However, it should be kept in mind that neither $\mathbf{A}$ nor $\bar{\mathbf{A}}$ are invertible, strictly speaking, since $\mathbf{A}$ is unbounded. A general criterion for invertibility is contained in the

Theorem 7.1.   If $\mathbf{A}$ is a linear operator on $\mathcal{H}$ whose domain $\mathcal{D}_\mathbf{A}$ and range $\mathcal{R}_\mathbf{A}$ are the entire space $\mathcal{H}$, then $\mathbf{A}$ is invertible if and only if there are positive constants m, M such that

$$m\|x\| \leq \|\mathbf{A}x\| \leq M\|x\| \tag{7.28}$$

for every x in $\mathcal{H}$.

---

[3]It is amusing to observe that $\mathbf{B}$ transforms the sum of a linear combination of orthogonal vectors into the sum of the components, with a unit vector included for appearance. This operator is often (too often!) used by first-year students in physics.

Proof. If $\mathbf{A}$ is invertible, then the inequalities in (7.28) are satisfied with $m = 1/\left|\mathbf{A}^{-1}\right|$ and $M = |\mathbf{A}|$. Conversely, if the inequalities are satisfied, then for any y in $\mathcal{H}$, we know that $y = \mathbf{A}x$ for some x in $\mathcal{H}$ (recall that $\mathcal{R}_\mathbf{A}$, the *range* of $\mathbf{A}$, is the set of all y such that $y = \mathbf{A}x$ for some x in $\mathcal{D}_\mathbf{A}$) and we can define a linear operator $\bar{\mathbf{A}}$ by

$$x \equiv \bar{\mathbf{A}}y \tag{7.29}$$

Evidently $\bar{\mathbf{A}}$ is defined on all of $\mathcal{H}$, with

$$\bar{\mathbf{A}}\mathbf{A} = \mathbf{A}\bar{\mathbf{A}} = 1 \tag{7.30}$$

and $\bar{\mathbf{A}}$ is bounded, with $\left|\bar{\mathbf{A}}\right| = 1/m$. Hence $\bar{\mathbf{A}} = \mathbf{A}^{-1}$. ∎

7.2.3   Compact Operators; Hilbert–Schmidt Operators

A stronger concept than continuity or boundedness of an operator is compactness.

Definition 7.5.   The linear operator $\mathbf{A}$ is *compact* (or *completely continuous*) if it transforms every weakly convergent sequence of vectors into a strongly convergent sequence, that is, if $\{x_n\} \rightharpoonup \theta$ implies $\{\mathbf{A}x_n\} \to \theta$. ∎

Thus, for example, if $\mathbf{A}$ is compact and $\{\phi_n\}$ is any orthonormal system, then $\{\mathbf{A}\phi_n\} \to \theta$. Every compact operator is continuous, but the converse is not true (the identity operator 1 is continuous, but not compact). The importance of compact operators becomes clear in the next section, where it will be seen that every compact normal operator has a complete orthonormal system of eigenvectors.

Definition 7.6.   Let $\mathbf{A}$ be a linear operator on $\mathcal{H}$, and $\{\phi_n\}$, $\{\psi_n\}$ a pair of complete orthonormal systems. The *norm* of $\mathbf{A}$ is defined by

$$\|\mathbf{A}\|^2 \equiv \operatorname{tr} \mathbf{A}^\dagger \mathbf{A} = \sum_{n=1}^\infty \|\mathbf{A}\phi_n\|^2 = \sum_{n,m=1}^\infty |(\mathbf{A}\phi_n, \psi_m)|^2 \tag{7.31}$$

$\mathbf{A}$ is a *Hilbert–Schmidt operator* if $\|\mathbf{A}\|$ is finite. Here $\operatorname{tr} \mathbf{X}$ denotes the trace of the operator $\mathbf{X}$, defined as usual as the sum of diagonal elements (the sum, if it exists, is independent of the choice of complete orthonormal system). ∎

The Hilbert–Schmidt operators themselves form a linear vector space $\mathcal{S}$, which is a Hilbert space with scalar product defined by

$$(\mathbf{A}, \mathbf{B}) \equiv \operatorname{tr} \mathbf{A}^\dagger \mathbf{B} = \sum_{n=1}^\infty (\mathbf{A}\phi_n, \mathbf{B}\phi_n) = \sum_{n,m=1}^\infty (\mathbf{A}\phi_n, \psi_m)(\psi_m, \mathbf{B}\phi_n) \tag{7.32}$$

This claim requires proof, especially of the property of closure, but details will be left to the problems. Any Hilbert–Schmidt operator is compact, but not conversely. For $\|\mathbf{A}\|$ to be finite, it is necessary that $\{\mathbf{A}\phi_n\} \to 0$, but $\{\mathbf{A}\phi_n\} \to 0$ does not guarantee convergence of the infinite series defining $\|\mathbf{A}\|^2$.

❑ **Example 7.4.**   Consider the function space $L_2(a, b)$ over the finite interval $I : a \leq t \leq b$, and suppose $K(t, u)$ is a function in $L_2(I \times I)$. Corresponding to $K(t, u)$ is the linear operator **K** in $L_2(a, b)$ defined by

$$(\mathbf{K}x)(t) = \int_a^b K(t, u)x(u)\, du \tag{7.33}$$

K is an *integral operator* of a type we will encounter again later in the chapter. Note that

$$\|\mathbf{K}\|^2 = \int_a^b \int_a^b |K(t, u)|^2\, dt\, du \tag{7.34}$$

is finite by assumption, so **K** is a Hilbert–Schmidt operator.   ∎

Finally, a linear operator **A** is of *finite rank* if the range of **A** ($\mathcal{R}_\mathbf{A}$, or im **A**) is finite dimensional. A projection operator onto a finite-dimensional linear manifold is of finite rank. The operator **B** defined by Eq. (7.24) is of rank 1. An operator of finite rank is certainly a Hilbert–Schmidt operator, hence compact and bounded.

Remark.   Every linear operator on a finite-dimensional vector space $\mathcal{V}^n$ is of finite rank. Hilbert–Schmidt and compact operators on $\mathcal{H}$ share many of the properties of operators of finite rank. For example, compact normal operators have a complete orthonormal system of eigenvectors belonging to discrete eigenvalues that in fact form a sequence converging to 0. Merely bounded operators are essentially infinite dimensional, with new characteristics not seen in finite dimensions, as illustrated by the examples in this and the preceding section. Unbounded operators can have even more unusual properties, since they have no finite-dimensional counterparts; it is for these that special care must be given to questions of domain, range, and closure.   □

7.2.4   Adjoint Operator

Definition 7.7.   The *adjoint* $\mathbf{A}^\dagger$ of the linear operator **A** is defined for the vectors y for which there exists a vector y_* such that

$$(y_*, x) = (y, \mathbf{A}x) \tag{7.35}$$

for every x in $\mathcal{D}_\mathbf{A}$. Then define

$$\mathbf{A}^\dagger y \equiv y_* \tag{7.36}$$

corresponding to the finite-dimensional definition.   ∎

Remark.   Note that $\mathcal{D}_\mathbf{A}$ must be dense in $\mathcal{H}$ to define $\mathbf{A}^\dagger$, for if there is a nonzero vector y_0 in $\mathcal{D}_\mathbf{A}^\perp$, then

$$(y_* + y_0, x) = (y_*, x) = (y, \mathbf{A}, x) \tag{7.37}$$

for every x in $\mathcal{D}_\mathbf{A}^\perp$, so that y_* is not uniquely defined.   □

To completely define $\mathbf{A}^\dagger$, it is necessary to specify the domain $\mathcal{D}_{\mathbf{A}^\dagger}$. If $\mathbf{A}$ is bounded, then $\mathbf{A}$ can be defined on all of $\mathcal{H}$, and $(y, \mathbf{A}x)$ is a bounded linear functional. Hence, by Theorem 2.2, there is a unique y_* satisfying (7.35), and thus $\mathcal{D}_{\mathbf{A}^\dagger}$ is all of $\mathcal{H}$. Furthermore, $\mathbf{A}^\dagger$ is bounded, and

$$\left|\mathbf{A}^\dagger\right| = |\mathbf{A}| \tag{7.38}$$

Even if $\mathbf{A}$ is not bounded, the adjoint $\mathbf{A}^\dagger$ is closed, since if $\{y_n\}$ is a sequence of vectors in $\mathcal{D}_{\mathbf{A}^\dagger}$ such that $\{y_n\} \to y$ and $\{\mathbf{A}^\dagger y_n\} \to y_*$, then

$$(y, \mathbf{A}x) = \lim_{n\to\infty} (y_n, \mathbf{A}x) \quad \text{and} \quad (y_*, x) = \lim_{n\to\infty} (\mathbf{A}^\dagger y_n, x) \tag{7.39}$$

for every x in $\mathcal{D}_{\mathbf{A}}$, so $\mathbf{A}^\dagger y = y_*$. Note that $\mathbf{A}^\dagger x = \theta$ if and only if $(x, \mathbf{A}y) = 0$ for every y in $\mathcal{D}_{\mathbf{A}}$; in short, $\ker \mathbf{A}^\dagger = \mathcal{R}_{\mathbf{A}}^\perp$.

The operator $(\mathbf{A}^\dagger)^\dagger$ is well defined if (and only if) $\mathcal{D}_{\mathbf{A}^\dagger}$ is dense in $\mathcal{H}$. This need not be the case, for if we consider the operator $\mathbf{B}$ defined by Eq. (7.24), we have

$$(\mathbf{B}^\dagger \phi_n, \phi_k) = (\phi_n, \mathbf{B}\phi_k) = (\phi_n, \phi_1) = 0 \tag{7.40}$$

and hence $\mathbf{B}^\dagger \phi_n = \theta$ for $n = 2, 3, \ldots$. However,

$$(\mathbf{B}^\dagger \phi_1, \phi_k) = (\phi_1, \mathbf{B}\phi_k) = 1 \tag{7.41}$$

for every k. Hence $\mathbf{B}^\dagger \phi_1$ is not defined, nor can $\mathbf{B}^\dagger \phi$ be defined on any vector ϕ with a nonzero component along ϕ_1. Thus ϕ_1 is in $\mathcal{D}_{\mathbf{B}^\dagger}^\perp$, and $\mathcal{D}_{\mathbf{B}^\dagger}$ cannot be dense in $\mathcal{H}$.

On the other hand, if the operator $(\mathbf{A}^\dagger)^\dagger$ does exist, then it is a closed extension (even the smallest closed extension) of $\mathbf{A}$, so that

$$\mathbf{A} \subset (\mathbf{A}^\dagger)^\dagger = [\mathbf{A}] \tag{7.42}$$

Definition 7.8. The linear operator $\mathbf{A}$ is *self-adjoint* if

$$\mathbf{A} = \mathbf{A}^\dagger \tag{7.43}$$

where equality includes the requirement $\mathcal{D}_{\mathbf{A}} = \mathcal{D}_{\mathbf{A}^\dagger}$.   ∎

In a finite-dimensional space $\mathcal{V}^n$, this is equivalent to the condition

$$(y, \mathbf{A}x) = (\mathbf{A}y, x) \tag{7.44}$$

for every x, y in $\mathcal{D}_{\mathbf{A}}$.

For an operator $\mathbf{A}$ on $\mathcal{H}$ that is bounded, it is also true that (7.43) and (7.44) are equivalent if $\mathcal{D}_{\mathbf{A}}$ is dense in $\mathcal{H}$, since in that case, we can always find a closed extension of $\mathbf{A}$ that is bounded. For unbounded operators, Eq. (7.44) defines a *symmetric*[4] operator, for which it is true that $\mathbf{A} \subset \mathbf{A}^\dagger$, but $\mathbf{A}$ may have no closed extension, in which case it cannot be self-adjoint. See Problem 12 for an important example from quantum mechanics.

---

[4]This terminology is suggested by Halmos. Equation (7.44) is often used in physics to characterize a *Hermitian* operator, but it is implicitly assumed that Eq. (7.43) is then automatically satisfied. It is not, but the examples seen here are too subtle to merit an extended explanatory digression in the standard exposition of quantum mechanics.

7.2.5   Unitary Operators; Isometric Operators

The operator $\mathbf{U}$ is *unitary* if $\mathbf{U}^\dagger = \mathbf{U}^{-1}$, which requires

$$\mathbf{U}^\dagger\mathbf{U} = \mathbf{U}\mathbf{U}^\dagger = 1 \tag{7.45}$$

In $\mathcal{V}^n$, $\mathbf{U}^\dagger\mathbf{U} = 1$ is sufficient to have $\mathbf{U}$ be unitary, but the example of the shift operator defined by Eq. (7.12) shows that this is not the case in $\mathcal{H}$; the two conditions in Eq. (7.45) are independent. The operator $\mathbf{U}$ is *isometric* if

$$\|\mathbf{U}x\| = \|x\| \tag{7.46}$$

for all x in $\mathcal{D}_\mathbf{U}$. The shift operator is isometric, but not unitary, since its range is not the entire space $\mathcal{H}$. In general, an isometric operator $\mathbf{U}$ in $\mathcal{H}$ maps an infinite-dimensional subspace $\mathcal{D}_\mathbf{U}$ into an infinite-dimensional subspace $\mathcal{R}_\mathbf{U}$. $\mathcal{H}$ is large enough that $\mathcal{D}_{\mathbf{U}\perp}$ and $\mathcal{R}_{\mathbf{U}\perp}$ can be non-trivial (they can even themselves be infinite-dimensional). If we let

$$m \equiv \dim\mathcal{D}_{\mathbf{U}\perp}\ ,\quad n \equiv \dim\mathcal{R}_{\mathbf{U}\perp} \tag{7.47}$$

then the ordered pair $[m, n]$ is the *deficiency index* (DI, for short) of $\mathbf{U}$.

❑ **Example 7.5.**   The deficiency index of the shift operator $\mathbf{U}$ is $[0, 1]$, and that of $\mathbf{U}^\dagger$ is $[1, 0]$. It is true in general that if $\mathbf{U}$ has DI=$[m, n]$, then $\mathbf{U}^\dagger$ has DI=$[n, m]$ (show this).   ∎

$\mathbf{U}$ is *maximal isometric* if it has no proper isometric extension. This will be the case if the deficiency index of $\mathbf{U}$ is of the form $[m, 0]$ with $m > 0$, or $[0, n]$ with $n > 0$. $\mathbf{U}$ is unitary if its DI is $[0, 0]$. It is *essentially unitary* if its DI is $[m, m]$, since we can then define a unitary extension of $\mathbf{U}$ by adjoining a unitary operator from $\mathcal{D}_\mathbf{U}^\perp$ onto $\mathcal{R}_\mathbf{U}^\perp$ (this extension is far from unique, since any m-dimensional unitary operator will do).

7.2.6   Convergence of Sequences of Operators in $\mathcal{H}$

For sequences of vectors in $\mathcal{H}$, there are two types of convergence, weak and strong. Convergence of sequences $\{\mathbf{A}_n\}$ of operators in $\mathcal{H}$ comes in three varieties:

   (i) *uniform convergence*. The sequence $\{\mathbf{A}_n\}$ converges *uniformly* to $\mathbf{0}$ ($\{\mathbf{A}_n\} \Rightarrow \mathbf{0}$) if $\{\|\mathbf{A}_n x\|\} \to 0$ uniformly in x, that is, if for every $\varepsilon > 0$ there is an N such that $\|\mathbf{A}x\| \le \varepsilon\|x\|$ for every $n > N$ and all x in $\mathcal{H}$.

   (ii) *strong convergence*. The sequence $\{\mathbf{A}_n\}$ converges *strongly* to $\mathbf{0}$ ($\{\mathbf{A}_n\} \to \mathbf{0}$) if $\{\|\mathbf{A}_n x\|\} \to 0$ for every x in $\mathcal{H}$.

   (iii) *weak convergence*. The sequence $\{\mathbf{A}_n\}$ converges *weakly* to $\mathbf{0}$ ($\{\mathbf{A}_n\} \rightharpoonup \mathbf{0}$) if $\{(y, \mathbf{A}_n x)\} \to 0$ for every y and x in $\mathcal{H}$.

Convergence to a limit is then defined in the obvious way: the sequence $\{\mathbf{A}_n\}$ converges to $\mathbf{A}$ in some sense, if the sequence $\{\mathbf{A}_n - \mathbf{A}\}$ converges to $\mathbf{0}$ in the same sense. If the limit is an unbounded operator, it must also be required that each of the $\mathbf{A}_n$ and the limit $\mathbf{A}$ can be defined on a common domain $\mathcal{D}_*$ that is dense in $\mathcal{H}$, and "for all x in $\mathcal{H}$" is to be understood as "for all x in $\mathcal{D}_*$."

These types of convergence are ordered, in the sense that uniform convergence implies strong convergence, and strong convergence implies weak convergence. In fact, only in infinite-dimensional space are these types of convergence distinct. In $\mathcal{V}^n$ the three types of convergence are equivalent, as are strong and weak convergence of sequences of vectors (the reader is invited to show this with the help of the following example).

❑ **Example 7.6.** Let $\{\phi_k\}$ be a complete orthonormal system, and $\mathbf{P}_k$ be the projection operator onto the linear manifold spanned by ϕ_k. Then the sequence $\{\mathbf{P}_k\} \to \mathbf{0}$ (strong, but not uniform convergence). The sequence $\{\mathbf{A}_n\}$, with

$$\mathbf{A}_n \equiv \sum_{k=1}^{n} \mathbf{P}_k \tag{7.48}$$

converges strongly, but not uniformly, to $\mathbf{1}$.   ∎

7.3 Spectrum of Linear Operators on Hilbert Space

The spectra of linear operators on an infinite-dimensional Hilbert space have a rich structure compared with the spectra of finite-dimensional operators, much as the theory of analytic functions of a complex variable is rich compared to the theory of polynomials. In this section, we will consider the spectra of normal operators[5] (recall $\mathbf{A}$ is normal if $\mathbf{A}\mathbf{A}^\dagger = \mathbf{A}^\dagger\mathbf{A}$). Since $\mathbf{A}$ is normal if (and only if) it has the form

$$\mathbf{A} = \mathbf{X} + i\mathbf{Y} \tag{7.49}$$

with $\mathbf{X}$ and $\mathbf{Y}$ self-adjoint, $[\mathbf{X}, \mathbf{Y}] = \mathbf{0}$ (and $\mathcal{D}_\mathbf{A} = \mathcal{D}_\mathbf{X} \cap \mathcal{D}_\mathbf{Y}$ dense in $\mathcal{H}$ if $\mathbf{A}$ is unbounded), it is actually sufficient to work out the theory for self-adjoint operators.

7.3.1 Spectrum of a Compact Self-Adjoint Operator

To begin, suppose $\mathbf{A}$ is a compact, self-adjoint operator. A compact operator converts a weakly convergent sequence into a strongly convergent one (by definition), so the influence of distant dimensions is reduced in some sense, and it turns that a compact, self-adjoint operator has a complete orthonormal system of eigenvectors with real eigenvalues, just as its finite-dimensional counterpart, and the eigenvalues define a sequence that converges to zero.

Since $\mathbf{A}$ is compact, it is bounded and closed, and $(x, \mathbf{A}x)$ is real for every x in $\mathcal{H}$, since $\mathbf{A}$ is self-adjoint. If we let

$$\lambda_+ = \sup_{\|x\|=1} (x, \mathbf{A}x), \quad \lambda_- = \inf_{\|x\|=1} (x, \mathbf{A}x) \tag{7.50}$$

then the bound of $\mathbf{A}$ is the larger of $|\lambda_+|$ and $|\lambda_-|$. Since λ_+ is the *least* upper bound of $(x, \mathbf{A}x)$ on the unit sphere ($\|x\| = 1$) in $\mathcal{H}$, there is a sequence $\{x_n\}$ of unit vectors such

---

[5]The shift operator introduced in Section 7.1 and other operators introduced in the problems, as well as any isometric operator that is not essentially unitary, are not normal. Those examples illustrate the additional spectral types that can occur with nonnormal operators.

that the sequence $\{(x_n, \mathbf{A}x_n)\}$ converges to λ_+. Since $\{x_n\}$ is bounded, it contains a subsequence $\{y_n\}$ that converges weakly to a limit y (see Section 6.1). Then the compactness of $\mathbf{A}$ implies that the sequence $\{\mathbf{A}y_n\}$ converges strongly to a limit $y_* = \mathbf{A}y$. Then we have

$$\|\mathbf{A}y_n - \lambda_+ y\|^2 = \|\mathbf{A}y_n\|^2 - 2\lambda_+(y, \mathbf{A}y_n) + \lambda_+^2 \to \|y_*\|^2 - \lambda_+^2 \leq 0 \qquad (7.51)$$

(note that $\{(y, \mathbf{A}y_n)\} \to \lambda_+$). But $\{\|\mathbf{A}y_n - \lambda_+ y\|\} \geq 0$, so the only consistent conclusion is that $\{\|\mathbf{A}y_n - \lambda_+ y\|\} \to 0$, which implies

$$\mathbf{A}y = \lambda_+ y \qquad (7.52)$$

Thus y is an eigenvector of $\mathbf{A}$ belonging to eigenvalue λ_+. The same argument shows that there is an eigenvector of $\mathbf{A}$ belonging to the eigenvalue λ_-. Thus every compact self-adjoint operator has at least one eigenvector belonging to a discrete eigenvalue.

Once some eigenvectors of $\mathbf{A}$ have been found, it is straightforward to find more. Suppose $\phi_1, \ldots, \phi_n$ is an orthonormal system of eigenvectors of $\mathbf{A}$ belonging to eigenvalues $\lambda_1, \ldots, \lambda_n$, that spans a linear manifold $\mathcal{M} = \mathcal{M}(\phi_1, \ldots, \phi_n)$. Then $\mathcal{M}(\phi_1, \ldots, \phi_n)$ is an invariant manifold of $\mathbf{A}$, and, since $\mathbf{A}$ is normal, it follows that so is $\mathcal{M}^\perp$. But $\mathbf{A}$ is a compact self-adjoint operator on $\mathcal{M}^\perp$, and thus has at least one eigenvector ϕ_{n+1} in $\mathcal{M}^\perp$ belonging to a discrete eigenvalue λ_{n+1}. Continuing this procedure leads to a complete orthonormal system $\{\phi_1, \phi_2, \ldots\}$ of eigenvectors belonging to discrete eigenvalues, and we have the

Theorem 7.2. Every compact normal linear operator on a Hilbert space $\mathcal{H}$ has a complete orthonormal system of eigenvectors belonging to discrete eigenvalues; the sequence $\{\lambda_1, \lambda_2, \ldots\}$ of eigenvalues converges to zero.

Proof. The convergence of the sequence of eigenvalues follows from that fact that the sequence $\{\phi_n\}$ of eigenvectors converges weakly to zero, and hence the sequence $\{\mathbf{A}\phi_n = \lambda_n\phi_n\} \to \theta$ strongly since $\mathbf{A}$ is compact. A corollary of this result is that the eigenmanifold $\mathcal{M}_\lambda$ belonging to an eigenvalue $\lambda \neq 0$ of $\mathbf{A}$ must be finite dimensional. ∎

7.3.2 Spectrum of Noncompact Normal Operators

Now suppose $\mathbf{A}$ is normal (but not necessarily compact, or even bounded), with a complete orthonormal system of eigenvectors and distinct eigenvalues $\lambda_1, \lambda_2, \ldots$ (such an operator is *separating*). If $\mathcal{M}_k$ is the eigenmanifold belonging to eigenvalue λ_k, with corresponding eigenprojector $\mathbf{P}_k$, then we have the spectral representation

$$\mathbf{A} = \sum_k \lambda_k \mathbf{P}_k \qquad (7.53)$$

as in the finite-dimensional case (compare with Eq. (2.167)), except that here the sum may be an infinite series. If the series (7.53) is actually infinite, then it converges uniformly if $\mathbf{A}$ is compact, strongly if $\mathbf{A}$ is bounded; if $\mathbf{A}$ is unbounded, the series converges only on the domain $\mathcal{D}_\mathbf{A}$ of $\mathbf{A}$.

Normal operators that are not compact need not have a complete orthonormal system of eigenvectors belonging to discrete eigenvalues. The concept of spectrum for such operators must be generalized to include a continuous spectrum.

Definition 7.9. The *spectrum* $\Sigma(\mathbf{A})$ of the normal operator $\mathbf{A}$ consists of the complex numbers λ such that for every $\varepsilon > 0$, there is a vector ϕ in $\mathcal{D}_{\mathbf{A}}$ for which

$$\|\mathbf{A}\phi - \lambda\phi\| < \varepsilon\|\phi\| \tag{7.54}$$

The *point spectrum* (or *discrete spectrum*) $\Pi(\mathbf{A})$ of $\mathbf{A}$ consists of the eigenvalues of $\mathbf{A}$. The elements of $\Sigma(\mathbf{A})$ spectrum of $\mathbf{A}$ not in $\Pi(\mathbf{A})$ form the *continuous spectrum* (or *approximate point spectrum*) $\Sigma_c(\mathbf{A})$ of $\mathbf{A}$.   ∎

❏ **Example 7.7.** Consider the operator $\mathbf{X}$ defined on the Hilbert space $L_2(-1, 1)$ by

$$\mathbf{X}f(x) = xf(x) \tag{7.55}$$

as in Example 7.2. As already noted, $\mathbf{X}$ has no point spectrum, but every x in the interval $-1 \leq x \leq 1$ is included in the continuous spectrum of $\mathbf{X}$. To see this, suppose $-1 < x_0 < 1$, and define the function $f_\varepsilon(x_0; x)$ by

$$f_\varepsilon(x_0; x) = \begin{cases} 1/\sqrt{2\varepsilon} & x_0 - \varepsilon \leq x \leq x_0 + \varepsilon \\ 0 & \text{otherwise} \end{cases} \tag{7.56}$$

We have $\|f_{\varepsilon(x_0)}\| = 1$, and

$$\|(\mathbf{X} - x_0\mathbf{1})f_\varepsilon(x_0)\|^2 < \tfrac{1}{3}\varepsilon^2 \tag{7.57}$$

This works for any $\varepsilon > 0$, so x_0 is in the continuous spectrum of $\mathbf{X}$. The discussion must be modified in an obvious way for $x_0 = \pm 1$, but the conclusion is the same. The only "eigenfunction" of $\mathbf{X}$ belonging to the eigenvalue x_0 is the Dirac δ-function $\delta(x - x_0)$, which is not in $L_2(-1, 1)$.   ∎

→ **Exercise 7.1.** Find the point spectrum and the continuous spectrum of the shift operator $\mathbf{U}$ defined by Eq. (7.12). Also, find the spectra of its adjoint $\mathbf{U}^\dagger$.   ▢

7.3.3   Resolution of the Identity

Equation (7.53) is a formal representation of the operator $\mathbf{A}$ in terms of its eigenvalues and eigenvectors. To generalize this representation to include operators with a continuous spectrum, we need to introduce the concept of a *resolution of the identity*. This is a family of projection operators $\{\mathbf{E}_\lambda\}$ defined for the real λ, with the properties:

(i) The projection manifold $\mathcal{M}_\lambda$ of $\mathbf{E}_\lambda$ shrinks to $\{\theta\}$ for $\lambda \to -\infty$ and grows to all of $\mathcal{H}$ for $\lambda \to +\infty$. or

$$\lim_{\lambda \to -\infty} \mathbf{E}_\lambda = 0 \quad \text{and} \quad \lim_{\lambda \to \infty} \mathbf{E}_\lambda = 1 \tag{7.58}$$

(ii) The projection manifold $\mathcal{M}_\lambda \subseteq \mathcal{M}_\mu$ if $\lambda < \mu$, so that $\mathbf{E}_\lambda \leq \mathbf{E}_\mu$ if $\lambda < \mu$. Thus $\mathbf{E}_\lambda$ is a nondecreasing function of λ that interpolates between 0 at $-\infty$ and 1 at ∞.

(iii) The limits

$$\lim_{\varepsilon \to 0^+} \mathbf{E}_{\lambda - \varepsilon} \equiv \mathbf{E}_\lambda^- \quad \text{and} \quad \lim_{\varepsilon \to 0^+} \mathbf{E}_{\lambda + \varepsilon} \equiv \mathbf{E}_\lambda^+ \tag{7.59}$$

exist for each λ. Note that while the limits $\mathbf{E}_\lambda^\pm$ must exist for each finite λ, they need not be equal, and the points where

$$\Delta \mathbf{E}_\lambda = \mathbf{E}_\lambda^+ - \mathbf{E}_\lambda^- \equiv \mathbf{P}_\lambda \neq 0 \tag{7.60}$$

are the *discontinuities* of $\mathbf{E}_\lambda$. $\mathbf{P}_\lambda$ is a projection operator onto a manifold of positive dimension, which may be finite or infinite. There may also be points at which $\mathbf{E}_\lambda$ is continuous, but

$$\mathbf{E}_{\lambda + \varepsilon} - \mathbf{E}_{\lambda - \varepsilon} \neq 0 \tag{7.61}$$

for every $\varepsilon > 0$. These are the *points of increase* of $\mathbf{E}_\lambda$.

If there is a finite a such that $\mathbf{E}_\lambda = 0$ for $\lambda < a$, but $\mathbf{E}_\lambda > 0$ for $\lambda > a$, then the family $\{\mathbf{E}_\lambda\}$ is bounded from below (with *lower bound* a). If there is a finite b such that $\mathbf{E}_\lambda = 1$ for $\lambda > b$, but $\mathbf{E}_\lambda < 1$ for $\lambda < b$, then $\{\mathbf{E}_\lambda\}$ is bounded from above (with *upper bound* b). If $\{\mathbf{E}_\lambda\}$ is bounded both above and below, it is *bounded* and $[a, b]$ is the *interval of variation* of $\{\mathbf{E}_\lambda\}$. If it is only bounded on one end, it is *semi-bounded*.

If $\mathbf{E}_\lambda$ is a resolution of the identity, then the operator $\mathbf{A}$ defined formally by

$$\mathbf{A} = \int \lambda \, d\mathbf{E}_\lambda \tag{7.62}$$

is self-adjoint, and bounded if and only if $\mathbf{E}_\lambda$ is bounded. The meaning of Eq. (7.62) is that[6]

$$(y, \mathbf{A}x) = \int \lambda \, d(y, \mathbf{E}_\lambda x) = \int \lambda \frac{d}{d\lambda}(y, \mathbf{E}_\lambda x) \, d\lambda \tag{7.66}$$

If $\mathbf{A}$ is unbounded, then the domain of $\mathbf{A}$ is the set of all vectors x for which the integral

$$\|\mathbf{A}x\|^2 = \int \lambda^2 d\|\mathbf{E}_\lambda x\|^2 < \infty \tag{7.67}$$

---

[6]We have the natural definition $\int df(x) = \int f'(x)dx$, which is known as a *Stieltjes integral*. If $f(x)$ is not strictly differentiable, this definition must be used with care, integrating by parts if necessary to give the integral

$$\int g(x) \, df(x) = \int g(x) f'(x) \, dx = -\int g'(x) f(x) \, dx \tag{7.63}$$

whenever $g(x)$ is differentiable. The Stieltjes integral also provides an alternative definition of the Dirac δ-function. If $\theta(x)$ is the step function defined by

$$\theta(x) = \begin{cases} 1 & x > 0 \\ 0 & \text{otherwise} \end{cases} \tag{7.64}$$

then we have

$$\int_{-\infty}^{\infty} g(x) \, d\theta(x) = g(0) = \int_{-\infty}^{\infty} g(x) \delta(x) \, dx \tag{7.65}$$

if $g(x)$ is continuous at $x = 0$.

Equation (7.62) shows how a resolution of the identity defines a self-adjoint operator on $\mathcal{H}$. The converse of this is Hilbert's fundamental theorem:

Hilbert's Fundamental Theorem. To every self-adjoint linear operator $\mathbf{A}$ on a Hilbert space $\mathcal{H}$, there is a unique resolution of the identity $\mathbf{E}_\lambda$ such that

$$\mathbf{A} = \int \lambda \, d\mathbf{E}_\lambda \tag{7.68}$$

Remark. This is the Hilbert space analog of the theorem that every self-adjoint operator has a complete orthonormal system of eigenvectors belonging to real eigenvalues.   □

❑ **Example 7.8.** To see how the resolution of the identity works, consider again the operator $\mathbf{X}$ defined by Eq. (7.9), but now let the function space be $L_2(a, b)$. If $a < \lambda < b$, we then let

$$\mathbf{E}_\lambda f(x) = \begin{cases} f(x) & a \le x \le \lambda \\ 0 & \lambda < x < b \end{cases} \tag{7.69}$$

or, more simply,

$$\mathbf{E}_\lambda f(x) = \theta(\lambda - x) f(x) \tag{7.70}$$

where $\theta(\xi)$ is the standard step function defined by Eq. (7.64). $\mathbf{E}_\lambda$ is evidently a resolution of the identity, and

$$\mathbf{X} = \int_a^b \lambda \, d\mathbf{E}_\lambda \tag{7.71}$$

since

$$d\mathbf{E}_\lambda f(x) = \delta(x - \lambda) f(\lambda) d\lambda \tag{7.72}$$

Here $\delta(x - \lambda)$ is the Dirac δ-function, and we have used the formal result $\theta'(\xi) = \delta(\xi)$. If a and b are finite, then $\mathbf{X}$ is a bounded operator. $\mathbf{X}$ is unbounded in the limits $a \to -\infty$ or $b \to \infty$, but the spectral representation with $\mathbf{E}_\lambda$ given by Eq. (7.70) is still valid.   ∎

➔ **Exercise 7.2.** The self-adjoint operator

$$\mathbf{P} \equiv \left[\frac{1}{i} \frac{d}{dx} \right]$$

on $L_2(-\infty, \infty)$ has the standard spectral representation

$$\mathbf{P} = \int_\infty^\infty \lambda \, d\mathbf{E}_\lambda$$

with the $\mathbf{E}_\lambda$ being a resolution of the identity. If $f(x)$ is in the domain of $\mathbf{P}$, find an expression for $\mathbf{E}_\lambda f(x)$. *Hint.* Consider the Fourier integral representation of $f(x)$.   □

Remark. In quantum mechanics, $\mathbf{P}$ is the momentum operator for a particle moving in one dimension.   □

7.3.4   Functions of a Self-Adjoint Operator

Functions of self-adjoint operators have an elegant definition in terms of the spectral representation: if $f(x)$ is defined for x in the spectrum of the self-adjoint operator $\mathbf{A}$ defined by Eq. (7.62), then

$$f(\mathbf{A}) = \int f(\lambda)\, d\mathbf{E}_\lambda \tag{7.73}$$

with domain $\mathcal{D}_{f(\mathbf{A})}$ containing those vectors x for which

$$\int |f(\lambda)|^2\, d\|\mathbf{E}_\lambda x\|^2 < \infty \tag{7.74}$$

The *resolvent* $R_\alpha(\mathbf{A})$ of the linear operator $\mathbf{A}$ is defined by

$$R_\alpha(\mathbf{A}) \equiv (\mathbf{A} - \alpha\mathbf{1})^{-1} \tag{7.75}$$

If $\mathbf{A}$ is a differential operator, the resolvent $R_\alpha(\mathbf{A})$ is often known as a *Green function*. The resolvent $R_\alpha(\mathbf{A})$ is bounded unless α is in the spectrum of $\mathbf{A}$. If α is in the point spectrum $\Pi(\mathbf{A})$ of $\mathbf{A}$ then $(\mathbf{A} - \alpha\mathbf{1})$ is singular, and $R_\alpha(\mathbf{A})$ can only be defined on $\mathcal{M}_\alpha^\perp$, where $\mathcal{M}_\alpha$ is the eigenmanifold of $\mathbf{A}$ belonging to eigenvalue α. If α is in the continuous spectrum $\Sigma_{\mathrm{c}}(\mathbf{A})$ of $\mathbf{A}$, then $R_\alpha(\mathbf{A})$ is unbounded, but defined on a dense subset of $\mathcal{H}$.

If $\mathbf{A}$ is self-adjoint, with resolution of the identity $\{\mathbf{E}_\lambda\}$, then the resolvent $R_\alpha(\mathbf{A})$ is given formally by

$$R_\alpha(\mathbf{A}) = \int \frac{1}{\lambda - \alpha}\, d\mathbf{E}_\lambda \tag{7.76}$$

This expression will be used to provide a more explicit construction of the resolvent operators for certain differential operators to appear soon.

If $\mathbf{A}$ is a normal operator, then it can be written in the form $\mathbf{A} = \mathbf{X} + i\mathbf{Y}$, with $\mathbf{X}$ and $\mathbf{Y}$ self-adjoint and $[\mathbf{X}, \mathbf{Y}] = \mathbf{0}$. If

$$\mathbf{X} = \int \lambda\, d\mathbf{E}_\lambda, \quad \mathbf{Y} = \int \mu\, d\mathbf{F}_\mu \tag{7.77}$$

then $[\mathbf{E}_\lambda, \mathbf{F}_\mu] = \mathbf{0}$ for every λ, μ. If $\mathbf{U}$ is a unitary operator, then there is a resolution of the identity $\mathbf{H}_\phi$ with interval of variation $\subseteq [0, 2\pi]$ such that

$$\mathbf{U} = \int e^{i\phi}\, d\mathbf{H}_\phi \tag{7.78}$$

Now that we have reviewed the abstract properties of linear operators on a Hilbert space, it is time to turn to the concrete examples of linear differential and integral operators that appear in practically every branch of physics.

7.4   Linear Differential Operators

7.4.1   Differential Operators and Boundary Conditions

Linear differential operators of the form

$$\mathbf{L} \equiv p_0(x) \frac{d^n}{dx^n} + p_1(x) \frac{d^{n-1}}{dx^{n-1}} + \cdots + p_{n-1}(x) \frac{d}{dx} + p_n(x) \tag{7.79}$$

were introduced in Chapter 5 (see Eq. (5.50)) and described there from an analytical point of view. Now we consider $\mathbf{L}$ as a linear operator on a Hilbert space of the type $L_2(a, b)$. Note that $\mathbf{L}$ is unbounded, since the derivative of a smooth function can be arbitrarily large. The formal domain of $\mathbf{L}$ as a differential operator includes the class $C^n(a, b)$ of functions that are continuous together with their first n derivatives on the interval (a, b), and $C^n(a, b)$ is everywhere dense in $L_2(a, b)$.

But to consider $\mathbf{L}$ as a Hilbert space operator, we must define the domain of $\mathbf{L}$ more precisely. In particular, we must carefully specify the boundary conditions to be imposed on functions included in the Hilbert space domain of $\mathbf{L}$ in order to verify properties, such as self-adjointness, that are critical to understanding the spectrum of $\mathbf{L}$. Indeed, we will see from the examples given here that the spectrum can be more sensitive to the boundary conditions than to the operator itself.

The simplest linear differential operator is the derivative $\mathbf{D}$ defined by

$$\mathbf{D}f(x) = \frac{d}{dx} f(x) \tag{7.80}$$

for functions that are differentiable on $a \le x \le b$, and extended by closure. Now integration by parts gives

$$
\begin{aligned}
(g, \mathbf{D}f) &= \int_a^b g^*(x) \frac{d}{dx} f(x) \, dx \\
&= [g^*(b)f(b) - g^*(a)f(a)] - \int_a^b f(x) \frac{d}{dx} g^*(x) \, dx \\
&= -(\mathbf{D}g, f) + B(g, f)
\end{aligned}
\tag{7.81}
$$

where the boundary term $B(g, f)$, also known as the *bilinear concomitant*,

$$B(g, f) \equiv g^*(b)f(b) - g^*(a)f(a) \tag{7.82}$$

contains the endpoint terms from the integration by parts.

In order to make a self-adjoint operator from $\mathbf{D}$, we first need to fix the minus sign on the right-hand side of Eq. (7.81). To do this, simply define

$$\mathbf{P} \equiv -i\mathbf{D} \tag{7.83}$$

(the factor $-i$ is chosen so that $\mathbf{P}$ can be identified with the usual momentum operator in quantum mechanics for appropriate boundary conditions). Then we need to choose boundary conditions on the values of the functions at the endpoints to make the boundary term $B(g, f)$ vanish identically. This can be arranged by imposing the *periodic boundary condition*

$$f(b) = f(a) \tag{7.84}$$

With this boundary condition, $\mathbf{P}$ is a self-adjoint operator on $L_2(a, b)$.

Remark.   More generally, it is sufficient to require

$$f(b) = \alpha f(a) \tag{7.85}$$

with $|\alpha| = 1$. Thus α has the form $\alpha = \exp(2\pi i\xi)$ and there is a continuous family $\mathbf{P}_\xi$ ($0 \le \xi \le 1$) of self-adjoint extensions of the linear operator $-i\mathbf{D}$. Which extension is appropriate for a particular physical problem must be determined by the physics. The boundary condition (7.84), corresponding to $\alpha = 1$, is most often physically correct, but there are fermionic systems for which (7.85) with $\alpha = -1$ is relevant.   □

With the periodic boundary condition (7.84), $\mathbf{P}$ has a complete orthonormal system $\{\phi_n(x)\}$ of eigenvectors given by

$$\phi_n(x) = \frac{1}{\sqrt{b-a}} \exp\left(\frac{2n\pi i x}{b-a}\right) \tag{7.86}$$

($n = 0, \pm 1, \pm 2, \ldots$) belonging to eigenvalues

$$\lambda_n = \frac{2n\pi}{b-a} \tag{7.87}$$

This analysis must be modified if $a \to -\infty$ or $b \to \infty$. $\mathbf{P}$ is already self-adjoint on $L_2(-\infty, \infty)$; the boundary term $B(g, f)$ vanishes automatically, since $f(x) \to 0$ for $x \to \pm\infty$. But the discrete eigenvalues merge into a continuous spectrum, and the complete orthonormal system (7.86) of eigenvectors must be replaced by a continuous resolution of the identity (see Exercise 7.2). On a semi-infinite interval ($L_2(0, \infty)$ for example) the behavior of $\mathbf{P}$ is more complicated; see Problem 12.

→ **Exercise 7.3.**   Consider the linear operator $\mathbf{P}_\xi$ defined by

$$\mathbf{P}_\xi \, \phi(x) \equiv \frac{1}{i} \frac{d}{dx} \phi(x)$$

for differentiable functions $\phi(x)$ in $L_2[-\pi, \pi]$ that satisfy the boundary condition

$$\phi(\pi) = e^{2\pi i\xi} \, \phi(-\pi)$$

with ξ real. Find the eigenvalues of $\mathbf{P}_\xi$, and construct a complete orthonormal system of eigenvectors. Then characterize the domain of the closure of $\mathbf{P}_\xi$ in terms of these eigenvectors. Is $\mathbf{P}_\xi$ self-adjoint for any (real) ξ?   □

7.4.2   Second-Order Linear Differential Operators

The general second-order linear differential operator on $L_2(a, b)$ has the form

$$\mathbf{L} = p(x)\frac{d^2}{dx^2} + q(x)\frac{d}{dx} + r(x) \tag{7.88}$$

Here integration by parts gives

$$(g, \mathbf{L}f) = \int_a^b g^*(x)\mathbf{L}[f(x)]dx = \int_a^b (\tilde{\mathbf{L}}[g(x)])^* f(x)dx + B(g, f) \tag{7.89}$$

where the *adjoint differential operator* $\tilde{\mathbf{L}}$ is defined by

$$\tilde{\mathbf{L}}[g(x)] = \frac{d^2}{dx^2}[p^*(x)g(x)] - \frac{d}{dx}[q^*(x)g(x)] + r^*(x)g(x) \tag{7.90}$$

and the boundary term is given by

$$B(g, f) = p(x)\left[g^*(x)f'(x) - g'^*(x)f(x)\right]\Big|_a^b + g^*(x)\left[q(x) - p'(x)\right]f(x)\Big|_a^b \tag{7.91}$$

For $\mathbf{L}$ to be symmetric ($\tilde{\mathbf{L}} = \mathbf{L}$) we have the requirements:
  (i) $p^*(x) = p(x)$ ($p(x)$ must be real),
  (ii) $p'(x) = \operatorname{Re} q(x)$ and
  (iii) $\operatorname{Im} q'(x) = 2\operatorname{Im} r(x)$.
If these conditions are satisfied, then $\mathbf{L}$ has the form

$$\mathbf{L} = \frac{d}{dx}\left[p(x)\frac{d}{dx}\right] + i\xi(x)\frac{d}{dx}\left[\xi(x)\cdot\right] + s(x) \tag{7.92}$$

with $p(x)$, $\xi(x) = \sqrt{\operatorname{Im} q(x)}$ and $s(x) = \operatorname{Re} r(x)$ real functions. The term containing $\xi(x)$ is absent if the operator $\mathbf{L}$ is real, which is often the case in practice.

To consider the boundary term let us suppose that $\mathbf{L}$ is real, to simplify the discussion. Then we have $p'(x) = q(x)$ and the boundary term is simply

$$B(g, f) = p(x)\left[g^*(x)f'(x) - g'^*(x)f(x)\right]\Bigg|_a^b \tag{7.93}$$

If $p(x) = 0$ at an endpoint, the differential equation is singular at the endpoint, but we can require that the solution be regular at that endpoint, which eliminates the contribution of that endpoint to the boundary term. If $p(x) \neq 0$ at the endpoint $x = \tau$ ($\tau = a, b$), then the endpoint contribution to $B(g, f)$ from $x = \tau$ will vanish if both $f(x)$ and $g(x)$ satisfy a boundary condition of the form

$$cf(\tau) + c'f'(\tau) = 0 \tag{7.94}$$

with c, c' being constants, one of which may vanish.

The operator **L** in the form

$$\mathbf{L} = \frac{d}{dx}\left[p(x)\frac{d}{dx}\right] + s(x) \tag{7.95}$$

is a *Sturm–Liouville* operator, and the boundary conditions required to make **L** self-adjoint are precisely those encountered in physical systems governed by the Laplace equation or the wave equation, as will be seen shortly.

Remark.   Note that the differential operator **L** can be multiplied by a function $\rho(x)$ without changing the solutions to the differential equation. Thus an operator that is not symmetric at first sight may in fact be converted to one which is by an inspired choice of multiplying factor $\rho(x)$. For examples, see Problems 13 and 14.   □

One important tool to analyze the spectrum of a Sturm–Liouville operator **L**, or any linear differential operator for that matter, is to find, or prove the existence of, a linear integral operator **K** such that $\mathbf{LK} = 1$. For if the integral operator **K** is compact (which it certainly will be if the domain of integration is bounded) and normal, then it has a complete orthonormal system of eigenvectors belonging to discrete eigenvalues λ_n, and these eigenvectors will also be eigenvectors of **L**, with eigenvalues $1/\lambda_n$. We now turn to the study of integral operators.

7.5   Linear Integral Operators; Green Functions

7.5.1   Compact Integral Operators

Consider the function space $L_2(\Omega)$ with Ω being a bounded region in $\mathbf{R}^n$ (often Ω is just an interval $[a, b]$ in one dimension). Suppose $K(x, y)$ is a function on $\Omega \times \Omega$ such that

$$\|\mathbf{K}\|^2 = \int_\Omega \int_\Omega |K(x,y)|^2 \, dx \, dy < \infty \tag{7.96}$$

Then corresponding to $K(x, y)$ is a linear operator **K** on $L_2(\Omega)$ defined by

$$g(x) \equiv (\mathbf{K}f)(x) = \int_\Omega K(x,y)f(y) \, dy \tag{7.97}$$

K is an *integral operator*; $K(x, y)$ is the *kernel*[7] corresponding to **K**. The condition (7.96) insures that **K** is a Hilbert–Schmidt operator, and hence compact.

Remark.   Other operators can be expressed as integral operators, even if they do not satisfy the condition (7.96). For example, the identity operator 1 can be represented as an integral operator with kernel

$$K(x,y) = \delta(x - y) \tag{7.98}$$

where $\delta(x-y)$ is the Dirac δ-function, even though this kernel $K(x, y)$ does not satisfy (7.96), and the representation is valid only on the linear manifold of functions that are continuous on

---

[7]The term "kernel" introduced here is not to be confused with the kernel of an operator introduced in Chapter 2 as the linear manifold transformed into the zero vector by the operator. The double usage is perhaps unfortunate, but confusion can be avoided by paying attention to context.

Ω. Note, however that this manifold is dense in $L_2(\Omega)$. Integral operators are also useful when defined on unbounded regions, when (7.96) need not be satisfied even by bounded continuous kernels $K(x, y)$.   □

The adjoint operator $\mathbf{K}^\dagger$ is given by

$$(\mathbf{K}^\dagger f)(x) = \int_\Omega K^*(y, x) f(y) \, dy \tag{7.99}$$

which corresponds to the expression for the matrix elements of the adjoint operator given in Eq. (2.2.88). Thus $\mathbf{K}$ is self-adjoint if and only if

$$K(y, x) = K^*(x, y) \tag{7.100}$$

almost everywhere.

If $\mathbf{K}$ is self-adjoint, and compact due to condition (7.96), then we know that it has a complete orthonormal system $\{\phi_n(x)\}$ of eigenvectors belonging to discrete real eigenvalues, and the sequence $\{\lambda_n\}$ of eigenvalues converges to zero. This leads to the formal representation

$$K(x, y) = \sum_n \lambda_n \phi_n(x) \phi_n^*(y) \tag{7.101}$$

which is valid in the sense of strong $L_2(\Omega \times \Omega)$ convergence.

Somewhat stronger results can be derived if the kernel $K(x, y)$ is continuous, or even if the *iterated kernel*

$$K^2(x, y) \equiv \int_\Omega K(x, z) \, K(z, y) \, dz \tag{7.102}$$

which corresponds to the operator $\mathbf{K}^2$, is a continuous function of x in Ω for fixed y, or even if $K^2(x, x)$ is continuous in x. In that case

$$K^2(x, y) = \sum_n \lambda_n^2 \phi_n(x) \phi_n^*(y) \tag{7.103}$$

is uniformly convergent in $\Omega \times \Omega$, and any function $g(x)$ in the range of $\mathbf{K}$ has an expansion

$$g(x) = \sum_n c_n \phi_n(x) \tag{7.104}$$

that is uniformly convergent in Ω.

These results follow from the continuity of $K^2(x, y)$, which implies that eigenvectors $\{\phi_n(x)\}$ of $\mathbf{K}$ belonging to nonzero eigenvalues are continuous functions of x. To see this, suppose $\mathbf{K}\phi = \lambda\phi$ with $\lambda \neq 0$. Then

$$\lambda\phi(x) = \int_\Omega K(x, y)\phi(y) \, dy \tag{7.105}$$

and

$$\lambda \left[\phi(x) - \phi(x')\right] = \int_\Omega \left[K(x, y) - K(x', y)\right] \phi(y) \, dy \tag{7.106}$$

The Schwarz inequality then gives

$$\lambda^2 \left| \phi(x) - \phi(x') \right|^2 \leq \|\phi\|^2 \int_\Omega |K(x, y) - K(x', y)|^2 \, dy \tag{7.107}$$

and the right-hand side is continuous, hence vanishes when $x \to x'$.

Equation (7.101) gives the expansion of $K(x, y)$ with respect to the complete orthonormal system $\{\phi_m(x)\phi_n^*(y)\}$ on $\Omega \times \Omega$. Expansion in terms of y alone has the form

$$K(x, y) = \sum_n f_n(x)\phi_n^*(y) \tag{7.108}$$

when $f_n(x) = \lambda_n \phi_n(x)$ follows from the fact that the $\{\phi_n(y)\}$ are eigenvectors of $\mathbf{K}$. Then also

$$K^2(x, x) = \int_\Omega |K(x, y)|^2 \, dy = \sum_n \lambda_n^2 |\phi_n(x)|^2 \tag{7.109}$$

and the convergence is uniform since the limit is continuous.[8] The uniform convergence in Eq. (7.103) then follows using the Schwarz inequality again.

7.5.2   Differential Operators and Green Functions

Now suppose $\mathbf{L}$ is a self-adjoint linear differential operator with boundary conditions that allow $\mathbf{L}$ to be defined as a self-adjoint operator on $L_2(\Omega)$. If λ is a real constant not in the spectrum of $\mathbf{L}$, consider the inhomogeneous equation

$$(\mathbf{L} - \lambda\mathbf{1})\, u(x) = f(x) \tag{7.110}$$

The solution to Eq. (7.110) has the general form

$$u(x) = \int_\Omega G_\lambda(x, y) f(y) \, dy \tag{7.111}$$

where $G_\lambda(x, y)$, the *Green function* for $(\mathbf{L} - \lambda\mathbf{1})$, is the solution of the formal differential equation

$$(\mathbf{L} - \lambda\mathbf{1})_x \, G_\lambda(x, y) = \delta(x - y) \tag{7.112}$$

satisfying the appropriate boundary conditions when x is on the boundary of Ω. That the solution has the form (7.111) is clear from the linearity of $\mathbf{L}$, and Eq. (7.112) is a formal expression of the linearity.

What is less obvious, but also true, is that $G_\lambda(x, y)$ satisfies the symmetry condition

$$G_\lambda(x, y) = G_\lambda^*(y, x) \tag{7.113}$$

---

[8]This is Dini's theorem in real analysis.

This is plausible, since $G_\lambda(x, y)$ is the kernel of an integral operator $\mathbf{G}_\lambda$ that is the inverse of a self-adjoint operator $(\mathbf{L} - \lambda\mathbf{1})$. Rather than give a more rigorous proof, we will illustrate the construction of the Green function $G_\lambda(x, y)$ for various differential operators in the examples and problems.

The main significance of the existence of the self-adjoint inverse $\mathbf{G}_\lambda$ of $\mathbf{L} - \lambda\mathbf{1}$ is that $\mathbf{G}_\lambda$ is compact so long as the region Ω is bounded. Hence it has a complete orthonormal system $\{\phi_n\}$ of eigenvectors, with corresponding discrete eigenvalues $\{\gamma_n\}$ forming a sequence converging to zero. Then the differential operator $\mathbf{L} - \lambda\mathbf{1}$ has the same eigenvectors, with

$$(\mathbf{L} - \lambda\mathbf{1})\phi_n = (1/\gamma_n)\phi_n \tag{7.114}$$

and $\mathbf{L}$ itself has the same eigenvectors $\{\phi_n\}$, with eigenvalues

$$\lambda_n = \lambda + \frac{1}{\gamma_n} \tag{7.115}$$

obtained directly from the eigenvalues of $\mathbf{G}_\lambda$. The consequences of this discussion are important enough to state them as a formal theorem:

Theorem 7.3. Let $\mathbf{L}$ be a self-adjoint linear differential operator on the space $L_2(\Omega)$, with Ω a bounded region. Then

1. $\mathbf{L}$ has a complete orthonormal system $\{\phi_n\}$ of eigenvectors belonging to discrete eigenvalues $\{\lambda_n\}$ such that the sequence $\{1/\lambda_n\}$ converges to zero; hence the sequence $\{\lambda_n\}$ is unbounded.

2. For every λ not in the spectrum of $\mathbf{L}$, the inhomogeneous equation (7.110) has the solution

$$f(x) = (\mathbf{L} - \lambda\mathbf{1})^{-1}u(x) = \int_\Omega G_\lambda(x, y)f(y)\, dy \tag{7.116}$$

with Green function $G_\lambda(x, y)$ in $L_2(\Omega \times \Omega)$. The Green function has the expansion

$$G_\lambda(x, y) = \sum_n \frac{\phi_n(x)\phi_n^*(y)}{\lambda_n - \lambda} \tag{7.117}$$

in terms of the normalized eigenvectors $\{\phi_n\}$ of $\mathbf{L}$.

❑ **Example 7.9.** Consider the differential operator

$$\Delta \equiv \frac{d^2}{dx^2} \tag{7.118}$$

on $L^2(0, 1)$, defined on functions $u(x)$ that satisfy the boundary conditions

$$u(0) = 0 = u(1) \tag{7.119}$$

These boundary conditions define Δ as a self-adjoint linear operator (other self-adjoint boundary conditions appear in Problem 15). Since $\lambda = 0$ is not an eigenvalue of Δ (show this), we can find the Green function $G_0(x, y)$ for Δ with these boundary conditions as a solution of

$$\frac{\partial^2}{\partial x^2} G_0(x, y) = -\delta(x - y)$$

(7.120)

(the minus sign here is conventional), with

$$G_0(0, y) = 0 = G_0(1, y)$$

(7.121)

For $x \neq y$, Eq. (7.120) together with symmetry requires

$$G(x, y) = \begin{cases} ax(1 - y) & x < y \\ a(1 - x)y & x > y \end{cases}$$

(7.122)

while as $x \to y$, it requires

$$\lim_{\varepsilon \to 0^+} \left\{ \left[\frac{\partial G_0}{\partial x} \right]_{x=y+\varepsilon} - \left[\frac{\partial G_0}{\partial x} \right]_{x=y-\varepsilon} \right\} = -1$$

(7.123)

Thus we need $a(1 - y) + ay = 1$, whence the constant $a = 1$.

For $\lambda = -k^2 \neq 0$, the Green function $G_\lambda(x, y)$ must satisfy

$$\left(\frac{\partial^2}{\partial x^2} + k^2 \right) G_\lambda(x, y) = -\delta(x - y)$$

(7.124)

with the same boundary conditions as above. Then we have

$$G_\lambda(x, y) = \begin{cases} A \sin kx \, \sin k(1 - y) & x < y \\ A \sin k(1 - x) \sin ky & x > y \end{cases}$$

(7.125)

and the condition (7.123) now requires

$$Ak \left[\cos ky \, \sin k(1 - y) + \cos k(1 - y) \, \sin ky \right] = Ak \sin k = 1$$

(7.126)

This uniquely determines A if $\sin k \neq 0$, and we have

$$G_\lambda(x, y) = \begin{cases} \dfrac{\sin kx \, \sin k(1 - y)}{k \sin k} & x < y \\ \dfrac{\sin k(1 - x) \sin ky}{k \sin k} & x > y \end{cases}$$

(7.127)

when $\sin \sqrt{-\lambda} \neq 0$.   ∎

Remark. Note that the eigenvalues of Δ have the form

$$\lambda_n = -n^2 \pi^2$$

(7.128)

with corresponding normalized eigenfunctions

$$\phi_n(x) = \sqrt{2}\sin n\pi x \tag{7.129}$$

$(n = 1, 2, \ldots)$. Thus $\sin k = 0$ corresponds to a value of λ in the spectrum of Δ, except for the case $\lambda = 0$ $(k = 0)$ which is *not* an eigenvalue for the boundary condition given here. Note also that

$$\lim_{\lambda \to 0} G_\lambda(x, y) = G_0(x, y) \tag{7.130}$$

The expansion (7.117) then implies

$$\sum_{n=1}^{\infty} \frac{\sin n\pi x \, \sin n\pi y}{n^2\pi^2 - k^2} = \frac{\sin kx \, \sin k(1 - y)}{2k\sin k} \tag{7.131}$$

for $x < y$, with the corresponding result for $x > y$ obtained by interchanging x and y. An independent verification of this result follows from the integral

$$\int_0^1 G_\lambda(x, y)\sin n\pi y \, dy = \frac{\sin n\pi x}{n^2\pi^2 - k^2} \tag{7.132}$$

which is obtained by a straightforward but slightly long calculation.   □

Bibliography and Notes

The book on Hilbert space by Halmos as well as others cited in Chapter 6 gives further details of the mathematical properties of linear operators on Hilbert space. The standard textbook by Byron and Fuller cited in Chapter 2 emphasizes linear vector spaces.

There are many modern books on quantum mechanics that discuss the essential connection between linear vector spaces and quantum mechanics. Two good introductory books are

David J. Griffiths, *Introduction to Quantum Mechanics* (2nd edition), Prentice-Hall (2004).

Ramamurti Shankar, *Principles of Quantum Mechanics* (2nd edition), Springer (2005)

Slightly more advanced but still introductory is

Eugen Merzbacher, *Quantum Mechanics* (3rd edition), Wiley (1997).

Two classics that strongly reflect the original viewpoints of their authors are

Paul A. M. Dirac, *The Principles of Quantum Mechanics* (4*th* edition), Clarendon Press, Oxford (1958)

John von Neumann, *Mathematical Foundations of Quantum Mechanics*, Princeton University Press (1955).

Dirac's work describes quantum mechanics as the mathematics flows from his own physical insight, while von Neumann presents an axiomatic formulation based on his deep understanding of Hilbert space theory. Both are important works for the student of the historical development of the quantum theory.

Problems

1. Let $\{\phi_n\}$ $(n = 1, 2, \ldots)$ be a complete orthonormal system in the (infinite-dimensional) Hilbert space $\mathcal{H}$. Consider the operators $\mathbf{U}_k$ defined by

   $$\mathbf{U}_k \phi_n = \phi_{n+k}$$

   $(k = 1, 2, \ldots)$.

   (i) Give an explicit form for $\mathbf{U}_k^\dagger$.

   (ii) Find the eigenvalues and eigenvectors of $\mathbf{U}_k$ and $\mathbf{U}_k^\dagger$.

   (iii) Discuss the convergence of the sequences $\{\mathbf{U}_k\}$, $\{\mathbf{U}_k^\dagger\}$, $\{\mathbf{U}_k^\dagger \mathbf{U}_k\}$, $\{\mathbf{U}_k \mathbf{U}_k^\dagger\}$.

2. Let $\{\phi_n\}$ $(n = 1, 2, \ldots)$ be a complete orthonormal system in $\mathcal{H}$, and define the linear operator $\mathbf{T}$ by

   $$\mathbf{T}\phi_n \equiv n\phi_{n+1} \quad (n = 1, 2, \ldots)$$

   (i) What is the domain of $\mathbf{T}$?

   (ii) How does $\mathbf{T}^\dagger$ act on $\{\phi_1, \phi_2, \ldots\}$? What is the domain of $\mathbf{T}^\dagger$?

   (iii) Find the eigenvalues and eigenvectors of $\mathbf{T}$.

   (iv) Find the eigenvalues and eigenvectors of $\mathbf{T}^\dagger$.

3. Let $\{\phi_n\}$ $(n = 0, \pm 1, \pm 2, \ldots)$ be a complete orthonormal set in the Hilbert space $\mathcal{H}$. Consider the operator $\mathbf{A}$ defined by

   $$\mathbf{A}\phi_n = a(\phi_{n+1} + \phi_{n-1})$$

   (i) Is $\mathbf{A}$ bounded? compact? self-adjoint?

   (ii) Find the spectrum of $\mathbf{A}$.

   Hint. Try to find an explicit representation of $\mathbf{A}$ on the function space $L^2(-\pi, \pi)$.

4. Suppose we add a "small" perturbation to the operator $\mathbf{A}$ in the preceding problem. Let

   $$\mathbf{B} \equiv \mathbf{A} + \mathbf{V} \equiv \mathbf{A} + \xi \mathbf{P}_0$$

   where $\mathbf{P}_0$ projects onto the linear manifold spanned by ϕ_0, and ξ is a real constant that may be either positive or negative.

   (i) Is $\mathbf{B}$ bounded? compact? self-adjoint?

   (ii) Find the spectrum of $\mathbf{B}$.

5. Let $\{\phi_n\}$ $(n = 0, \pm 1, \pm 2, \ldots)$ be a complete orthonormal set in the Hilbert space $\mathcal{H}$. Consider the operators $\mathbf{U}_N$ defined by

   $$\mathbf{U}_N \phi_n = \phi_{n+N}$$

($N = \pm 1, \pm 2, \dots$). These are *not* the same operators as in Problem 1, since the range of indices on the complete orthonormal system is different!

(i) Give an explicit form for $\mathbf{U}_N^\dagger$.

(ii) Is $\mathbf{U}_N$ bounded, compact, unitary?

(iii) Find the spectra of $\mathbf{U}_N$ and $\mathbf{U}_N^\dagger$.

(iv) Find a unitary operator $\mathbf{S}_N$ such that $\mathbf{U}_N^\dagger = \mathbf{S}_N \mathbf{U}_N \mathbf{S}_N^\dagger$. Is $\mathbf{S}_N$ unique?

(v) Discuss the convergence of the sequences $\{\mathbf{U}_N\}$, $\{\mathbf{U}_N^\dagger\}$, $\{\mathbf{U}_N^\dagger \mathbf{U}_N\}$ and $\{\mathbf{U}_N \mathbf{U}_N^\dagger\}$ in the limit $N \to \infty$.

Hint. Again, try to find an explicit representation of $\mathbf{U}$ on the function space $L^2(-\pi, \pi)$.

6. Consider the three-dimensional Lie algebra spanned by the linear operators $\bar{\mathbf{A}}$, $\mathbf{A}$ and $\mathbf{1}$, satisfying the commutation relations

$$[\mathbf{A}, \mathbf{1}] = \mathbf{0} = [\bar{\mathbf{A}}, \mathbf{1}], \quad [\mathbf{A}, \bar{\mathbf{A}}] = \mathbf{1}$$

Remark. Strictly speaking, the last commutator should be written as $[\mathbf{A}, \bar{\mathbf{A}}] \subset \mathbf{1}$ since the domain of the commutator cannot be the entire space $\mathcal{H}$. There is a general theorem, which we will not prove here, that if $[\mathbf{B}, \mathbf{C}]$ is a constant, then $\mathbf{B}$ and $\mathbf{C}$ must be unbounded. Note that the results found here are consistent with the theorem. □

(i) Show that the operator $\mathbf{N} \equiv \bar{\mathbf{A}}\mathbf{A}$ satisfies the commutation relations

$$[\mathbf{N}, \mathbf{A}] = -\mathbf{A}, \quad [\mathbf{N}, \bar{\mathbf{A}}] = \bar{\mathbf{A}}$$

(ii) Suppose ϕ_λ is an eigenstate of $\mathbf{N}$ with $\mathbf{N}\phi_\lambda = \lambda\phi_\lambda$. Show that

$$\mathbf{N}(\mathbf{A}\phi_\lambda) = (\lambda - 1)\mathbf{A}\phi_\lambda$$

$$\mathbf{N}(\bar{\mathbf{A}}\phi_\lambda) = (\lambda + 1)\bar{\mathbf{A}}\phi_\lambda$$

Remark. Thus $\bar{\mathbf{A}}$ and $\mathbf{A}$ are *ladder operators* (or *raising* and *lowering operators* for $\mathbf{N}$, since they raise ($\bar{\mathbf{A}}$) or lower ($\mathbf{A}$) the eigenvalue of $\mathbf{N}$ by 1. In a quantum mechanical context, $\mathbf{N}$ is an operator representing the number of quanta of some harmonic oscillator or normal mode of a field (the electromagnetic field, for example), in which case the operators represent *creation* ($\bar{\mathbf{A}}$) and *annihilation* ($\mathbf{A}$) *operators* for the quanta. □

(iii) Since $\mathbf{N} \geq 0$, the lowest eigenvalue of $\mathbf{N}$ must be $\lambda = 0$, and thus the eigenvalues of $\mathbf{N}$ are $\lambda = 0, 1, 2, \dots$, with unit eigenvectors ϕ_n corresponding to $\lambda = n$. Show that

$$\|\mathbf{A}\phi_n\| = \sqrt{n}, \quad \|\bar{\mathbf{A}}\phi_n\| = \sqrt{n+1}$$

and then that phases can be chosen so that

$$\mathbf{A}\phi_n = \sqrt{n}\,\phi_{n-1}, \quad \bar{\mathbf{A}}\phi_n = \sqrt{n+1}\,\phi_{n+1}$$

Remark. The creation and annihilation operators introduced in this problem satisfy commutation rules, and the quanta they create obey Bose–Einstein statistics. Hence they are known as *bosons*. In our world, there are also spin-$\frac{1}{2}$ particles (electrons, protons, neutrons, quarks, etc.) for which the number of quanta present in a specific state can only be zero or 1 (this is often stated as the *Pauli exclusion principle*). These particles obey Fermi–Dirac statistics, and hence are known as *fermions*. A formalism that incorporates the Pauli principle introduces creation and annihilation operators for fermions that satisfy anticommutation rules, rather than the commutation rules for boson operators—see Problem 8. □

7. Suppose $\mathbf{A}$ is a linear operator whose commutator with its adjoint is given by

$$\left[\mathbf{A}, \mathbf{A}^\dagger\right] \equiv \mathbf{A}\mathbf{A}^\dagger - \mathbf{A}^\dagger\mathbf{A} = 1$$

as in the preceding problem (where $\bar{\mathbf{A}}$ is identified with $\mathbf{A}^\dagger$).

(i) Show that $\mathbf{A}$ has a normalized eigenvector ψ_α for any complex number α, and find the expansion of the ψ_α in terms of the ϕ_n.

(ii) Show that any vector x in $\mathcal{H}$ can be expanded in terms of the normalized ϕ_α as

$$x = \int \psi_\alpha \left(\psi_\alpha, x\right) \rho(\alpha) \, d^2\alpha$$

for suitable weight function $\rho(\alpha)$, and find $\rho(\alpha)$.

Remark. As noted above, $\mathbf{N}$ corresponds in quantum theory to the number of quanta of some harmonic oscillator, and $\mathbf{A}$ and $\mathbf{A}^\dagger$ act as *annihilation* and *creation operators* for the quanta. The eigenvectors ψ_α are *coherent states* of the oscillator, corresponding in some sense to classical oscillations with complex amplitude α. See the quantum mechanics books cited in the notes for more discussion of these states. □

(iii) Find the eigenvectors of $\mathbf{A}^\dagger$.

8. Let $a_\sigma^\dagger$, a_σ be the creation and annihilation operators for a spin-$\frac{1}{2}$ fermion in spin state σ [$\sigma = \pm\frac{1}{2}$, or spin up ($\uparrow$) and spin down ($\downarrow$)]. These satisfy the anticommutation rules

$$\{a_\alpha^\dagger, a_\beta\} = a_\alpha^\dagger a_\beta + a_\beta a_\alpha^\dagger = \delta_{\alpha\beta}\mathbf{1}$$

$$\{a_\alpha^\dagger, a_\beta^\dagger\} = 0 = \{a_\alpha, a_\beta\}$$

(i) Show that the number operator

$$\mathbf{N}_\alpha \equiv a_\alpha^\dagger a_\alpha$$

satisfies

$$\left[\mathbf{N}_\alpha, a^\dagger\right] = -\delta_{\alpha\beta}a_\beta, \quad \left[\mathbf{N}_\alpha, a_\beta^\dagger\right] = \delta_{\alpha\beta}a_\beta^\dagger$$

(ii) Show that $\mathbf{N}_\alpha$ has eigenvalues $0, 1$.

Remark. Hence the Pauli exclusion principle follows from the anticommutation rules. □

(iii) From (ii) it follows that a basis for the fermion states is the set

$$|0\rangle \qquad |\uparrow\rangle \equiv a_\uparrow^\dagger |0\rangle \qquad |\downarrow\rangle \equiv a_\downarrow^\dagger |0\rangle \qquad |\uparrow\downarrow\rangle \equiv a_\uparrow^\dagger |\downarrow\rangle$$

Write down the matrices representing the operators a_α, $a_\alpha^\dagger$, and N_α in this basis.

Remark.   Both here and in Problem 6, the creation and annihilation operators are those for a single mode of the relevant particle. For physical particles, there are many modes, labeled by momentum, energy and other possible quantum labels, and we will have creation and annihilation operators defined for each such mode.   □

9. Let $\{A_k\}$ and $\{B_k\}$ be two sequences of operators on the (infinite-dimensional) Hilbert space $\mathcal{H}$ such that

$$\{A_k\} \to A \text{ and } \{B_k\} \to B$$

(i) Show that $\{A_k B_k\} \to AB$.

(ii) If $\{A_k\} \to A$ and $\{B_k\} \rightharpoonup B$, what can you say about the convergence of the sequence $\{A_k B_k\}$?

(iii) If $\{A_k\} \rightharpoonup A$ and $\{B_k\} \rightharpoonup B$, what can you say about the convergence of the sequence $\{A_k B_k\}$?

10. Consider a pair of operators Q and P that satisfy the commutation rules

$$[Q, P] \equiv QP - PQ = \varepsilon 1$$

Remark.   Again, this should be written $[Q, P] \subset \varepsilon 1$ to be precise.   □

(i) Show that

$$[Q, P^n] = n\varepsilon P^{n-1}$$

$$[Q^n, P] = n\varepsilon Q^{n-1}$$

(ii) Show that if A, B are two operators that are polynomials in Q and P, then we have

$$[A, B] = \varepsilon \left\{ \frac{\partial A}{\partial Q} \frac{\partial B}{\partial P} - \frac{\partial B}{\partial Q} \frac{\partial A}{\partial P} \right\}$$

to lowest order in ε.

11. Consider the linear operator

$$P \equiv \left[\frac{1}{i} \frac{d}{dx} \right]$$

defined on $L^2(-\pi, \pi)$ (here the bracket [ ] denotes the closure of the operator). The functions $\phi_n(x) \equiv e^{inx}/\sqrt{2\pi}$ $(n = 0, \pm 1, \pm 2, \ldots)$ are eigenvectors of $\mathbf{P}$, with

$$\mathbf{P}\,\phi_n(x) = n\phi_n(x)$$

However, it also appears that the functions

$$\phi_\lambda(x) \equiv \frac{1}{\sqrt{2\pi}}\, e^{i\lambda x}$$

are eigenvectors of $\mathbf{P}$ for any complex λ. This seems puzzling, since we know from the theory of Fourier series that the $\{\phi_n(x)\}$ with integer n form a complete orthonormal system of eigenvectors of $\mathbf{P}$. To clarify the puzzle, first calculate the coefficients $c_n(\lambda)$ in the expansion

$$\phi_\lambda(x) = \sum_{n=-\infty}^{\infty} c_n(\lambda)\phi_n(x)$$

Then find the expansion of $\mathbf{P}\,\phi_\lambda(x)$. Is this the expansion of a vector in the Hilbert space? Explain and discuss.

12. The linear operator

$$\mathbf{A} = \left[\frac{1}{i}\frac{d}{dx}\right]$$

is symmetric on $L_2(0, \infty)$ when restricted to functions $f(x)$ for which $f(0) = 0$. However, $\mathbf{A}^\dagger$ is defined on a larger domain, whence $\mathbf{A} \subset \mathbf{A}^\dagger$, and $\mathbf{A}$ need not be self-adjoint.

(i) Show that $\mathbf{A}$ has eigenvectors $f_\alpha(x) = \exp(i\alpha x)$ that are in $L_2(0, \infty)$ for any α in the upper half α-plane (Im $\alpha > 0$).

Remark. This shows that $\mathbf{A}$ cannot be self-adjoint, since a self-adjoint operator has only real eigenvalues. □

(ii) Find formal eigenvectors of the operator

$$\mathbf{A}_\kappa \equiv \mathbf{A} - \frac{i\kappa}{x}$$

corresponding to eigenvalues α in the upper half α-plane. For what values of κ are these eigenvectors actually in $L_2(0, \infty)$?

Remark. The operators $\mathbf{A}_\kappa$ satisfy the commutation rule $[\mathbf{A}_\kappa, x] \subset 1$ for any κ. This commutator is required in quantum mechanics of an operator corresponding to the momentum conjugate to the coordinate x, which might be a radial coordinate in spherical or cylindrical coordinates, for example. Quantum mechanical operators corresponding to observables must also be strictly self-adjoint, and not merely symmetric. Hence the naive choice of $\mathbf{A} = \mathbf{A}_0$ as the momentum conjugate to the radial coordinate does not work, and a suitable $\mathbf{A}_\kappa$ must be used instead. The reader is invited to decide which $\mathbf{A}_\kappa$ might be appropriate in spherical or cylindrical coordinates. □

13. (i) Under what conditions on the parameters a, b, c is the differential operator

$$\mathbf{L} = x(x-1)\frac{d^2}{dx^2} + [(a+b+1)x - c]\frac{d}{dx} + ab$$

in the hypergeometric equation (5.5.A13) symmetric, and self-adjoint in $L_2(0,1)$?

(ii) Under what conditions is there a function $\rho(x)$ such that $\rho(x)\mathbf{L}$ is symmetric, and self-adjoint in $L_2(0,1)$?

14. (i) Under what conditions on the parameters a and c is the differential operator

$$\mathbf{L} = x\frac{d^2}{dx^2} + (c-x)\frac{d}{dx} - a$$

in the confluent hypergeometric equation (5.5.B38) symmetric, and self-adjoint in $L_2(0,\infty)$?

(ii) Under what conditions is there a function $\rho(x)$ such that $\rho(x)\mathbf{L}$ is symmetric, and self-adjoint in $L_2(0,\infty)$?

15. Consider the operator

$$\Delta \equiv \frac{d^2}{dx^2}$$

defined on the (complex) function space $L^2(-1,1)$.

(i) If f and g are two functions that are twice differentiable, compute

$$B(f,g) \equiv (f, \Delta g) - (\Delta f, g)$$

in terms of the boundary values of f and g at $x = \pm 1$.

(ii) Find the complete class of boundary conditions on f and g that lead to a self-adjoint extension of the operator Δ to a maximal domain.

(iii) The functions $\exp(\alpha x)$ are formal eigenfunctions of Δ for any complex α. What values of α are consistent with each of the boundary conditions introduced in part (ii)? In other words, find the spectrum of Δ for each boundary condition that defines Δ as a self-adjoint operator on $L^2(-1,1)$.

(iv) Find the Green function $G_\lambda(x,y) = (\Delta - \lambda 1)^{-1}$ for each of the self-adjoint boundary conditions introduced in parts (ii) and (iii).

16. Consider the linear differential operator

$$K \equiv -\frac{d^2}{dx^2} + x^2$$

defined on $L_2(-\infty, \infty)$.

(i) Show that K is self-adjoint.

(ii) Show that the Hermite functions

$$h_n(x) = C_n H_n(x) e^{-\frac{1}{2}x^2}$$

defined in Problem 6.20 are eigenfunctions of K, and find the corresponding eigenvalues. Here the $H_n(x)$ are the Hermite polynomials introduced in Chapter 6.

(iii) Find the normalization constants C_n.

Remark. The operator H is obtained by change of variables from the Hamiltonian

$$H = -\frac{\hbar^2}{2m} \frac{d^2}{dx^2} + \frac{m\omega^2}{2} x^2$$

for the one-dimensional quantum mechanical harmonic oscillator (frequency ω). □

17. Consider the linear differential operator

$$K \equiv -\frac{d^2}{dx^2} + \frac{U}{\cosh^2 x}$$

defined on $L_2(-\infty, \infty)$.

(i) Show that K is self-adjoint.

(ii) Introduce the variable $\xi = \tanh x$, and show that the eigenvalue equation for K is the same as the differential equation for associated Legendre functions introduced in Exercise 6.8.

(iii) What can you say about the spectrum of K? Consider the cases $U > 0$ and $U < 0$ separately.

Remark. The operator K is related to the Hamiltonian for a particle in a potential

$$V(x) = \frac{V_0}{\cosh^2 x}$$

Thus the discrete spectrum of K is related to the existence of bound states in this potential. Note that there is always at least one bound state in the potential for $V_0 < 0$. This potential also plays a role of the theory of the KdV equation discussed in Chapter 8. □

18. The integral operator $\mathbf{K}$ on $L_2(\Omega)$ is *separable* if its kernel $K(x,y)$ has the form

$$K(x,y) = \alpha\, u(x)v(y)$$

with α being a constant, and u, v functions in $L_2(\Omega)$.

(i) Show that $\mathbf{K}$ has at most one nonzero eigenvalue. Find that eigenvalue and a corresponding eigenvector.

(ii) Under what conditions on α, u, v is $\mathbf{K}$ self-adjoint?

More generally, $\mathbf{K}$ is *degenerate* if $K(x,y)$ has the form

$$K(x,y) = \sum_{k=1}^{n} \alpha_k\, u_k(x)v_k(y)$$

with $\alpha_1, \ldots, \alpha_n$ constants, $\{u_1, \ldots, u_n\}$ and $\{v_1, \ldots, v_n\}$ each sets of linearly independent functions in $L_2(\Omega)$.

(iii) Show that $\mathbf{K}$ is of finite rank, and characterize the range $\mathbf{R_K}$.

(iv) Show that the nonzero eigenvalues of $\mathbf{K}$ are determined by finding the eigenvalues of a finite-dimensional matrix; give explicit expressions for the elements of this matrix.

8 Partial Differential Equations

Ordinary differential equations describe the evolution of a curve in a manifold as a variable, often understood as time, increases. A common example is the evolution of the coordinates in the phase space of a Hamiltonian dynamical system according to Hamilton's equations of motion. However, there are many physical variables that are described by functions, often called *fields*, defined on a manifold of space and time coordinates. The evolution of these variables is described by equations of motion that involve not only time derivatives, but also spatial derivatives of the variables. Such equations involving derivatives with respect to more than one variable are partial differential equations.

A linear first order partial differential equation of the form

$$\sum_{k=1}^{n} v^k(x)\, \frac{\partial u}{\partial x^k} = \vec{v} \cdot \vec{\nabla} u = f(x)$$

can be analyzed using geometrical methods as introduced in Chapter 3. The equation determines the evolution of the function $u(x)$ along the lines of flow of the vector field v, the *characteristics* of the equation. A particular solution is defined by specifying the values of the function $u(x)$ on a surface that intersects each of the lines of flow of the vector field v exactly once.

Many equations of physics are second-order linear equations for which the Hilbert space methods introduced in Chapter 7 are especially useful. Maxwell's equations for the electromagnetic field and the nonrelativistic Schrödinger equation for a particle moving in a potential are two examples of such equations that involve the Laplacian operator

$$\Delta \equiv \nabla^2 \equiv \left(\frac{\partial^2}{\partial x^2} + \frac{\partial^2}{\partial y^2} + \frac{\partial^2}{\partial z^2} \right)$$

The boundary conditions needed to make the Laplacian a self-adjoint linear operator in a Hilbert space are derived, and the spectrum of the Laplacian determined for some examples.

Green functions, representing the solution due to a point source with homogeneous boundary conditions on the surface of some region are introduced. They are then used to construct formal solutions of *Laplace's equation*

$$\Delta u = 0$$

with inhomogeneous boundary conditions, and the related *Poisson's equation*

$$\Delta u = -4\pi\rho$$

Introduction to Mathematical Physics. Michael T. Vaughn
Copyright © 2007 WILEY-VCH Verlag GmbH & Co. KGaA, Weinheim
ISBN: 978-3-527-40627-2

To construct the Green functions requires knowledge of solutions to the eigenvalue equation

$$\Delta u = \lambda u$$

(and thus of the spectrum of the Laplacian). This equation is closely related to the *Helmholtz equation*

$$(\Delta + \kappa^2)u = -4\pi\rho$$

that appears in the analysis of wave equations for waves with a definite frequency. The method of choice for finding the spectrum of the Laplacian is the method of separation of variables, which is feasible in various coordinate systems. This method is based on seeking solutions to the eigenvalue equation that are expressed as products of functions, each depending on only one of the coordinates. These functions then satisfy ordinary differential equations in a singe coordinate. For the coordinate systems in which Laplace's equation is separable, many of the ordinary differential equations that arises are closely related to the hypergeometric and confluent hypergeometric equations studied in Chapter 5.

Equations of motion that involve both time derivatives and the Laplacian are introduced. The *diffusion equation* (or *heat equation*)

$$\frac{\partial \chi}{\partial t} = a\nabla^2 \chi$$

describes the evolution of the temperature distribution in a heat conducting material, for example. A general solution to this equation, given a fixed set of initial conditions, is based on the eigenvalues and eigenfunctions of the Laplacian; one key result is that solutions to this equation decrease exponentially in time with time constant depending on the spectrum of the Laplacian.

The *Schrödinger equation*

$$i\hbar \frac{\partial \psi}{\partial t} = \mathbf{H}\psi$$

governs the time development of the state vector ψ of a quantum mechanical system with Hamiltonian $\mathbf{H}$. States of definite energy E satisfy the eigenvalue equation $\mathbf{H}\psi = E\psi$; hence finding the spectrum of the Hamiltonian is a critical problem. We discuss the solutions for the problem of a charged particle (for example, an electron) moving in the Coulomb field of a point charge (an atomic nucleus, for example).

Two second-order equations of interest are the *wave equation*

$$\left(\nabla^2 - \frac{1}{c^2}\frac{\partial^2}{\partial t^2}\right)\phi = -4\pi\rho$$

satisfied by the amplitude ϕ of a wave from a source ρ propagating with speed c. We look at Green functions corresponding to retarded and advanced boundary conditions; these are obtained from the same Fourier transform by choosing different contours in the complex frequency plane for the Fourier transform that returns from frequency space to real time. We also look at the multipole expansion for the radiation from a known source.

Finally, we introduce the *Klein–Gordon equation*

$$\left[\frac{1}{c^2}\frac{\partial^2}{\partial t^2} - \nabla^2 + \left(\frac{mc}{\hbar}\right)^2\right]\phi = 0$$

which is a relativistic wave equation for a free particle of mass m, although we pay more attention to nonlinear variations of this equation than to the Klein–Gordon equation itself.

Nonlinear partial differential equations are increasingly important as linear systems have been more thoroughly studied. We begin with a look at a quasilinear first-order equation

$$\frac{\partial u}{\partial t} + s(u)\frac{\partial u}{\partial x} = 0$$

that describes wave propagation with speed $s(u)$ that depends on the amplitude of the wave. We analyze this equation using the method of characteristics, and see how shock waves can arise when characteristics join together. Addition of a dispersive term proportional to the third derivative of the wave amplitude leads to the Kortweg–deVries (KdV) equation

$$\frac{\partial u}{\partial t} + 6u\frac{\partial u}{\partial x} + \frac{\partial^3 u}{\partial x^3} = 0$$

This equation has solutions that can be characterized as *solitary waves*, or *solitons*—these are sharply peaked pulses that propagate with speed proportional to their height, and also with a width that decreases for taller pulses. Even more remarkable, though we do not derive the results here, is the fact that these pulses can propagate, passing through other pulses, with form unchanged after the interaction. There is also an infinite set of conservation laws associated with solutions to the KdV equation; again we refer the reader to other sources for the details.

The Klein–Gordon equation augmented by a cubic nonlinear term corresponds to an equation of motion for an interacting classical relativistic field. While the equation is generally difficult to study, it happens that in one space dimension there are solitary waves that appear to be similar to the solitons of the KdV equation. But the shape of these pulses is altered by interaction with other pulses, and there is no infinite set of conservation laws.

Another variation on the Klein–Gordon equation is derived from a potential

$$U(\phi) = \lambda v^4 \left[1 - \cos\left(\frac{\phi}{v}\right)\right] = 2\lambda v^4 \sin^2\left(\frac{\phi}{2v}\right)$$

This leads to the equation of motion

$$\frac{\partial^2 \phi}{\partial t^2} - \frac{\partial^2 \phi}{\partial x^2} = -\lambda v^3 \sin\left(\frac{\phi}{v}\right)$$

that is known as the *sine-Gordon equation*. This equation appears in several branches of physics. Here we note that it has shares many features with the KdV equation; it has solitonic solutions that can pass through each other without distortion, and there is an infinite set of conservation laws.

In Appendix A, we describe the basic ideas of how classical field theories are derived from a Lagrangian density, with symmetries such as Lorentz invariance incorporated in the Lagrangian.

8.1   Linear First-Order Equations

The general linear first order partial differential equation on an n-dimensional manifold $\mathcal{M}$ has the form

$$\sum_{k=1}^{n} v^k(x)\, \frac{\partial u}{\partial x^k} = \vec{v} \cdot \vec{\nabla} = f(x) \tag{8.1}$$

If the coefficients $v^k(x)$ are well-behaved functions on $\mathcal{M}$, they form the components of a vector field v, and Eq. (8.1) defines the rate of change of $u(x)$ along the integral curves of v. Thus Eq. (8.1) becomes

$$\frac{du}{d\lambda} = f(x(\lambda)) \tag{8.2}$$

along each integral curve of v. To generate a solution of Eq. (8.1) valid on the entire manifold $\mathcal{M}$ except possibly at singular points of v, we need to specify the values of u on an $(n-1)$-dimensional surface through which every integral curve of v passes exactly once.

Remark.   The integral curves of v are *characteristics* of Eq. (8.1), and the method outlined here of finding solutions to Eq. (8.1) is the *method of characteristics*. This method transforms the partial differential equation into a set of ordinary differential equations—first to determine the characteristic curves, and then to evaluate the projection of the solution along each characteristic curve.   □

❑ **Example 8.1.**   The simplest linear first-order equation is

$$\frac{\partial u}{\partial t} + s\, \frac{\partial u}{\partial x} = 0 \tag{8.3}$$

where s is a constant. Here $u = u(x,t)$ describes a wave that propagates in one space dimension with speed s. Equation (8.3) is obtained from the usual second-order wave equation by restricting solutions to waves that propagate only in one direction, here the positive X-direction.

If we begin with the amplitude

$$u(x,0) = \psi(x) \tag{8.4}$$

at $t = 0$, then the solution to Eq. (8.3) that develops from this initial condition is exactly

$$u(x,t) = \psi(x - st) \tag{8.5}$$

Thus Eq. (8.3) describes propagation of the initial wave form $\psi(x)$ with constant speed s in the X-direction.

The characteristic curves of Eq. (8.3) are straight lines of the form

$$x - st = \xi \tag{8.6}$$

and a typical surface on which to specify the value of u is $t = 0$. With coordinates $\xi = x - st$ and $\tau = t$, Eq. (8.3) is simply

$$\frac{du}{d\tau} = 0 \tag{8.7}$$

with solution $u(\xi, \tau) = \psi(\xi)$ as already noted.

The inhomogeneous equation corresponding to Eq. (8.3) is

$$\frac{\partial u}{\partial t} + s \frac{\partial u}{\partial x} = f(x, t) \tag{8.8}$$

In terms of the variables ξ, τ this equation is

$$\frac{\partial u}{\partial \tau} = f(\xi + s\tau, \tau) \tag{8.9}$$

and the solution with $u(x, 0) = \psi(x)$ is

$$u(x, t) = \psi(x - st) + \int_0^t f[x - s(t - \tau), \tau] \, d\tau \tag{8.10}$$

Thus the initial condition is propagated along each characteristic curve; there is also a contribution from sources at points looking backward in time along the characteristic.   ∎

→ **Exercise 8.1.**   Consider the linear partial differential equation

$$\frac{\partial u}{\partial t} + \sum_{k=1}^{n} s^k \frac{\partial u}{\partial x^k} = f(x, t) \tag{8.11}$$

with constant coefficients $\{s^k\}$, on a manifold with n spatial coordinates $x = (x^1, \ldots, x^n)$ and a time coordinate t. Find the solution to this equation corresponding to the initial condition $u(x, 0) = \psi(x)$, and give a physical interpretation of this solution.   □

If we can express the initial condition in Example 8.1 as a Fourier integral

$$\psi(x) = \int_{-\infty}^{\infty} \chi(k) e^{ikx} \, dk \tag{8.12}$$

then the solution (8.5) is given by

$$u(x, t) = \int_{-\infty}^{\infty} \chi(k) e^{ikx - i\omega t} \, dk \tag{8.13}$$

Here

$$\omega = 2\pi f = ks \tag{8.14}$$

is the (angular) frequency (the usual frequency is f) corresponding to wave vector $k = 2\pi/\lambda$ (λ is the wavelength). The relation (8.14) between ω and k is generally determined by the physics leading to Eq. (8.3); it is often called a *dispersion relation*. A strictly linear relation between ω and k, corresponding to a speed of propagation independent of ω, is unusual— except for electromagnetic waves propagating in vacuum, ω cannot be exactly linear in k. However, the speed of propagation may vary slowly enough over the range of k contained in the initial amplitude that the approximation of constant speed is reasonable.

To go beyond the approximation of constant speed, we can expand the expression for $\omega = \omega(k)$ as a power series,

$$\omega = ks(1 - \alpha^2 k^2 + \cdots) \tag{8.15}$$

where α is a constant. Note that the dispersion relation (8.15) can only be valid for $k\alpha < 1$; it is a long-wavelength ($\lambda > 2\pi\alpha$) approximation. However, it leads to a wave equation

$$\frac{\partial u}{\partial t} + s\,\frac{\partial u}{\partial x} + s\alpha^2\,\frac{\partial^3 u}{\partial x^3} = 0 \tag{8.16}$$

that involves a third order partial derivative. Nevertheless, the solution is straightforward; starting from an initial condition (8.12), we have the solution given by a Fourier integral

$$u(x,t) = \int_{-\infty}^{\infty} \chi(k)e^{ikx - i\omega(k)t}\,dk \tag{8.17}$$

with $\omega(k)$ given by Eq. (8.15). Equation (8.16) is a linear version of the *Kortweg–deVries equation* that will appear in nonlinear form in Section 8.4.

Equation (8.17) is the most general expression for a linear wave propagating in one direction, although it does not arise from a linear partial differential equation unless the dispersion relation for $\omega(k)$ is a polynomial in k. The energy associated with such a wave is typically quadratic in the amplitude $u(x,t)$ and its derivatives, or proportional to an integral of $|\chi(k)|^2$ times a polynomial in k^2. Thus a solution of finite energy cannot be a pure plane wave, but must be a superposition of plane waves integrated over some finite interval in k.

Suppose $|\chi(k)|^2$ is sharply peaked around some value k_0, so that we can expand

$$\omega(k) \simeq \omega(k_0) + \left.\frac{d\omega}{dk}\right|_{k=k_0} (k - k_0) \tag{8.18}$$

Then the solution (8.17) can be approximated as

$$u(x,t) \simeq e^{ik_0 x - i\omega_0 t} \int_{-\infty}^{\infty} \chi(k)e^{i(k-k_0)(x-v_g t)}\,dk \tag{8.19}$$

Such a solution is called a *wave packet*. It can be understood as a phase factor propagating with speed $v_{ph} = \omega_0/k_0$ (the *phase velocity*) times the integral, which propagates with speed

$$v_g = \left.\frac{d\omega}{dk}\right|_{k=k_0} \tag{8.20}$$

Here v_g is the *group velocity*; it is the speed with which energy and information in the wave are transmitted.

Linear waves described by the standard second-order wave equation are treated later in Section 8.3.2. Nonlinear waves are described by generalizations of Eq. (8.3) in which the wave speed depends also on the amplitude of the wave. Some examples of these are analyzed in Section 8.4.

8.2   The Laplacian and Linear Second-Order Equations

8.2.1   Laplacian and Boundary Conditions

Much of the physics of the last two centuries is expressed mathematically in the form of linear partial differential equations. *Laplace's equation*

$$\Delta u \equiv \nabla^2 u = \left(\frac{\partial^2}{\partial x^2} + \frac{\partial^2}{\partial y^2} + \frac{\partial^2}{\partial z^2} \right) u = 0 \tag{8.21}$$

and its inhomogeneous version (*Poisson's equation*)

$$\nabla^2 u = -4\pi\rho \tag{8.22}$$

describe many physical quantities—two examples are the electrostatic potential $u(\vec{r})$ due to a fixed distribution $\rho(\vec{r})$ of charge and the temperature distribution $u(\vec{r})$ of a system in thermal equilibrium with its surroundings. The factor 4π is conventional.

The linear operator Δ (or ∇^2) is the *Laplace operator*, or simply *Laplacian*. Is the Laplacian a self-adjoint linear operator? From the discussion of differential operators in Section 7.5.2, we expect the answer depends critically on the boundary conditions imposed on the solutions to Eqs. (8.21) or (8.22). Suppose we seek solutions in a region Ω bounded by a closed surface S. Integration by parts gives

$$(v, \Delta u) = \int_\Omega v^* \left(\nabla^2 u \right) d\Omega = - \int_\Omega \left(\vec{\nabla} v^* \right) \cdot \left(\vec{\nabla} u \right) d\Omega + \int_S v^* \, \hat{\mathbf{n}} \cdot \left(\vec{\nabla} u \right) dS \tag{8.23}$$

where $\hat{\mathbf{n}}$ is the normal to the surface S directed outward from the region Ω. A second integration by parts gives

$$\int_\Omega v^* \nabla^2 u \, d\Omega = \int_\Omega \left(\nabla^2 v^* \right) u \, d\Omega + \int_S \left[v^* \, \hat{\mathbf{n}} \cdot \left(\vec{\nabla} u \right) - u \, \hat{\mathbf{n}} \cdot \left(\vec{\nabla} v^* \right) \right] dS \tag{8.24}$$

Equation (8.24) is *Green's theorem*. It follows that Δ is a self-adjoint operator if the boundary conditions on the solutions insure that the surface integral in Eq. (8.24) vanishes automatically.

One boundary condition that makes Δ self-adjoint is the *Dirichlet* condition

$$u(\vec{r}) = v(\vec{r}) = 0 \quad \text{on } S. \tag{8.25}$$

For example, this boundary condition is appropriate for the electrostatic potential on a conducting surface, in order to have no current flow along the surface. Another boundary condition that makes Δ self-adjoint is the *Neumann* condition

$$\hat{\mathbf{n}} \cdot \vec{\nabla} u(\vec{r}) = \hat{\mathbf{n}} \cdot \vec{\nabla} v(\vec{r}) = 0 \quad \text{on } S. \tag{8.26}$$

This condition is physically relevant for the temperature distribution for a system bounded by an insulating surface, for example; it prohibits heat flow across the boundary. A general self-adjoint boundary condition for Δ is the *mixed* boundary condition

$$\alpha(\vec{r}) u(\vec{r}) + \beta(\vec{r}) \hat{\mathbf{n}} \cdot \vec{\nabla} u(\vec{r}) = \alpha(\vec{r}) v(\vec{r}) + \beta(\vec{r}) \hat{\mathbf{n}} \cdot \vec{\nabla} v(\vec{r}) = 0 \quad \text{on } S. \tag{8.27}$$

with $\alpha(\vec{r})$ and $\beta(\vec{r})$ are real functions on S (see Problem 1). All these boundary conditions are homogeneous. However, solutions that satisfy inhomogeneous boundary conditions can generally be found using the Green function with homogeneous boundary conditions.

8.2.2   Green Functions for Laplace's Equation

The solution to the inhomogeneous equation (8.22) is obtained formally by finding a solution to the equation

$$\nabla_r^2 G(\vec{r}, \vec{s}) = -4\pi\delta(\vec{r} - \vec{s}) \tag{8.28}$$

with appropriate homogeneous boundary conditions. The solution $G(\vec{r}, \vec{s})$ is the solution with the prescribed boundary conditions for a unit point source at $\vec{s}$; it is the *Green function* for the Laplacian with the given boundary conditions.

The solution to Eq. (8.22) with an arbitrary source $\rho(\vec{r})$ and homogeneous boundary condition is then given by

$$u(\vec{r}) = \int_\Omega G(\vec{r}, \vec{s})\rho(\vec{s})\, d\Omega_s \tag{8.29}$$

The Green function also provides a solution to Laplace's equation with inhomogeneous boundary conditions. Green's theorem (8.24) with $v(\vec{s}) = G(\vec{r}, \vec{s})$ gives the general result

$$u(\vec{r}) = \int_S \left[G^*(\vec{r}, \vec{s})\,\hat{\mathbf{n}} \cdot \vec{\nabla}_s\, u(\vec{s}) - u(\vec{s})\,\hat{\mathbf{n}} \cdot \vec{\nabla}_s\, G^*(\vec{r}, \vec{s}) \right] dS_s \tag{8.30}$$

This can be simplified in a particular problem using the (homogeneous) boundary condition satisfied by $G(\vec{r}, \vec{s})$.

One important consequence of the results (8.29) and (8.30) is that there are no nontrivial solutions to Laplace's equation in a region Ω that satisfy homogeneous boundary conditions on the boundary of Ω. There must either be explicit sources within Ω that lead to the solution in Eq. (8.29), or sources on the boundary of Ω whose presence is implied by inhomogeneous boundary conditions, as seen in Eq. (8.30). Of course both types of sources may be present, in which case the full solution is obtained as a linear combination of the two solutions.

A simple example is the construction of the Green function for the Laplacian on the entire n-dimensional space $\mathbf{R}^n$, with boundary condition simply that the Green function vanish far from the source. In this case, we expect that the Green function should depend only on the relative coordinate of source and observer, so that

$$G(\vec{r}, \vec{s}) = G(\vec{r} - \vec{s}) \tag{8.31}$$

Introduce the Fourier integral

$$G(\vec{r} - \vec{s}) = \frac{1}{(2\pi)^n} \int e^{i\vec{k}\cdot(\vec{r}-\vec{s})}\widetilde{G}(\vec{k})\, d^n k \tag{8.32}$$

and note that the Fourier integral theorem can be expressed formally as

$$\int e^{i\vec{k}\cdot(\vec{r}-\vec{s})}\, d^n k = (2\pi)^n\,\delta(\vec{r} - \vec{s}) \tag{8.33}$$

Poisson's equation (8.28) requires $k^2\widetilde{G}(\vec{k}) = 4\pi$, so that

$$\widetilde{G}(\vec{k}) = \frac{4\pi}{k^2} \tag{8.34}$$

The Green function in coordinate space is then given by

$$G(\vec{r}, \vec{s}) = \frac{4\pi}{(2\pi)^n} \int \frac{e^{i\vec{k}\cdot(\vec{r}-\vec{s})}}{k^2} d^n k \tag{8.35}$$

In three dimensions, this integral can be evaluated in spherical coordinates. We have

$$G(\vec{r}, \vec{s}) = \frac{4\pi}{(2\pi)^3} \int \frac{e^{i\vec{k}\cdot(\vec{r}-\vec{s})}}{k^2} d^3 k = \frac{8\pi^2}{(2\pi)^3} \int_0^\infty \int_{-1}^1 e^{ik\rho\mu} d\mu \, dk \tag{8.36}$$

with $\rho \equiv |\vec{r} - \vec{s}|$. Now

$$\int_{-1}^1 e^{ik\rho\mu} d\mu = \frac{1}{ik\rho} \left(e^{ik\rho} - e^{-ik\rho} \right) = \frac{\sin k\rho}{k\rho} \tag{8.37}$$

and

$$\int_0^\infty \frac{\sin k\rho}{k\rho} dk = \frac{\pi}{2\rho} \tag{8.38}$$

which leads to the final result

$$G(\vec{r}, \vec{s}) = \frac{1}{|\vec{r} - \vec{s}|} \tag{8.39}$$

The reader will recognize this as the Coulomb potential due to a point charge (in Gaussian units obtained from SI units by setting $4\pi\varepsilon_0 = 1$).

If we imagine Eq. (8.22) to be Poisson's equation for the Coulomb potential $u(\vec{r})$ due to a charge distribution $\rho(\vec{r})$, then the solution (8.29) has the expected form

$$u(\vec{r}) = \int \frac{\rho(\vec{s})}{|\vec{r} - \vec{s}|} d^3 s \tag{8.40}$$

apart from the factor 4π due to the unrationalized units. If the charge distribution is of finite extent, we can use the generating function (5.137) for the Legendre polynomials to expand

$$\frac{1}{|\vec{r} - \vec{s}|} = \sum_{n=0}^\infty \frac{s^n}{r^{n+1}} P_n(\cos\Theta) \tag{8.41}$$

outside the charge distribution, where $r > s$ for all s in the integral. Here Θ is the angle between $\vec{r}$ and $\vec{s}$, and we have

$$P_n(\cos\Theta) = \frac{4\pi}{2n+1} \sum_{m=-n}^n Y_{nm}^*(\theta_s, \phi_s) Y_{nm}(\theta, \phi) \tag{8.42}$$

from the spherical harmonic addition theorem (Eq. (6.159)).

Finally, we have

$$\phi(\vec{r}) = \sum_{n=0}^\infty \frac{4\pi}{2n+1} \frac{1}{r^{n+1}} \sum_{m=-n}^n q_{nm} Y_{nm}(\theta, \phi) \tag{8.43}$$

for $r \to \infty$. This is the *multipole expansion* of the potential; the coefficients

$$q_{nm} = \int Y_{nm}^*(\theta_s, \phi_s) \, s^n \rho(\vec{s}) \, d^3 s \qquad (8.44)$$

are the *multipole moments* of the charge distribution.

Remark. The multipole expansion (8.43) is also valid as an asymptotic series even if the charge density decreases exponentially for $r \to \infty$, rather than having a sharp cutoff at some finite distance. The expansion is limited only by the requirement that it can only include terms for which the multipole moments are finite.   □

Remark. The normalization of the multipole moments is somewhat arbitrary, and various conventions are used in the literature. For example, the quadrupole moment of an atomic nucleus is defined by

$$Q = \int (3z^2 - r^2) \rho(\vec{r}) \, d^3 r$$

where $\rho(\vec{r})$ is the nuclear charge density. More precisely, if the nucleus has angular momentum J, then $\rho(\vec{r})$ is the charge density in the state with $J_z = J$, where J_z is the Z-component of the angular momentum.   □

→ **Exercise 8.2.** A localized charge distribution has charge density

$$\rho(\vec{r}) = \gamma r^2 e^{-\alpha r} \cos^2 \theta$$

in terms of spherical coordinates.

(i) Find the total charge q of the distribution in terms of γ and α.

(ii) Find the quadrupole moment of the charge distribution, using the nuclear convention introduced above, and find the asymptotic expansion (for $r \to \infty$) of the potential due to this distribution in terms of q and α.   □

The Green function (8.35) is similar to the eigenfunction expansion (7.117) of the Green function for an ordinary differential operator introduced in Chapter 7. However, the Laplacian defined on all of $\mathbf{R}^n$ has a continuous spectrum that covers the entire negative real axis. There are *continuum eigenvectors*, or *almost eigenvectors*, of the form

$$\phi_{\vec{k}}(\vec{r}) = C_k \, e^{i\vec{k}\cdot\vec{r}} \qquad (8.45)$$

for every real vector $\vec{k}$, with C_k a normalization constant, and

$$\Delta \phi_{\vec{k}} = -k^2 \phi_{\vec{k}} \qquad (8.46)$$

The sum in Eq. (7.117) is replaced in Eq. (8.35) by an integral over the continuous spectrum of Δ, with C_k taken to be $1/(2\pi)^{\frac{n}{2}}$.

8.2.3   Spectrum of the Laplacian

In order to analyze the various partial differential equations that involve the Laplacian operator, we need to understand the spectrum of the Laplacian. This spectrum depends both on the region Ω on which the operator is defined, and on the boundary conditions imposed on the boundary of Ω. For any function u satisfying boundary conditions that make Δ self-adjoint, integration by parts gives

$$(u, \Delta u) = \int_\Omega u^* \left(\nabla^2 u \right) d\Omega = - \int_\Omega \left(\vec{\nabla} u^* \right) \cdot \left(\vec{\nabla} u \right) d\Omega < 0 \tag{8.47}$$

Note that the surface integral omitted here is zero for function that satisfy self-adjoint boundary conditions; it is precisely the vanishing of the surface terms that is required for Δ to be self-adjoint.

If the region Ω is bounded, then the Green function $G(\vec{r}, \vec{s})$ is a Hilbert–Schmidt operator, and Δ has a discrete spectrum as in the one-dimensional examples studied in Chapter 7. Then the Green function can be expressed, formally at least, as an infinite series of the form (7.117). If Ω is unbounded, then Δ has a continuous spectrum, and the infinite series is replaced by an integral over the continuous spectrum.

The eigenvalue equation for the Laplacian

$$\Delta u = \lambda u \tag{8.48}$$

is the same as the homogeneous version of the *Helmholtz equation*

$$(\Delta + k^2)u = -\rho \tag{8.49}$$

that arises from equations of motion such as the wave equation or the heat equation after factoring out an exponential time dependence. Thus determining the eigenvalues and eigenvectors of the Laplacian also generates Green functions for the Helmholtz equation and related equations, examples of which we will soon see.

The importance of the spectrum of the Laplacian, and of related Hilbert space operators that include the Laplacian, has led to the development of a variety of methods, both analytic and approximate, for determining this spectrum for various regions and boundary conditions. The book by Morse and Feshbach cited in the bibliography gives a good sample of standard analytical methods. Beyond this, the widespread availability of modern high-speed computers has stimulated the analysis of a broad range of numerical approximation methods that can generate highly accurate results for specific problems of practical interest.

One classical method for finding the spectrum of the Laplacian works is the method of *separation of variables*. This method is most useful for problems in which the boundaries of the region Ω coincide with surfaces of constant coordinate in some special coordinate system. Here we give examples that use cartesian coordinates and spherical coordinates; some other useful coordinate systems are noted in the problems (see also Chapter 3). The book by Morse and Feshbach has a comprehensive list of coordinate systems in which Laplace's equation is separable, as well as detailed discussions of solutions to both Laplace's equation and the Helmholtz equation in these coordinate systems.

❑ **Example 8.2.**   Consider the eigenvalue equation

$$\Delta u = \lambda u \tag{8.50}$$

in a rectangular box characterized by

$$0 \le x \le a \qquad 0 \le y \le b \qquad 0 \le z \le c \tag{8.51}$$

with boundary condition $u(\vec{r}) = 0$ on the surface of the box. Since the boundaries of the box coincide with surfaces of constant coordinate, it is plausible to look for solutions of the form

$$u(\vec{r}) = X(x)Y(y)Z(z) \tag{8.52}$$

that is, solutions that are products of functions of one coordinate only. Then the eigenvalue equation (8.50) is reduced to three eigenvalue equations

$$\frac{d^2 X}{dx^2} = \lambda_x X \quad \frac{d^2 Y}{dy^2} = \lambda_y Y \quad \frac{d^2 Z}{dz^2} = \lambda_z Z \tag{8.53}$$

with boundary conditions

$$X(0) = 0 = X(a) \quad Y(0) = 0 = Y(b) \quad Z(0) = 0 = Z(c) \tag{8.54}$$

whose solutions will produce an eigenvector of Δ with eigenvalue

$$\lambda = \lambda_x + \lambda_y + \lambda_z \tag{8.55}$$

The solutions to the one-dimensional problems are the (unnormalized) eigenvectors

$$X_k(x) = \sin \frac{l\pi x}{a} \qquad Y_l(y) = \sin \frac{m\pi y}{b} \qquad Z_m(z) = \sin \frac{n\pi z}{c} \tag{8.56}$$

$(k, l, m = 1, 2, 3, \ldots)$ with corresponding eigenvalues

$$\lambda_x = -\left(\frac{l\pi x}{a}\right)^2 \qquad \lambda_y = -\left(\frac{m\pi y}{b}\right)^2 \qquad \lambda_z = -\left(\frac{n\pi z}{c}\right)^2 \tag{8.57}$$

These solutions generate the three-dimensional eigenvectors

$$u_{lmn}(\vec{r}) = \sin \frac{l\pi x}{a} \sin \frac{m\pi y}{b} \sin \frac{n\pi z}{c} \tag{8.58}$$

$(k, l, m = 1, 2, 3, \ldots)$ with corresponding eigenvalues

$$\lambda_{lmn} = -\left\{ \left(\frac{l\pi x}{a}\right)^2 + \left(\frac{m\pi y}{b}\right)^2 + \left(\frac{n\pi z}{c}\right)^2 \right\} \tag{8.59}$$

The spectrum with periodic boundary conditions is considered in Problem 4.   ∎

❑ **Example 8.3.** We now want to find the eigenvalues λ and eigenvectors $u(\vec{r})$ of the Laplacian defined inside a sphere of radius R, with the boundary condition

$$u(\vec{r}) = 0 \qquad \text{for} \quad r = R \tag{8.60}$$

on the surface of the sphere. To this end, we recall from Chapters 3 and 6 that the Laplacian in spherical coordinates is given by

$$\Delta = \frac{1}{r^2} \left\{ \frac{\partial}{\partial r} \left(r^2 \frac{\partial}{\partial r} \right) + \frac{1}{\sin\theta} \frac{\partial}{\partial \theta} \left(\sin\theta \frac{\partial}{\partial \theta} \right) + \frac{1}{\sin^2\theta} \frac{\partial^2}{\partial \phi^2} \right\} \tag{3.193}$$

We then look for solutions of the eigenvalue equation (8.50) that have the form

$$u(\vec{r}) = R(r)Y(\theta, \phi) \tag{8.61}$$

The angular part of the Laplacian has already been discussed in Section 6.5.3. There we found angular functions, the spherical harmonics $Y_{nm}(\theta, \phi)$ with $n = 0, 1, 2, \ldots$ and $m = n, n-1, \ldots, -n+1, -n$, that satisfy

$$\left\{ \frac{1}{\sin\theta} \frac{\partial}{\partial \theta} \left(\sin\theta \frac{\partial}{\partial \theta} \right) + \frac{1}{\sin^2\theta} \frac{\partial^2}{\partial \phi^2} \right\} Y_{nm}(\theta, \phi) = -n(n+1)Y_{nm}(\theta, \phi) \tag{8.62}$$

The radial equation for the corresponding function $R_n(r)$ then has the form

$$\frac{1}{r^2} \left\{ \frac{d}{dr} \left(r^2 \frac{d}{dr} \right) - \frac{n(n+1)}{r^2} \right\} R_n(r) = \lambda R_n(r) \tag{8.63}$$

Now let $\lambda = -\kappa^2$, and introduce the variable $x = \kappa r$. Then, with $R_n(r) = X_n(\kappa r)$, the function $X_n(x)$ must satisfy the equation

$$X_n''(x) + \frac{2}{x} X_n'(x) + \left[1 - \frac{n(n+1)}{x^2} \right] X_n(x) = 0 \tag{8.64}$$

which is same as the differential equation satisfied by the spherical Bessel function $j_n(x)$ (see Eq. (5.180)).

To satisfy the boundary condition $u(R) = 0$, we need to have

$$j_n(\kappa R) = 0 \tag{8.65}$$

Now each of the functions $j_n(x)$ has an infinite sequence $\{x_{n,1}, x_{n,2}, \ldots\}$ of zeroes, with corresponding values $\kappa_{n,q} = x_{n,q}/R$ of κ. Thus we have a set of eigenvectors of the Laplacian that have the form

$$u_{pnm}(\vec{r}) = A_{p,n} j_n(\kappa_{n,p} r) Y_{nm}(\theta, \phi) \tag{8.66}$$

with eigenvalues related to zeroes of the spherical Bessel functions by

$$\lambda_{p,n} = -\kappa_{n,p}^2 \tag{8.67}$$

Note that the orthogonality of the $j_n(\kappa_{n,p} r)$ for fixed n, different p follows from general principles, but can also be proved directly from Bessel's equation (see Problem 5.23). ∎

8.3   Time-Dependent Partial Differential Equations

Time-dependent systems are described by equations such as
   (i) the *diffusion equation*

$$\frac{\partial \chi}{\partial t} = a \nabla^2 \chi \tag{8.68}$$

with diffusion constant $a > 0$. In this equation, χ might be the concentration of a solvent in a solution, or the temperature of a system (in the latter context, the equation is also known as the *heat equation*.
   (ii) the time-dependent *Schrödinger equation*

$$i\hbar \frac{\partial \psi}{\partial t} = \mathbf{H}\psi \tag{8.69}$$

for the wave function ψ of a quantum mechanical system, where $\mathbf{H}$ is a self-adjoint operator, the *Hamiltonian*, of the system.
   (i) the *wave equation*

$$\left(\frac{1}{c^2} \frac{\partial^2}{\partial t^2} - \nabla^2 \right) \phi = 4\pi\rho \tag{8.70}$$

This describes the amplitude ϕ of a wave propagating with speed c emitted by a source ρ— electromagnetic waves and sound waves are two examples. The equation also describes the transverse oscillations of a vibrating string.
   (iv) the *Klein–Gordon equation*

$$\left(\frac{1}{c^2} \frac{\partial^2}{\partial t^2} - \nabla^2 + a^2 \right) \phi = 0 \tag{8.71}$$

that describes a wave with a nonlinear dispersion relation

$$\omega^2 = (k^2 + a^2)c^2$$

The equation appears as a relativistic equation of motion for a scalar field ϕ whose quanta are particles of mass m in a quantum theory (in this context we have $a = mc/\hbar$).
   One typical problem is to find the time evolution of the variable $\phi = \phi(\vec{r}, t)$ (or χ or ψ) in a spatial region Ω, starting from an initial condition

$$\phi(\vec{r}, t = 0) = \phi_0(\vec{r}) \tag{8.72}$$

and, for an equation that is second order in time,

$$\frac{\partial}{\partial t}\phi(\vec{r}, t) \bigg|_{t=0} = \pi_0(\vec{r}) \tag{8.73}$$

In addition, ϕ must satisfy boundary conditions on the boundary S of Ω. These boundary conditions may be inhomogeneous, but they must be of a type that would make Δ self-adjoint if they were homogeneous.

8.3.1   The Diffusion Equation

Consider the diffusion equation (8.68)

$$\frac{\partial \chi}{\partial t} = a\nabla^2 \chi \tag{8.68}$$

with diffusion constant $a > 0$. This equation describes diffusion phenomena in liquids and gases. It can also describe heat flow, and thus temperature changes, in a heat-conducting medium. If κ is the thermal conductivity, ρ the mass density and C the specific heat of the medium, then the temperature distribution satisfies Eq. (8.68) with

$$a = \kappa C/\rho \tag{8.74}$$

Now suppose we have an infinite medium with initial temperature distribution $\chi_0(\vec{r})$. Then we introduce the Fourier transform $\tilde{\chi}(\vec{k}, t)$ by

$$\chi(\vec{r}, t) = \int \tilde{\chi}(\vec{k}, t) e^{i\vec{k}\cdot\vec{r}} \, d^3k \tag{8.75}$$

The diffusion equation for the Fourier transform $\tilde{\chi}(\vec{k}, t)$ is simply

$$\frac{\partial \tilde{\chi}(\vec{k}, t)}{\partial t} = -ak^2 \tilde{\chi}(\vec{k}, t) \tag{8.76}$$

with the solution

$$\tilde{\chi}(\vec{k}, t) = e^{-ak^2 t} \tilde{\chi}(\vec{k}, 0) \tag{8.77}$$

Since $\tilde{\chi}(\vec{k}, 0)$ is simply the Fourier transform of the initial distribution,

$$\tilde{\chi}(\vec{k}, 0) = \frac{1}{(2\pi)^3} \int e^{-i\vec{k}\cdot\vec{s}} \chi_0(\vec{s}) \, d^3s \tag{8.78}$$

we then have

$$\chi(\vec{r}, t) = \frac{1}{(2\pi)^3} \int e^{-ak^2 t} e^{i\vec{k}\cdot(\vec{r}-\vec{s})} \chi_0(\vec{s}) \, d^3s \, d^3k \tag{8.79}$$

The integral over d^3k can be done by completing the square in the exponent and using the standard Gaussian integral

$$\int_{-\infty}^{\infty} e^{-\alpha x^2} \, dx = \sqrt{\frac{\pi}{\alpha}} \tag{8.80}$$

This leads to the final result

$$\chi(\vec{r}, t) = \left(\frac{1}{4\pi a t}\right)^{3/2} \int \chi_0(\vec{s}) \exp\left(-\frac{(\vec{r}-\vec{s})^2}{4at}\right) d^3s \tag{8.81}$$

What this result means is that any disturbance in $\chi(\vec{r})$ decreases in time, spreading out on a distance scale proportional to $\ell = \sqrt{at}$ and thus over a volume proportional to ℓ^3. That the distance is proportional to $\sqrt{t}$ rather than to t is similar to a random walk process, in which the mean distance from the starting point increases with time only as $\sqrt{t}$.

Now consider the diffusion equation in a finite region Ω bounded by a closed surface S. A typical boundary condition might be that the temperature is held fixed with some distribution on the surface S, or that the normal derivative of the temperature should vanish on the surface so that no heat flows across the surface.

The homogeneous version of these and other boundary conditions on S will guarantee that Δ is a self-adjoint operator. Since the region Ω is bounded, there will be a complete orthonormal system $\{\phi_n\}$ of eigenvectors of Δ, with corresponding eigenvalues $\lambda_n = -\kappa_n^2 < 0$. The general solution $\chi(\vec{r}, t)$ of the diffusion equation can then be expanded as

$$\chi(\vec{r}, t) = \sum_{n=1}^{\infty} c_n(t)\phi_n(\vec{r}) \tag{8.82}$$

with time-dependent coefficients $c_n(t)$ that must satisfy the equations

$$\frac{d}{dt} c_n(t) = a\lambda_n c_n(t) = -a\kappa_n^2 c_n(t) \tag{8.83}$$

These equations have the elementary solutions

$$c_n(t) = c_n(0)e^{-a\kappa_n^2 t} \tag{8.84}$$

and the initial values $c_n(0)$ are determined simply from

$$c_n(0) = (\phi_n, \chi_0) \tag{8.85}$$

The formal solution to the original problem is then

$$\chi(\vec{r}, t) = \sum_{n=1}^{\infty} c_n(0)\phi_n(\vec{r})e^{-a\kappa_n^2 t} \tag{8.86}$$

Note that the convergence is exponential in time; it is an important property of the diffusion equation that solutions approach their asymptotic limit exponentially in time.

Remark. If $\chi(\vec{r}, t)$ satisfies an inhomogeneous boundary condition $\chi(\vec{r}, t) = \chi_s(\vec{r})$ on S, then a formal solution $\overline{\chi}(\vec{r})$ to Laplace's equation that satisfies these boundary conditions can be obtained from the Green function as in Eq. (8.30). Here we have

$$\overline{\chi}(\vec{r}) = -\int_S \chi_s(\vec{r})\,\hat{\mathbf{n}} \cdot \vec{\nabla}_s\, G^*(\vec{r}, \vec{s})\, dS_s \tag{8.87}$$

$\overline{\chi}(\vec{r})$ is the steady-state solution; the difference $\chi(\vec{r}, t) - \overline{\chi}(\vec{r})$ then has the form of Eq. (8.86) with the initial condition

$$\chi(\vec{r}, t = 0) = \chi_0(\vec{r}) - \overline{\chi}(\vec{r}) \tag{8.88}$$

and homogeneous boundary conditions on S. $\square$

8.3.2 Inhomogeneous Wave Equation: Advanced and Retarded Green Functions

Another typical problem is to find the wave amplitude due to a known source $\rho(\vec{r}, t)$. Here the relevant Green function $G(\vec{r}, t; \vec{s}, u)$ must satisfy the equation

$$\left(\nabla_r^2 - \frac{1}{c^2}\frac{\partial^2}{\partial t^2}\right) G(\vec{r}, t; \vec{s}, u) = -4\pi\delta(\vec{r} - \vec{s})\,\delta(t - u) \tag{8.89}$$

with appropriate conditions on the boundary of the region Ω in which the waves propagate. In addition, $G(\vec{r}, t; \vec{s}, u)$ must satisfy suitable initial conditions. In classical physics, the most relevant Green function is the *retarded* Green function $G_{\text{ret}}(\vec{r}, t; \vec{s}, u)$ defined by the condition

$$G_{\text{ret}}(\vec{r}, t; \vec{s}, u) = 0 \quad \text{for } t \leq u. \tag{8.90}$$

This condition is consistent with the intuitive notion of causality, which requires that the response appear later in time than the source that produces it. The wave amplitude due to the source $\rho(\vec{r}, t)$ is then given by

$$\phi(\vec{r}, t) = \int_{-\infty}^{\infty} \int_{\Omega} G_{\text{ret}}(\vec{r}, t; \vec{s}, u)\rho(\vec{s}, u)\,d\Omega_s\,du \tag{8.91}$$

If we consider the wave equation in n space dimensions and require that the amplitude vanish far from the source, the retarded Green function should depend only on the relative coordinates of the source and the observer, so that it can be expressed as a Fourier integral

$$G_{\text{ret}}(\vec{r}, t; \vec{s}, u) = \frac{4\pi}{(2\pi)^{n+1}} \int \widetilde{G}_{\text{ret}}(\vec{k}, \omega) e^{i\vec{k}\cdot(\vec{r}-\vec{s})} e^{-i\omega(t-u)}\,d^n k\,d\omega \tag{8.92}$$

where the signs in the exponent of the Fourier integral are chosen so that the results have a natural interpretation for electromagnetic waves. The inhomogeneous wave equation then requires

$$\left(\frac{\omega^2}{c^2} - k^2\right) \widetilde{G}_{\text{ret}}(\vec{k}, \omega) = -4\pi \tag{8.93}$$

so that

$$\widetilde{G}_{\text{ret}}(\vec{k}, \omega) = -\frac{4\pi c^2}{\omega^2 - k^2 c^2} \tag{8.94}$$

and

$$G_{\text{ret}}(\vec{r}, t; \vec{s}, u) = -\frac{4\pi c^2}{(2\pi)^{n+1}} \int \frac{1}{\omega^2 - k^2 c^2} e^{i\vec{k}\cdot(\vec{r}-\vec{s})} e^{-i\omega(t-u)}\,d^n k\,d\omega \tag{8.95}$$

The integration over ω requires a choice of contour in the complex ω-plane to avoid the singularities of the integrand at $\omega = \pm kc$. This choice of contour can be used to satisfy the initial conditions. For example, the contour C_{ret} in Fig. 8.1 is correct for the retarded

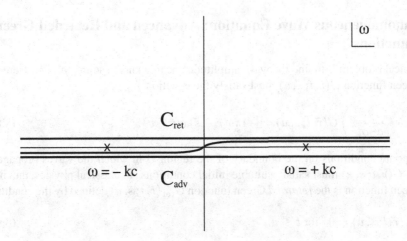

Figure 8.1: Contours of integration in the complex ω-plane used to evaluate various Green functions for the wave equation. The upper contour $C_{\rm ret}$ gives the retarded Green function; the lower contour $C_{\rm adv}$ the advanced Green function. The middle contour gives the Feynman Green function, which is the photon 'propagator' in quantum field theory.

condition (8.90). For $t < 0$ the contour can be closed in the upper-half plane, where there are no singularities of the integrand, while for $t > 0$ the contour must be closed in the lower half-plane, and the integral picks up contributions from both poles of the integrand. Then for $t > 0$, we have

$$G_{\rm ret}(\vec{r}, t; \vec{s}, u) = \frac{4\pi c}{(2\pi)^n} \int \frac{\sin kc\tau}{k} e^{i\vec{k}\cdot\vec{\rho}} d^n k \tag{8.96}$$

where $\vec{\rho} \equiv \vec{r} - \vec{s}$ and $\tau \equiv t - u$.

In three dimensions, the integral can be evaluated in spherical coordinates to give

$$G_{\rm ret}(\vec{\rho}, \tau) = \frac{4\pi c}{(2\pi)^2} \int_0^\infty \int_{-1}^1 \frac{\sin kc\tau}{k} e^{ik\rho\mu} d\mu\, k^2\, dk \tag{8.97}$$

$$= \frac{2c}{\pi\rho} \int_0^\infty \sin k\rho \sin kc\tau\, dk$$

Here the Fourier integral theorem gives

$$\int_0^\infty \sin k\rho \sin kc\tau\, dk = \frac{\pi}{2} \left[\delta(\rho - c\tau) - \delta(\rho + c\tau)\right] \tag{8.98}$$

where here $\delta(\rho + c\tau) = 0$ since $\rho > 0$ and $\tau > 0$. Thus, finally,

$$G_{\rm ret}(\vec{\rho}, \tau) = \frac{c}{\rho} \delta(\rho - c\tau) = \frac{1}{\rho} \delta\left(\tau - \frac{\rho}{c}\right) \tag{8.99}$$

This result has the natural interpretation that the wave propagates outward from its source with speed c, with amplitude inversely proportional to the distance from the source.

Remark.   The retarded Green function has this sharp form only in an odd number of spatial dimensions. In an even number of dimensions there is a trailing wave in addition to the sharp wave front propagating outward from the source. The interested reader is invited to work this out for $n = 2$, which is realized physically by waves emitted from a long linear source.   □

The *advanced* Green function $G_{\text{adv}}(\vec{r}, t; \vec{s}, u)$ is defined by the condition

$$G_{\text{adv}}(\vec{r}, t; \vec{s}, u) = 0 \quad \text{for} \quad t \geq u. \tag{8.100}$$

An appropriate contour for this Green function is also shown in Fig. 8.1. The contour between C_{ret} and C_{adv} defines the *Feynman*, or *causal* Green function. This contour gives a contribution to the Green function from the pole at $\omega = -kc$ for $t < 0$, and from the pole at $\omega = +kc$ for $t > 0$. This Green function is especially important since it is the "propagator" of the photon in quantum electrodynamics. A derivation of this propagator can be found in most textbooks on quantum field theory.

If the radiation source is oscillating at a single frequency, so that

$$\rho(\vec{r}, t) = \rho_\omega(\vec{r}) e^{-i\omega t} \tag{8.101}$$

then the wave amplitude also has the same frequency, so that

$$\phi(\vec{r}, t) = \phi_\omega(\vec{r}) e^{-i\omega t} \tag{8.102}$$

and $\phi_\omega(\vec{r})$ must satisfy

$$\left(\Delta + \frac{\omega^2}{c^2} \right) \phi_\omega(\vec{r}) = -4\pi \rho_\omega(\vec{r}) \tag{8.103}$$

which is the scalar *Helmholtz equation*.

The Green function for this equation is a resolvent of the Laplacian and can be formally expressed in terms of the eigenfunctions and eigenvalues of Δ. The boundary condition on the Green function is that the waves should look asymptotically like waves radiating *outward* from the source; this corresponds to the retarded condition (8.90) on the full Green function.

❏ **Example 8.4.**   In three dimensions, the wave amplitude ϕ_ω is given by

$$\phi_\omega(\vec{r}) e^{-i\omega t} = \int \frac{\rho_\omega(\vec{s})}{|\vec{r} - \vec{s}|} \delta\left(t - u - \frac{|\vec{r} - \vec{s}|}{c} \right) e^{-i\omega u} \, d^3s \, du \tag{8.104}$$

using the retarded Green function (8.99), so that

$$\phi_\omega(\vec{r}) = \int \frac{e^{ik|\vec{r} - \vec{s}|}}{|\vec{r} - \vec{s}|} \rho_\omega(\vec{s}) \, d^3s \tag{8.105}$$

$(k = \omega/c)$, corresponding to the Green function

$$G_\omega(\vec{r}, \vec{s}) = \frac{e^{ik|\vec{r} - \vec{s}|}}{|\vec{r} - \vec{s}|} \tag{8.106}$$

Note that far from the source (which supposes that the source is of finite extent), the amplitude has the asymptotic form

$$\phi_\omega(\vec{r}) \overset{r \to \infty}{\longrightarrow} \frac{e^{ikr}}{r} \int e^{-i\vec{k}\cdot\vec{s}}\rho_\omega(\vec{s})\, d^3s \tag{8.107}$$

where $\vec{k} = k\hat{r}$ points radially outward from the source to the point of observation. This corresponds to a wave radiating outward from the source, with amplitude depending on direction through the spatial Fourier transform of the source density.

From the result of Problem 5.14, we have the expansion

$$e^{-i\vec{k}\cdot\vec{s}} = \sum_{n=0}^{\infty} (-i)^n (2n+1) j_n(ks) P_n \cos\theta \tag{8.108}$$

where the $j_n(ks)$ are spherical Bessel functions and the P_n are Legendre polynomials. Using the spherical harmonic addition theorem, we have the asymptotic expansion

$$\phi_\omega(\vec{r}) \overset{r \to \infty}{\longrightarrow} 4\pi \frac{e^{ikr}}{r} \sum_{n=0}^{\infty} (-i)^n \sum_{m=-n}^{n} q_{nm}(\omega) Y_{nm}(\theta, \phi) \tag{8.109}$$

where the multipole moments $q_{nm}(\omega)$ are here given by

$$q_{nm}(\omega) = \int Y_{nm}^*(\theta_s, \phi_s) j_n(ks) \rho_\omega(\vec{s})\, d^3s \tag{8.110}$$

As with the multipole expansion for the Coulomb potential, there are various normalization conventions in use. ∎

Remark. The full Green function (8.106) for the scalar Helmholtz equation also has a multipole expansion; for $r < s$, we have

$$\frac{e^{ik|\vec{r}-\vec{s}|}}{abs\vec{r}-\vec{s}} = 4\pi ik \sum_{n=0}^{\infty} (-i)^n j_n(ks)\, h_n(kr)_> \sum_{m=-n}^{n} Y_{nm}(\theta, \phi) Y_{nm}^*(\theta_s, \phi_s) \tag{8.111}$$

It is left to the reader to verify this expansion. □

Analysis of the Klein–Gordon equation and its Green functions proceeds in a similar way. The details are left to Problem 10. See also Section 8.4.3 for a Klein–Gordon equation with an additional nonlinear term.

8.3.3   The Schrödinger Equation

The time-dependent Schrödinger equation for a quantum mechanical system with Hamiltonian **H** is

$$i\hbar \frac{\partial \Psi(t)}{\partial t} = \mathbf{H}\Psi(t) \tag{8.112}$$

The Hamiltonian may be constructed by correspondence with the classical Hamiltonian of the system (see Chapter 3), or by other physical principles for a system, such as a spin system, with no classical analog. In any case, physics requires **H** to be a self-adjoint operator; the eigenstates of the Hamiltonian are states of definite energy of the system, and only energies in the spectrum of **H** are possible for the system. A formal solution to Eq. (8.112) can be given in terms of the resolution of the identity of the Hamiltonian—see Problem 8—but it is purely formal.

An eigenstate ψ of **H** satisfies the *time-independent* Schrödinger equation

$$\mathbf{H}\psi = E\psi \tag{8.113}$$

Then $\Psi(t) = \psi \exp(i\omega t)$ is a solution of the time-dependent Schrödinger equation (8.112), where

$$E = \hbar \omega \tag{8.114}$$

and $\hbar$ is Planck's constant; Eq. (8.114) is essentially the relation between energy and frequency suggested by Planck. Since $\Psi(t)$ changes in time only by a phase factor $\exp(i\omega t)$, all matrix elements $(\Psi, \mathbf{A}\Psi) = (\psi, \mathbf{A}\psi)$ are independent of time; hence ψ is a *stationary* state of the system. Equation (8.113) is also known as the (time-independent) Schrödinger equation.

For a particle of mass μ moving in a potential $V(\vec{r})$, the Hamiltonian is

$$\mathbf{H} = -\frac{\hbar^2}{2\mu}\Delta + V(\vec{r}) \tag{8.115}$$

Equation (8.113) is a (partial) differential equation for the particle wave function $\psi(\vec{r})$. The physical boundary conditions on the wave function are such that If $V(\vec{r})$ is real, then **H** is a self-adjoint operator whose spectrum is determined by the properties of the solutions of Eq. (8.113).

❑ **Example 8.5.**   For an electron in the Coulomb potential of a point nucleus with charge Ze, we have $V(r) = -Ze^2/r$ (again we set $4\pi\varepsilon_0 = 1$). The time-independent Schrödinger equation is then

$$\mathbf{H}\psi(\vec{r}) = \left\{ -\frac{\hbar^2}{2m_e}\Delta - \frac{Ze^2}{r} \right\} \psi(\vec{r}) = E\psi(\vec{r}) \tag{8.116}$$

where m_e is the mass of the electron. We can separate variables in spherical coordinates, and look for solutions of the form

$$\psi(\vec{r}) = R_{E\ell}(r)Y_{\ell m}(\theta, \phi) \tag{8.117}$$

where $Y_{\ell m}(\theta, \phi)$ is a spherical harmonic as introduced in Section 6.5.3. We note that ℓ is related to the angular momentum of the electron, as explained further in Chapter 10.

The radial function $R_{E\ell}(r)$ must then satisfy the equation

$$-\frac{\hbar^2}{2m_e r^2}\frac{d}{dr}\left(r^2\frac{dR_{E\ell}}{dr}\right) + \left[\frac{\ell(\ell+1)\hbar^2}{2m_e r^2} - \frac{Ze^2}{r}\right]R_{E\ell} = ER_{E\ell} \tag{8.118}$$

This equation has a regular singular point at $r = 0$ and irregular singular point at ∞, suggesting that it might be related to the confluent hypergeometric equation introduced in Chapter 5. To proceed further, we introduce the parameter $a_Z = \hbar^2/Zm_e e^2$ to set the length scale (note that for $Z = 1$, $a = \hbar^2/m_e e^2 = 52.9$ pm is the Bohr radius). Then introduce the dimensionless variable $\rho = 2r/a_Z$ (the reason for the factor two will become apparent soon), and let

$$R_{E\ell}(r) = u(\rho) \tag{8.119}$$

(we drop the subscripts E and ℓ for now). The radial equation (8.118) then becomes

$$u''(\rho) + \frac{2}{\rho}u'(\rho) + \left[\frac{1}{\rho} - \frac{\ell(\ell+1)}{\rho^2}\right]u(\rho) = -\tfrac{1}{4}\lambda u(\rho) \tag{8.120}$$

Here $\lambda = E/E_Z$, with energy scale E_Z defined by

$$E_Z = \frac{Ze^2}{2a_Z} = \frac{Z^2 m_e e^4}{2\hbar^2} = \tfrac{1}{2}Z^2\alpha^2 m_e c^2 \tag{8.121}$$

where $\alpha = e^2/\hbar c$ is the dimensionless fine structure constant and c is the speed of light. We will soon see that E_Z is the binding energy of the ground state of the electron.

The indices of Eq. (8.120) at $\rho = 0$ are ℓ and $-(\ell + 1)$, so there are solutions proportional to ρ^ℓ and $\rho^{-\ell-1}$ for $\rho \to 0$. The second solution is singular at $\rho = 0$ and must not be present. Only the first solution is allowed. Thus we let

$$u(\rho) = \rho^\ell v(\rho) \tag{8.122}$$

Then $v(\rho)$ must be a solution of the equation

$$\rho v''(\rho) + 2(\ell + 1)v'(\rho) + (\tfrac{1}{4}\lambda\rho + 1)v(\rho) = 0 \tag{8.123}$$

For $\rho \to \infty$, this equation has the asymptotic form

$$v'' + \tfrac{1}{4}\lambda v = 0 \tag{8.124}$$

For positive energy E, we can let $\lambda = \kappa^2$ and there are solutions $v_\pm(\rho)$ that have the asymptotic behavior $\exp(\pm\tfrac{1}{2}i\kappa\rho)$ for $\rho \to \infty$, corresponding to incoming $(-)$ or outgoing $(+)$ spherical waves. These solutions have logarithmic phase factors in addition to the usual $\exp(ikr)$ and $\exp(ikz)$ forms, arising from the long-range nature of the Coulomb potential. These subtleties are best seen in parabolic coordinates introduced in Problem 3.9. We leave them for the reader to work out in Problem 9.

For $E < 0$, we let $\lambda = -\beta^2$. Then there are solutions with asymptotic behavior $\exp(\pm\frac{1}{2}\beta\rho)$ for $\rho \to \infty$. Only the exponentially decreasing solution, corresponding to a bound state of the electron, is allowed. In general, the solution $v(\rho)$ that is analytic at $r = 0$ will not be exponentially decreasing as $\rho \to \infty$; there will be a discrete set of values of β for which this is the case, and the corresponding values of energy $E < 0$ define the discrete spectrum of the Hamiltonian (8.116). If we now let

$$v(\rho) = e^{-\frac{1}{2}\beta\rho}w(\rho) \tag{8.125}$$

then $w(\rho)$ must satisfy

$$\rho w''(\rho) + [2(\ell + 1) - \beta\rho]w'(\rho) + [1 - \beta(\ell + 1)]w(\rho) = 0 \tag{8.126}$$

To transform this to the confluent hypergeometric equation, we now let $x = \beta\rho$ and $f(x) = w(\rho)$. Then $f(x)$ satisfies the equation

$$xf''(x) + [2(\ell + 1) - x]f'(x) + (\beta^{-1} - \ell - 1)f(x) = 0 \tag{8.127}$$

This is the confluent hypergeometric equation (5.B38), with

$$a = \ell + 1 - \beta^{-1} \quad \text{and} \quad c = 2(\ell + 1)$$

The solution $F(a|c|x)$ that is analytic at $x = 0$ has a part that grows exponentially, unless a is zero or a negative integer, when there is a polynomial solution. Here that requirement means that we must have

$$\frac{1}{\beta} = k + \ell + 1 \equiv n \tag{8.128}$$

with $k = 0, 1, 2, \ldots$, or $n = \ell, \ell + 1, \ell + 2, \ldots$. In this case, the solution is a polynomial of degree k, proportional to an associated Laguerre polynomial defined in Appendix A. Explicitly, we have a solution

$$f(x) = F(\ell + 1 - n|2(\ell + 1)|\beta\rho) = AL^{2\ell+1}_{n-\ell-1}(\beta\rho) \tag{8.129}$$

where A is a normalization constant. Returning to the original variables, we can write the wave functions $\psi_{n\ell m}(\vec{r})$ corresponding to energy

$$E_n = -\frac{Ze^2}{n^2 a_Z} \tag{8.130}$$

as

$$\psi_{n\ell m}(\vec{r}) = AL^{2\ell+1}_{n-\ell-1}\left(\frac{2r}{na_Z}\right)\exp\left(-\frac{r}{na_Z}\right)Y_{\ell m}(\theta, \phi) \tag{8.131}$$

with $n = 1, 2, \ldots, \ell = 0, 1, \ldots, n - 1$ and $m = \ell, \ell - 1, \ldots, -\ell$.   ∎

8.4   Nonlinear Partial Differential Equations

8.4.1   Quasilinear First-Order Equations

In Section 8.1, we studied a linear first order partial differential equation that describes a wave propagating in one space dimension with speed s. Suppose now that the speed s depends on the amplitude of the wave, so that we have a *nonlinear* wave equation

$$\frac{\partial u}{\partial t} + s(u)\frac{\partial u}{\partial x} = 0 \tag{8.132}$$

This equation can arise from a conservation law for a system described by a density $u(x,t)$ and a flux density Φ, such that

$$\frac{d}{dt}\int_a^b u(x,t)\,dx = \Phi(a) - \Phi(b) = -\int_a^b \frac{\partial \Phi(x,t)}{dx}\,dx \tag{8.133}$$

for any a and b. From this equation it follows that

$$\frac{\partial u}{\partial t} + \frac{\partial \Phi}{\partial x} = 0 \tag{8.134}$$

Equation (8.134) is a equation of continuity similar to Eq. (3.138); it expresses a local version of conservation of a charge q defined by

$$q = \int_{-\infty}^{\infty} u(x,t)\,dx \tag{8.135}$$

since Eq. (8.133) means that charge can leave the interval $[a,b]$ only if there is a flux across the endpoints of the interval.

❑ **Example 8.6.**   In the book by Billingham and King cited at the end of the chapter, this equation appears as a model to describe the flow of traffic on a highway. $u(x,t)$ is the density of cars, and the flux $\Phi(x,t)$ is expressed as $vu(x,t)$, where v is the speed of the cars. With the empirical assumption that v depends only on the density—there is data that is consistent with this assumption, at least as a first approximation—the conservation law leads to Eq. (8.132) with

$$s(u) = v(u) + uv'(u) \tag{8.136}$$

Note here that $s(u)$ is not the speed of the cars, but the speed of propagation of fluctuations in the density of cars.   ∎

In general, if the flux Φ is a function of u only, with no explicit dependence on x or t, Eq. (8.134) can be written as

$$\frac{\partial u}{\partial t} + \Phi'(u)\frac{\partial u}{\partial x} = 0 \tag{8.137}$$

which is the original Eq. (8.132) with speed $s(u) = \Phi'(u)$.

We want to find solutions to Eq. (8.132) starting from an initial condition

$$u(x,0) = u_0(x) \tag{8.138}$$

As with the linear equation with constant s, we can construct characteristics of the partial differential equation as curves $x = x(t)$ on which the wave amplitude u is a constant. Along a characteristic, we have

$$du = \frac{\partial u}{\partial x}\, dx + \frac{\partial u}{\partial t}\, dt = \frac{\partial u}{\partial x}\, [dx - s(u)dt] = 0 \tag{8.139}$$

Thus the characteristics are straight lines, and the characteristic passing through the point x at $t = 0$ is determined from its slope

$$\frac{dx}{dt} = s_0(x) = s[u_0(x)] \tag{8.140}$$

Then to find the amplitude u at a later time $t > 0$, we need to solve the implicit equation

$$u(x,t) = u(x - s(u_0)t, 0) \tag{8.141}$$

as we trace the solution along its characteristic so that

$$u(x + s(u_0)t, t) = u(x,0) = u_0 \tag{8.142}$$

However, the solution is not as simple as that in general. If the wave speed depends on the amplitude, then the characteristics are not parallel, and two characteristics will meet, either in the future or in the past. Since waves are propagating to the right along the X-axis, it is clear that a faster wave coming from the left will catch up with a slower wave starting to the right of the fast wave. Thus some characteristics will meet at some time after $t = 0$ unless $s(u_0)$ is a nondecreasing function of x. In that case, the wave is described as a *rarefaction wave*, since there are no faster waves coming from the left; all the intersections of the characteristics lie on the past (i.e., for $t < 0$).

What happens, then, when two characteristics meet? Consider the characteristics coming from $x = a$ and $x = b$, and suppose the initial conditions are such that

$$u(a,0) = u_a \quad \text{and} \quad u(b,0) = u_b$$

If $s_a = s(u_a)$ and $s_b = s(u_b)$, then the characteristics from a and b will meet

$$\text{at} \quad x = \frac{a+b}{2} + \frac{s_a + s_b}{2}\frac{a-b}{s_b - s_a} \quad \text{when} \quad t = \frac{a-b}{s_b - s_a}$$

The characteristics cannot continue beyond this point, and if they have not already encountered other characteristics, they will form a point of discontinuity in the solution of the partial differential equation (8.132). The discontinuities formed by merging characteristics will form a line that is understood as a *shock wave* or *shock front*, across which the solution $u(x,t)$ has a discontinuity in x.

We can compute the discontinuity of the solution across a shock using the original partial differential equation. If the discontinuity of $u(x,t)$ is at $x = X(t)$, then we can integrate Eq. (8.132) from $a < X$ to $b > X$. This gives

$$\frac{\partial}{\partial t} \int_a^b u(x,t)\, dx = \Phi(a) - \Phi(b) \tag{8.143}$$

With $u^{\pm} = \lim_{\varepsilon \to 0^+} u(X \pm \varepsilon, t)$, we then have

$$\frac{\partial}{\partial t} \left[(b - X)u^+ + (X - a)u^- \right] = \Phi(a) - \Phi(b) \tag{8.144}$$

The velocity of the shock wave is $V = dX/dt$; if we pass to the limit $\varepsilon \to 0$ we find a relation

$$V(u^- - u^+) = \Phi(u^-) - \Phi(u^+) \tag{8.145}$$

between the discontinuities and the velocity of the shock wave. Most often this allows us to compute V as

$$V = \frac{\Phi(u^-) - \Phi(u^+)}{u^- - u^+} \tag{8.146}$$

Problem 12 gives an explicit example of such a shock wave.

8.4.2   KdV Equation

The linear wave equation for a wave with a dispersion relation $\omega = ks(1 - \alpha^2 k^2)$ was given in Eq. (8.16). Now we are interested in the waves that result when the speed of propagation also depends on the amplitude u of the wave as in the preceding section. Thus we start with a general equation of the form

$$\frac{\partial u}{\partial t} + s(u)\frac{\partial u}{\partial x} + s\alpha^2 \frac{\partial^3 u}{\partial x^3} = 0 \tag{8.147}$$

If we consider only the equation when $s(u)$ is a linear function of u, then linear transformations of the variables (u, x and t) allow Eq. (8.147) to be cast in the form

$$\frac{\partial u}{\partial t} + 6u\frac{\partial u}{\partial x} + \frac{\partial^3 u}{\partial x^3} = 0 \tag{8.148}$$

This is a standard form of the *Kortweg–deVries* or *KdV equation*.

→ **Exercise 8.3.**   Find a set of linear transformations of the variables that change Eq. (8.147) to Eq. (8.148).   □

One important property of this equation is that it has travelling wave solutions of the form

$$u(x,t) = U(x - st) \equiv U(\xi) \tag{8.149}$$

that travel to the right with speed $s = \kappa^2$; here we have introduced the variable $\xi = x - st$. The KdV equation requires $U(\xi)$ to satisfy the equation

$$U'''(\xi) + 6U(\xi)U'(\xi) - \kappa^2 U'(\xi) = 0 \tag{8.150}$$

Equation (8.150) can be immediately integrated to give

$$U''(\xi) + 3[U(\xi)]^2 - \kappa^2 U(\xi) = 0 \tag{8.151}$$

where a constant of integration on the right-hand side has been set to zero so that $U(\xi)$ tends to zero for $|\xi| \to \infty$. Then multiply by $U'(\xi)$ and integrate once more to give

$$\tfrac{1}{2}[U'(\xi)]^2 + [U(\xi)]^3 - \tfrac{1}{2}\kappa^2[U(\xi)]^2 = 0 \tag{8.152}$$

where another integration constant has been set to zero so that $U(\xi)$ will vanish for $|\xi| \to \infty$. We now have the differential equation

$$\frac{dU}{d\xi} = U(\xi)\sqrt{2U(\xi) - \kappa^2} \tag{8.153}$$

The integral of this differential equation is not immediately obvious, but it is not too difficult to verify that the solution

$$U(\xi) = \frac{\kappa^2}{2}\operatorname{sech}^2(\tfrac{1}{2}\kappa\xi) = \frac{\kappa^2}{2\cosh^2(\tfrac{1}{2}\kappa\xi)} \tag{8.154}$$

actually satisfies Eq. (8.153).

The solution (8.154) is sharply peaked around $\xi = 0$ and vanishes exponentially for $\xi \to \infty$. The one parameter κ characterizes both the height and width of the peak, as well as the speed of the wave—larger κ means a more rapid wave, with a higher and narrower peak. The shape of the wave is constant in time, in contrast to linear waves with no dispersion, i.e., a speed of propagation independent of frequency. The solution is called a *solitary wave*, or *soliton*. However, we note that in modern terminology, the term soliton actually implies further properties of the equations—especially that there are multisoliton solutions that correspond to solitons that pass through each other maintaining their shape both before and after interacting. The KdV solitons have all these properties, but we do not discuss them further here—see the books cited in the bibliography, as well as many others, for further details.

➜ **Exercise 8.4.**   One property of the KdV equation is that it admits an infinite set of conservation laws. Show that it follows from the equations of motion that the following quantities are conserved:

$$m = \int_{-\infty}^{\infty} u\, dx$$

$$P = \int_{-\infty}^{\infty} u^2\, dx$$

$$E = \int_{-\infty}^{\infty} (\tfrac{1}{2}u_x^2 - u^3)\, dx$$

Assume that u and its derivatives vanish as $x \to \pm\infty$. We note that these three laws correspond to conservation of mass, momentum and energy, and are quite general. It is the laws beyond this that are special for the KdV equation.   □

8.4.3   Scalar Field in $1 + 1$ Dimensions

The Klein–Gordon equation (8.71) for a scalar field $\phi(x, t)$ in one space (+ one time) dimension can be derived from a Lagrangian density

$$\mathcal{L} = \tfrac{1}{2} \left(\phi_t^2 - \phi_k^2 - m^2 \phi^2 \right) \tag{8.155}$$

as explained in Appendix A. Here $\phi_t = \partial \phi / \partial t$ and $\phi_t = \partial \phi / \partial x$. As already noted, the Klein–Gordon equation describes the propagation of waves with a dispersion relation

$$\omega^2 = k^2 + m^2 \tag{8.156}$$

that is associated waves with a free relativistic particle m in units with $\hbar = c = 1$. The dispersion relation also describes propagation of electromagnetic waves in an ideal plasma, where m is the plasma frequency.

The Lagrangian (8.155) is a special case of the Lagrangian

$$\mathcal{L} = \tfrac{1}{2} \left(\phi_t^2 - \phi_k^2 \right) - U(\phi) \tag{8.157}$$

with potential $U(\phi) = \tfrac{1}{2} m^2 \phi^2$, for which the equation of motion is

$$\frac{\partial^2 \phi}{\partial t^2} - \frac{\partial^2 \phi}{\partial x^2} = -\frac{dU(\phi)}{d\phi} \tag{8.158}$$

as derived in Appendix A. Also noted there was the energy of the field, which is given by

$$E[\phi] = \int_{-\infty}^{\infty} \left[\tfrac{1}{2} \left(\phi_t^2 + \phi_x^2 \right) + U(\phi) \right] dx \tag{8.159}$$

Now assume that the potential $U(\phi)$ is non-negative, and that $U(\phi) = 0$ only for a discrete set of values $\phi = v_1, v_2, \ldots, v_N$. Then these values of ϕ are absolute minima of $U(\phi)$, and

$$E[\phi] = 0 \text{ if and only if } \phi(x, t) = v_k$$

for some $k = 1, \ldots, N$. These (trivial) static solutions minimize the energy $\mathcal{E}[\phi]$.

If the potential has more than one minimum, there may also be nontrivial static solutions of the field equation (8.158) that have finite total energy. These solutions must satisfy the equation

$$\frac{d^2 \phi}{dx^2} = \frac{dU(\phi)}{d\phi} \tag{8.160}$$

This equation has an integral

$$W = \frac{1}{2} \left(\frac{d\phi}{dx} \right)^2 - U(\phi) \tag{8.161}$$

W must be constant, independent of x, for any solution of Eq. (8.160). For the solution to have finite energy, it is necessary that

$$E[\phi] = \int_{-\infty}^{\infty} dx \left[\frac{1}{2} \left(\frac{d\phi}{dx} \right)^2 + U(\phi) \right] < \infty \tag{8.162}$$

This requires

$$\lim_{x \to \pm\infty} \phi(x) \equiv v^{\pm} \tag{8.163}$$

with v^+ and v^- among the $\{v_1, \ldots, v_N\}$, since Eq. (8.163) implies $U(\phi) \to 0$ as well as $W \to 0$ for $x \to \pm\infty$. Then also $W = 0$ for all x. However, if $N > 1$, we need not have $v^+ = v^-$, and there may be nontrivial solutions of $W = 0$ that interpolate between two minima of $U(\phi)$, as well as solutions for which $v^+ = v^-$.

Remark. Equation (8.160) has the same form as Newton's equation of motion for a particle with potential energy $-U(x)$. A finite energy static solution of the field equation corresponds to a finite action zero-energy solution of the particle equation of motion, in which the particle moves between two adjacent zeros of U. □

Remark. Static finite-energy solutions to the field equations derived from the Lagrangian (8.157) can only exist in one space dimension (this result is known as *Derrick's theorem*). To see this, consider a field configuration $\phi_\alpha(x) \equiv \phi(\alpha x)$ with rescaled spatial coordinates. The energy associated with this configuration in q space dimensions is

$$E_\alpha \equiv E[\phi_\alpha] = \int_{-\infty}^{\infty} d^q x \left[\tfrac{1}{2}(\vec{\nabla}\phi_\alpha \cdot \vec{\nabla}\phi_\alpha) + U(\phi_\alpha) \right] \equiv K_\alpha + V_\alpha$$

$$= \int_{-\infty}^{\infty} d^q \xi \left[\frac{1}{2}\alpha^{2-q}(\vec{\nabla}\phi_\alpha \cdot \vec{\nabla}\phi_\alpha) + \alpha^{-q}U(\phi) \right] = \alpha^{2-q}K + \alpha^{-q}V \tag{8.164}$$

with $\xi = \alpha x$. Here K, V are the values of the integrals for $\alpha = 1$. Since E_α must have an extremum at $\alpha = 1$, it follows that

$$(2 - q)K = qV \tag{8.165}$$

which is consistent with positivity of K and V only for $0 < q < 2$. □

The equation $W = 0$ can be solved by quadrature to give

$$x - x_0 = \pm \int_{\phi_0}^{\phi} \frac{1}{\sqrt{U(\psi)}} \, d\psi \tag{8.166}$$

with $\phi_0 = \phi(x_0)$. This shows how the solutions interpolate between minima of $U(\phi)$, which by assumption have $U(\phi) = 0$. If $U(\phi)$ is $O(\phi - v_k)^2$ near the minimum at $\phi = v_k$, then the integral diverges as the minimum is approached, corresponding to $x \to \pm\infty$.

❑ **Example 8.7.** For the $(1 + 1)$-dimensional ϕ^4 field theory, the potential is

$$U(\phi) = \frac{1}{2}\lambda \left(\phi^2 - \frac{m^2}{\lambda} \right)^2 \tag{8.167}$$

with $U(\phi) = 0$ at the degenerate minima $\phi = \pm v \equiv v_\pm$, where $v^2 = m^2/\lambda$. Static solutions of the field equations satisfy

$$\frac{d^2\phi}{dx^2} = 2\lambda\phi^3 - 2m^2\phi \tag{8.168}$$

whence

$$x - x_0 = \pm\sqrt{\frac{1}{\lambda}} \int_{\phi_0}^{\phi} \frac{1}{\psi^2 - v^2}\, d\psi \tag{8.169}$$

This equation can be inverted to give

$$\phi(x) = \pm v \tanh\left[m(x - x_0)\right] \equiv \phi_{\pm}(x; x_0) \tag{8.170}$$

The solution $\phi_+(x; x_0)$ $[\phi_-(x; x_0)]$ interpolates between $-v$ $[+v]$ as $x \to -\infty$ to $+v$ $[-v]$ as $x \to +\infty$, while vanishing at x_0. Hence ϕ_+ $[\phi_-]$ is a known as a *kink* [*antikink*]. The energy density of the kink (or antikink) solution is given by

$$\mathcal{E}(x) = 2U(\phi) = \frac{m^4}{\lambda} \frac{1}{\cosh^4[m(x - x_0)]} \tag{8.171}$$

which is sharply localized near $x = x_0$. The total energy of the kink solution (the *classical mass M_**) of the kink is

$$E[\phi_{\pm}] \equiv M_* = \int_{-\infty}^{\infty} \mathcal{E}(x)dx = \frac{4m}{3\lambda} \tag{8.172}$$

where we have used the integral

$$\int_0^{\infty} \frac{d\xi}{\cosh^4 \xi} = \frac{1}{3}\left[\frac{\sinh \xi}{\cosh^3 \xi} + 2\frac{\sinh \xi}{\cosh \xi}\right]_0^{\infty} = \frac{2}{3} \tag{8.173}$$

Since Eq. (8.158) is invariant under Lorentz transformations (in $1 + 1$ dimensions), we can transform of the kink and antikink solutions to find solutions that correspond to moving kinks or antikinks. We have

$$\phi_{\pm}^u(x; x_0) = \pm v \tanh\left[\frac{m(x - x_0 - ut)}{\sqrt{(1 - u^2)}}\right] \tag{8.174}$$

with energy

$$E[\phi_{\pm}^u] = M_*/\sqrt{1 - u^2} \tag{8.175}$$

appropriate to a particle of mass M_* with speed u (note that we have set $c = 1$). ∎

Remark.   For any field ϕ in $1 + 1$ dimensions, there is a *topological current density*

$$K^{\mu} \equiv \varepsilon^{\mu\nu} \frac{\partial \phi}{\partial x^{\nu}} \tag{8.176}$$

that automatically satisfies $\partial K^{\mu}/\partial x^{\mu} = 0$ for nonsingular fields. Hence the *topological charge*

$$Q \equiv \int_{-\infty}^{\infty} K_0\, dx = v^+ - v^- \tag{8.177}$$

is conserved. Then the space of nonsingular finite-energy solutions can be divided into *topological sectors* associated with distinct values of the topological charge.   □

➡ **Exercise 8.5.**   Find the topological charge (in units of $2v$) for the static kink and antikink solutions. Is this charge the same for the corresponding moving solitons?   □

8.4.4 Sine-Gordon Equation

A special potential $U(\phi)$ which has applications in nonlinear optics, in addition to its interest as a toy model, is

$$U(\phi) = \lambda v^4 \left[1 - \cos\left(\frac{\phi}{v}\right) \right] = 2\lambda v^4 \sin^2\left(\frac{\phi}{2v}\right) \tag{8.178}$$

where v is a dimensional constant which we can identify with $m/\sqrt{\lambda}$ of the ϕ^4 theory. This potential is bounded, and has zeros for

$$\phi = 2n\pi v \equiv \phi_n \tag{8.179}$$

$(n = 0, \pm 1, \pm 2, \ldots)$. The field equation for this potential is

$$\frac{\partial^2 \phi}{\partial t^2} - \frac{\partial^2 \phi}{\partial x^2} = -\lambda v^3 \sin\left(\frac{\phi}{v}\right) \tag{8.180}$$

For small ϕ, the right-hand side is approximated by $-m^2\phi$, and the equation looks like the Klein–Gordon equation in this limit. This has led to the slightly whimsical name *sine-Gordon equation* for (8.180).

Equation (8.180) has static finite energy solutions obtained from

$$x - x_0 = \pm \int_{\phi_0}^{\phi} \frac{1}{2\sqrt{\lambda} v^2 \sin(\psi/2v)} \, d\psi \tag{8.181}$$

This can be evaluated using the integral

$$\int \frac{d\xi}{\sin \xi} = \ln\left(\tan \tfrac{1}{2}\xi\right) \tag{8.182}$$

to give

$$\ln\left(\tan \frac{\phi}{4v}\right) = \ln\left(\tan \frac{\phi_0}{4v}\right) \pm m(x - x_0) \tag{8.183}$$

If we choose x_0 such that

$$\phi_0 = \phi(x_0) = (2n+1)\pi v \tag{8.184}$$

$(n = 0, \pm 1, \pm 2, \ldots)$, then we have

$$\frac{\phi}{4v} = (-1)^n \tan^{-1}\left[e^{\pm m(x - x_0)}\right] \tag{8.185}$$

where each choice of branch of the arctangent leads to a solution which interpolates between a different pair of adjacent zeros of $U(\phi)$.

→ **Exercise 8.6.** If we choose the branch of the arctangent for which $\tan^{-1}(0) = n\pi$, find the limits

$$\phi_\pm = \lim_{x \to \pm\infty} \phi(x)$$

for each of the choices of sign in the exponent in Eq. (8.185).   □

A Lagrangian Field Theory

In Section 3.5.3, we described the Lagrangian and Hamiltonian dynamics of systems with a finite number of degrees of freedom. Here we present the extension of that theory to systems of fields, i.e., by functions of space and time coordinates defined on some spacetime manifold. There are some subtleties in that extension, especially when dealing with fields such as the electromagnetic field with its gauge invariance. However, we will not be concerned with such points here, as we simply present the basic ideas of classical Lagrangian field theory.

Consider first a real scalar field $\phi(x, t)$, whose dynamics is to be described by a Lagrangian density $\mathcal{L}$ that depends on ϕ and its first derivatives, which we denote by ϕ_t for the time derivative $\partial\phi/\partial t$ and by ϕ_k for the spatial derivatives $\partial\phi/\partial x^k$. The dynamics is based on an extension of Hamilton's principle (see Section 3.5.3) to require that the action

$$S[\phi(x)] = \int \mathcal{L}(\phi, \phi_k, \phi_t) \, d^n x \, dt \tag{8.A1}$$

be an extremum relative to nearby fields (n is the number of space dimensions). This leads to a generalization of the Euler–Lagrange equations of motion,

$$\frac{\partial}{\partial t} \frac{\partial \mathcal{L}}{\partial \phi_t} + \frac{\partial}{\partial x^k} \frac{\partial \mathcal{L}}{\partial \phi_k} - \frac{\partial \mathcal{L}}{\partial \phi} = 0 \tag{8.A2}$$

(summation over k understood) that provides equations of motion for the fields.

Canonical momenta for the field can be introduced by

$$\pi(x, t) \equiv \frac{\partial \mathcal{L}}{\partial \phi_t(x, t)} \tag{8.A3}$$

The Hamiltonian density $\mathcal{H}$ defined by

$$\mathcal{H} = \pi(x, t) \frac{\partial \mathcal{L}}{\partial \phi_t(x, t)} - \mathcal{L}(\phi, \phi_k, \phi_t) \tag{8.A4}$$

can often be identified as an energy density of the field. The total field energy is then given by

$$H = E[\phi] = \int \left(\pi(x, t) \frac{\partial \mathcal{L}}{\partial \phi_t(x, t)} - \mathcal{L}(\phi, \phi_k, \phi_t) \right) d^n x \tag{8.A5}$$

The wave equation for the scalar field $\phi(x, t)$ can be derived from the Lagrangian density

$$\mathcal{L} = \frac{1}{2} \left(\frac{1}{c^2} \phi_t^2 - \phi_k^2 \right) \tag{8.A6}$$

(again with implied summation over k), The canonical momentum density is then given by $\pi(x, t) = \phi_t(x, t)$ and the Hamiltonian density is

$$\mathcal{H} = \frac{1}{2} \left\{ [\pi(x, t)]^2 + [\phi_k(x, t)]^2 \right\} \tag{8.A7}$$

Symmetry principles can be enforced by constraining the Lagrangian density to be invariant under the desired symmetry transformations. This is especially useful in constructing theories of elementary particles that are supposed to possess certain symmetries, as mentioned in Chapter 10. Here we note that a relativistic field theory can be constructed by requiring $\mathcal{L}$ to be invariant under Lorentz transformations.

❑ **Example 8.8.** For a relativistic scalar field $\phi(x, t)$ in one space dimension, the principle of invariance under Lorentz transformations requires Lagrangian density to have the form

$$\mathcal{L} = \frac{1}{2}\left(\frac{1}{c^2}\phi_t^2 - \phi_x^2\right) - U(\phi) \qquad\qquad 8.157$$

as already noted in Section 8.4.3. in order to be invariant under Lorentz transformations. Here $U(\phi)$ is a functional of ϕ (and not its derivatives) that is often referred to as the *potential* for the field ϕ. Stability requires $U(\phi)$ to be bounded from below, and we will generally assume that $U(\phi) \geq 0$ for all ϕ. The equation of motion for the field is

$$\frac{1}{c^2}\frac{\partial^2\phi}{\partial t^2} - \frac{\partial^2\phi}{\partial x^2} = -\frac{dU(\phi)}{d\phi} \qquad\qquad (8.A8)$$

The conserved energy functional for the field is

$$E[\phi] = \int_{-\infty}^{\infty} \left[\tfrac{1}{2}\left(\phi_t^2 + \phi_x^2\right) + U(\phi)\right] dx \qquad\qquad (8.A9)$$

The Klein–Gordon equation is the equation of motion derived from a Lagrangian density with potential

$$U(\phi) = \tfrac{1}{2}m^2\phi \qquad\qquad (8.A10)$$

Other potentials $U(\phi)$ lead to interesting phenomena, some of which are described in Sections 8.4.3 and 8.4.4.   ∎

For a scalar field in n space dimensions, the Lorentz invariant Lagrangian density is

$$\mathcal{L} = \tfrac{1}{2}(\phi_t^2 - \phi_k^2) - U(\phi) \qquad\qquad (8.A11)$$

with $U(\phi)$ a potential that is fairly arbitrary for a classical field, but more constrained in a quantum theory. The canonical momentum is $\pi = \phi_t$ leading to the Hamiltonian density

$$\mathcal{H} = \tfrac{1}{2}\pi^2 + \tfrac{1}{2}\nabla\phi \cdot \nabla\phi + U(\phi) \qquad\qquad (8.A12)$$

The momentum and energy can be derived from the stress-energy tensor $\mathcal{T}_{\mu\nu}$ defined by

$$\mathcal{T}_{\mu\nu} \equiv \frac{\partial\phi}{\partial x^\mu}\frac{\partial\phi}{\partial x^\nu} - \mathcal{L}g_{\mu\nu} \qquad\qquad (8.A13)$$

We then have

$$P_\mu = \int \mathcal{T}_{0\mu}\, d^n x \qquad\qquad (8.A14)$$

Conservation of four-momentum follows from the equations of motion (show this).

Bibliography and Notes

A monumental treatise with an extensive discussion of methods to find both exact and approximate solutions to linear partial differential equations involving the Laplace operator is

> Philip M. Morse and Herman Feshbach, *Methods of Theoretical Physics*, (two volumes) McGraw-Hill (1953)

Laplace's and Poisson's equations, the scalar Helmholtz equation, the wave equation, the Schrödinger equation, and the diffusion equation are all discussed at length. The method of separation of variables, the ordinary differential equations that arise after separating variables, and the special functions that appear as solutions to these equations are thoroughly analyzed. The coordinate systems in which Laplace's equation and the scalar Helmholtz equation are separable are analyzed at length. Complex variable methods are thoroughly described, but abstract linear vector space methods are presented only in a very archaic form.

A handbook that has many useful formulas—differential equations, generating functions, recursion relations, integrals and more—is

> Milton Abramowitz and Irene Stegun, *Handbook of Mathematical Functions*, Dover (1972).

This book has extensive graphs and tables of values of the functions, as well as the formulas. While the tables are less important with the availability of high-level programming systems such as Matlab and Mathematica, the graphs are still of some use for orientation. It is also available online as a free download; it was originally created as a project of the (U. S.) National Bureau of Standards and thus not subject to copyright.

A classic text from the famous Sommerfeld lecture series is

> Arnold Sommerfeld, *Lectures on Theoretical Physics VI: Partial Differential Equations in Physics*, Academic Press (1964)

This book is an outstanding survey of methods for analyzing partial differential equations that were available in the 1920s and 1930s. These methods go far beyond the discussion of separation of variables given here, and cover many applications that are still interesting today.

A modern survey of various topics in linear and nonlinear wave motion is

> J. Billingham and A. C. King, *Wave Motion*, Cambridge University Press (2000).

This book starts an elementary level, but moves on to describe elastic waves in solids, water waves in various limiting cases both linear and nonlinear, electromagnetic waves, and waves in chemical and biological systems.

Another introduction that emphasized nonlinear partial differential equations that have solitonic solutions is

> P. G. Drazin and R. S. Johnson, *Solitons: An Introduction*, Cambridge University Press (1989).

This book treats the Kortweg-de Vries equation in great detail, but considers other nonlinear equations as well.

Problems

1. Consider the operator

$$\Delta \equiv \nabla^2 = \left(\frac{\partial^2}{\partial x^2} + \frac{\partial^2}{\partial y^2} + \frac{\partial^2}{\partial z^2} \right)$$

defined on the (complex) function space $L^2(\Omega)$, where Ω is three-dimensional region bounded by a closed surface S. Show that this operator is self-adjoint when defined on the subspace of twice differentiable functions that satisfy the mixed boundary condition (see Eq. (8.27))

$$\alpha(\mathbf{r})\, u(\mathbf{r}) + \beta(\mathbf{r})\, \hat{\mathbf{n}} \cdot \vec{\nabla}\, u(\mathbf{r}) = \alpha(\mathbf{r})\, v(\mathbf{r}) + \beta(\mathbf{r})\, \hat{\mathbf{n}} \cdot \vec{\nabla}\, v(\mathbf{r}) = 0 \text{ on } S.$$

with $\alpha(\mathbf{r})$ and $\beta(\mathbf{r})$ real functions on S.

2. Evaluate the Green function for the Laplacian in two dimensions, with boundary condition that $\hat{\mathbf{n}} \cdot \nabla_r G(\vec{r}, \vec{s})$ vanish far from the source, from its Fourier transform

$$G(\vec{r}, \vec{s}) = \frac{1}{(2\pi)^2} \int \frac{e^{i\vec{k}\cdot(\vec{r}-\vec{s})}}{k^2}\, d^2 k$$

and compare this result with the standard result for the electrostatic potential due to a line charge. (*Note.* The evaluation is tricky. With $\rho \equiv |\vec{r} - \vec{s}|$, you might want to first evaluate $dG/d\rho$.)

3. Use the continuum eigenvectors defined in Eq. (8.45) to construct an explicit representation of the operators $\mathbf{E}_\lambda$ that define a resolution of the identity (Section 7.3.3) for the operator Δ on $\mathbf{R}^n$.

4. Consider the eigenvalue equation

$$\Delta u = \lambda u$$

in a rectangular box characterized by

$$0 \le x \le a \quad 0 \le y \le b \quad 0 \le z \le c$$

with *periodic boundary conditions*

$$u(\vec{r} + a\hat{\mathbf{e}}_x) = u(\vec{r}) \quad u(\vec{r} + b\hat{\mathbf{e}}_y) = u(\vec{r}) \quad u(\vec{r} + c\hat{\mathbf{e}}_z) = u(\vec{r})$$

(here $\hat{\mathbf{e}}_x$, $\hat{\mathbf{e}}_y$, and $\hat{\mathbf{e}}_z$ are unit vectors along the coordinate axes, which are parallel to the edges of the box).

(i) Show that Δ is a self-adjoint operator with the periodic boundary conditions.

(ii) Find a complete orthonormal system of eigenvectors of Δ with periodic boundary conditions. What are the corresponding eigenvalues?

5. Consider the Laplacian Δ inside a cylinder of radius R and length L.

(i) Express Δ in terms of partial derivatives with respect to the standard cylindrical coordinates ρ, ϕ, and z (see Eq. (3.194)).

(ii) Consider solutions to the eigenvalue equation

$$\Delta u = \lambda u$$

of the form

$$u(\rho, \phi, z) = G(\rho)F(\phi)Z(z)$$

Find the differential equations that must be satisfied by the functions $G(\rho)$, $F(\phi)$, and $Z(z)$, introducing additional constants as needed.

(iii) Find solutions of these equations that are single valued inside the cylinder and that vanish on the surface of the cylinder.

(iv) Then describe the spectrum of the Laplacian in the cylinder with these boundary conditions, and find the corresponding complete orthonormal system of eigenvectors.

6. (i) Show that the retarded Green function for the wave equation (8.70) can be written as

$$G_{\mathrm{ret}}(\vec{\rho}, \tau) = -\frac{c^2}{(2\pi)^{n+1}} \lim_{\varepsilon \to 0+} \int \frac{1}{(\omega + i\varepsilon)^2 - k^2 c^2} e^{i\vec{k}\cdot\vec{\rho} - i\omega\tau} \, d^n k \, d\omega$$

with $\vec{\rho} \equiv \vec{r} - \vec{s}$ and $\tau \equiv t - u$.

(ii) Show that the Feynman Green function G_F can be written as

$$G_F(\vec{\rho}, \tau) = -\frac{c^2}{(2\pi)^{n+1}} \lim_{\varepsilon \to 0+} \int \frac{1}{\omega^2 - k^2 c^2 + i\varepsilon} e^{i\vec{k}\cdot\vec{\rho} - i\omega\tau} \, d^n k \, d\omega$$

Remark. The problem here is simply to show that each so-called "$i\varepsilon$ prescription" given here is equivalent to the corresponding contour in Fig. 8.1.   □

7. Show that the solution (8.99) for the retarded Green function in three dimensions leads to the potential

$$\phi(\vec{r}, t) = \frac{1}{4\pi} \int \frac{1}{|\vec{r} - \vec{s}|} \rho\left(\vec{s}, t - \frac{|\vec{r} - \vec{s}|}{c}\right) d^3 s$$

Remark. This potential is the *retarded potential*. The potential at the point $P = (\vec{r}, t)$ is determined by the sources on the "backward light cone" from P. Draw a picture to explain this statement.   □

8. (i) Show that a formal solution to the Schrödinger equation (8.69)

$$i\hbar \frac{\partial \psi}{\partial t} = \mathbf{H}\psi$$

is given by

$$\psi(t) = \exp(-\frac{i}{\hbar}\mathbf{H}t)\psi(0) \equiv \mathbf{U}(t)\psi(0) \qquad (*)$$

(ii) If the Hamiltonian $\mathbf{H}$ has a resolution of the identity $\mathbf{E}_\omega$ such that

$$\mathbf{H} = \hbar \int \omega d\mathbf{E}_\omega$$

then

$$\psi(t) = \int e^{-i\omega t} d\mathbf{E}_\omega \, \psi(0)$$

(iii) For a particle of mass m with potential energy $V(\vec{r})$ the Hamiltonian is

$$\mathbf{H} = -\frac{\hbar^2}{2m}\nabla^2 + V(\vec{r})$$

The formal solution $(*)$ can be expressed as

$$\psi(\vec{r}, t) = \int G(\vec{r}, t; \vec{s}, 0)\psi(\vec{s}, 0) \, d^3s$$

where $G(\vec{r}, t; \vec{s}, 0)$ is the Green function for this Schrödinger equation.

(iv) Evaluate the Green function $G_0(\vec{r}, t; \vec{s}, 0)$ for a free particle $(V = 0)$.

9. Parabolic coordinates ξ, η, ϕ were introduced in Problem 3.9 in terms of Cartesian coordinates x, y, z.

(i) Show that ξ, η can be expressed in terms of the spherical coordinates r and θ as

$$\xi = r(1 - \cos\theta) = r - z \qquad \eta = r(1 + \cos\theta) = r + z$$

(ii) Show the Laplacian Δ in parabolic coordinates is

$$\Delta = \frac{4}{\xi + \eta}\left[\frac{\partial}{\partial\xi}\left(\xi\frac{\partial}{\partial\xi}\right) + \frac{\partial}{\partial\eta}\left(\eta\frac{\partial}{\partial\eta}\right)\right] + \frac{1}{\xi\eta}\frac{\partial^2}{\partial\phi^2}$$

(iii) Show that the Schrödinger equation (8.116) can be separated in parabolic coordinates.

(iv) Find scattering solutions corresponding to an incoming plane wave plus outgoing scattered wave, including the logarithmic corrections to the phase that are related to the long-range character of the Coulomb potential.

10. (i) Show that the Green function for the Klein–Gordon equation (8.71) in n space dimensions has the general form

$$G(\vec{\rho}, \tau) = -\frac{c^2}{(2\pi)^{n+1}}\int \frac{1}{\omega^2 - k^2c^2 - (mc^2/\hbar)^2} e^{i\vec{k}\cdot\vec{\rho} - i\omega\tau} \, d^n k \, d\omega$$

(again $\vec{\rho} \equiv \vec{r} - \vec{s}$ and $\tau \equiv t - u$) where, as for the wave equation, the contour of integration in the complex ω-plane must be chosen so that G satisfies the desired initial conditions.

(ii) Show that the contour C_{ret} in Fig. 8.1 again gives a retarded Green function.

(iii) Show that doing the integral over ω gives

$$G_{ret}(\vec{\rho}, \tau) = \frac{c^2}{(2\pi)^n} \int \frac{\sin \omega_k \tau}{\omega_k} e^{i \vec{k} \cdot \vec{\rho}} d^n k$$

with $\omega_k = \sqrt{k^2 c^2 + (mc^2/\hbar)^2}$.

11. Show that the Schrödinger equation for a particle of mass m moving in a potential $V(\vec{r})$ can be derived from a Lagrangian

$$\mathcal{L} = \frac{i}{2} \left(\psi^* \psi_t - \psi_t^* \psi \right) - \frac{\hbar^2}{2m} \psi_k^* \psi_k - \psi^* V(\vec{r}) \psi$$

Treat ψ and ψ^* as independent fields.

12. Consider the partial differential equation

$$\frac{\partial u}{\partial t} + (1 + u) \frac{\partial u}{\partial x} = 0$$

with $u(x, 0)$ given by a "hat" function

$$u(x, 0) = \begin{cases} u_0(1 + x) & -1 < x \leq 0 \\ u_0(1 - x) & 0 < t < 1 \\ 0 & \text{otherwise} \end{cases}$$

(i) Draw a sketch in the x–t plane of the characteristics of this equation starting from the X-axis at $t = 0$.

(ii) Find the trajectory of the shock wave associated with these initial conditions.

(iii) Then find the complete solution $u(x, t)$ of the partial differential equation with these initial conditions.

13. Derive Eq. (8.175) for the energy of the moving kink solution (8.174).

9 Finite Groups

Symmetries and invariance principles lead to conservation laws that are at the foundation of physical theory. The idea that physical laws should be independent of the coordinate system used to describe spacetime leads to the fundamental conservation laws of classical physics. That laws are independent of the choice of origin of the spatial coordinate system (*translation invariance*) is equivalent to conservation of momentum; independence of the choice of initial time is equivalent to conservation of energy. Rotational invariance of physical laws is equivalent to conservation of angular momentum. Even in systems such as solids, where the full symmetries of spacetime are not present, there are discrete rotational and translational symmetries of the lattice that have consequences for the physical properties of such systems.

Systems of identical particles have special properties in quantum mechanics. All known elementary particles are classified as either bosons (with integer spin) or fermions (spin $\frac{1}{2}$). The requirement that any state of a system of bosons (fermions) be symmetric (antisymmetric) under the exchange of any pair of particles leads naturally to study of the properties of permutations. Antisymmetry of many-electron states under permutations is the basis of the Pauli exclusion principle, which leads to an elementary qualitative picture of the periodic table of the elements. Moreover, the theory of atomic and molecular spectra requires knowledge of the constraints on the allowed states of atoms and molecules imposed by the Pauli exclusion principle. The allowed states of atomic nuclei are also restricted by the Pauli principle applied to the constituent protons and neutrons.

These and other symmetries are described by a mathematical structure known as a *group*, and there is a highly developed theory of groups with physical applications. In this chapter, we introduce some general properties of groups derived from the group axioms that were introduced in Chapter 1, and give examples of both finite and continuous groups. These include symmetry groups of simple geometric objects such as polygons and polyhedra that appear in molecules and crystal lattices, as well as the group of permutations of N objects, the *symmetric group* S_N. Space–time symmetries such as rotations and translations are described by continuous groups, also known as *Lie groups*, as are the more abstract symmetries associated with conservation of electric charge and with the fundamental interactions of quarks and leptons. Lie groups are described in detail in Chapter 10.

In this chapter, we study groups with a finite number n of elements (n is the *order* of the group). An important element of this study is to find the (*conjugacy*) *classes* of $\mathcal{G}$. Two group elements a and b are *conjugate* ($a \sim b$) if there is an element g of $\mathcal{G}$ such that $b = gag^{-1}$. A *class* of $\mathcal{G}$ contains all the group elements conjugate to any one element of the class. A group of finite order n has p ($\leq n$) classes with $h_1, \ldots, h_p$ elements, with $\sum h_k = n$.

Introduction to Mathematical Physics. Michael T. Vaughn
Copyright © 2007 WILEY-VCH Verlag GmbH & Co. KGaA, Weinheim
ISBN: 978-3-527-40627-2

In physical applications, we are most often interested in the representation of a group by linear operators on a linear vector space, such as the coordinate space of a system of coupled oscillators, or the state space of a quantum mechanical system. A representation is a map $g \to \mathbf{D}(g)$ such that the group multiplication is preserved, i.e.,

$$g = g_1 g_2 \qquad \text{if and only if} \qquad \mathbf{D}(g) = \mathbf{D}(g_1)\mathbf{D}(g_2)$$

Of special interest are the *irreducible representations*, in which the group acts on a vector space that has no proper subspace invariant under the action of the entire group. For a finite group, any irreducible representation is equivalent to a representation by unitary matrices.

The theory of irreducible representations of finite groups starts from orthogonality relations based on two lemmas due to Schur. The first lemma states that any operator that commutes with all the operators in an irreducible representation is a multiple of the identity operator. The second states that if Γ $[g \to \mathbf{D}(g)]$ and Γ' $[g \to \mathbf{D}'(g)]$ are two inequivalent irreducible representations and $\mathbf{A}$ is a linear operator such that

$$\mathbf{A}\mathbf{D}(g) = \mathbf{D}'(g)\mathbf{A}$$

for all g, then $\mathbf{A} = \mathbf{0}$. From these lemmas, we derive the fundamental theorem on representations of finite groups, which states that a finite group $\mathcal{G}$ with p classes has exactly p distinct inequivalent irreducible representations, of dimensions $m_1, \ldots, m_p$ such that

$$\sum_{k=1}^{p} m_k^2 = n$$

We give special attention to the symmetric groups $\mathcal{S}_N$, as these are important both for the quantum-mechanical description of systems of identical particles and for the representation theory of continuous groups. Each class of $\mathcal{S}_N$ is associated with one of the $\pi(N)$ partitions of N and its corresponding *Young diagram*, as every permutation with a cyclic structure described by one partition belongs to the same class of $\mathcal{S}_N$. Then $\mathcal{S}_N$ also has $\pi(N)$ inequivalent irreducible representations, each of which can also be associated with a partition of N and its Young diagram.

There is a remarkable generating function, due to Frobenius, for the simple characters $\chi^{(\lambda)}_{(m)}$ of $\mathcal{S}_N$. This function, derived in Appendix B, allows us to develop graphical methods based on Young diagrams for computing the characters, for the reduction of tensor products of irreducible representations of $\mathcal{S}_N$, and for the reduction of outer products $\Gamma^{(\mu)} \circ \Gamma^{(\nu)}$ of irreducible representations of $\mathcal{S}_m$ and $\mathcal{S}_n$. The outer products are representations of $\mathcal{S}_{m+n}$ induced by representations of the subgroup $\mathcal{S}_m \otimes \mathcal{S}_m$ using a standard procedure for obtaining representations of a group from those of its subgroups. These graphical methods can be used to compute various properties not only of $\mathcal{S}_N$, but also of Lie groups and their Lie algebras, as explained in Chapter 10.

9.1 General Properties of Groups

9.1.1 Group Axioms

The essential properties of a group are that (i) multiplication is defined for any two elements of the group, (ii) there is an identity element that leaves any element unchanged under multiplication, and (iii) for every element of the group, there is an inverse under multiplication that brings the group element back to the identity under multiplication. These properties are formalized in the following definition.

Definition 9.1. A *group* $\mathcal{G}$ is a set of elements with a law of composition (*multiplication*) that associates with any ordered pair (g, g') of elements of $\mathcal{G}$ a unique elements gg', the *product* of g and g', such that

(i) (associative law) for every g, g', g'' in $\mathcal{G}$, it is true that

$$(gg')g'' = g(g'g''); \tag{9.1}$$

(ii) there is a unique element in $\mathcal{G}$ (the *identity*) denoted by 1 or by e, such that $eg = g$ for every g in $\mathcal{G}$;

(iii) for every g in $\mathcal{G}$, there is a (unique) element g^{-1}, the *inverse* of g, such that

$$g^{-1}g = e = gg^{-1}. \tag{9.2}$$

The group $\mathcal{G}$ is *finite* if it contains a finite number of elements, otherwise *infinite*. If $\mathcal{G}$ is finite, the number of elements of $\mathcal{G}$ is the *order* of $\mathcal{G}$. A subset $\mathcal{H}$ of $\mathcal{G}$ that is a group under multiplication in $\mathcal{G}$ is a *subgroup* of $\mathcal{G}$, *proper* unless $\mathcal{H} = \mathcal{G}$ or $\mathcal{H} = \{e\}$. ∎

Definition 9.2. The group $\mathcal{G}$ is *Abelian* if $gg' = g'g$ for every pair g and g' in $\mathcal{G}$. The law of composition of an Abelian group is often called *addition*; the group is *additive*. ∎

❏ **Example 9.1.** The set $\{1, -1\}$ is a group (of order 2) under ordinary multiplication; it is evidently Abelian. This group is denoted by $\mathbf{Z}_2$. ∎

❏ **Example 9.2.** The integers $\mathbf{Z} = \{0, \pm 1, \pm 2, \ldots\}$ form an (infinite) Abelian group under addition. ∎

❏ **Example 9.3.** The permutations of N distinct objects form a group S_N, the *symmetric group*, since the result of applying two permutations of the N objects is another permutation. S_N is of order $N!$ (there are $N!$ distinct permutations of N objects). ∎

❏ **Example 9.4.** The complex numbers z on the unit circle ($|z| = 1$) form an Abelian group under the usual complex multiplication. The elements of this group depend on a single real parameter θ [with $z = \exp(i\theta)$] in the range $0 \le \theta < 2\pi$. ∎

❏ **Example 9.5.** The $n \times n$ unitary matrices $\mathbf{U}$ form the *unitary group* $U(n)$ (recall from Chapter 2 that the product of unitary matrices is unitary). $U(n)$ is non-Abelian for $n > 1$; $U(1)$ is equivalent to the group of complex numbers on the unit circle. The subgroup of $U(n)$ containing those matrices $\mathbf{U}$ with det $\mathbf{U} = 1$ is the *special unitary group* $SU(n)$. ∎

❏ **Example 9.6.** The rotations in the plane depicted in Fig. 2.3 form the group $SO(2)$, which is equivalent to the group $U(1)$ in the preceding example. The three-dimensional rotations depicted in Fig. 2.4 form the group $SO(3)$. ∎

❏ **Example 9.7.** The orthogonal (real unitary) linear operators in an n-dimensional real vector space form a group $O(n)$. $O(n)$ has a subgroup $SO(n)$ containing the rotations that are connected continuously to the identity transformation; reflections are excluded from $SO(n)$. ∎

Two groups appearing in different contexts may have the same abstract structure, and are thus equivalent from a mathematical viewpoint. This equivalence is called *isomorphism*. For computational purposes, we do not need to distinguish between groups that are isomorphic.

Definition 9.3. The groups $\mathcal{G}$ and $\mathcal{G}'$ are *isomorphic* ($\mathcal{G} \cong \mathcal{G}'$) if there is a one-to-one mapping $\mathcal{G} \leftrightarrow \mathcal{G}'$ that preserves multiplication, so that if a, b in $\mathcal{G}$ are mapped to a', b' in $\mathcal{G}'$, then

$$a'b' = ab$$ ∎

Definition 9.4. A group element g is of *order* n if n is the smallest integer for which $g^n = e$. If no such integer exists, then g is of infinite order. The group $\mathcal{G}$ is *periodic* if every element is of finite order. $\mathcal{G}$ is *cyclic* if there is an element a in $\mathcal{G}$ such that every g in $\mathcal{G}$ can be expressed as $g = a^m$ for some integer m. The *cyclic group* of order n is denoted by $\mathbf{Z}_n$. It contains elements $e, a, a^2, \ldots, a^{n-1}$. ∎

❏ **Example 9.8.** $\mathbf{Z}_n$ and $\mathbf{Z}$ are cyclic groups. $\mathbf{Z}$ is of infinite order. ∎

❏ **Example 9.9.** The smallest non-Abelian group is the group D_3 of order 6 whose elements $\{e, a, b, c, d, d^{-1}\}$ have the properties

$$a^2 = b^2 = c^2 = e \qquad ab = bc = ca = d \qquad ac = cb = ba = d^{-1}$$

The elements a, b, c are evidently of order 2. ∎

→ **Exercise 9.1.**   (i) For the group D_3, show that

$$aba = c = bab \qquad bcb = a = cbc \qquad aca = b = cac$$

(ii) What is the order of d? of d^{-1}?

(iii) Show that the elements d and d^{-1} of D_3 can be identified with rotations in a plane through angles $\pm 2\pi/3$, while the elements a, b, and c can be identified with reflections through three axes making angles of $2\pi/3$ with each other.

(iv) Finally, show that these elements contain all six permutations of the three vertices of an equilateral triangle.

Remark. This exercise shows that the group D_3 is isomorphic to S_3. D_3 is the smallest of a class known as *dihedral groups* (see Section 9.2.2). ☐

Definition 9.5. The *direct product* $\mathcal{G}_1 \otimes \mathcal{G}_2$ of the groups $\mathcal{G}_1$ and $\mathcal{G}_2$ contains the ordered pairs (g_1, g_1) with g_1 from $\mathcal{G}_1$ and g_2 from $\mathcal{G}_2$. With a_1, b_1 from $\mathcal{G}_1$ and a_2, b_2 from $\mathcal{G}_2$, multiplication is defined by

$$(a_1, a_2)(b_1, b_2) = (a_1 b_1, a_2 b_2) \tag{9.3}$$

If $\mathcal{G} \cong \mathcal{G}_1 \otimes \mathcal{G}_2$, then $\mathcal{G}$ is *decomposed* into the direct product of $\mathcal{G}_1$ and $\mathcal{G}_2$. ∎

➙ **Exercise 9.2.** Consider the cyclic groups $\mathbf{Z}_2 = \{e, a\}$ and $\mathbf{Z}_3 = \{e, b, b^2\}$. The direct product $\mathbf{Z}_2 \otimes \mathbf{Z}_3$ is a group of order 6, with elements

$$1 = (e, e), (a, e), (e, b), (e, b^2), (a, b), (a, b^2) \tag{9.4}$$

Show that $\mathbf{Z}_2 \otimes \mathbf{Z}_3 \cong \mathbf{Z}_6$. *Hint.* Consider the element $g = (a, b)$ and its powers. ☐

9.1.2 Cosets and Classes

Definition 9.6. Suppose $\mathcal{G}$ is a group with subgroup $\mathcal{H}$, and g an element of $\mathcal{G}$. The set $g\mathcal{H}$ consisting of the elements of $\mathcal{G}$ of the form gh with h in $\mathcal{H}$, is a *(left) coset* of $\mathcal{H}$ in $\mathcal{G}$. Similarly, the set $\mathcal{H}g$ is a *(right) coset* of $\mathcal{H}$ in $\mathcal{G}$. ∎

An important property of cosets is that if g_1 and g_2 are two elements of $\mathcal{G}$, then the cosets $g_1 \mathcal{H}$ and $g_2 \mathcal{H}$ are either (i) identical, or (ii) contain no common element. To show this, suppose $g_1 \mathcal{H}$ and $g_2 \mathcal{H}$ have a common element. Then there are elements h_1 and h_2 in $\mathcal{H}$ such that $g_1 h_1 = g_2 h_2$. Thus $g_2 = g_1 h_1 h_2^{-1}$ is in $g_1 \mathcal{H}$, and so is $g_2 h$ for any h in $\mathcal{H}$. Thus $g_1 \mathcal{H} = g_2 \mathcal{H}$.

It follows that if $\mathcal{G}$ is a group of finite order n with subgroup $\mathcal{H}$ of order m, we can find group elements $g_1 = e, g_2, \ldots, g_t$ such that the cosets $g_1 \mathcal{H}, g_2 \mathcal{H}, \ldots, g_t \mathcal{H}$ are disjoint, but every element of $\mathcal{G}$ is in exactly one of these cosets. Then $n = mt$, so that the order of $\mathcal{H}$ is a divisor of the order of $\mathcal{G}$ (*Lagrange's theorem*). The integer t is the *index* of $\mathcal{H}$ in $\mathcal{G}$.

An important corollary of Lagrange's theorem is that if g is an element of order m in the group $\mathcal{G}$ of finite order n, then m is a divisor of n. Thus, for example, every element g in a group of prime order p has order p (except the identity element, of course), and the group must be isomorphic to the cyclic group $\mathbf{Z}_p$ with elements $e, g, g^2, \ldots, g^{p-1}$.

Definition 9.7. Two elements a and b of a group $\mathcal{G}$ are *conjugate* ($a \sim b$) if there is an element g of $\mathcal{G}$ such that

$$b = gag^{-1} \tag{9.5}$$

If $gag^{-1} = a$ for every g in $\mathcal{G}$, then a is *self-conjugate*. ∎

A group can be divided into (conjugacy) *classes*, such that each class contains the group elements conjugate to one member of the class. The identity element is in a class by itself. In an Abelian group, every element is in a class by itself. Every self-conjugate element is in class by itself. All the elements of a class have the same order (show this).

Remark. Understanding the class structure of a finite group $\mathcal{G}$ is especially important, since the number of inequivalent irreducible representations of $\mathcal{G}$ is equal to the number of distinct classes of $\mathcal{G}$, as will soon be shown. ☐

❑ **Example 9.10.** In D_3, the elements a, b, c form a class, as do the rotations d, d^{-1}. ∎

❑ **Example 9.11.** The class structure of the rotation group $SO(3)$ is determined by noting that a rotation in $SO(3)$ can be characterized by a unit vector $\mathbf{n}$ defining an axis of rotation and an angle Φ of rotation ($0 \leq \Phi \leq \pi$). Two rotations are conjugate if and only if they have the same angle of rotation (see Problem 2.13). Thus the classes of $SO(3)$ are characterized by a rotation angle Φ (in $[0, \pi]$). ∎

Definition 9.8. If a is self-conjugate, then $ga = ag$ for every g in $\mathcal{G}$, so that self-conjugate elements of $\mathcal{G}$ form an Abelian subgroup of $\mathcal{G}$, the *center* $Z_\mathcal{G}$ of $\mathcal{G}$. ∎

❑ **Example 9.12.** The group $U(n)$ of $n \times n$ unitary matrices has as its center the group of matrices of the form $\mathbf{U} = \exp(i\alpha)\mathbf{1}$, which is isomorphic to the group $U(1)$. ∎

Definition 9.9. Two subgroups $\mathcal{H}$ and $\mathcal{H}'$ of the group $\mathcal{G}$ are *conjugate* if there is an element g of $\mathcal{G}$ such that

$$\mathcal{H}' = g\mathcal{H}g^{-1} \tag{9.6}$$

The subgroup $\mathcal{H}$ is a *invariant subgroup*, or *normal subgroup*, if $g\mathcal{H}g^{-1} = \mathcal{H}$ for every element g of $\mathcal{G}$. The invariant subgroup $\mathcal{H}$ of $\mathcal{G}$ is *maximal* if it is not an (invariant) subgroup of any proper invariant subgroup of $\mathcal{G}$, *minimal* if it contains no proper subgroup that is an invariant subgroup of $\mathcal{G}$. $\mathcal{G}$ is *simple* if it contains no proper invariant subgroups, *semisimple* if it contains no Abelian invariant subgroups. ∎

❑ **Example 9.13.** The center $Z_\mathcal{G}$ of $\mathcal{G}$ is an Abelian invariant subgroup of $\mathcal{G}$. ∎

If $\mathcal{H}$ is an invariant subgroup of $\mathcal{G}$, then the cosets of $\mathcal{H}$ in $\mathcal{G}$ form a group under multiplication, since we then have

$$g\mathcal{H} = (g\mathcal{H}g^{-1})g = \mathcal{H}g \tag{9.7}$$

(the left cosets and the right cosets of $\mathcal{H}$ coincide), and then

$$g_1\mathcal{H}g_2\mathcal{H} = g_1g_2\mathcal{H} \tag{9.8}$$

This group defined by the coset multiplication is the *factor group*, or *quotient group*, denoted by $\mathcal{G}/\mathcal{H}$ (read $\mathcal{G}$ *mod* $\mathcal{H}$).

❑ **Example 9.14.** The group $\mathcal{H} \equiv \{e, d, d^{-1}\} \cong \mathbf{Z}_3$ is an (Abelian) invariant subgroup of D_3, but d, d^{-1} do not commute with every element of $\mathcal{G}$, so not every Abelian invariant subgroup of a group $\mathcal{G}$ is in the center $Z_\mathcal{G}$ of $\mathcal{G}$. The factor group $D_3/\mathbf{Z}_3 \cong \mathbf{Z}_2$. Note that it is *not* true that $D_3 \cong \mathbf{Z}_2 \otimes \mathbf{Z}_3$. ∎

❑ **Example 9.15.** The group $2\mathbf{Z} \equiv \{0, \pm2, \pm4, \ldots\}$ of even integers is an invariant subgroup of $\mathbf{Z}$. Note that $2\mathbf{Z}$ is isomorphic to $\mathbf{Z}$ (!) Again, the factor group $\mathbf{Z}/(2\mathbf{Z}) = \mathbf{Z}_2$, but it is *not* true that $\mathbf{Z} \cong \mathbf{Z}_2 \otimes \mathbf{Z}$. ∎

Definition 9.10. A mapping f of a group $\mathcal{G}$ onto a subgroup $\mathcal{H}$ that preserves multiplication is a *homomorphism*, written as $\mathcal{G} \xrightarrow{f} \mathcal{H}$. The elements g in $\mathcal{G}$ that are mapped into the identity on $\mathcal{H}$ form a group ker f, the *kernel* of f. The elements h of $\mathcal{H}$ such that $g \xmapsto{f} h$ for some g in $\mathcal{G}$ form a group im f, the *image* of $\mathcal{G}$ (under f). ∎

Remark. Compare these definitions of kernel and image with those in Chapter 2. Note that a linear vector space $\mathcal{V}$ is an Abelian group under vector addition, with identity element θ (the zero vector), and a linear operator can be described as a homomorphism of $\mathcal{V}$ into itself. □

➜ **Exercise 9.3.** Show that the kernel of the homomorphism f is an invariant subgroup of $\mathcal{G}$, and im $f \cong \mathcal{G}/\ker f$. □

➜ **Exercise 9.4.** Show that if $\mathcal{H}$ is an invariant subgroup of $\mathcal{G}$, then there is a homomorphism f with ker $f = \mathcal{H}$ such that $\mathcal{G} \xrightarrow{f} \mathcal{G}/\mathcal{H}$. □

Definition 9.11. Let $\mathcal{G}$ be a group with invariant subgroup $\mathcal{H}$, and $\mathcal{K} = \mathcal{G}/\mathcal{H}$. Then $\mathcal{G}$ is an *extension* of $\mathcal{K}$ by the group $\mathcal{H}$. If $\mathcal{H}$ is in the center of $\mathcal{G}$, then $\mathcal{G}$ is a *covering group* of $\mathcal{K}$, or *central extension* of $\mathcal{K}$ by $\mathcal{H}$. ∎

Remark. The concept of covering group is important in the discussion of the global properties of Lie groups. □

9.1.3 Algebras; Group Algebra

A linear vector space is an Abelian group under addition of vectors, but it has more structure in the form of multiplication by scalars, and the existence of norm (and scalar product in unitary spaces). Still more structure is obtained, if there is a rule for multiplication of two vectors to produce a third vector; such a space is called an algebra. We have the formal

Definition 9.12. An *algebra* is a linear vector space $\mathcal{V}$ on which is defined, in addition to the usual vector space operations, an operation ∘, *multiplication* of vectors, that for every pair (x, y) of vectors defines a unique vector $x \circ y$, the *product* of x and y. This multiplication must satisfy the axioms.

(i) For every scalar α and every pair of vectors x, y,

$$(\alpha x) \circ y = \alpha (x \circ y) = x \circ (\alpha y) \tag{9.9}$$

(ii) For every triple of vectors x, y, and z,

$$\begin{aligned} (x + y) \circ z &= x \circ z + y \circ z \\ x \circ (y + z) &= x \circ y + x \circ z \end{aligned} \tag{9.10}$$

In addition to these mandatory distributive laws, special types of algebras can be defined by further axioms. ∎

❑ **Example 9.16.** The linear operators on a linear vector space $\mathcal{V}$ form an algebra $\mathcal{O}(\mathcal{V})$ with multiplication defined as operator multiplication, the *operator algebra* of $\mathcal{V}$. If $\mathcal{V}$ is of finite dimension n, then $\mathcal{O}(\mathcal{V})$ has dimension n^2. ∎

❑ **Example 9.17.** The linear operators on a linear vector space form another algebra $\mathcal{L}(\mathcal{V})$, the *Lie algebra* of $\mathcal{V}$, with multiplication of two operators $\mathbf{A}$ and $\mathbf{B}$ defined as the commutator,

$$\mathbf{A} \circ \mathbf{B} \equiv [\mathbf{A}, \mathbf{B}] = \mathbf{AB} - \mathbf{BA} \tag{9.11}$$

Lie algebras play an important role in the theory of continuous groups.   ∎

❑ **Example 9.18.** The linear operators on a linear vector space form yet another algebra $\mathbf{J}(\mathcal{V})$ with multiplication of two operators $\mathbf{A}$ and $\mathbf{B}$ defined as the anticommutator,

$$\mathbf{A} \circ \mathbf{B} \equiv \tfrac{1}{2}\{\mathbf{A}, \mathbf{B}\} = \tfrac{1}{2}(\mathbf{AB} + \mathbf{BA}) \tag{9.12}$$

$\mathbf{J}(\mathcal{V})$ is a *Jordan algebra*. While Jordan algebras are encountered in axiomatic formulations of quantum mechanics, they are not discussed at length here.   ∎

Definition 9.13. An algebra is *commutative* if for every pair of vectors x, y we have

$$x \circ y = y \circ x$$

  ∎

❑ **Example 9.19.** $\mathbf{J}(\mathcal{V})$ is commutative; $\mathcal{O}(\mathcal{V})$ and $\mathcal{L}(\mathcal{V})$ are not.   ∎

Definition 9.14.
    An algebra is *associative* if for every triple of vectors x, y, and z we have

$$x \circ (y \circ z) = (x \circ y) \circ z$$

  ∎

❑ **Example 9.20.** $\mathcal{O}(\mathcal{V})$ is associative; $\mathcal{L}(\mathcal{V})$ and $\mathbf{J}(\mathcal{V})$ are not.   ∎

Definition 9.15. An algebra has a *unit* if there is an element $\mathbf{1}$ of the algebra such that for every vector x, we have

$$\mathbf{1} \circ x = c = x \circ \mathbf{1}$$

  ∎

❑ **Example 9.21.** $\mathcal{O}(\mathcal{V})$ and $\mathbf{J}(\mathcal{V})$ have a unit; $\mathcal{L}(\mathcal{V})$ does not.   ∎

If $\mathcal{G}$ is a finite group, the *group algebra* $\mathcal{A}(\mathcal{G})$ consists of linear combinations of the group elements with scalar coefficients (complex numbers in general). Multiplication of vectors is simply defined by the group multiplication table, extended by the distributive laws (9.9) and (9.10). If $\mathcal{G}$ is of finite order n, then $\mathcal{A}(\mathcal{G})$ is of dimension n. $\mathcal{A}(\mathcal{G})$ has a unit, the identity element of $\mathcal{G}$. It is associative since the group multiplication is associative; it is commutative if and only if the group is Abelian. We use the group algebra of the symmetric group $\mathcal{S}_N$ to construct projection operators onto irreducible representations of $\mathcal{S}_N$. The group algebra can also be used to construct a more abstract theory of representations than required here.

9.2 Some Finite Groups

9.2.1 Cyclic Groups

The *cyclic group* $\mathbf{Z}_N$ is generated from a single element a; it has elements $1, a, \ldots, a^{N-1}$ with $a^N \equiv 1$. A concrete realization of the group $\mathbf{Z}_N$ is obtained with

$$a = \exp\left(\frac{2\pi i}{N}\right) \qquad (9.13)$$

using the ordinary rules of complex multiplication.

→ **Exercise 9.5.** Show that if p and q are prime (even relatively prime), the cyclic group $\mathbf{Z}_{pq}$ can be factorized into the direct product $\mathbf{Z}_p \otimes \mathbf{Z}_q$. □

As already noted, the only group of prime order p is the cyclic group $\mathbf{Z}_p$, which is Abelian. If the integer N is a product of distinct primes $p_1, \ldots, p_m$, then the only Abelian group of order N is the cyclic group $\mathbf{Z}_N$, for in that case we have

$$\mathbf{Z}_N \cong \mathbf{Z}_{p_1} \otimes \mathbf{Z}_{p_2} \otimes \cdots \otimes \mathbf{Z}_{p_m} \qquad (9.14)$$

However, if p is prime, then the cyclic group $\mathbf{Z}_{p^m}$ is not factorizable into the product of smaller groups. Hence there are distinct Abelian groups of order p^m of the form

$$\mathbf{Z}_{p^{m_1}} \otimes \mathbf{Z}_{p^{m_2}} \otimes \cdots \otimes \mathbf{Z}_{p^{m_q}}$$

where $m_1, m_2, \ldots, m_q$ is a partition of m into q parts (see the discussion of permutations below for a definition of partitions). The number of such distinct groups is evidently the number $\pi(m)$ of partitions of m.

❑ **Example 9.22.** The group $\mathbf{Z}_4$ has elements $\{1, a, a^2, a^3\}$ with $a^4 = 1$. The elements a, a^3 have order 4, while the element a^2 is of order 2. On the other hand, the group $\mathbf{Z}_2 \otimes \mathbf{Z}_2$ has elements $\{1, a, b, ab = ba\}$, and the elements a, b, ab are each of order 2. ∎

9.2.2 Dihedral Groups

The rotations in the plane that leave a regular N-sided polygon invariant form a group isomorphic to the cyclic group $\mathbf{Z}_N$. Further transformations that leave the polygon invariant are (i) inversion σ of the axis normal to the plane of the polygon, (ii) rotation π_k through angle π about any of N symmetry axes $v_1, \ldots, v_N$ in the plane of the polygon, and (iii) inversions σ_k of an axis in the plane normal to one of the v_k.

Each of the transformations σ, σ_k, and π_k has order 2, and

$$\sigma \pi_k = \sigma_k = \pi_k \sigma \qquad (9.15)$$

$(k - 1, \ldots, N)$. If a denotes rotation in the plane through angle $2\pi/N$, then also

$$\sigma_k a \sigma_k = a^{-1} = \pi_k a \pi_k \qquad \sigma a \sigma = a \qquad (9.16)$$

If we let ρ denote any one of the rotations π_k about a symmetry axis, then each of the transformations $b = \rho, b_1 = \rho a, \ldots, b_{N-1} = \rho a^{N-1}$ is of order 2 (show this). The elements $\mathbf{1}$, a, $\ldots$, $a^{N-1}, b = \rho, b_1 = \rho a, \ldots, b_{N-1}$ form a group of order $2N$, the *dihedral group* D_N.

Remark. Note that if N is odd, the symmetry axes in the plane pass through one vertex of the polygon and bisect the opposite edge. On the other hand, if $N = 2m$ is even, then m of the axes join opposite vertices, and the other m axes bisect opposite edges of the polygon. $\square$

➜ **Exercise 9.6.** Analyze the class structure of the dihedral group D_N. Find the classes, and the number of elements belonging to each class. Explain geometrically the difference between the structure when N is even and when N is odd. $\square$

➜ **Exercise 9.7.** Show that $\mathbf{Z}_N$ is an invariant subgroup of the dihedral group D_N. What is the quotient group D_N/Z_N? $\square$

9.2.3 Tetrahedral Group

The tetrahedral group $\mathcal{T}$ is the group of rotations that transform a regular tetrahedron into itself. Evidently this group contains rotations through angles $\pm 2\pi/3$ about any of the four axes $\mathcal{X}_1, \mathcal{X}_2, \mathcal{X}_3, \mathcal{X}_4$ that pass through the center and one of the four vertices of the tetrahedron. These rotations generate three further transformations in which the vertices are interchanged pairwise. Thus $\mathcal{T}$ is of order 12; it is in fact isomorphic to the group $\mathcal{A}_4$ of even permutations of the four vertices (see Problem 8).

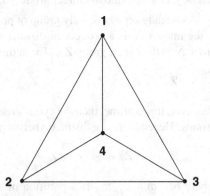

Figure 9.1: A regular tetrahedron viewed from above one vertex, here labeled "4."

The tetrahedron has additional symmetry if reflections are included. There are six planes that contain one edge of the tetrahedron and bisect the opposite edge. The tetrahedron in invariant under reflections in each of these planes. Each reflection exchanges a pair of vertices of the tetrahedron; denote the reflection exchanging vertices j and k by $\mathbf{R}_{jk}$. Then more symmetries are generated by following the reflection $\mathbf{R}_{jk}$ by a rotation through angles $\pm 2\pi/3$ about either of the axes j and k. Only six of these combined reflection–rotation transformations are distinct, so there are a total of 12 new symmetry transformations including the reflections. Thus the complete symmetry group $\mathcal{T}_d$ of the tetrahedron is of order 24, which is the same as the order of the group S_4 of permutations of the four vertices. This suggests that $\mathcal{T}_d \cong S_4$; it is left to Problem 8 to show that this is actually the case.

Remark. In addition to the tetrahedral groups, the symmetry groups of the three-dimensional cube are important in the theory of solids. These are the octahedral groups $\mathcal{O}$ and $\mathcal{O}_d$, which are analyzed in Problems 9 and 10. $\square$

9.3   The Symmetric Group S_N

9.3.1   Permutations and the Symmetric Group S_N

Permutations arise directly in the quantum description of systems of identical particles, since the states of such systems are required to be either symmetric (*Bose–Einstein* statistics) or antisymmetric (*Fermi–Dirac* statistics) under permutations of the particles. Also, the classification of symmetry types is important in the analysis of representations of other groups, especially the classical Lie groups (orthogonal, unitary, and symplectic) described in the next chapter.

Definition 9.16.   A *permutation* (of *degree* N) is a one-to-one mapping of a set Ω_N of N elements onto itself. The permutation is a *rearrangement*, or *reordering*, of Ω_N. If the elements of Ω_N are labeled $1, 2, \ldots, N$, and the permutation $\mathbf{P}$ maps the elements by

$$\mathbf{P} : 1 \to i_1, 2 \to i_2, \ldots, N \to i_N \tag{9.17}$$

then we can write

$$\mathbf{P} = (i_1 i_2 \cdots i_N) \tag{9.18}$$

A longer notation that is useful when considering the product of permutations is to write

$$\mathbf{P} = \begin{pmatrix} 1 & 2 & \cdots & N \\ i_1 & i_2 & \cdots & i_N \end{pmatrix} = \begin{pmatrix} \alpha_1 & \alpha_2 & \cdots & \alpha_N \\ i_{\alpha_1} & i_{\alpha_2} & \cdots & i_{\alpha_N} \end{pmatrix} \tag{9.19}$$

where $\alpha_1, \alpha_2, \ldots, \alpha_N$ is an arbitrary reordering of $1, 2, \ldots, N$.   ∎

With this notation, the inverse of the permutation (9.19) is evidently given by

$$\mathbf{P}^{-1} = \begin{pmatrix} i_1 & i_2 & \cdots & i_N \\ 1 & 2 & \cdots & N \end{pmatrix} \tag{9.20}$$

Also, given two permutations

$$\mathbf{P} = \begin{pmatrix} 1 & 2 & \cdots & N \\ i_1 & i_2 & \cdots & i_N \end{pmatrix} \qquad \mathbf{Q} = \begin{pmatrix} i_1 & i_2 & \cdots & i_N \\ j_1 & j_2 & \cdots & j_N \end{pmatrix} \tag{9.21}$$

we have the product

$$\mathbf{QP} = \begin{pmatrix} 1 & 2 & \cdots & N \\ j_1 & j_2 & \cdots & j_N \end{pmatrix} \tag{9.22}$$

Thus the permutations of Ω_N form a group, the *symmetric group* S_N (of *degree* N). S_N is of order $N!$, since there are $N!$ distinct permutations of N objects.

Definition 9.17.   A permutation $\mathbf{P}_{jk}$ that interchanges elements (j, k) of Ω_N, leaving the other elements in place, is a *transposition*. A transposition is *elementary* if the transposed elements are adjacent, i.e., if $k \equiv j \pm 1 \pmod{N}$.   ∎

Every permutation $\mathbf{P}$ can be expressed as a product of transpositions, even of elementary transpositions. If the permutation $\mathbf{P}$ is the product of n transpositions, then the *parity* of $\mathbf{P}$ is defined by $\varepsilon_{\mathbf{P}} = (-1)^n$. $\mathbf{P}$ is *even* if $\varepsilon_{\mathbf{P}} = +1$, *odd* if $\varepsilon_{\mathbf{P}} = -1$.

The expression of a permutation $\mathbf{P}$ as a product of transpositions is not unique. However, the parity $\varepsilon_{\mathbf{P}}$ of $\mathbf{P}$ *is* unique. To show this, consider the *alternant* $A(x_1, \ldots, x_N)$ of the N variables $x_1, \ldots, x_N$, defined by

$$A(x_1, \ldots, x_N) \equiv \prod_{j<k} (x_j - x_k) \tag{9.23}$$

Under a permutation $\mathbf{P} = (i_1 \cdots i_N)$ of the $x_1, \ldots, x_N$, the alternant transforms according to

$$A(x_1, \ldots, x_N) \to A(x_{i_1}, \ldots x_{i_N}) \equiv \varepsilon_{\mathbf{P}} A(x_1, \ldots, x_N) \tag{9.24}$$

In fact, Eq. (9.24) is an alternate definition of $\varepsilon_{\mathbf{P}}$. It shows that the parity of the product of two permutations $\mathbf{P}_1$, $\mathbf{P}_2$ is

$$\varepsilon_{\mathbf{P}_1\mathbf{P}_2} = \varepsilon_{\mathbf{P}_1}\varepsilon_{\mathbf{P}_2} = \varepsilon_{\mathbf{P}_2\mathbf{P}_1} \tag{9.25}$$

It follows from this that the even permutations form a subgroup of $\mathcal{S}_N$, (since the product of two even permutations is an even permutation) the *alternating group* $\mathcal{A}_N$. It is an invariant subgroup as well, since if $\mathbf{P}$ is an even permutation, so is $\mathbf{QPQ}^{-1}$ for any permutation $\mathbf{Q}$ (even *or* odd).

❏ **Example 9.23.**   In the group $\mathcal{S}_3$, the transpositions are $\mathbf{P}_{12} = (213)$, $\mathbf{P}_{13} = (321)$, $\mathbf{P}_{23} = (132)$. The remaining two permutations (apart from $\mathbf{1}$) are

$$\begin{aligned}
(231) &= \mathbf{P}_{13}\mathbf{P}_{12} = \mathbf{P}_{12}\mathbf{P}_{23} = \mathbf{P}_{23}\mathbf{P}_{13} \\
(312) &= \mathbf{P}_{12}\mathbf{P}_{13} = \mathbf{P}_{23}\mathbf{P}_{12} = \mathbf{P}_{13}\mathbf{P}_{23}
\end{aligned} \tag{9.26}$$

The transpositions are odd, while $\mathbf{1}$, (231), and (312) are even. Note that here the alternating group $\mathcal{A}_3 \cong \mathbf{Z}_3$ is cyclic, and $\mathcal{S}_3/\mathcal{A}_3 \cong \mathbf{Z}_2$. Nevertheless, $\mathcal{S}_3$ is *not* a direct product $\mathbf{Z}_2 \otimes \mathbf{Z}_3$.   ∎

→ **Exercise 9.8.**   Express the permutations in $\mathcal{S}_4$ as products of the elementary transpositions $\mathbf{P}_{12}, \mathbf{P}_{23}, \mathbf{P}_{34}, \mathbf{P}_{41}$.   ❏

→ **Exercise 9.9.**   Show that $\mathcal{S}_N/\mathcal{A}_N \cong \mathbf{Z}_2$.   ❏

Definition 9.18.   Let $\mathbf{P} = (i_1 \cdots i_N)$ be a permutation of degree N. Associated with $\mathbf{P}$ is the $N \times N$ *permutation matrix* $\mathbf{A} = \mathbf{A}(\mathbf{P})$ with elements

$$A_{jk} = \delta_{ji_k} \tag{9.27}$$

→ **Exercise 9.10.**   Show that the matrix $\mathbf{A}$ is orthogonal, and that $\det \mathbf{A}(\mathbf{P}) = \varepsilon_{\mathbf{P}}$.   ❏

Definition 9.19.   A permutation $\mathbf{P}$ mapping the elements $a_1, a_2, \ldots, a_p$ of Ω_N by

$$a_1 \to a_2 \,, \; a_2 \to a_3 \,, \; \ldots \,, \; a_p \to a_1 \tag{9.28}$$

leaving the other elements fixed, is a *cycle* of *length* p, or p-cycle. Such a cycle can be represented by the abbreviated notation

$$\mathbf{P} = (a_1\, a_2 \, \cdots \, a_p) = (a_2 \, \cdots \, a_p\, a_1) = \cdots = (a_p\, a_1\, a_2 \, \cdots \, a_{p-1}) \qquad\quad ∎$$

→ **Exercise 9.11.**   If $\mathbf{P}$ is a p-cycle, then $\varepsilon_{\mathbf{P}} = (-1)^{p+1}$.   □

Every permutation can be expressed as a product of disjoint cycles, uniquely, apart from ordering of the cycles. For we can choose an element a_1 of Ω_N and follow the chain $a_1 \rightarrow a_2 \rightarrow \cdots \rightarrow a_p \rightarrow a_1$ to the end of its cycle. Next we choose an element b_1 not in the previous cycle, and follow the chain $b_1 \rightarrow b_2 \rightarrow \cdots \rightarrow b_q \rightarrow b_1$ to the end of its cycle. Then we end up with $\mathbf{P}$ expressed as

$$\mathbf{P} = (a_1\, a_2\, \cdots\, a_p)(b_1\, b_2\, \cdots\, b_q) \cdots \tag{9.29}$$

where, if the degree of $\mathbf{P}$ is clear, the 1-cycles can be omitted.

❑ **Example 9.24.**   In cyclic notation, the transpositions of the group S_3 are

$$\mathbf{P}_{12} = (12) \qquad \mathbf{P}_{13} = (13) \qquad \mathbf{P}_{23} = (23) \tag{9.30}$$

and the 3-cycles are

$$\begin{aligned}
(123) &= (13)(12) = (12)(23) = (23)(13) \\
(321) &= (12)(13) = (23)(12) = (13)(23)
\end{aligned} \tag{9.31}$$

Note that the 3-cycles are even, since they are products of two transpositions. Be careful to distinguish the cyclic notation here and the notation in Example 9.23.   ∎

→ **Exercise 9.12.**   Find the cycle structure for each of the permutations in S_4.   □

→ **Exercise 9.13.**   If the permutation $\mathbf{P}$ in S_N is a product of p disjoint cycles, then

$$\varepsilon_{\mathbf{P}} = (-1)^{N+p} \tag{9.32}$$

Note explicitly how this works for S_3 and S_4.   □

The classes of S_N are determined by the cycle structure of permutations, since two permutations with the same cycle structure belong to the same class of S_N. To see this, note that if the permutation $\mathbf{P}$ is expressed as a product of disjoint cycles,

$$\mathbf{P} = (a_1\, a_2\, \cdots\, a_p)(b_1\, b_2\, \cdots\, b_q) \cdots \tag{9.33}$$

and $\mathbf{Q}$ is the permutation

$$\mathbf{Q} = \begin{pmatrix} a_1 & a_2 & \cdots & a_p & b_1 & b_2 & \cdots & b_q & \cdots \\ a_1' & a_2' & \cdots & a_p' & b_1' & b_2' & \cdots & b_q' & \cdots \end{pmatrix} \tag{9.34}$$

then

$$\mathbf{P}' \equiv \mathbf{Q}\mathbf{P}\mathbf{Q}^{-1} = (a_1'\, a_2'\, \cdots\, a_p')(b_1'\, b_2'\, \cdots\, b_q') \cdots \tag{9.35}$$

is a permutation with the same cycle structure as $\mathbf{P}$. Conversely, if $\mathbf{P}_1$ and $\mathbf{P}_2$ are two permutations of degree N with the same cycle structure, then there is a (unique) permutation $\mathbf{Q}$ such that

$$\mathbf{P}_2 = \mathbf{Q}\mathbf{P}_1\mathbf{Q}^{-1} \tag{9.36}$$

Thus any two permutations with the same cycle structure belong to the same class of S_N.

9.3.2 Permutations and Partitions

The cycle structure of a permutation, as noted in Eq. (9.29), is defined by a set $m_1, m_2, \ldots, m_p$ of (positive) integers such that

$$\text{(i)} \ m_1 \geq m_2 \geq \cdots \geq m_p > 0 \qquad \text{and} \qquad \text{(ii)} \ m_1 + m_2 + \cdots + m_p = N \quad (9.37)$$

A set $(m) = (m_1 m_2 \cdots m_p)$ of positive integers that satisfies (i) and (ii) is a *partition* of N (into p parts). We have just shown a one-to-one correspondence between classes of $\mathcal{S}_N$ and partitions of N; let $K_{(m)}$ denote the class whose permutations have cycle structure (m).

Remark. An integer q that is repeated r times in the partition (m) can be expressed as q^r. Thus the partitions of 2 are (2) and $(11) = (1^2)$, for example.   □

❑ **Example 9.25.** The partitions of 3 are (3), (21), and (1^3). The partitions of 4 are (4), (31), (2^2), (21^2), and (1^4).   ∎

To each partition $(m) = (m_1 m_2 \cdots m_p)$ of N into p parts corresponds a *Young diagram* $\mathcal{Y}^{(m)}$ constructed with N boxes placed into p rows, such that m_1 boxes are in the first row, m_2 boxes in the second row, and so on.

❑ **Example 9.26.** The Young diagrams

$$(2) = \boxed{} \qquad (1^2) = \boxed{} \qquad\qquad\qquad\qquad\qquad (9.38)$$

correspond to the partitions of 2. The diagrams

$$(3) = \boxed{} \qquad (21) = \boxed{} \qquad (1^3) = \boxed{} \qquad\qquad (9.39)$$

correspond to the partitions of 3, and the diagrams

$$(4) = \boxed{} \qquad (31) = \boxed{} \qquad (2^2) = \boxed{} \qquad (21^2) = \boxed{} \qquad (1^4) = \boxed{}$$

correspond to the partitions of 4.   ∎

Partitions of N can be ordered by the (dictionary) rule:

$$(m) = (m_1 m_2 \cdots m_p) \text{ precedes } (m') = (m_1' m_2' \cdots m_q')$$

[or simply $(m) < (m')$] if the first nonzero integer in the sequence $\{m_1 - m_1', m_2 - m_2', \ldots\}$ is positive. The partitions of $N = 2$, $N = 3$, and $N = 4$ in the preceding example have been given in this order.

→ **Exercise 9.14.** Enumerate the partitions of $N = 5$ and $N = 6$ in dictionary order, and draw the corresponding Young diagrams.   □

To each partition (m) of N is associated a *conjugate partition* $(\widetilde{m})$ whose diagram is obtained from that of (m) by interchanging rows and columns.

❏ **Example 9.27.** $(\widetilde{n}) = (1^n)$ $(\widetilde{21}) = (21)$ $(\widetilde{22}) = (22)$ $(\widetilde{31}) = (21^2)$.   ∎

Remark.   The partitions (21) and (22) are *self-conjugate*.   □

If $(m) = (m_1 m_2 \cdots m_p)$ is a partition of N, let $\nu_k \equiv m_k - m_{k+1}$ $(k = 1, \ldots, p)$, with $m_{p+1} \equiv 0$. Then the conjugate partition $(\widetilde{m})$ can be written in the form $(\widetilde{m}) = (p^{\nu_p} \cdots 2^{\nu_2} 1^{\nu_1})$, omitting the term k^{ν_k} if $\nu_k = 0$. Conversely, if $(m) = (q^{\nu_q} \cdots 2^{\nu_2} 1^{\nu_1})$ is a partition of N, then the conjugate partition is $(\widetilde{m}) = (\widetilde{m}_1 \widetilde{m}_2 \cdots \widetilde{m}_q)$, with

$$
\begin{aligned}
\widetilde{m}_1 &= \nu_1 + \nu_2 + \cdots + \nu_q \\
\widetilde{m}_2 &= \nu_2 + \cdots + \nu_q \\
&\;\;\vdots \\
\widetilde{m}_q &= \nu_q
\end{aligned}
\tag{9.40}
$$

These relations are easily verified by drawing a few diagrams.

Associated with a Young diagram corresponding to a partition (m) of N is a set of $d[(m)]$ *regular* (or *standard*) tableaux obtained by assigning the numbers $1, 2, \ldots, N$ to the boxes of the Young diagram such that the numbers increase (i) across each row and (ii) down each column of the diagram. For the diagrams with either a single row or a single column, there is associated a unique regular tableau, since rules (i) and (ii) require the numbers $1, 2, \ldots, N$ to appear in order across the row or down the column of the diagram. However, $d[(m)] > 1$ for other Young diagrams, and we shall see that $d[(m)]$ is the dimension of the irreducible representation of the symmetric group S_N corresponding to the partition (m).

❏ **Example 9.28.** $d[(21)] = 2$, since

Similarly, $d[(22)] = 2$, since

$d[(31)] = 3$, since

Computation of the $d(m)$ for partitions of $N > 4$ is left to the exercises.   ∎

→ **Exercise 9.15.**   Show that $d[(m)] = d[(\widetilde{m})]$ for any conjugate pair (m), $(\widetilde{m})$.   □

There is also a normal (dictionary) ordering of the standard tableaux associated with a partition. Comparing elements of two standard tableaux while reading across the rows in order, we place the tableau with the first smaller number ahead of the comparison tableau. The tableaux in the examples above have been written in normal order.

→ **Exercise 9.16.**   Enumerate the regular tableaux (in normal order) associated with each partition (m) of $N = 5, 6$. Then find the dimension $d[(m)]$ for each (m).   □

9.4   Group Representations

9.4.1   Group Representations by Linear Operators

Of prime importance for physical applications is the study of group representations on a linear vector space $\mathcal{V}$, especially in quantum mechanics, where $\mathcal{V}$ is often the space of states of a physical system. In this and the next section, we develop the general theory of representations of finite groups; in Section 9.5 we work out the theory of representations of $\mathcal{S}_N$ in great detail.

Definition 9.20. If $\mathcal{G}$ is a group and Γ a group of linear operators on a linear vector space $\mathcal{V}$, with a homomorphism $\mathcal{G} \to \Gamma$, then Γ is a *representation* of $\mathcal{G}$. $\mathcal{V}$ is the *representation space*, dim $\mathcal{V}$ the *dimension* of the representation.   ∎

In other words, to each element g of $\mathcal{G}$, there corresponds a linear operator $\mathbf{D}^{\Gamma}(g)$ on $\mathcal{V}$ such that

$$\mathbf{D}^{\Gamma}(g_2)\mathbf{D}^{\Gamma}(g_1) = \mathbf{D}^{\Gamma}(g_2 g_1) \tag{9.41}$$

so that the $\mathbf{D}^{\Gamma}(g)$ follow the multiplication law for the group $\mathcal{G}$. For every group $\mathcal{G}$, the map $g \to 1$ for every g in $\mathcal{G}$ is a representation, the *trivial representation*, or *identity representation*. The representation Γ is *faithful* if every element of $\mathcal{G}$ is represented by a distinct linear operator, so that

$$\mathbf{D}^{\Gamma}(g_2) = \mathbf{D}^{\Gamma}(g_1) \tag{9.42}$$

if and only if $g_2 = g_1$.

Definition 9.21. Two representations Γ_1 and Γ_2 of $\mathcal{G}$ on $\mathcal{V}$ are *equivalent* ($\Gamma_1 \sim \Gamma_2$) if there is a nonsingular linear operator $\mathbf{S}$ on $\mathcal{V}$ such that

$$\mathbf{D}^{\Gamma_2}(g) = \mathbf{S}\mathbf{D}^{\Gamma_1}(g)\mathbf{S}^{-1} \tag{9.43}$$

for every g in $\mathcal{G}$; if this is the case, Γ_1 and Γ_2 differ only by a change of basis in $\mathcal{V}$.   ∎

Definition 9.22. If $\mathcal{V}$ contains a subspace $\mathcal{M}$ invariant under Γ, so that $\mathcal{M}$ is an invariant manifold of $\mathbf{D}^{\Gamma}(g)$ for every group element g, then Γ is *reducible*. It is *fully reducible* if $\mathcal{M}^{\perp}$ is also invariant under Γ; in this case the representation Γ can be split into two parts acting on $\mathcal{M}$ and $\mathcal{M}^{\perp}$ with no connection between the subspaces. If there is no subspace invariant under Γ, then Γ is *irreducible*.   ∎

A general problem for any group $\mathcal{G}$ is to find all the possible inequivalent irreducible representations of $\mathcal{G}$. We will find the solution to this problem for finite groups in Theorem 9.4.

❑ **Example 9.29.**   A representation Γ^m of the cyclic group $\mathbf{Z}_N = \{1, a, \dots, a^{N-1}\}$ is defined by setting

$$a = \exp\left(\frac{2\pi i m}{N}\right) \tag{9.44}$$

for any $m = 0, 1, \dots, N - 1$. Each Γ^m is one-dimensional, hence irreducible, and the N different values of m correspond to inequivalent representations. Γ^m is faithful unless either $m = 0$ (identity representation) or m is a divisor of N. Furthermore, these representations provide a complete construction of the irreducible representations of Abelian groups, since we have seen in Section 9.2.1 that every Abelian group can be expressed as a direct product of cyclic groups whose order is a prime number raised to some power, ∎

❑ **Example 9.30.** The symmetric group $\mathcal{S}_N$ has two one-dimensional representations, the *symmetric* representation Γ_s with $\mathbf{P} \to 1$ for every $\mathbf{P}$, and the *antisymmetric* representation Γ_a with $\mathbf{P} \to \varepsilon_{\mathbf{P}}$ for every $\mathbf{P}$. These representations are inequivalent. ∎

❑ **Example 9.31.** The action of the permutations of three objects can be represented on $\mathbf{C}^3$ by 3×3 matrices,

$$\mathbf{P}_{12} = \begin{pmatrix} 0 & 1 & 0 \\ 1 & 0 & 0 \\ 0 & 0 & 1 \end{pmatrix} \qquad \mathbf{P}_{13} = \begin{pmatrix} 0 & 0 & 1 \\ 0 & 1 & 0 \\ 1 & 0 & 0 \end{pmatrix} \qquad \mathbf{P}_{23} = \begin{pmatrix} 1 & 0 & 0 \\ 0 & 0 & 1 \\ 0 & 1 & 0 \end{pmatrix} \qquad (9.45)$$

$$(123) = \begin{pmatrix} 0 & 0 & 1 \\ 1 & 0 & 0 \\ 0 & 1 & 0 \end{pmatrix} \qquad (321) = \begin{pmatrix} 0 & 1 & 0 \\ 0 & 0 & 1 \\ 1 & 0 & 0 \end{pmatrix} \qquad \mathbf{1} = \begin{pmatrix} 1 & 0 & 0 \\ 0 & 1 & 0 \\ 0 & 0 & 1 \end{pmatrix} \qquad (9.46)$$

that permute the basis vectors of $\mathbf{C}^3$. This representation of $\mathcal{S}_3$ is reducible, since the vector $\phi_0 = (1, 1, 1)$ is an eigenvector (eigenvalue $+1$) of each permutation in $\mathcal{S}_3$. Thus $\mathcal{M}(\phi_0)$ is invariant under the representation. However, the representation restricted to $\mathcal{M}^{\perp}(\phi_0)$ *is irreducible*. ∎

➜ **Exercise 9.17.** In the three-dimensional space of the preceding example, let

$$\psi_1 \equiv \sqrt{\tfrac{1}{2}}(1, -1, 0) \qquad \psi_2 \equiv \sqrt{\tfrac{1}{6}}(1, 1, -2) \qquad (9.47)$$

(i) Show that $\mathcal{M}(\psi_1, \psi_2) = \mathcal{M}^{\perp}(\phi_0)$.

(ii) Construct the 2×2 matrices representing $\mathcal{S}_3$ on $\mathcal{M}(\psi_1, \psi_2)$ in the basis $\{\psi_1, \psi_2\}$.

(iii) Show that this representation is irreducible. ❑

Definition 9.23. If Γ $[g \to \mathbf{D}(g)]$ is a representation of the group $\mathcal{G}$, then $g \to \mathbf{D}^*(g)$ defines a representation Γ^* of $\mathcal{G}$, the *complex conjugate* of Γ. There are three possibilities:

1. Γ is equivalent to a real representation (Γ is *real*);

2. $\Gamma \sim \Gamma^*$, but Γ cannot be transformed to a real representation (Γ is *pseudoreal*);

3. Γ is not equivalent to Γ^* (Γ is *complex*). ∎

Remark. This classification of representations is especially useful in quantum physics, where complex conjugation is related to time reversal and charge conjugation. ❑

Definition 9.24. If Γ $[g \to \mathbf{D}(g)]$ is a representation of the group $\mathcal{G}$, then $g \to \mathbf{D}^{\dagger}(g^{-1})$ defines a representation $\overline{\Gamma}$ of $\mathcal{G}$, the *dual* of Γ. If the $\mathbf{D}(g)$ are unitary, then $\Gamma = \overline{\Gamma}$; in general the representation Γ is *unitary* if it is equivalent to a representation by unitary operators. ∎

➜ **Exercise 9.18.** Verify that $\overline{\Gamma}$ actually is a representation. ❑

Theorem 9.1. Any finite-dimensional representation of a group of finite order is equivalent to a unitary representation.

Proof. Suppose $\mathcal{G}$ is a group of order n, and Γ a representation of $\mathcal{G}$ with $g \to \mathbf{D}(g)$. Let

$$\mathbf{H} \equiv \sum_{k=1}^{n} \mathbf{D}(g_k)\mathbf{D}^\dagger(g_k) \tag{9.48}$$

Then $\mathbf{H}$ is positive definite, and if the $\mathbf{D}(g)$ are unitary, then $\mathbf{H}$ is simply n times the unit matrix. In any case, however, the matrix $\Delta = \mathbf{H}^{1/2}$ can be chosen to be positive definite. If we now define

$$\mathbf{U}(g) \equiv \Delta^{-1}\mathbf{D}(g)\Delta \tag{9.49}$$

then the $\mathbf{U}(g)$ define a representation of $\mathcal{G}$ equivalent to Γ. Now

$$\mathbf{U}(g)\mathbf{U}^\dagger(g) = \Delta^{-1}\mathbf{D}(g)\mathbf{H}\mathbf{D}^\dagger(g)\Delta^{-1} \tag{9.50}$$

However,

$$\mathbf{D}(g)\mathbf{H}\mathbf{D}^\dagger(g) = \mathbf{D}(g)\sum_{k=1}^{n} \mathbf{D}(g_k)\mathbf{D}^\dagger(g_k)\mathbf{D}^\dagger(g) = \sum_{k=1}^{n} \mathbf{D}(gg_k)\mathbf{D}^\dagger(gg_k) \tag{9.51}$$

and as k runs from 1 to n, the $\{gg_k\}$ range over the entire group, so that

$$\sum_{k=1}^{n} \mathbf{D}(gg_k)\mathbf{D}^\dagger(gg_k) = \sum_{\ell=1}^{n} \mathbf{D}(g_\ell)\mathbf{D}^\dagger(g_\ell) = \mathbf{H} \tag{9.52}$$

It follows that $\mathbf{U}$ is unitary, since

$$\mathbf{U}(g)\mathbf{U}^\dagger(g) = \Delta^{-1}\mathbf{D}(g)\mathbf{H}\mathbf{D}^\dagger(g)\Delta^{-1} = \Delta^{-1}\mathbf{H}\Delta^{-1} = 1 \qquad\blacksquare$$

Remark. Since every representation of a finite group is equivalent to a representation by unitary operators, it is no loss of generality to assume that the linear operators are actually unitary, unless explicitly stated otherwise.   □

Definition 9.25. Let Γ be a representation of the group $\mathcal{G}$ with $g \to \mathbf{D}^{(\Gamma)}(g)$. The *character* $\chi^{(\Gamma)}(g)$ of g in Γ is

$$\chi^{(\Gamma)}(g) \equiv \operatorname{tr}\mathbf{D}^{(\Gamma)}(g) \tag{9.53}$$

If g_1 and g_2 are in the same class of $\mathcal{G}$, then $\chi^{(\Gamma)}(g_1) = \chi^{(\Gamma)}(g_2)$ in every representation Γ of $\mathcal{G}$, so the character is a class function. If K is a class in $\mathcal{G}$, then the character $\chi^{(\Gamma)}(K)$ of K in Γ is the character of any element of K in Γ. The set $\{\chi^{(\Gamma)}(K)\}$ of class characters is the *character* of the representation Γ, *simple* if Γ is irreducible, otherwise *compound*.   ∎

❏ **Example 9.32.**   Since all the irreducible representations of an Abelian group are one-dimensional, the character of a group element represents the element itself in an irreducible representation.   ∎

❑ **Example 9.33.**   The characters of the classes of $\mathcal{S}_3$ (labeled by partitions of 3) in the reducible three-dimensional representation of Example 9.31 are

$$\chi^{(\Gamma)}(3) = 0 \qquad \chi^{(\Gamma)}(21) = 1 \qquad \chi^{(\Gamma)}(1^3) = 3 \tag{9.54}$$

As an exercise, find these characters in the two-dimensional irreducible representation defined on $\mathcal{M}^\perp(\phi_0)$ (see also Exercise 9.17).   ∎

❑ **Example 9.34.**   The character of a rotation through angle θ in the defining representation of $SO(2)$ is

$$\chi(\theta) = 2\cos\theta \tag{9.55}$$

since the matrix for rotation through angle θ is

$$\mathbf{R}_z(\theta) = \begin{pmatrix} \cos\theta & -\sin\theta \\ \sin\theta & \cos\theta \end{pmatrix} \tag{9.56}$$

as given in Eq. (2.107).   ∎

❑ **Example 9.35.**   The character of a rotation through angle θ in the defining representation of $SO(3)$ is

$$\chi(\theta) = 1 + 2\cos\theta \tag{9.57}$$

To see this, recall the matrix for rotation through angle θ about the Z-axis,

$$\mathbf{R}_z(\theta) = \begin{pmatrix} \cos\theta & -\sin\theta & 0 \\ \sin\theta & \cos\theta & 0 \\ 0 & 0 & 1 \end{pmatrix} \tag{9.58}$$

and note that rotations through angle θ about any axis belong to the same class of $SO(3)$ (see Exercise 2.13).   ∎

One observation that can be useful for constructing representations is that if the group $\mathcal{G}$ has an invariant subgroup $\mathcal{H}$, with factor group $\mathcal{F} \equiv \mathcal{G}/\mathcal{H}$, then every representation of $\mathcal{F}$ is also a representation of $\mathcal{G}$ in which $h \to \mathbf{D}(h) = 1$ for every h in $\mathcal{H}$. Other representations of $\mathcal{G}$ can be constructed from nontrivial representations of $\mathcal{H}$; this is discussed later on.

❑ **Example 9.36.**   The alternating group $\mathcal{A}_N$ is an invariant subgroup of the symmetric group $\mathcal{S}_N$, and $\mathcal{S}_N/\mathcal{A}_N = \mathbf{Z}_2$. The factor group $\mathbf{Z}_2$ has two inequivalent irreducible representations; the corresponding irreducible representations of $\mathcal{S}_N$ are the symmetric ($\mathbf{P} \to 1$) and antisymmetric ($\mathbf{P} \to \varepsilon_\mathbf{P}$) representations.   ∎

9.4.2   Schur's Lemmas and Orthogonality Relations

The basic properties of finite-dimensional unitary representations of groups are derived from two fundamental theorems, both due to Schur.

Theorem 9.2. (Schur's Lemma I) Let Γ [$g \to \mathbf{D}(g)$] be an irreducible representation of the group $\mathcal{G}$ by unitary operators on the linear vector space $\mathcal{V}$ and suppose $\mathbf{A}$ is a bounded linear operator on $\mathcal{V}$ such that

$$\mathbf{A}\mathbf{D}(g) = \mathbf{D}(g)\mathbf{A} \tag{9.59}$$

for every g in $\mathcal{G}$. Then $\mathbf{A} = \alpha \mathbf{1}$ for some scalar α.

Remark.   In other words, any operator that commutes with every matrix in an irreducible unitary representation of a group is a multiple of the identity. The restriction to bounded operators ensures that the theorem also works for infinite-dimensional representations.   □

Proof. If $\mathbf{A}$ commutes with $\mathbf{D}(g)$ for every g, so does $\mathbf{A}^\dagger$, since Eq. (9.59) implies

$$\mathbf{A}^\dagger \mathbf{D}^\dagger(g) = \mathbf{D}^\dagger(g)\mathbf{A}^\dagger \tag{9.60}$$

and unitarity means that $\mathbf{D}^\dagger(g) = \mathbf{D}(g^{-1})$ is in Γ for every g. Hence we can take $\mathbf{A}$ to be self-adjoint, with a spectral resolution. If the spectrum of $\mathbf{A}$ contains more than one point, then $\mathcal{V}$ can be decomposed into a direct sum $\mathcal{V}_1 \oplus \mathcal{V}_2$ such that the spectra of $\mathbf{A}$ on $\mathcal{V}_1$ and $\mathcal{V}_2$ are disjoint. But $\mathcal{V}_1$ and $\mathcal{V}_2$ are invariant under Γ, so that Γ would be reducible. Since it is not, the spectrum of $\mathbf{A}$ can contain only one point, i.e., $\mathbf{A} = \alpha\mathbf{1}$ for some scalar α.   ■

Theorem 9.3. (Schur's Lemma II) Let Γ [$g \to \mathbf{D}(g)$] and Γ' [$g \to \mathbf{D}'(g)$] be inequivalent unitary irreducible representations of the group $\mathcal{G}$ on linear vector spaces $\mathcal{V}$ and $\mathcal{V}'$ of finite dimensions m and m', respectively. Let $\mathbf{A}$ be an $m \times m'$ matrix mapping $\mathcal{V}'$ to $\mathcal{V}$ such that

$$\mathbf{D}(g)\mathbf{A} = \mathbf{A}\mathbf{D}'(g) \tag{9.61}$$

for every g in $\mathcal{G}$. Then $\mathbf{A} = \mathbf{0}$.

Proof. Suppose $m > m'$. Then $\mathbf{A}\mathcal{V}'$ defines a linear manifold $\mathcal{M}_\mathbf{A}$ in $\mathcal{V}$ of dimension $\dim \mathcal{M}_\mathbf{A} \leq m' < m$. But $\mathcal{M}_\mathbf{A}$ is invariant under Γ if $\mathbf{D}(g)\mathbf{A} = \mathbf{A}\mathbf{D}'(g)$, and thus Γ is reducible if $\dim \mathcal{M}_\mathbf{A} \neq 0$. Hence $\mathbf{A} = \mathbf{0}$. If Γ and Γ' are irreducible representations of the same dimension, then *either* $\dim \mathcal{M}_\mathbf{A} = 0$, in which case $\mathbf{A} = \mathbf{0}$, *or* $\dim \mathcal{M}_\mathbf{A} = m$, in which case Eq. (9.61) implies $\mathbf{D}(g) = \mathbf{A}\mathbf{D}'(g)\mathbf{A}^{-1}$ and thus $\Gamma \sim \Gamma'$. The latter is contrary to hypothesis, hence $\mathbf{A} = \mathbf{0}$.   ■

Remark.   The statements of the two lemmas are valid for an arbitrary group, not necessarily finite. However, the lemmas apply to only to finite-dimensional unitary representations, which may, or may not, exist for an infinite group.   □

Two corollaries of Schur's lemmas lead to a set of orthogonality relations for the representation matrices of finite groups. These relations lead to a fairly complete theory of representations of finite groups, part of which is outlined here.

Corollary 1. Let $\mathcal{G}$ be a finite group of order n, Γ [$g \to \mathbf{D}(g)$] an irreducible representation of $\mathcal{G}$ on the m-dimensional linear vector space $\mathcal{V}$. If $\mathbf{X}$ is any linear operator on $\mathcal{V}$, then there is a scalar $\alpha = \alpha(\mathbf{X})$ such that

$$A(\mathbf{X}) \equiv \sum_{k=1}^{n} \mathbf{D}(g_k)\mathbf{X}\mathbf{D}(g_k^{-1}) = \alpha(\mathbf{X})\,\mathbf{1} \tag{9.62}$$

Proof. We have

$$A(\mathbf{X})\mathbf{D}(g) = \sum_{k=1}^{n} \mathbf{D}(g_k)\mathbf{X}\mathbf{D}(g_k^{-1})\mathbf{D}(g) = \mathbf{D}(g)\sum_{k=1}^{n} \mathbf{D}(g^{-1}g_k)\mathbf{X}\mathbf{D}(g_k^{-1}g) \tag{9.63}$$

But, as noted earlier, the sum over the elements $\{gg_k\}$ is equivalent to a sum over all the elements of $\mathcal{G}$, so we have

$$A(\mathbf{X})\mathbf{D}(g) = \mathbf{D}(g)A(\mathbf{X}) \tag{9.64}$$

whence $A(\mathbf{X})$ must be a multiple of the identity by Schur's Lemma I.   ∎

Corollary 2. Let $\Gamma\,[g \rightarrow \mathbf{D}(g)]$ and $\Gamma'\,[g \rightarrow \mathbf{D}'(g)]$ be inequivalent irreducible representations of the group $\mathcal{G}$ of order n on the linear vector spaces $\mathcal{V}$ and $\mathcal{V}'$ of dimension m and m', respectively. Let $\mathbf{X}$ be an $m \times m'$ matrix mapping $\mathcal{V}'$ to $\mathcal{V}$. Then

$$\mathcal{F}(\mathbf{X}) \equiv \sum_{k=1}^{n} \mathbf{D}(g_k)\mathbf{X}\mathbf{D}'(g_k^{-1}) = 0 \tag{9.65}$$

Proof. By the same argument used in the preceding proof, we have

$$\mathbf{D}(g)\mathcal{F}(\mathbf{X}) = \mathcal{F}(\mathbf{X})\mathbf{D}'(g) \tag{9.66}$$

for every g in $\mathcal{G}$. Hence $\mathcal{F}(\mathbf{X}) = 0$ by Schur's Lemma II.   ∎

Since the results (9.62) and (9.65) are true for any matrix $\mathbf{X}$, we can use the special matrices $\mathbf{X}^{jk}$ with a single nonvanishing matrix element,

$$\left(\mathbf{X}^{jk}\right)_{j'k'} \equiv \delta_{j'}^{j}\delta_{k'}^{k} \tag{9.67}$$

to derive some useful properties of the representation matrices. In particular, if $\mathcal{G}$ is a finite group of order n with an m-dimensional representation $\Gamma\,[g \rightarrow \mathbf{D}(g)]$, we can treat each set of matrix elements $D_{jk}(g_\ell)$ ($\ell = 1,\ldots,n$) as the components of a vector in an n-dimensional vector space $\mathcal{V}^n$. Here $\mathcal{V}^n$ is exactly the vector space underlying the group algebra $\mathcal{A}(\mathcal{G})$ introduced in Section 9.1.3.

In particular, suppose $\Gamma\,[g \rightarrow \mathbf{D}(g)]$ is an m-dimensional irreducible representation of $\mathcal{G}$. Then from Eq. (9.62), we have

$$A(\mathbf{X}^{kk'}) = \sum_{\ell=1}^{n} \mathbf{D}(g_\ell)\mathbf{X}^{kk'}\mathbf{D}(g_\ell^{-1}) = \lambda_{kk'}\mathbf{1} \tag{9.68}$$

for some scalar $\lambda_{kk'}$, and, since the $\mathbf{D}(g)$ are unitary, we then have

$$\sum_{\ell=1}^{n} D_{jk}(g_\ell)D_{j'k'}^{*}(g_\ell) = \lambda_{kk'}\delta_{jj'} \tag{9.69}$$

The scalar $\lambda_{kk'}$ is evaluated by taking the trace of this equation to give

$$m\lambda_{kk'} = \sum_{\ell=1}^{n} \text{tr}\left\{\mathbf{D}(g_\ell)\mathbf{X}^{kk'}\mathbf{D}(g_\ell^{-1})\right\} = \sum_{\ell=1}^{n} \text{tr}\left\{D_{k'k}(g_\ell^{-1}g_\ell)\right\} = n\delta_{kk'} \qquad (9.70)$$

so that we have, finally,

$$\sum_{\ell=1}^{n} D_{jk}(g_\ell)D_{j'k'}^{*}(g_\ell) = \frac{n}{m}\delta_{jj'}\delta_{kk'} \qquad (9.71)$$

Thus the D_{jk} form a set of m^2 orthogonal vectors in $\mathcal{V}^n$.

Furthermore, if Γ^a $[g \rightarrow \mathbf{D}^a(g)]$ and Γ^b $[g \rightarrow \mathbf{D}^b(g)]$ are two inequivalent finite-dimensional unitary irreducible representations of $\mathcal{G}$, then Eq. (9.65) tells us that

$$\mathcal{F}(\mathbf{X}^{kk'}) = \sum_{k=1}^{n} \mathbf{D}(g_k)\mathbf{X}^{kk'}\mathbf{D}'(g_k^{-1}) = 0 \qquad (9.72)$$

and then

$$\sum_{\ell=1}^{n} D_{jk}^{a}(g_\ell)D_{j'k'}^{b*}(g_\ell) = 0 \qquad (9.73)$$

Thus the $D_{jk}^{a}(g_\ell)$, considered as components of vectors D_{jk}^{a} in $\mathcal{V}^n$, define an orthogonal system in $\mathcal{V}^n$. We can then sum over all the inequivalent irreducible representations Γ^a of $\mathcal{G}$ to obtain

$$\sum_{a} m_a^2 \leq n \qquad (9.74)$$

(m_a is the dimension of Γ^a), since there are at most n orthogonal vectors in $\mathcal{V}^n$, In fact, the equality is always true, as shown in Theorem 9.9.4.

Equations (9.71) and (9.73) can be expressed simply in terms of the characters of the representations: If $\chi^a(g)$, $\chi^b(g)$ denote the characters of g in the inequivalent irreducible finite-dimensional representations Γ^a and Γ^b of $\mathcal{G}$, then

$$\sum_{\ell=1}^{n} \chi^a(g_\ell)\chi^{b*}(g_\ell) = n\,\delta^{ab} \qquad (9.75)$$

The character of g depends only on the class of $\mathcal{G}$ to which g belongs. If $K_1 = \{e\}, K_2, \ldots, K_p$ are the classes of $\mathcal{G}$, with $h_1 = 1, h_2, \ldots, h_p$ elements, and that χ_k^a is the character of the class K_k in the irreducible representation Γ^a. Then Eq. (9.75) is equivalent to

$$\sum_{k=1}^{p} h_k \chi_k^a \chi_k^{b*} = n\,\delta^{ab} \qquad (9.76)$$

Thus the p-dimensional vectors $\mathbf{v}^a$ with components

$$v_k^a \equiv \sqrt{\frac{h_k}{n}}\, \chi_k^a \tag{9.77}$$

$(k = 1, \ldots, p)$ form an orthonormal system, so there are at most p inequivalent finite-dimensional irreducible representations of $\mathcal{G}$.

Remark. There are *exactly* p inequivalent finite-dimensional irreducible representations of $\mathcal{G}$, and the $p \times p$ matrix $\mathbf{V} = (v_k^a)$ is unitary, but more work is needed to derive that result. $\square$

Now suppose Γ is a finite-dimensional representation of $\mathcal{G}$. In general, Γ is reducible, but we can express it as a direct sum of inequivalent irreducible representations of $\mathcal{G}$,

$$\Gamma = \oplus_a c_a^\Gamma \, \Gamma^a \tag{9.78}$$

with nonnegative integer coefficients c_a^Γ. If χ_k^Γ is the character of the class K_k of $\mathcal{G}$ in Γ, then we also have the expansion

$$\chi_k^\Gamma = \sum_a c_a^\Gamma \chi_k^a \tag{9.79}$$

in terms of the characters of the χ_k^a irreducible representations of $\mathcal{G}$. The orthogonality relation (9.76) then gives

$$\sum_{k=1}^p h_k \chi_k^{a*} \chi_k^\Gamma = n c_a^\Gamma \quad \text{and} \quad \sum_{k=1}^p h_k \left| \chi_k^\Gamma \right|^2 = n \sum_a \left| c_a^\Gamma \right|^2 \tag{9.80}$$

Hence the representation Γ is irreducible if and only if

$$\sum_{k=1}^p h_k \left| \chi_k^\Gamma \right|^2 = n \tag{9.81}$$

Remark. Thus we can reduce any representation Γ of $\mathcal{G}$ to a direct sum over the inequivalent irreducible representations using the characters χ_k^a once we have the characters χ_k^Γ of Γ. $\square$

Definition 9.26. If $\mathcal{G}$ is a group of finite order n, then every element g of $\mathcal{G}$ defines a permutation $\mathbf{P}_g$ $(g_k \to g g_k, \; k = 1, \ldots, n)$. The map $g \to \mathbf{P}_g$ is a *permutation representation* of $\mathcal{G}$. If $\mathbf{A}(\mathbf{P}_g)$ is the permutation matrix associated with $\mathbf{P}_g$ (see Eq. (9.27)), then the representation $\Gamma^R \, [g \to \mathbf{A}(\mathbf{P}_g)]$ is a faithful representation, the *regular representation*, of $\mathcal{G}$. $\blacksquare$

The characters of Γ^R are given by $\chi_1^R = n$, $\chi_k^R = 0$ $(k = 2, \ldots, p)$. From Eq. (9.80) we have $c_a^R = m_a$, where m_a is the dimension of the irreducible representation Γ^a. Thus Γ^R is expressed as a direct sum of the inequivalent irreducible representations $\Gamma^1, \ldots, \Gamma^q$ of $\mathcal{G}$ as

$$\Gamma^R = \oplus_{a=1}^q m_a \Gamma^a \tag{9.82}$$

Then also

$$\sum_{a=1}^q m_a^2 = n \quad \text{and} \quad \sum_{a=1}^q m_a \chi_k^a = 0 \tag{9.83}$$

Now suppose $\Gamma = \Gamma^a$ $[g \to \mathbf{D}(g)]$ is an irreducible representation of $\mathcal{G}$ (dimension m_a), and K_k is a class of $\mathcal{G}$ with h_k elements. Then the *class matrix* $\mathbf{D}_k^\Gamma$ of K_k in Γ is defined by

$$\mathbf{D}_k^\Gamma \equiv \sum_{g \text{ in } K_k} \mathbf{D}(g) \tag{9.84}$$

Since $g(K_k) = (K_k)g$ follows directly from the definition of a class, the class matrix $\mathbf{D}_k^\Gamma$ commutes with every g in $\mathcal{G}$. Since $\Gamma = \Gamma^a$ is irreducible, we must have

$$\mathbf{D}_k^\Gamma \equiv \mathbf{D}_k^a = \lambda_k^a \mathbf{1} \tag{9.85}$$

by Schur's Lemma I; here λ_k^a is computed by taking traces on both sides to give

$$m_a \lambda_k^a = h_k \chi_k^a \tag{9.86}$$

In general, the class matrices satisfy

$$\mathbf{D}_k^\Gamma \mathbf{D}_k^\Gamma = \sum_{\ell=1}^p c_{jk}^\ell \mathbf{D}_\ell^\Gamma \tag{9.87}$$

where the c_{jk}^ℓ are the class multiplication coefficients of $\mathcal{G}$ (see Problem 2). If $\Gamma = \Gamma^a$, then

$$h_j h_k \, \chi_j^a \chi_k^a = m_a \sum_{\ell=1}^p c_{jk}^\ell h_\ell \chi_\ell^a \tag{9.88}$$

If we now sum Eq. (9.88) over the inequivalent irreducible representations of $\mathcal{G}$, we obtain

$$h_j h_k \sum_{a=1}^q \chi_j^a \chi_k^a = \sum_{\ell=1}^p c_{jk}^\ell h_\ell \sum_{a=1}^q m_a \chi_\ell^a = n c_{jk}^1 \tag{9.89}$$

in view of Eq. (9.83). Now c_{jk}^1 is the number of times the identity element appears in the product of the classes K_j and K_k. Clearly $c_{jk}^1 = 0$ unless K_j is the class $K_{\bar{k}}$ inverse to K_k; in that case, c_{jk}^1 must be the number h_k of group elements in the class K_k. Thus we have

$$c_{jk}^1 = h_k \delta_{j\bar{k}} \tag{9.90}$$

Since Γ^a is unitary, we also have $\chi_{\bar{k}}^a = \chi_k^{a*}$, and then

$$h_k \sum_{a=1}^q \chi_j^{a*} \chi_k^a = n \delta_{jk} \tag{9.91}$$

Thus the q-dimensional vectors $\mathbf{u}_k$ with components $u_k^a = \chi_k^a$ ($a = 1, \ldots, q$) form an orthogonal system. Hence there are at most q classes of $\mathcal{G}$.

The preceding results taken together form the fundamental theorem on the inequivalent irreducible representations of a finite group.

Theorem 9.4. (Fundamental Representation Theorem) Suppose $\mathcal{G}$ is a group of finite order n containing the classes $K_1 = \{e\}, K_2, \ldots, K_p$ with $h_1 = 1, h_2, \ldots, h_p$ elements, respectively. Let $\Gamma^1, \ldots, \Gamma^q$ be the inequivalent irreducible representations of $\mathcal{G}$, and let χ_k^a be the character of the class K_k in the irreducible representation Γ^a. Then

(i) $q = p$, and

(ii) the $p \times p$ matrix $\mathbf{V} = (v_k^a)$ with matrix elements

$$v_k^a \equiv \sqrt{\frac{h_k}{n}} \, \chi_k^a \tag{9.92}$$

is unitary.

Remark. The characters of the group can then be presented as a $p \times p$ matrix, the *character table* of the group, as illustrated in the examples below. □

Thus the number of inequivalent irreducible representations of $\mathcal{G}$ is equal to the number of classes of $\mathcal{G}$. The characters form an orthonormal system (with appropriately chosen weights) considered as vectors with components either along the classes of $\mathcal{G}$, or along the inequivalent irreducible representations of $\mathcal{G}$. Equations (9.76) and (9.88), together with Eq. (9.83), are enough to completely determine the characters and even the representations for small groups, and lead to general methods for computing the characters and representations of larger groups.

❑ **Example 9.37.** The symmetry group of a square consists of rotations in the plane through an integer multiple of $\pi/2$, and any of these rotations combined with a rotation ρ through π about a diagonal of the square. This group has elements that we can denote by $e, a, a^2, a^3, \rho, \rho a, \rho a^2, \rho a^3$, with

$$a^4 = e \qquad \rho^2 = e \qquad \rho a \rho = a^3 = a^{-1} \tag{9.93}$$

The group is in fact isomorphic to the dihedral group D_4 introduced in Section 9.2 (show this). The classes of the group are easily identified:

$$K_1 = \{e\} \quad K_2 = \{a, a^3\} \quad K_3 = \{a^2\}$$

$$K_4 = \{\rho, \rho a^2\} \qquad K_5 = \{\rho a, \rho a^3\}$$

With five classes, there are five inequivalent irreducible representations. It is clear from Eq. (9.83) that of these, four are one-dimensional and one is two-dimensional. For the one-dimensional representations, we can identify $a = \pm 1$ and $\rho = \pm 1$ independently. For the two-dimensional representation, the requirement $\rho^2 = e$ allows

	K_1	K_2	K_3	K_4	K_5
h	1	2	1	2	2
Γ^1	1	1	1	1	1
Γ^2	1	1	1	-1	-1
Γ^3	1	-1	1	1	-1
Γ^4	1	-1	1	-1	1
Γ^5	2	0	-2	0	0

Character table for D_4.

us to identify $\rho = \sigma_1$; then $\rho a \rho = a^{-1}$ leads to the choice $a = i\sigma_2$, unique up to sign, to obtain a real representation (it is also possible to start with $\rho = \sigma_3$). The characters of the group are then easily computed and can be arranged in the character table shown at the right (check the orthogonality relations as an exercise). ∎

❑ **Example 9.38.** The alternating group $\mathcal{A}_4$ contains the 12 even permutations of degree 4. These permutations have cycle structure (1^4), (22), and (4), but from Eq. (9.83) we can be certain that at least one of these classes of $\mathcal{S}_4$ must divide into smaller classes in $\mathcal{A}_4$. In fact, a short computation shows that the classes of $\mathcal{A}_4$ are

$$K_1 = \{e\} \qquad K_2 = \{(12)(34), (13)(24), (14)(23)\}$$

$$K_3 = \{(123), (134), (421), (432)\}$$

$$K_{\bar{3}} = \{(321), (431), (124), (234)\}$$

Thus there must be three one-dimensional and one three-dimensional irreducible representation. Since the 3-cycles are of order 3, they must be represented in the one-dimensional representations by cube roots of unity, which are 1, ω, and ω^*, where

$$\omega = \exp\left(\frac{2\pi i}{3}\right)$$

	K_1	K_2	K_3	$K_{\bar{3}}$
h	1	3	4	4
Γ^1	1	1	1	1
Γ^2	1	1	ω	ω^*
Γ^3	1	1	ω^*	ω
Γ^4	3	-1	0	0

Character table for $\mathcal{A}_4$.

Note that the classes K_1 and K_2 form an invariant subgroup of $\mathcal{A}_3$ isomorphic to $\mathbf{Z}_2 \otimes \mathbf{Z}_2$, with factor group $\mathbf{Z}_3$. Hence we expect three one-dimensional representations corresponding to those of $\mathbf{Z}_3$. Note also that K_3 and $K_{\bar{3}}$ are inverse classes, so $\chi(K_{\bar{3}}) = \chi^*(K_3)$ in any representation. To construct the three-dimensional representation, start with the fundamental permutation representation on $\mathbf{C}^4$ and consider the three-dimensional subspace spanned by the orthonormal system

$$\psi_1 = \tfrac{1}{2}(1, 1, -1, -1) \qquad \psi_2 = \tfrac{1}{2}(1, -1, 1, -1) \qquad \psi_3 = \tfrac{1}{2}(1, -1, -1, 1) \quad (9.94)$$

On this subspace, the elements of the class K_2 of $\mathcal{A}_4$ are represented by

$$(12)(34) = \begin{pmatrix} 1 & 0 & 0 \\ 0 & -1 & 0 \\ 0 & 0 & -1 \end{pmatrix} \qquad (13)(24) = \begin{pmatrix} -1 & 0 & 0 \\ 0 & 1 & 0 \\ 0 & 0 & -1 \end{pmatrix}$$

$$(14)(23) = \begin{pmatrix} -1 & 0 & 0 \\ 0 & -1 & 0 \\ 0 & 0 & 1 \end{pmatrix} \qquad (9.95)$$

while the 3-cycles in the classes K_3 and $K_{\bar{3}}$ are represented by

$$(123) = \begin{pmatrix} 0 & 1 & 0 \\ 0 & 0 & -1 \\ -1 & 0 & 0 \end{pmatrix} = \widetilde{(321)} \qquad (421) = \begin{pmatrix} 0 & -1 & 0 \\ 0 & 0 & -1 \\ 1 & 0 & 0 \end{pmatrix} = \widetilde{(124)}$$

$$(9.96)$$

$$(134) = \begin{pmatrix} 0 & -1 & 0 \\ 0 & 0 & 1 \\ -1 & 0 & 0 \end{pmatrix} = \widetilde{(431)} \qquad (432) = \begin{pmatrix} 0 & 1 & 0 \\ 0 & 0 & 1 \\ 1 & 0 & 0 \end{pmatrix} = \widetilde{(234)}$$

This is the irreducible representation Γ^4 of $\mathcal{A}_4$. ∎

9.4.3   Kronecker Product of Representations

There are many systems in which we encounter a space that is a tensor product of group representation spaces. In quantum mechanics, for example, we consider separately the symmetry of a two-electron state under exchange of the spatial and spin coordinates of the two electrons. The combined state must be antisymmetric according to the Pauli principle, but this can be achieved either with symmetric space and antisymmetric spin states, or *vice versa*. This example is simple enough, but when more particles are involved, the implementation of the Pauli principle requires further analysis.

Definition 9.27. Let Γ^a $[g \rightarrow \mathbf{D}^a(g)]$ and Γ^b $[g \rightarrow \mathbf{D}^b(g)]$ be irreducible representations of $\mathcal{G}$ on $\mathcal{V}^a$ and on $\mathcal{V}^b$. Then the representation $\Gamma^a \times \Gamma^b$ of $\mathcal{G}$ on $\mathcal{V}^a \otimes \mathcal{V}^b$ defined by

$$\Gamma^a \times \Gamma^b: \quad g \rightarrow \mathbf{D}^{a \times b} \equiv \mathbf{D}^a \otimes \mathbf{D}^b \tag{9.97}$$

is the *Kronecker product* (or *tensor product*) of Γ^a and Γ^b.   ∎

The Kronecker product can be reduced to a sum of irreducible representations of the form

$$\Gamma^a \times \Gamma^b = \oplus_c C_c^{ab} \, \Gamma^c \tag{9.98}$$

The coefficients C_c^{ab} in this reduction are called the *coefficients of composition* of $\mathcal{G}$.

The character of $\Gamma^a \times \Gamma^b$ is simply the product of the characters of Γ^a and Γ^b,

$$\chi_k^{a \times b} = \chi_k^a \chi_k^b \tag{9.99}$$

It follows from the orthogonality relation (9.76) that

$$nC_c^{ab} = \sum_{k=1}^{p} h_k \chi_k^a \chi_k^b \chi_k^{c*} = \sum_{k=1}^{p} h_k \chi_k^{a*} \chi_k^{b*} \chi_k^c \tag{9.100}$$

Thus the C_c^{ab} can be computed directly from the character table.

One useful general result can be obtained if we let Γ^c be the identity representation. Then Eq. (9.100) becomes

$$nC_1^{ab} = \sum_{k=1}^{p} h_k \chi_k^a \chi_k^b = n\delta^{\bar{a}b} \tag{9.101}$$

where $\Gamma^{\bar{a}} = \Gamma^{a*}$ is the complex conjugate of Γ^a, and Γ_1 denotes the identity representation. Hence Γ_1 appears only in the Kronecker product of an irreducible representation with its complex conjugate.

There are also symmetries of the C_c^{ab} that follow directly from Eq. (9.100). For example,

$$C_c^{ab} = C_c^{ba} = C_{\bar{c}}^{\bar{a}\bar{b}} = C_{\bar{b}}^{a\bar{c}} \tag{9.102}$$

These symmetry relations simplify the evaluation of many Kronecker products.

❑ **Example 9.39.**   Consider the group D_4 with character table given in Example 9.37. For the one-dimensional representations, Kronecker products can be read off directly from the character table to give

$$\Gamma^2 \times \Gamma^3 = \Gamma^4 \qquad \Gamma^2 \times \Gamma^4 = \Gamma^3 \qquad \Gamma^3 \times \Gamma^4 = \Gamma^2 \tag{9.103}$$

Note that the characters of the one-dimensional representations themselves form an Abelian group, the *character group*, here isomorphic to $\mathbf{Z}_2 \otimes \mathbf{Z}_2$. Also,

$$\Gamma^a \times \Gamma^5 = \Gamma^5 \tag{9.104}$$

$(a = 1, \ldots, 4)$, and finally,

$$\Gamma^5 \times \Gamma^5 = \Gamma^1 \oplus \Gamma^2 \oplus \Gamma^3 \oplus \Gamma^4 \tag{9.105}$$

This result follows directly from the symmetry relations (9.102), though it can also be obtained by calculation using Eq. (9.100) and the character tables.   ∎

→ **Exercise 9.19.** Compute the reduction of the Kronecker product $\Gamma^4 \times \Gamma^4$ in the group $\mathcal{A}_4$ from the character table given in Example 9.38.   □

9.4.4 Permutation Representations

In Section 9.4.2, we introduced the representation of a group $\mathcal{G}$ of order n by the permutations of the n group elements associated with group multiplication, i.e., if $g_1 = e, g_2, \ldots, g_n$ are the elements of $\mathcal{G}$, then

$$g g_k \rightarrow g_{i_k} \tag{9.106}$$

defines a permutation $\mathbf{P}_g = (i_1, \ldots, i_n)$ of n. The regular representation Γ^R of $\mathcal{G}$ was then introduced (Definition 9.26) as the representation $[g \rightarrow \mathbf{A}(\mathbf{P}_g)]$ of $\mathcal{G}$ by the permutation matrices $\mathbf{A}(\mathbf{P})$ corresponding to these permutations.

There are other useful representations of groups by permutations and their associated matrices. Every element $\mathbf{P}$ of the symmetric group $\mathcal{S}_N$ corresponds to an $N \times N$ permutation matrix $\mathbf{A}(\mathbf{P})$; this is the *fundamental* (or *defining*) *representation* of $\mathcal{S}_N$. The fundamental representation is reducible, since the vector $\xi_0 = (1, 1, \ldots, 1)$ is transformed into itself by every permutation of the basis vectors. However, the representation of $\mathcal{S}_N$ on the $(N-1)$-dimensional manifold $\mathcal{M}^\perp(\xi_0)$ actually is irreducible, as we now show for the case of $\mathcal{S}_3$.

❑ **Example 9.40.** For the group $\mathcal{S}_3$ of permutations of three objects, the fundamental three-dimensional representation was introduced in Example 9.31, with representation matrices given in Eq. (9.46). This representation of $\mathcal{S}_3$ is reducible, since one-dimensional subspace $\mathcal{M}(\xi_0)$ is invariant under $\mathcal{S}_3$. We have

$$\mathbf{A}(\mathbf{P})\xi_0 = \xi_0 \tag{9.107}$$

for every permutation $\mathbf{P}$ in $\mathcal{S}_3$. This defines the identity (symmetric) representation of $\mathcal{S}_3$. To construct a two-dimensional representation on $\mathcal{M}^\perp(\xi_0)$, note that the vectors

$$\psi_1 \equiv \sqrt{\tfrac{1}{2}}(1, -1, 0) \qquad \psi_2 \equiv \sqrt{\tfrac{1}{6}}(1, 1, -2) \tag{9.108}$$

form a complete orthonormal system on $\mathcal{M}^\perp(\xi_0)$. These vectors are chosen so that (i) they are orthogonal to ξ_0 and (ii) they are either symmetric (ψ_2) or antisymmetric (ψ_1)

under the transposition of ϕ_1 and ϕ_2. A short calculation shows that in this basis, the transpositions are represented by the 2×2 matrices

$$(12) = \begin{pmatrix} -1 & 0 \\ 0 & 1 \end{pmatrix} \qquad (13) = \frac{1}{2} \begin{pmatrix} 1 & -\sqrt{3} \\ -\sqrt{3} & -1 \end{pmatrix} \qquad (23) = \frac{1}{2} \begin{pmatrix} 1 & \sqrt{3} \\ \sqrt{3} & -1 \end{pmatrix}$$

$$(9.109)$$

Then also

$$(123) = (13)(12) = \frac{1}{2} \begin{pmatrix} -1 & -\sqrt{3} \\ \sqrt{3} & -1 \end{pmatrix} \qquad (321) = (12)(13) = \frac{1}{2} \begin{pmatrix} -1 & \sqrt{3} \\ -\sqrt{3} & -1 \end{pmatrix}$$

$$(9.110)$$

This representation is irreducible, since the matrices do not commute with each other, and thus cannot have common eigenvectors. ∎

Remark.   The group $\mathcal{S}_3$ thus has three inequivalent irreducible representations: Two one-dimensional representations, symmetric ($\mathbf{P} \rightarrow \mathbf{1}$) and antisymmetric ($\mathbf{P} \rightarrow \varepsilon_\mathbf{P}$), and the two-dimensional representation found here, the *mixed symmetry* representation. □

If a group $\mathcal{G}$ of order n has a subgroup $\mathcal{H}$ of order m, index $t = n/m$, then permutations of the cosets of $\mathcal{H}$ in $\mathcal{G}$ associated with the group multiplication define a permutation representation $\mathcal{P}(\mathcal{G}, \mathcal{H})$ of $\mathcal{G}$, in which $\mathbf{P}(g, \mathcal{H})$ is the permutation of the cosets of $\mathcal{H}$ in $\mathcal{G}$ associated with multiplication by the group element g. To understand this representation, note that there are elements $g_1 = e, g_2, \ldots, g_t$ of $\mathcal{G}$ such that $g_1\mathcal{H}, g_2\mathcal{H}, \ldots, g_t\mathcal{H}$ are the disjoint cosets of $\mathcal{H}$ in $\mathcal{G}$. Then we can form the $t \times t$ matrices $\sigma(g, h)$ such that

$$\sigma_{\alpha\beta}(g, h) = \begin{cases} 1, & \text{if } gg_\beta = g_\alpha h \\ 0 & \text{otherwise} \end{cases}$$

$$(9.111)$$

These matrices have at most one nonzero element in any row or column, since given g in $\mathcal{G}$, and $\beta = 1, \ldots, t$, the element gg_β belongs to exactly one coset of $\mathcal{H}$, say $g_\alpha\mathcal{H}$. Then let

$$\pi(g, \mathcal{H}) \equiv \sum_{h \in \mathcal{H}} \sigma(g, h) = \mathbf{A}(\mathbf{P}(g, \mathcal{H}))$$

$$(9.112)$$

where $\mathbf{A}(\mathbf{P}(g, \mathcal{H}))$ is the permutation matrix associated with the permutation $\mathbf{P}(g, \mathcal{H})$.

→ Exercise 9.20.   Show that

$$\pi(g_1, \mathcal{H})\pi(g_2, \mathcal{H}) = \pi(g_1 g_2, \mathcal{H})$$

so that the $\pi(g, \mathcal{H})$ actually form a representation of $\mathcal{G}$. □

The $\pi(g, \mathcal{H})$ define the *principal representation* of $\mathcal{G}$ induced by $\mathcal{H}$. Other representations of $\mathcal{G}$ can be constructed from those of $\mathcal{H}$ using the matrices $\sigma(g, h)$. If $\Delta [h \rightarrow \mathbf{D}^\Delta(h)]$ is a representation of $\mathcal{H}$, then

$$\mathbf{D}^\Gamma(g) \equiv \sum_{h \text{ in } \mathcal{H}} \sigma(g, h) \otimes \mathbf{D}^\Delta(h)$$

$$(9.113)$$

is a representation $\Gamma \equiv \Delta_{\text{ind}}$ of $\mathcal{G}$, the representation of $\mathcal{G}$ *induced* by Δ. The principal representation $[g \to \pi(g, \mathcal{H})]$ of $\mathcal{G}$ is evidently induced by the identity representation of $\mathcal{H}$.

❑ **Example 9.41.**   Consider the alternating group $\mathcal{A}_4$ introduced in Example 9.38. This group has an Abelian subgroup $\mathcal{H}$ isomorphic to $\mathbf{Z}_2 \otimes \mathbf{Z}_2$ with elements

$$e \quad a = (12)(34) \qquad b = (13)(24) \qquad c = (14)(23) \tag{9.114}$$

with

$$a^2 = b^2 = c^2 = e \quad \text{and} \quad abc = e \tag{9.115}$$

$\mathcal{H}$ has four inequivalent irreducible representations: The identity representation Γ^0, and three representations labeled Γ^a, Γ^b, and Γ^c, in which the elements a, b, c, respectively, are represented by $+1$ and the other two elements of order 2 by -1. The cosets of $\mathcal{H}$ are $\mathcal{H} = \{K_1 \cup K_2\}$, K_3, and $K_{\bar{3}}$, using the notation of Example 9.38, and we can take the coset representatives to be e, (123) and (321).

In the principal representation $\Gamma^{\mathcal{H}}$ of $\mathcal{A}_4$ induced by $\mathcal{H}$, all the elements of each coset are represented by the same matrix, and we have

$$\mathcal{H} \to \begin{pmatrix} 1 & 0 & 0 \\ 0 & 1 & 0 \\ 0 & 0 & 1 \end{pmatrix} \qquad K_3 \to \begin{pmatrix} 0 & 1 & 0 \\ 0 & 0 & 1 \\ 1 & 0 & 0 \end{pmatrix} \qquad K_{\bar{3}} \to \begin{pmatrix} 0 & 0 & 1 \\ 1 & 0 & 0 \\ 0 & 1 & 0 \end{pmatrix} \tag{9.116}$$

These matrices have common eigenvectors $\psi_\lambda = (1, \lambda, \lambda^2)$ with $\lambda^3 = 1$, so that $\Gamma^{\mathcal{H}}$ is reducible; in fact,

$$\Gamma^{\mathcal{H}} = \Gamma^1 \oplus \Gamma^2 \oplus \Gamma^3 \tag{9.117}$$

where the $\Gamma^{1,2,3}$ are the irreducible representations of $\mathcal{A}_4$ constructed in Example 9.38.

The three-dimensional irreducible representation Γ^4 of $\mathcal{A}_4$ appears only in the representations of $\mathcal{A}_4$ induced by one of the representations Γ^a, Γ^b, Γ^c of $\mathcal{H}$. Since these lead to equivalent representations of $\mathcal{A}_4$, we can choose Γ^a. We then find solutions to the following equation

$$g g_\beta = g_\alpha h \tag{9.118}$$

as g runs through $\mathcal{A}_4$ and g_β runs through the set $\{e, (123), (321)\}$. The results are shown in the table, together with the corresponding matrices in the induced representation. The matrices for the 3-cycles in the class $K_{\bar{3}}$ of $\mathcal{A}_4$ are not given explicitly; the matrix such a 3-cycle is simply the transpose of the matrix for its inverse in the class K_3.   ∎

Table 9.1: Multiplication table for generating representations of $\mathcal{A}_4$ induced by an irreducible representation of its $\mathbf{Z}_2 \otimes \mathbf{Z}_2$ subgroup. The rightmost column gives the matrices for the representation of $\mathcal{A}_4$ induced by the representation of $\mathbf{Z}_2 \otimes \mathbf{Z}_2$ in which the elements e and a are represented by $+1$, the elements b and c by -1.

g	g_β	g_α	h	$\displaystyle\sum_h \sigma_{\alpha\beta}\, \mathbf{D}(h)$
a	e	e	a	$\begin{pmatrix} 1 & 0 & 0 \\ 0 & -1 & 0 \\ 0 & 0 & -1 \end{pmatrix}$
a	(123)	(123)	b	
a	(321)	(321)	c	
b	e	e	b	$\begin{pmatrix} -1 & 0 & 0 \\ 0 & -1 & 0 \\ 0 & 0 & 1 \end{pmatrix}$
b	(123)	(123)	c	
b	(321)	(321)	a	
c	e	e	c	$\begin{pmatrix} -1 & 0 & 0 \\ 0 & 1 & 0 \\ 0 & 0 & -1 \end{pmatrix}$
c	(123)	(123)	a	
c	(321)	(321)	b	
(123)	e	(123)	e	$\begin{pmatrix} 0 & 0 & 1 \\ 1 & 0 & 0 \\ 0 & 1 & 0 \end{pmatrix}$
(123)	(123)	(321)	e	
(123)	(321)	e	e	
(134)	e	(123)	a	$\begin{pmatrix} 0 & 0 & -1 \\ 1 & 0 & 0 \\ 0 & -1 & 0 \end{pmatrix}$
(134)	(123)	(321)	b	
(134)	(321)	e	c	
(421)	e	(123)	c	$\begin{pmatrix} 0 & 0 & -1 \\ -1 & 0 & 0 \\ 0 & 1 & 0 \end{pmatrix}$
(421)	(123)	(321)	a	
(421)	(321)	e	b	
(432)	e	(123)	b	$\begin{pmatrix} 0 & 0 & 1 \\ -1 & 0 & 0 \\ 0 & -1 & 0 \end{pmatrix}$
(432)	(123)	(321)	c	
(432)	(321)	e	a	

Remark. The matrices in the table are not exactly the same as those given in Example 9.38. However, the two representations are in fact equivalent, since one can be transformed into the other by a rotation through $\pi/2$ in the ψ_2-ψ_3 plane. The details are left to the reader. $\qquad\square$

9.4.5   Representations of Groups and Subgroups

Any representation of a group $\mathcal{G}$ is a representation of a subgroup $\mathcal{H}$, but in general, an irreducible representation of $\mathcal{G}$ is reducible when restricted to $\mathcal{H}$. As explained in the preceding
section, a representation of $\mathcal{H}$ induces a representation of $\mathcal{G}$, which is also reducible in general.
The Frobenius reciprocity theorem, explained below and derived in Appendix A, provides a
fundamental relation between the reduction of irreducible representations of $\mathcal{G}$ restricted to a
subgroup $\mathcal{H}$ and the representations of $\mathcal{G}$ induced by irreducible representations of $\mathcal{H}$.

Suppose $\mathcal{G}$ is a group of order n, with inequivalent irreducible representations $\Gamma^1, \ldots, \Gamma^p$,
of dimensions $m_1, \ldots, m_p$, respectively. Also, suppose $\mathcal{H}$ is a subgroup of $\mathcal{G}$ of order m with
inequivalent irreducible representations denoted by $\Delta^1, \ldots, \Delta^q$, with μ_c the dimension of Δ^c
($c = 1, \ldots, q$). A representation Γ of $\mathcal{G}$ restricted to $\mathcal{H}$ defines the representation Γ_{sub} of $\mathcal{H}$
subduced by Γ. The irreducible representation Γ^a of $\mathcal{G}$ subduces a representation Γ_{sub}^a of $\mathcal{H}$
that can be reduced to a sum of irreducible representations of $\mathcal{H}$,

$$\Gamma_{\text{sub}}^a = \oplus_c \alpha_{ca} \Delta^c \tag{9.119}$$

with nonnegative integer coefficients α_{ca}. On the other hand, the irreducible representation
Δ^c of $\mathcal{H}$ induces a representation Δ_{ind}^c of $\mathcal{G}$, also reducible in general,

$$\Delta_{\text{ind}}^c = \oplus_a \beta_{ca} \Gamma^a \tag{9.120}$$

with nonnegative integer coefficients β_{ca}. The *Frobenius reciprocity theorem* states that

$$\alpha_{ca} = \beta_{ca} \tag{9.121}$$

In words, the coefficients that appear in the reduction of an irreducible representation of $\mathcal{G}$
restricted to a subgroup $\mathcal{H}$ are equal to the coefficients in the reduction of the representation of
$\mathcal{G}$ induced by an irreducible representation of $\mathcal{H}$. A proof of this crucial theorem is presented
in Appendix A for the curious reader.

An important corollary that is extremely useful in practical computations is

$$\Gamma^a \times \Delta_{\text{ind}}^c = [\Gamma_{\text{sub}}^a \times \Delta^c]_{\text{ind}} \tag{9.122}$$

This allows Kronecker products obtained for a subgroup $\mathcal{H}$ to be used to reduce those in $\mathcal{G}$.
This result will be applied to the the symmetric group in the next section.

Representations of direct product groups are often useful in constructing representations
of larger groups. If $\mathcal{G}_1$ and $\mathcal{G}_2$ are groups with representations $\Gamma(1)$ $[g_1 \rightarrow \mathbf{D}_{(1)}(g_1)]$ and $\Gamma(2)$
$[g_2 \rightarrow \mathbf{D}_{(2)}(g_2)]$, respectively, then

$$(g_1, g_2) \rightarrow \mathbf{D}(g_1, g_2) \equiv \mathbf{D}_{(1)}(g_1) \otimes \mathbf{D}_{(2)}(g_2) \tag{9.123}$$

is a representation $\Gamma(1) \otimes \Gamma(2)$ of $\mathcal{G}_1 \otimes \mathcal{G}_2$, irreducible if and only if both $\Gamma(1)$ and $\Gamma(2)$
are irreducible. If $\{\Gamma^a(1)\}$ and $\{\Gamma^c(2)\}$ are the inequivalent irreducible representations of $\mathcal{G}_1$
and $\mathcal{G}_2$, then the inequivalent irreducible representations of $\mathcal{G}_1 \otimes \mathcal{G}_2$ have the form

$$\Gamma^{(a,c)} \equiv \Gamma^a(1) \otimes \Gamma^c(2) \tag{9.124}$$

The representations of $\mathcal{G}$ induced by those of $\mathcal{G}_1 \otimes \mathcal{G}_2$ have a special name:

Definition 9.28. If $\mathcal{G}$ is a group with subgroup $\mathcal{G}_1 \otimes \mathcal{G}_2$, then the representation of $\mathcal{G}$ induced by the representation $\Gamma(1) \otimes \Gamma(2)$ of $\mathcal{G}_1 \otimes \mathcal{G}_2$, using the construction of Eq. (9.113), is the *outer product* $[\Gamma(1) \circ \Gamma(2)]_{\mathcal{G}}$ of $\Gamma(1)$ and $\Gamma(2)$ (with respect to $\mathcal{G}$). The subscript $\mathcal{G}$ may be omitted if the group $\mathcal{G}$ is understood.   ∎

❏ **Example 9.42.** A rather simple illustration of these ideas can be extracted from the discussion of the alternating group $\mathcal{A}_4$ and its $\mathbf{Z}_2 \otimes \mathbf{Z}_2$ subgroup in examples 9.38 and 9.41. Denote the irreducible representations of $\mathbf{Z}_2$ by

$$\Gamma^S = \{1, 1\} \quad \text{and} \quad \Gamma^A = \{1, -1\}$$

(S = symmetric, A = antisymmetric), those of $\mathbf{Z}_2 \otimes \mathbf{Z}_2$ by

$$\Gamma^0 = \Gamma^S \otimes \Gamma^S \qquad \Gamma^a = \Gamma^S \otimes \Gamma^A \qquad \Gamma^b = \Gamma^A \otimes \Gamma^S \qquad \Gamma^c = \Gamma^A \otimes \Gamma^A$$

as in Example 9.41, and those of $\mathcal{A}_4$ by $\Gamma^{1,2,3,4}$ as in Example 9.38. Under the restriction of $\mathcal{A}_4$ to its $\mathbf{Z}_2 \otimes \mathbf{Z}_2$ subgroup, we have

$$\Gamma^{1,2,3} \to \Gamma^0 \qquad \Gamma^4 \to \Gamma^a \oplus \Gamma^b \oplus \Gamma^c \tag{9.125}$$

The Frobenius reciprocity theorem then requires that the representations of $\mathcal{A}_4$ induced by those of $\mathbf{Z}_2 \otimes \mathbf{Z}_2$ satisfy

$$\Gamma^0_{\text{ind}} = \Gamma^S \circ \Gamma^S = \Gamma^1 \oplus \Gamma^2 \oplus \Gamma^3 \tag{9.126}$$

$$\Gamma^{a,b,c}_{\text{ind}} = \Gamma^S \circ \Gamma^A = \Gamma^A \circ \Gamma^S = \Gamma^A \circ \Gamma^A = \Gamma^4 \tag{9.127}$$

as already derived in Example 9.41.   ∎

Outer products of representations of $\mathcal{S}_N$ are especially important, as they play a central role in the reduction of Kronecker products of Lie groups. As we shall see in the next chapter, irreducible representations of Lie groups are associated with irreducible representations of $\mathcal{S}_N$. For example, consider the group $U(n)$ of unitary operators on an n-dimensional linear vector space $\mathcal{V}^n$. Tensors of rank N are defined on the tensor product $\otimes_N \mathcal{V}^n$ of N identical copies of $\mathcal{V}^n$. These tensors define a reducible representation of $U(n)$; irreducible tensors are those that transform as irreducible representations of $\mathcal{S}_N$ under permutations of the N indices.

If Γ_a and Γ_b are representations of a Lie group corresponding to representations $\mathcal{S}_M$ and $\mathcal{S}_N$, respectively, then the Kronecker product of these representations corresponds to the representation of $\mathcal{S}_{M+N}$ defined by the outer product $\Gamma_a \circ \Gamma_b$. Thus we are interested in finding the irreducible representations contained the reduction of outer products of irreducible representations of $\mathcal{S}_N$. Some elegant graphical techniques are available to perform this reduction; these are described in the next section.

9.5   Representations of the Symmetric Group $\mathcal{S}_N$

9.5.1   Irreducible Representations of $\mathcal{S}_N$

The symmetric group $\mathcal{S}_N$ of permutations of N elements has classes corresponding to partitions of N, as described in Section 9.3. Each permutation has a cycle structure that can be identified with a partition (ξ) of N, and the class $K_{(\xi)}$ of $\mathcal{S}_N$ consists of all the permutations in $\mathcal{S}_N$ with cycle structure characterized by (ξ). Thus there are $\pi(N)$ classes in $\mathcal{S}_N$. It follows that there are also $\pi(N)$ inequivalent irreducible representations of $\mathcal{S}_N$, and these can also be associated with partitions of N; denote the irreducible representation of $\mathcal{S}_N$ associated with the partition (λ) of N by $\Gamma^{(\lambda)}$.

The simple characters of $\mathcal{S}_N$ are given by a remarkable generating function due to Frobenius. This generating function (Eq. (9.B42)) and the graphical methods that can be used to actually compute the characters are derived in detail in Appendix B. Here we simply note that the dimension $d(\lambda)$ of $\Gamma^{(\lambda)}$ is given, as already noted in Section 9.3.2, by the number of regular Young tableaux that can be associated with the Young diagram $\mathcal{Y}^{(\lambda)}$, or by the elegant hook formula (9.B48).

The regular tableaux can be used to define projection operators that act on a general function of N variables to project out functions transforming among themselves under an irreducible representation on $\mathcal{S}_N$ under permutations of the variables. For example, the operators $\mathbf{S}$ and $\mathbf{A}$ defined by

$$\mathbf{S} = \sum_{P \in \mathcal{S}_N} \mathbf{P} \quad \text{and} \quad \mathbf{A} = \sum_{P \in \mathcal{S}_N} \varepsilon_{\mathbf{P}} \mathbf{P} \tag{9.128}$$

project out the symmetric and antisymmetric components of a function of N variables. These components may not be present, of course, in which case the projection gives zero.

For a general regular tableau t associated with a partition (λ) of N, we let H_t (V_t) denote the subgroup of $\mathcal{S}_N$ that leaves the rows (columns) of the tableau unchanged. Then define

$$\mathbf{S}_t = \sum_{P \in H_t} \mathbf{P} \quad \text{and} \quad \mathbf{A}_t = \sum_{P \in V_t} \varepsilon_{\mathbf{P}} \mathbf{P} \tag{9.129}$$

$\mathbf{S}_t$ $(\mathbf{A}_t)$ projects out components of a function that are symmetric (antisymmetric) under permutations of variables in the same row (column) of t. We then define the *Young symmetrizer* associated with t as

$$\mathbf{Y}_t = \mathbf{A}_t \mathbf{S}_t \tag{9.130}$$

Note that if t and t' are two distinct regular tableaux, there is necessarily at least one pair of elements in some row of t' that appears in the same column of t. Then $\mathbf{S}_{t'} \mathbf{A}_t = \mathbf{0}$, since $\mathbf{A}_t$ projects out the component antisymmetric under exchange of this pair, which is then annihilated by $\mathbf{S}_{t'}$. Thus

$$\mathbf{Y}_{t'} \mathbf{Y}_t = 0 \tag{9.131}$$

❑ **Example 9.43.** For the symmetric group S_3, the operators **S** and **A** are given by

$$\mathbf{S} = e + (12) + (13) + (23) + (123) + (321)$$
$$\mathbf{A} = e - (12) - (13) - (23) + (123) + (321)$$

(9.132)

(here e is the identity permutation), with $\mathbf{S}^2 = 6\mathbf{S}$ and $\mathbf{A}^2 = 6\mathbf{A}$ so that the actual projection operators are $\mathbf{S}/6$ and $\mathbf{A}/6$. As noted in Example 9.28, there are two regular tableaux associated with the partition [21] of 3 (here we use square brackets to denote partitions to avoid confusion with permutations). These are

(a) $\begin{array}{|c|c|}\hline 1 & 2 \\\hline 3 \\\cline{1-1}\end{array}$ and   (b) $\begin{array}{|c|c|}\hline 1 & 3 \\\hline 2 \\\cline{1-1}\end{array}$

For the tableau (a), we have

$$\mathbf{S}_a = e + (12) \qquad \mathbf{A}_a = e - (13)$$

(9.133)

and then

$$\mathbf{Y}_a = e + (12) - (13) - (123)$$

(9.134)

while for the tableau (b), we have

$$\mathbf{S}_b = e + (13) \qquad \mathbf{A}_b = e - (12)$$

(9.135)

and then

$$\mathbf{Y}_b = e - (12) + (13) - (321)$$

(9.136)

Now $\mathbf{Y}_a\mathbf{Y}_b = 0$ as expected, and also $\mathbf{Y}_a^2 = 3\mathbf{Y}_a$ and $\mathbf{Y}_b^2 = 3\mathbf{Y}_b$ so that the actual projection operators here are $\mathbf{Y}_a/3$ and $\mathbf{Y}_b/3$.   ∎

Note that the projection operators in the example satisfy

$$\tfrac{1}{6}(\mathbf{S} + \mathbf{A}) + \tfrac{1}{3}(\mathbf{Y}_a + \mathbf{Y}_b) = e$$

(9.137)

thus defining a resolution of the identity on the six-dimensional space of arrangements of three objects. **S** and **A** project onto the one-dimensional spaces of symmetric and antisymmetric combinations, while $\mathbf{Y}_a$ and $\mathbf{Y}_b$ project onto two-dimensional manifolds in which permutations are represented by an irreducible representation $\Gamma^{(21)}$ of S_3, so the total dimension of the projection manifolds is six, as required.

However, if we consider the projection operators as elements of the group algebra introduced in Section 9.1.3, there are only four projection operators, while the group algebra is six-dimensional. Thus we need two more independent vectors to define a basis of the group algebra. We can choose these to be $(23)\mathbf{Y}_a$ and $(23)\mathbf{Y}_b$. It can be verified by direct calculation that these vectors are independent of the projection operators; hence, we have a set of basis vectors for the group algebra.

Now recall the regular representation of a group introduced in Definition 9.26; it is equivalent to representing the group by its action as a linear operator on the group algebra. Introducing the Young symmetrizers allows us to reduce the regular representation of $\mathcal{S}_N$ on its own group algebra into its irreducible components, as here the group algebra is spanned by the one-dimensional vector spaces $\mathcal{V}_S$ and $\mathcal{V}_A$ defined by $\mathbf{S}$ and $\mathbf{A}$, together with the two-dimensional spaces $\mathcal{V}_a$ and $\mathcal{V}_b$ spanned by $\{\mathbf{Y}_a, (23)\mathbf{Y}_a\}$ and $\{\mathbf{Y}_b, (23)\mathbf{Y}_b\}$, span the entire group algebra.

→ **Exercise 9.21.** Show that

$$(23)\mathbf{Y}_a = \mathbf{Y}_b(23) \quad \text{and} \quad (23)\mathbf{Y}_b = \mathbf{Y}_a(23)$$

How are these identities related to the fact that the Young tableaux (a) and (b) are transformed into each other by interchanging 2 and 3?   □

Young symmetrizers can be constructed for any irreducible representation of any $\mathcal{S}_N$ using the standard tableaux. They provide a separation of a general function of N variables into parts that have a definite symmetry, i.e., parts that transform according to definite irreducible representations of $\mathcal{S}_N$ under permutations of the variables. There are various methods for constructing explicit representation matrices if these are actually needed. We have given the matrices for the mixed [(21)] representation of $\mathcal{S}_3$ in Example 9.40, and invite the reader to find matrices for the irreducible representations of $\mathcal{S}_4$ in Problem 16. General methods are described in the book by Hamermesh cited in the bibliography, among many others.

9.5.2   Outer Products of Representations of $\mathcal{S}_m \otimes \mathcal{S}_n$

One avenue to study the irreducible representations of a symmetric group is to understand the relations between representations of the group and those of its subgroups, as described in general in Section 9.4.5. Of particular interest here are the representations of a symmetric group $\mathcal{S}_{m+n}$ induced by irreducible representations $\Gamma^{(\mu)} \otimes \Gamma^{(\nu)}$ of a subgroup $\mathcal{S}_m \otimes \mathcal{S}_n$; these representations are the outer products defined in Section 9.4.5 (see Definition 9.28). The outer products of symmetric groups are also related directly to the Kronecker products of representations of Lie groups, as we will soon see in Chapter 10. The outer product $\Gamma^{(\mu)} \circ \Gamma^{(\nu)}$ is in general reducible; we have

$$\Gamma^{(\mu)} \circ \Gamma^{(\nu)} = \sum_{(\lambda)} K\{(\mu)(\nu)|(\lambda)\}\Gamma^{(\lambda)} \tag{9.138}$$

where the summation is over partitions (λ) of $m + n$. The expansion coefficients $K\{(\mu)(\nu)|(\lambda)\}$ can be evaluated by graphical methods that we describe here; some derivations are given in Appendix B.3.

We first consider the case where the partition $(\mu) = (m)$, i.e., the partition has a single part, corresponding to a Young diagram with m nodes in a single row. Then, as derived in Appendix B, the coefficient $K\{(m)(\nu)|(\lambda)\}$ is equal to one if and only if the diagram $\mathcal{Y}^{(\lambda)}$ can be constructed from $\mathcal{Y}^{(\nu)}$ by the addition of m nodes, no two of which appear in the same column of $\mathcal{Y}^{(\lambda)}$. Otherwise, the coefficient is zero.

❑ **Example 9.44.**   For $\mathcal{S}_2 \otimes \mathcal{S}_2$, we have the products

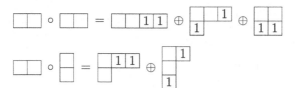

For $\mathcal{S}_2 \otimes \mathcal{S}_3$, we have the products

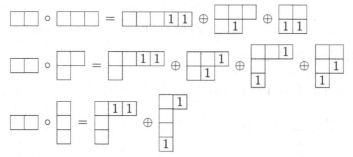

These should show the general picture.   ∎

➔ **Exercise 9.22.**   Show that if $m \leq n$, then we have

$$\Gamma^{(m)} \circ \Gamma^{(n)} = \sum_{k=0}^{m} \Gamma^{(m+n-k\ k)}$$

Draw a picture of this in diagrams.   ❑

It remains to reduce the outer product $\Gamma^{(\mu)} \circ \Gamma^{(\nu)}$ of a general pair of irreducible representations of $\mathcal{S}_m$ and $\mathcal{S}_n$. An algebraic expression is given in Appendix B (Eq. (9.B53)), but a graphical rule that is more useful in practice is also given there (Theorem 9.5). This rule states that if $(\mu) = (\mu_1 \mu_2 \dots \mu_q)$ is a partition of m and (ν) is a partition of n, then the coefficient $K\{(\mu)(\nu)|(\lambda)\}$ in the reduction (9.138) is equal to the number of distinct constructions of the diagram $\mathcal{Y}^{(\lambda)}$ from the diagram $\mathcal{Y}^{(\nu)}$ by successive application of μ_1 nodes labeled 1, μ_2 nodes labeled 2, ..., μ_q nodes labeled q such that

(N) no two nodes with the same label appear in the same column of $\mathcal{Y}^{(\lambda)}$, *and*

(R) if we scan $\mathcal{Y}^{(\lambda)}$ from left to right across the rows from top to bottom, then the ith node labeled $k + 1$ appears in a lower row of $\mathcal{Y}^{(\lambda)}$ than the ith node labeled k.

❑ **Example 9.45.**   A very simple example is

where the latter two diagrams are excluded by rule (R). This result also agrees with the outer product in the previous example.   ∎

❑ **Example 9.46.** We have

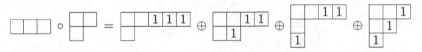

The same product in reverse order is given by

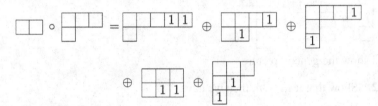

Thus the outer product does not depend on the order of the terms in the product. ∎

→ **Exercise 9.23.** Evaluate the outer product $\Gamma^{(21)} \circ \Gamma^{(21)}$.   ☐

❑ **Example 9.47.** As a final example, we have

The evaluation of $\Gamma^{(2)} \circ \Gamma^{(22)}$ is left to the reader. ∎

9.5.3 Kronecker Products of Irreducible Representations of $\mathcal{S}_N$

Consider now the Kronecker product

$$\Gamma^{(\mu)} \times \Gamma^{(\nu)} = \sum_{(\lambda)} C\{(\mu)(\nu)|(\lambda)\}\Gamma^{(\lambda)} \tag{9.139}$$

of the irreducible representations $\Gamma^{(\mu)}$ and $\Gamma^{(\nu)}$ of $\mathcal{S}_N$. A formal expression for the coefficients $C\{(\mu)(\nu)|(\lambda)\}$ is

$$C\{(\mu)(\nu)|(\lambda)\} = \sum_{(m)} \frac{N_{(m)}}{N!} \chi^{(\mu)}_{(m)} \chi^{(\nu)}_{(m)} \chi^{(\lambda)}_{(m)} \tag{9.140}$$

which shows that the coefficients $C\{(\mu)(\nu)|(\lambda)\}$ are symmetric in (μ), (ν) and (λ) (the characters of $\mathcal{S}_N$ are real). Also,

$$C\{(\tilde{\mu})(\tilde{\nu})|(\lambda)\} = C\{(\mu)(\nu)|(\lambda)\} \tag{9.141}$$

so that we can conjugate any two of the partitions in $C\{(\mu)(\nu)|(\lambda)\}$ and have the same result. Finally,

$$C\{(\mu)(\nu)|(N)\} = \delta_{(\mu)(\nu)} \quad \text{and} \quad C\{(\mu)(\nu)|(1^N)\} = \delta_{(\mu)(\tilde{\nu})} \tag{9.142}$$

Thus the Kronecker product of an irreducible representation with itself, and only with itself, contains the identity representation, while the antisymmetric representation is contained only the product of an irreducible representation with the irreducible representation belonging to the conjugate partition.

Actual computation of the coefficients $C\{(\mu)(\nu)|(\lambda)\}$, or of the complete reduction (9.139), is aided by the symmetry of the coefficients and the dimension check

$$d^{(\mu)}d^{(\nu)} = \sum_{(\lambda)} C\{(\mu)(\nu)|(\lambda)\}\, d^{(\lambda)} \tag{9.143}$$

The full calculation of the reduction can be done either with the character tables, or graphically with the aid of Eq. (9.122). If $(\xi) = (\xi_1\,\xi_2\,\dots\,\xi_q)$ is a partition of N, let

$$\mathcal{G}(\xi) \equiv \mathcal{S}_{\xi_1} \otimes \mathcal{S}_{\xi_2} \otimes \cdots \otimes \mathcal{S}_{\xi_q}$$

and let $\Gamma_{(\xi)}$ be the representation of $\mathcal{S}_N$ induced by the identity representation of $\mathcal{G}(\xi)$. If (μ) another partition of N, then Eq. (9.122) has the form

$$\Gamma^{(\mu)} \times \Gamma_{(\xi)} = [\Gamma_{(\xi)}^{(\mu)}]_{\text{ind}} \tag{9.144}$$

Here $\Gamma_{(\xi)}^{(\mu)}$ is the (reducible) representation of $\mathcal{G}(\xi)$ subduced by $\Gamma^{(\mu)}$. By the Frobenius reciprocity theorem, this representation contains all the representations

$$\Gamma^{(\xi_1)} \otimes \Gamma^{(\xi_2)} \otimes \cdots \otimes \Gamma^{(\xi_q)}$$

for which $\Gamma(\mu)$ is contained in the outer product

$$\Gamma^{(\xi_1)} \circ \Gamma^{(\xi_2)} \circ \cdots \circ \Gamma^{(\xi_q)}$$

To compute all the coefficients $C\{(\mu)(\nu)|(\lambda)\}$ for fixed N, we can proceed through the partitions (ξ) of N in order. The representation $\Gamma_{(\xi)}^{(\mu)}$ can be reduced with the aid of graphical rules and the Frobenius reciprocity theorem, and the reduction of $[\Gamma_{(\xi)}^{(\mu)}]_{\text{ind}}$ is done with the graphical rules for outer products. Since the reduction of $\Gamma_{(\xi)}$ involves only (ξ) and partitions of N that precede (ξ), the product $\Gamma^{(\mu)} \times \Gamma^{(\xi)}$ can then be obtained by subtraction.

❑ **Example 9.48.** The Kronecker products in $\mathcal{S}_3$ are straightforward. The nontrivial result

follows directly from counting dimensions and the general result (9.142).   ∎

❑ **Example 9.49.** For $\mathcal{S}_4$, the Kronecker products of the identity representation are trivial. Next, note that since

$$\Gamma^{(31)}_{(31)} = (\,\square\square\square \otimes \square\,) \oplus \left(\,\begin{array}{c}\square\square\\\square\end{array} \otimes \square\,\right)$$

is computed by noting all the ways in which a single node can be removed from the diagram $\mathcal{Y}^{(31)}$. Then note that

$$\square\square\square \circ \square = \square\square\square\square \oplus \begin{array}{c}\square\square\square\\\square\end{array} \quad , \quad \begin{array}{c}\square\square\\\square\end{array} \circ \square = \begin{array}{c}\square\square\square\\\square\end{array} \oplus \begin{array}{c}\square\square\\\square\square\end{array} \oplus \begin{array}{c}\square\square\\\square\\\square\end{array}$$

Since

$$\begin{array}{c}\square\square\\\square\end{array} = \square\square\square \circ \square - \square\square\square\square$$

we then have

$$\begin{array}{c}\square\square\\\square\end{array} \times \begin{array}{c}\square\square\\\square\end{array} = \square\square\square\square \oplus \begin{array}{c}\square\square\square\\\square\end{array} \oplus \begin{array}{c}\square\square\\\square\square\end{array} \oplus \begin{array}{c}\square\square\\\square\\\square\end{array}$$

Note that the dimension check is satisfied. Also

$$\begin{array}{c}\square\square\\\square\square\end{array} \times (\,\square\square\square \circ \square\,) = \left[\Gamma^{(31)}_{(22)}\right]_{\text{ind}}$$

Now $\Gamma^{(31)}_{(22)}$ contains the representations of $\mathcal{S}_3 \otimes \mathcal{S}_1$ obtained by removing a single node from the diagram $\mathcal{Y}^{(22)}$. Thus

$$\Gamma^{(31)}_{(22)} = \begin{array}{c}\square\square\\\square\end{array} \otimes \square$$

and then

$$\left[\Gamma^{(31)}_{(22)}\right]_{\text{ind}} = \begin{array}{c}\square\square\square\\\square\end{array} \oplus \begin{array}{c}\square\square\\\square\square\end{array} \oplus \begin{array}{c}\square\square\\\square\\\square\end{array}$$

so that

$$\begin{array}{c}\square\square\\\square\square\end{array} \times \begin{array}{c}\square\square\\\square\end{array} = \begin{array}{c}\square\square\square\\\square\end{array} \oplus \begin{array}{c}\square\square\\\square\\\square\end{array}$$

This actually could have been deduced from the dimension check alone, using the symmetry relation (9.142).

$\blacksquare$

→ **Exercise 9.24.** Reduce the Kronecker products

$$\begin{array}{c}\square\square\\\square\\\square\end{array} \times \begin{array}{c}\square\square\\\square\end{array} \quad \text{and} \quad \begin{array}{c}\square\square\\\square\square\end{array} \times \begin{array}{c}\square\square\\\square\square\end{array}$$

using the graphical methods given here.

$\square$

9.6   Discrete Infinite Groups

The simplest discrete infinite group is the group $\mathbf{Z}$ of integers under addition, which corresponds to the group of translations of a one-dimensional lattice. If we let $\mathbf{T}$ denote the translation of the lattice by one unit, then the group consists of the elements $\mathbf{T}^n$ ($n = 0, \pm 1, \pm 2, \ldots$). For every complex number z, there is an irreducible representation $\Gamma(z)$ of the group, in which

$$\mathbf{T} \to z \quad \text{and} \quad \mathbf{T}^n \to z^n \tag{9.145}$$

($n = 0, \pm 1, \pm 2, \ldots$). The representation is unitary if and only if $|z| = 1$, in which case let $z = \exp(ik)$ and then

$$\mathbf{T}^n \to e^{ikn} \tag{9.146}$$

Remark. This example shows that representations of an infinite group need not be unitary, unlike those of a finite group.   □

For an N-dimensional lattice, there are N fundamental translations $\mathbf{T}_1, \ldots, \mathbf{T}_N$, and the group elements have the form

$$\mathbf{T}(\mathbf{x}) = \mathbf{T}_1^{x_1} \cdots \mathbf{T}_N^{x_N} \tag{9.147}$$

with $\mathbf{x} = (x_1, \ldots, x_N)$ and $x_m = 0, \pm 1, \pm 2, \ldots$ ($m = 1, \ldots, N$). There is an irreducible representation $\Gamma(z_1, \ldots, z_N)$ for every N-tuple $(z_1, \ldots, z_N)$ of complex numbers, with

$$\mathbf{T}(\mathbf{x}) \to z_1^{x_1} \cdots z_N^{x_N} \tag{9.148}$$

The representation is unitary if and only if $|z_m| = 1$ ($m = 1, \ldots, N$), in which case let $z_m = \exp(ik_m)$ and then

$$\mathbf{T}(\mathbf{x}) \to e^{i\mathbf{k} \cdot \mathbf{x}} \tag{9.149}$$

with $\mathbf{k} = (k_1, \ldots, k_N)$, and $\mathbf{k} \cdot \mathbf{x}$ the usual scalar product of $\mathbf{k}$ and $\mathbf{x}$. This representation is also denoted as $\Gamma(\mathbf{k})$.

Remark. The constraint $|z_m| = 1$ means that k_m is only defined *modulo* 2π. For the infinite lattice, this constrains the vector $\mathbf{k}$ to lie on an N-dimensional torus with period $2\pi/L$ along any coordinate axis. Thus in three dimensions, we can restrict $\mathbf{k}$ to the cube

$$-\frac{\pi}{L} \le k_x \le \frac{\pi}{L} \qquad -\frac{\pi}{L} \le k_y \le \frac{\pi}{L} \qquad -\frac{\pi}{L} \le k_z \le \frac{\pi}{L} \tag{9.150}$$

with points on opposite faces of the cube identified with the same representation $\Gamma(\mathbf{k})$. The constraints (9.150) define the (first) *Brillouin zone* for the vector $\mathbf{k}$.   □

Remark. The vector $\mathbf{k}$ is the *wave vector* of the representation $\Gamma(\mathbf{k})$, since the function $\exp(i\mathbf{k} \cdot \mathbf{x})$ corresponds to a wave propagating in the direction of $\mathbf{k}$ with wavelength $2\pi/|\mathbf{k}|$ in units of the lattice spacing. In a quantum-mechanical system on the lattice, $\hbar \mathbf{k}$ is also interpreted as the *quasi-momentum* associated with the representation.   □

Remark. An alternative to using an infinite lattice is to consider a very large lattice with L^N points and impose periodic boundary conditions, which is equivalent to the requirement $\mathbf{T}^N = 1$. Then the components k_m are constrained to have the values of the form

$$k_m = \frac{2\pi \kappa_m}{L} \tag{9.151}$$

with κ_m an integer ($\kappa_m = 1, \ldots, N$ or $\kappa_m = -\frac{1}{2}N, -\frac{1}{2}N + 1, \ldots, \frac{1}{2}N - 1, \frac{1}{2}N$ with $\kappa_m = \pm \frac{1}{2}N$ identified as the same point). For practical purposes, this lattice is equivalent to the infinite lattice if $L \gg 1$. $\qquad\square$

The full symmetry group of a cubic lattice, denoted here by $\mathcal{G}_C$, includes not only translations of the lattice, but also the group $\mathcal{O}_d$ of rotations and reflections that leave the cube invariant. The group $\mathcal{G}_C$ is not a direct product, since the rotations do not commute with the translations. In fact, if R is an element of $\mathcal{O}_d$, then

$$R\,\mathbf{T}(\mathbf{x})R^{-1} = \mathbf{T}(\mathbf{Rx}) \;\Rightarrow\; R\,\mathbf{T}(\mathbf{x}) = \mathbf{T}(\mathbf{Rx})R \tag{9.152}$$

where $\mathbf{Rx}$ is the image of $\mathbf{x}$ under the rotation R. Equation (9.152) shows that the translations are an invariant subgroup of $\mathcal{G}_C$, and if $|\mathbf{k}\rangle$ is an eigenvector of $\mathbf{T}(\mathbf{x})$ with

$$\mathbf{T}(\mathbf{x}) = |\mathbf{k}\rangle \tag{9.153}$$

then $\mathbf{D}(R)|\mathbf{k}\rangle$ will be an eigenvector of $\mathbf{T}(\mathbf{x})$ with eigenvalue $\mathbf{Rk}$. This observation leads to a method for constructing irreducible representations of $\mathcal{G}_C$ that can be generalized to other groups, notably the group of inhomogeneous Lorentz transformations, also known as the *Poincaré group*, whose irreducible representations are constructed in Chapter 10.

We start with fixed $\mathbf{k}$ and let $\mathcal{G}_\mathbf{k}$ denote the subgroup of rotations of the cube that leave the vector $\mathbf{k}$ invariant. $\mathcal{G}_\mathbf{k}$ is the *isotropy group*, or *little group*, at $\mathbf{k}$. For a general $\mathbf{k}$, $\mathcal{G}_\mathbf{k}$ is the trivial group, but if two (or all three) components of $\mathbf{k}$ have the same magnitude, there will be nontrivial rotations that leave $\mathbf{k}$ invariant. There is even more symmetry if one or more of the components of $\mathbf{k}$ is equal to $\pm\pi/L$ (i.e., if $\mathbf{k}$ lies on the boundary of the Brillouin zone).

❏ **Example 9.50.** If $\mathbf{k}$ is in the direction of one of the coordinate axes (so that two components of $\mathbf{k}$ vanish), then $\mathcal{G}_\mathbf{k} \cong D_4$ is the dihedral group generated by (i) a rotation ρ through $\pi/2$ about the direction of $\mathbf{k}$, and (ii) reflection σ of an axis perpendicular to $\mathbf{k}$, which can be represented as space inversion $\mathbf{P}$ combined with rotation through π about the axis of reflection. If further $|\mathbf{k}| = \pi/L$, so that $\mathbf{k}$ is on the boundary of the Brillouin zone, then $\mathbf{k}$ and $-\mathbf{k}$ are equivalent. In this case, both $\mathbf{P}$ and rotations through π about any coordinate axis perpendicular to $\mathbf{k}$ each leave $\mathbf{k}$ invariant, and the isotropy group is enlarged to $D_4 \otimes \mathbf{Z}_2$. This group is the largest isotropy group for any nonzero $\mathbf{k}$. ∎

❏ **Example 9.51.** If $\mathbf{k}$ is along one of the main diagonals of the cube,

$$\mathbf{k} = \kappa\,(\pm\mathbf{e}_x \pm \mathbf{e}_y \pm \mathbf{e}_z)$$

then $\mathcal{G}_\mathbf{k} \cong \mathbf{Z}_3$ contains the rotations through $\pm\pi/3$ about the direction of $\mathbf{k}$. If $\kappa = \pi/L$ (so that $\mathbf{k}$ is on the boundary of the Brillouin zone), then $\mathcal{G}_\mathbf{k}$ contains space inversion as well, and the isotropy group is $\mathbf{Z}_6 \cong \mathbf{Z}_3 \otimes \mathbf{Z}_2$. ∎

→ **Exercise 9.25.**   Other classes of **k** that have nontrivial isotropy groups are

$$(i)\ \ \mathbf{k}_1 = \pm\mathbf{k}_2 \quad \mathbf{k}_3 = 0 \qquad \text{and} \qquad (ii)\ \ \mathbf{k}_1 = \pm\mathbf{k}_2 \quad \mathbf{k}_3 \neq 0$$

Find the largest isotropy group (including reflections) for each of these classes. Describe the extra symmetry if **k** is on a boundary of the Brillouin zone.   □

Next choose an irreducible representation of $\mathcal{G}_\mathbf{k}$ of dimension d, say, and let $|\mathbf{k}, \alpha\rangle$ ($\alpha = 1, \dots, d$) be a basis for this representation. Then if R is an element of $\mathcal{G}_\mathbf{k}$,

$$\mathbf{D}(R)|\mathbf{k}, \alpha\rangle = \sum_\beta D_{\beta\alpha}(R)|\mathbf{k}, \beta\rangle \tag{9.154}$$

where $\mathbf{D}(R)$ is the linear operator representing R in the chosen irreducible representation of $\mathcal{G}_\mathbf{k}$. If R is not in $\mathcal{G}_\mathbf{k}$, then

$$\mathbf{D}(R)|\mathbf{k}, \alpha\rangle = \sum_\beta C_{\beta\alpha}(R)|\mathbf{R}\mathbf{k}, \beta\rangle \tag{9.155}$$

with matrix elements $C_{\beta\alpha}(R)$ constructed to be consistent with the group multiplication rules. The set of vectors $\{\mathbf{R}\mathbf{k}\}$ as R runs through the full point symmetry group $\mathcal{O}_d$ of the cube is the *orbit* $\Lambda(\mathbf{k})$ of **k** under $\mathcal{O}_d$.

❑ **Example 9.52.**   Consider the case $\mathbf{k} = k\mathbf{e}_z$ for which the isotropy group $\mathcal{G}_\mathbf{k}$ is the dihedral group D_4 generated by the rotation a through $\pi/2$ about the Z-axis and the reflection σ of the X-axis. Note that

$$a^4 = e \qquad \sigma^2 = e \qquad \sigma a \sigma = a^{-1} \tag{9.156}$$

As shown in Section 9.4, this group has four one-dimensional representations, in which

$$a \to \pm 1 \equiv \alpha \qquad \sigma \to \pm 1 \equiv \eta \tag{9.157}$$

and a two-dimensional irreducible representation with

$$a \to i\sigma_y = \begin{pmatrix} 0 & 1 \\ -1 & 0 \end{pmatrix} \qquad \sigma \to \sigma_z = \begin{pmatrix} 1 & 0 \\ 0 & -1 \end{pmatrix} \tag{9.158}$$

The orbit of $\Lambda(\mathbf{k})$ under $\mathcal{O}_d$ contains the vectors $\pm k\mathbf{e}_x, \pm k\mathbf{e}_y, \pm k\mathbf{e}_z$. Thus if we start with a one-dimensional representation of the isotropy group, we need to consider a set of six basis vectors

$$|\pm k\mathbf{e}_x\rangle \qquad |\pm k\mathbf{e}_y\rangle \qquad |\pm k\mathbf{e}_z\rangle \tag{9.159}$$

We then have

$$\mathbf{P}|\mathbf{k}\rangle = \eta|-\mathbf{k}\rangle \tag{9.160}$$

while the $\mathcal{O}_d$ rotations $R_x(\pi/2)$, $R_y(\pi/2)$, $R_z(\pi/2)$ act on these states as shown in the table at the right. The remaining elements of $\mathcal{O}_d$ can be expressed as products of $\mathbf{P}$ and these rotations. Starting

	$|k\mathbf{e}_x\rangle$	$|k\mathbf{e}_y\rangle$	$|k\mathbf{e}_z\rangle$
$R_x(\pi/2)$	$\alpha|k\mathbf{e}_x\rangle$	$\alpha|-k\mathbf{e}_z\rangle$	$\alpha|k\mathbf{e}_y\rangle$
$R_y(\pi/2)$	$\alpha|k\mathbf{e}_z\rangle$	$\alpha|k\mathbf{e}_y\rangle$	$\alpha|-k\mathbf{e}_x\rangle$
$R_z(\pi/2)$	$\alpha|-k\mathbf{e}_y\rangle$	$\alpha|k\mathbf{e}_x\rangle$	$\alpha|k\mathbf{e}_z\rangle$

Action of rotations on basis vectors.

with the two-dimensional irreducible representation of the little group, acting on basis states $|k\mathbf{e}_z, \pm\rangle$, leads to a twelve-dimensional representation of $\mathcal{O}_d$ on a space with basis vectors

$$| \pm k\mathbf{e}_x, \pm\rangle \qquad | \pm k\mathbf{e}_y, \pm\rangle \qquad | \pm k\mathbf{e}_z, \pm\rangle \tag{9.161}$$

Equation (9.158) corresponds to

$$R_z\left(\frac{\pi}{2}\right)|k\mathbf{e}_z, \pm\rangle = \mp|k\mathbf{e}_z, \mp\rangle \quad \text{and} \quad \mathbf{P}R_x(\pi)|k\mathbf{e}_z, \pm\rangle = \pm|k\mathbf{e}_z, \pm\rangle \tag{9.162}$$

whence also

$$R_z(\pi)|k\mathbf{e}_z, \pm\rangle = -|k\mathbf{e}_z, \pm\rangle \tag{9.163}$$

For the action of $\mathbf{P}$ and $R_x(\pi)$ separately, it is convenient to take

$$\mathbf{P}|k\mathbf{e}_z, \pm\rangle = \pm| - k\mathbf{e}_z, \pm\rangle \tag{9.164}$$

$$R_x(\pi)|k\mathbf{e}_z, \pm\rangle = | - k\mathbf{e}_z, \pm\rangle \tag{9.165}$$

(this is no more than a phase convention for the basis vector $| - k\mathbf{e}_z, -\rangle$). From the group multiplication $R_x(\pi)R_y(\pi) = R_z(\pi)$, it then follows that

$$R_y(\pi)|k\mathbf{e}_z, \pm\rangle = -| - k\mathbf{e}_z, \pm\rangle \tag{9.166}$$

It is consistent with Eq. (9.165) to choose

$$R_x\left(\frac{\pi}{2}\right)|k\mathbf{e}_z, \pm\rangle = |k\mathbf{e}_y, \mp\rangle \qquad R_x\left(\frac{\pi}{2}\right)|k\mathbf{e}_y, \pm\rangle = | - k\mathbf{e}_z, \mp\rangle \tag{9.167}$$

It is further consistent with Eq. (9.166) to choose

$$R_y\left(\frac{\pi}{2}\right)|k\mathbf{e}_z, \pm\rangle = \pm| - k\mathbf{e}_x, \mp\rangle \qquad R_y\left(\frac{\pi}{2}\right)| - k\mathbf{e}_x, \pm\rangle = \pm| - k\mathbf{e}_z, \mp\rangle \tag{9.168}$$

The group multiplication rules then lead to a complete construction of the twelve-dimensional representation of $\mathcal{O}_d$. The full representation follows from the preceding, and the additional results

$$R_x\left(\frac{\pi}{2}\right)|k\mathbf{e}_x, \pm\rangle = \mp|k\mathbf{e}_x, \mp\rangle \qquad R_y\left(\frac{\pi}{2}\right)|k\mathbf{e}_y, \pm\rangle = \mp|k\mathbf{e}_y, \mp\rangle \tag{9.169}$$

derived from the group multiplication rule

$$R_x\left(\frac{\pi}{2}\right)R_y\left(\frac{\pi}{2}\right)R_x\left(-\frac{\pi}{2}\right) = R_z\left(\frac{\pi}{2}\right) \tag{9.170}$$

It is an instructive exercise to reduce this six-dimensional representation of $\mathcal{O}_d$ into irreducible components. ∎

A   Frobenius Reciprocity Theorem

In Section 9.4.5, we presented the *Frobenius reciprocity theorem*, which states that the co-efficients that appear in the reduction of an irreducible representation of $\mathcal{G}$ restricted to a subgroup $\mathcal{H}$ are equal to the coefficients in the reduction of the representation of $\mathcal{G}$ induced by an irreducible representation of $\mathcal{H}$. Here we present a proof of the theorem and an important corollary.

Suppose $\mathcal{G}$ is a group of finite order n whose classes $K_1 = \{e\}, K_2, \ldots, K_p$ contain $h_1 = 1, h_2, \ldots, h_p$ elements, respectively. Denote the inequivalent irreducible representations of $\mathcal{G}$ by $\Gamma^1, \ldots, \Gamma^p$, with m_a the dimension of Γ^a and χ_k^a the character of the class K_k in Γ^a. The matrix representing g_ℓ in Γ^a is $\mathbf{D}^a(g_\ell)$, with matrix elements $D_{jk}^a(g_\ell)$.

Now let $\mathcal{H}$ be a subgroup of $\mathcal{G}$ of order m, index $t = n/m$. The classes K_k of $\mathcal{G}$ split into classes $L_{1k}, \ldots, L_{n_k k}$ of $\mathcal{H}$, with ℓ_{jk} elements in the class L_{jk} ($j = 1, \ldots, n_k$). Some classes of $\mathcal{G}$ may not be in $\mathcal{H}$, so that $k = 1, \ldots, s \leq p$, and others may be only partially included in $\mathcal{H}$, so that

$$\sum_{j=1}^{n_k} \ell_{jk} \leq h_k \tag{9.A1}$$

in general. The total number of classes of $\mathcal{H}$ is given by

$$q = \sum_{k=1}^{s} n_k \tag{9.A2}$$

Then we can denote the inequivalent irreducible representations of $\mathcal{H}$ by $\Delta^1, \ldots, \Delta^q$, the dimension of Δ^c by μ_c ($c = 1, \ldots, q$), and the character of L_{jk} in Δ^c by ϕ_{jk}^c.

The irreducible representation Γ^a of $\mathcal{G}$ subduces a representation Γ_{sub}^a of $\mathcal{H}$, which can be reduced according to

$$\Gamma_{\mathrm{sub}}^a = \oplus_c \alpha_{ca} \Delta^c \tag{9.A3}$$

with nonnegative integer coefficients α_{ca}. The character χ_k^a can be expressed in terms of the characters in $\mathcal{H}$ as

$$\chi_k^a = \sum_{c=1}^{q} \alpha_{ca} \phi_{jk}^c \tag{9.A4}$$

it follows from the orthogonality relation (9.75) that

$$m\alpha_{ca} = \sum_{k} \sum_{j=1}^{n_k} \ell_{jk} \phi_{jk}^{c*} \chi_k^a \tag{9.A5}$$

Thus the integers α_{ca} are determined from the characters of $\mathcal{H}$ and $\mathcal{G}$, as expected. But using the second orthogonality relation (9.91), we have

$$\sum_{a=1}^{p} \alpha_{ca} \chi_k^a = \begin{cases} \displaystyle\sum_{j=1}^{n_k} \left(\frac{n\ell_{jk}}{mh_k}\right) \phi_{jk}^c & j = 1, \ldots, s \\ 0 & \text{otherwise} \end{cases} \tag{9.A6}$$

This defines the character of a (reducible) representation Γ of $\mathcal{G}$ in terms of the character ϕ_{jk}^c of an irreducible representation of $\mathcal{H}$. We shall see that Γ is precisely the representation of $\mathcal{G}$ induced by Γ^c of $\mathcal{H}$ in the construction of Eq. (9.113).

Now suppose Δ^c is an irreducible representation of $\mathcal{H}$. Then the representation Δ_{ind}^c of $\mathcal{G}$ induced by Δ^c is reducible in general,

$$\Delta_{\text{ind}}^c = \oplus_a \, \beta_{ca} \Gamma^a \tag{9.A7}$$

with nonnegative integer coefficients β_{ca}. The character ξ_k^c of the class K_k of $\mathcal{G}$ in Δ_{ind}^c is

$$\xi_k^c = \sum_{a=1}^{p} \beta_{ca} \chi_k^a \tag{9.A8}$$

Now recall the definition (Eq. (9.111))

$$\sigma_{\alpha\beta}(g, h) = \begin{cases} 1, & \text{if } g g_\beta = g_\alpha h \\ 0 & \text{otherwise} \end{cases} \tag{9.A9}$$

and note that $\sigma_{\alpha\alpha}(g, h) = 1$ if and only if $g = g_\alpha h g_\alpha^{-1}$, i.e., if and only if g and h belong to the same class of $\mathcal{G}$. Thus

$$\sum_{g \text{ in } K_k} \text{tr}\,\sigma(g, h) = \begin{cases} t \, (= n/m), & \text{if } h \text{ in } K_k \\ 0 & \text{otherwise} \end{cases} \tag{9.A10}$$

Also, recall the definition (Eq. (9.113))

$$\mathbf{D}^\Gamma(g) \equiv \sum_{h \text{ in } \mathcal{H}} \sigma(g, h) \otimes \mathbf{D}^\Delta(h) \tag{9.A11}$$

Then we have

$$h_k \xi_k^c = \sum_{g \text{ in } K_k} \xi^c(g) = \sum_{g \text{ in } K_k} \sum_{h \text{ in } \mathcal{H}} [\text{tr}\,\sigma(g, h)] \, \phi^c(h) = t \sum_{j=1}^{n_k} \ell_{jk} \phi_{jk}^c \tag{9.A12}$$

and then

$$\xi_k^c = \sum_{a=1}^{p} \beta_{ca} \chi_k^a = \sum_{j=1}^{n_k} \left(\frac{n \ell_{jk}}{m h_k} \right) \phi_{jk}^c = \sum_{a=1}^{p} \alpha_{ca} \chi_k^a \tag{9.A13}$$

where the last equality follows from Eq. (9.A6). Since Eq. (9.A13) is true for all $k = 1, \ldots, p$, it must be that

$$\alpha_{ca} = \beta_{ca} \tag{9.A14}$$

which is the *Frobenius Reciprocity Theorem*.

A corollary of this result is

$$\Gamma^a \times \Delta^c_{\text{ind}} = [\Gamma^a_{\text{sub}} \times \Delta^c]_{\text{ind}} \tag{9.A15}$$

as noted in Eq. (9.122), so that Kronecker products in $\mathcal{H}$ can be used to reduce those in $\mathcal{G}$. This corollary is quite useful in practice. To derive the result, note that the character of the class K_k of $\mathcal{G}$ is given in the representation $\Gamma^a \times \Delta^c_{\text{ind}}$ by

$$\xi^c_k \chi^a_k = \left[\sum_{j=1}^{n_k} \left(\frac{n\ell_{jk}}{h_k m} \right) \phi^c_{jk} \right] \chi^a_k = \sum_{j=1}^{n_k} \left(\frac{n\ell_{jk}}{h_k m} \right) [\phi^c_{jk} \chi^a_k] \tag{9.A16}$$

But the last expression is the same character of the class K_k in the induced representation $[\Gamma^a_{\text{sub}} \times \Delta^c]_{\text{ind}}$. Hence the two representations must be the same.

B   *S*-Functions and Irreducible Representations of $\mathcal{S}_N$

B.1   Frobenius Generating Function for the Simple Characters of $\mathcal{S}_N$

In Eq. (9.24), we introduced the alternant $A(x_1, x_2, \ldots, x_N)$ as an antisymmetric function of the N variables $x_1, x_2, \ldots, x_N$. Now we can also write

$$A(x_1, x_2, \ldots, x_N) = \prod_{j<k} (x_j - x_k) = \sum_{\mathbf{P}=(i_1 i_2 \cdots i_N)} \varepsilon_{\mathbf{P}} \, x^{N-1}_{i_1} x^{N-2}_{i_2} \cdots x_{i_{N-1}} \tag{9.B17}$$

where the sum is over all permutations $\mathbf{P} = (i_1 i_2 \cdots i_N)$ of N. This form shows that the alternant is actually the determinant of the matrix $\mathbf{A}$ defined by

$$\mathbf{A} = \begin{pmatrix} 1 & x_1 & x_1^2 & \cdots & x_1^{N-1} \\ 1 & x_2 & x_2^2 & \cdots & x_2^{N-1} \\ \vdots & \vdots & \vdots & \ddots & \vdots \\ 1 & x_N & x_N^2 & \cdots & x_N^{N-1} \end{pmatrix} \tag{9.B18}$$

Note that the alternant is homogeneous of degree $\frac{1}{2}N(N-1)$ in the $x_1, x_2, \ldots, x_N$.

We can also define an antisymmetric polynomial $A^{(\lambda)}(x_1, x_2, \ldots, x_N)$ associated with a partition $(\lambda) = (\lambda_1 \cdots \lambda_s)$ of N by

$$A^{(\lambda)}(x_1, x_2, \ldots, x_N) \equiv \sum_{\mathbf{P}=(i_1 i_2 \cdots i_N)} \varepsilon_{\mathbf{P}} \, x^{\lambda_1+N-1}_{i_1} x^{\lambda_2+N-2}_{i_2} \cdots x^{\lambda_N}_{i_N} \tag{9.B19}$$

$A^{(\lambda)}(x_1, x_2, \ldots, x_N)$ is homogeneous of degree $d_N = \frac{1}{2}N(N+1)$ in the $x_1, x_2, \ldots, x_N$. The homogeneous antisymmetric polynomials of degree d_N form a linear vector space $\mathcal{V}$, and the exponents of every monomial in such a polynomial must correspond to the exponents in Eq. (9.B19) for some partition (λ) of N. Thus the $A^{(\lambda)}$ define a basis of $\mathcal{V}$, and $\mathcal{V}$ has dimension $\pi(N)$, the number of distinct partitions of N.

→ **Exercise 9.B1.**   Show that if $(\lambda) = (\lambda_1 \cdots \lambda_s)$ is a partition of N, then the function $A^{(\lambda)}(x_1, x_2, \ldots, x_N)$ can be expressed as the determinant of the matrix $\mathbf{A}^{(\lambda)}$ defined by

$$
\mathbf{A} =
\begin{pmatrix}
x_1^{\lambda_N} & x_1^{\lambda_{N-1}+1} & x_1^{\lambda_{N-2}+2} & \cdots & x_1^{\lambda_1+N-1} \\
x_2^{\lambda_N} & x_2^{\lambda_{N-1}+1} & x_2^{\lambda_{N-2}+2} & \cdots & x_2^{\lambda_1+N-1} \\
\vdots & \vdots & \vdots & \ddots & \vdots \\
x_N^{\lambda_N} & x_N^{\lambda_{N-1}+1} & x_N^{\lambda_{N-2}+2} & \cdots & x_N^{\lambda_1+N-1}
\end{pmatrix}
$$

(again, if $s < N$, define $\lambda_{s+1} = \cdots = \lambda_N = 0$).   □

Now define the symmetric functions

$$
s_k = s_k(x_1, x_2, \ldots, x_N) \equiv \sum_{i=1}^{N} (x_i)^k \tag{9.B20}
$$

The functions $s_1, s_2, \ldots, s_N$ are functionally independent, since the Jacobian determinant is

$$
\det\left(\frac{\partial s_k}{\partial x_i}\right) = N! \sum_{\mathbf{P}=(i_1 i_2 \cdots i_N)} \varepsilon_{\mathbf{P}} \, x_{i_1}^{N-1} x_{i_2}^{N-2} \cdots x_{i_{N-1}} = N! \, A(x_1, x_2, \ldots, x_N)
$$

$$
\tag{9.B21}
$$

where the sum is over permutations of $(1 \cdots N)$, and the alternant $A(x_1, x_2, \ldots, x_N)$ vanishes only if two of the variables are equal. Note that powers of the $\{s_k\}$ are expressed as

$$
(s_k)^\nu = \sum_{(\nu)} \sum_{\mathbf{P}=(i_1 i_2 \cdots i_N)} \frac{\nu!}{\nu_1! \, \nu_2! \cdots \nu_N!} x_{i_1}^{k\nu_1} x_{i_2}^{k\nu_2} \cdots x_{i_N}^{k\nu_N} \tag{9.B22}
$$

where the summation is over partitions (ν) of ν; if the number of parts q is less than N, we define $\nu_{q+1} = \cdots = \nu_N = 0$. Then if $(m) = (m_1 m_2 \cdots m_t) = (p^{\nu_p} \cdots 2^{\nu_2} 1^{\nu_1})$ is a partition of N, define

$$
S_{(m)} = S_{(m)}(x_1, x_2, \ldots, x_N) = s_{m_1} s_{m_2} \cdots s_{m_t} = \prod_{k=1}^{p} [s_k(x_1, x_2, \ldots, x_N)]^{\nu_k} \tag{9.B23}
$$

Evidently $S_{(m)}(x_1, x_2, \ldots, x_N)$ is a symmetric, homogeneous polynomial of degree N in the $x_1, x_2, \ldots, x_N$. The $S_{(m)}$ are known as *Schur functions*, or simply *S-functions*.

Consider now the product $S_{(m)}(x_1, \ldots, x_N) A(x_1, \ldots, x_N)$. It is a homogeneous polynomial of degree in the variables $x_1, \ldots, x_N$, antisymmetric under interchange of any pair of variables. Thus it must be a linear combination of the polynomials $A^{(\lambda)}(x_1, \ldots, x_N)$, and we can write

$$
S_{(m)}(x_1, \ldots, x_N) A(x_1, \ldots, x_N) = \sum_{(\lambda)} \chi_{(m)}^{(\lambda)} A^{(\lambda)}(x_1, \ldots, x_N) \tag{9.B24}
$$

Remarkably, the coefficients $\chi_{(m)}^{(\lambda)}$ are exactly the simple characters of $\mathcal{S}_N$.

To prove this, we will show that the $\chi_{(m)}^{(\lambda)}$ satisfy the orthogonality relations

$$\sum_{(m)} N_{(m)} \chi_{(m)}^{(\lambda)} \chi_{(m)}^{(\xi)} = N! \, \delta^{(\lambda)(\xi)} \tag{9.B25}$$

where $N_{(m)}$ is the number of elements in the class $K_{(m)}$.

Remark. In Section 9.3, we identified the classes of $\mathcal{S}_N$ with the $\pi(N)$ partitions of N, and denoted the class whose cycle structure belongs to the partition $(m) = (m_1 m_2 \cdots m_t) = (p^{\nu_p} \cdots 2^{\nu_2} 1^{\nu_1}$ by $K_{(m)}$. The number of elements in the class $K_{(m)}$ is given by

$$N_{(m)} = \frac{N!}{(1^{\nu_1} \, 2^{\nu_2} \, \cdots \, p^{\nu_p}) \, (\nu_1! \, \nu_2! \, \cdots \, \nu_p!)} \equiv N(\nu_1 \, \nu_2 \, \cdots \, \nu_p) \tag{9.B26}$$

To see this, note that there are $N!$ arrangements of $1, \ldots, N$ in the given cycle structure. But there are n equivalent expressions of each n-cycle, and $\nu_n!$ arrangements of the ν_n n-cycles, each of which corresponds to the same permutation in $\mathcal{S}_N$.   □

Remark. Equation (9.B24) serves to define the association of the partition (λ) with a particular irreducible representation. In addition to the orthogonality relations, we need to show that

$$\chi_{(1^N)}^{(\lambda)} > 0 \tag{9.B27}$$

since that character is the dimension of the representation $\Gamma^{(\lambda)}$. We also need to verify the identification of (N) with the symmetric, and (1^N) with the antisymmetric, representation. □

Now introduce a second set $(y_1, \ldots, y_N)$ of N variables, and note that

$$\sum_{(m)} N_{(m)} S_{(m)}(x_1, \ldots, x_N) S_{(m)}(y_1, \ldots, y_N)$$

$$= N! \sum_{\{(\nu_k)| \sum_k k\nu_k = N\}} \prod_{k=1}^{p} \frac{[s_k(x) s_k(y)]^{\nu_k}}{k^{\nu_k} \nu_k!} \tag{9.B28}$$

where $N_{(m)}$ is the number of elements in the class $K_{(m)}$ of $\mathcal{S}_N$ given by Eq. (9.B26). Then keep the number of variables fixed at n, and sum over N to determine a function $S(x, y)$ defined by

$$S(x, y) \equiv \sum_{N=0}^{\infty} \sum_{(m)} \frac{N_{(m)}}{N!} S_{(m)}(x_1, \ldots, x_n) S_{(m)}(y_1, \ldots, y_n)$$

$$= \prod_{k=1}^{\infty} \sum_{\nu_k=0}^{\infty} \frac{1}{\nu_k!} \left[\frac{s_k(x) s_k(y)}{k}\right]^{\nu_k} = \exp\left\{\sum_{k=1}^{\infty} \frac{s_k(x) s_k(y)}{k}\right\} \tag{9.B29}$$

The series in the exponential can be summed in closed form to give

$$\sum_{k=1}^{\infty} \frac{s_k(x) s_k(y)}{k} = \sum_{i=1}^{n} \sum_{j=1}^{n} \sum_{k=1}^{\infty} \frac{(x_i y_j)^k}{k} = -\sum_{i=1}^{n} \sum_{j=1}^{n} \ln(1 - x_i y_j) \tag{9.B30}$$

and then

$$S(x, y) = \exp\left\{ -\sum_{i=1}^{n} \sum_{j=1}^{n} \ln(1 - x_i y_j) \right\} = \prod_{i=1}^{n} \prod_{j=1}^{n} \left(\frac{1}{1 - x_i y_j} \right) \qquad (9.\text{B}31)$$

Next define the $n \times n$ matrix

$$\mathbf{M} = (M_{ij}) = \left(\frac{1}{1 - x_i y_j} \right) \qquad (9.\text{B}32)$$

The determinant of $\mathbf{M}$ can be evaluated formally by expanding the denominators and using the definition of the determinant to obtain

$$\det \mathbf{M} = \sum_{\mu_1=0}^{\infty} \cdots \sum_{\mu_n=0}^{\infty} \sum_{\mathbf{P}=(i_1 i_2 \cdots i_n)} \varepsilon_{\mathbf{P}} \, (x_{i_1} y_1)^{\mu_1} (x_{i_2} y_2)^{\mu_2} \cdots (x_{i_n} y_n)^{\mu_n} \qquad (9.\text{B}33)$$

Note that the exponents $\mu_1, \mu_2, \ldots, \mu_n$ in a single monomial in this series must all be distinct, due to the antisymmetry in the x and y variables enforced by the factor $\varepsilon_{\mathbf{P}}$. We can reorder the exponents in each term are so that we always have

$$\mu_1 > \mu_2 > \cdots > \mu_n$$

and the last equation can be written as

$$\det \mathbf{M} = \sum_{N=0}^{\infty} \sum_{(\lambda)} \sum_{\mathbf{P}=(i)} \sum_{\mathbf{Q}=(j)} \varepsilon_{\mathbf{P}} \, \varepsilon_{\mathbf{Q}} (x_{i_1} y_{j_1})^{\lambda_1+n-1} (x_{i_2} y_{j_2})^{\lambda_2+n-2} \cdots (x_{i_n} y_{j_n})^{\lambda_n}$$

$$(9.\text{B}34)$$

where for each N, the sums on $\mathbf{P}$ and $\mathbf{Q}$ are over permutations $i_1, i_2, \ldots, i_N$ and $j_1, j_2, \ldots, j_N$ of N, and the sum on (λ) is over the partitions of N. Strictly speaking, the (λ) are partitions into not more than n parts, but the proof of the result (9.B41) for a particular N only requires $n \geq N$, which can always be arranged since n is arbitrary. From the definition of the $A^{(\lambda)}(x)$ in Eq. (9.B19), it follows that Eq. (9.B34) is equivalent to

$$\det \mathbf{M} = \sum_{N=0}^{\infty} \sum_{(\lambda)} A^{(\lambda)}(x) A^{(\lambda)}(y) \qquad (9.\text{B}35)$$

On the other hand, we also have

$$\det \mathbf{M} = \left\| \frac{1}{1 - x_i y_j} \right\| = A(x_1, \ldots, x_n) \, A(y_1, \ldots, y_n) \prod_{i=1}^{n} \prod_{j=1}^{n} \left(\frac{1}{1 - x_i y_j} \right) \qquad (9.\text{B}36)$$

This follows from the observation that after extracting the product of the factors $1/(1 - x_i y_j)$ from the determinant, there remains a numerator that is a polynomial in the x and y antisymmetric under exchange of any two x or any two y variables, since this exchange corresponds to exchanging two rows or two columns, respectively, of the matrix $\mathbf{M}$. To evaluate this polynomial, consider first the 2×2 determinant, given by

$$\left(\frac{1}{1-x_1y_1}\right)\left(\frac{1}{1-x_2y_2}\right) - \left(\frac{1}{1-x_1y_2}\right)\left(\frac{1}{1-x_2y_1}\right)$$

$$= \frac{(x_1-x_2)(y_1-y_2)}{(1-x_1y_1)(1-x_1y_2)(1-x_2y_1)(1-x_2y_2)}$$

(9.B37)

To generalize this formula to arbitrary n, subtract the ith row of the determinant from the first row, and then subtract the jth column of the resulting determinant from the first column. The result will be a product of the form

$$\det \mathbf{M} = \frac{1}{1-x_1y_1}\left(\prod_{i=2}^{n}\frac{x_i-x_1}{1-x_iy_1}\right)\left(\prod_{j=2}^{n}\frac{y_j-y_1}{1-x_1y_j}\right)\det \mathbf{M}'$$

(9.B38)

where $\mathbf{M}'$ is the $(n-1) \times (n-1)$ matrix obtained from $\mathbf{M}$ by removing the first row and the first column. Thus Eq. (9.B36) is true by induction on n.

Using the result (9.B31) in Eq. (9.B36), and then the definition in Eq. (9.B29), we have

$$\det \mathbf{M} = S(x,y)A(x)A(y) = \sum_{N=0}^{\infty}\sum_{(m)}\frac{N_{(m)}}{N!}S_{(m)}(x)A(x)S_{(m)}(y)A(y)$$

(9.B39)

From definition (9.B24) of the $\chi_{(m)}^{(\lambda)}$, we then have

$$\det \mathbf{M} = \sum_{N=0}^{\infty}\sum_{(\lambda)}\sum_{(\xi)}\sum_{(m)}\frac{N_{(m)}}{N!}\chi_{(m)}^{(\lambda)}\chi_{(m)}^{(\xi)}A^{(\lambda)}(x)A^{(\xi)}(y)$$

(9.B40)

Comparing Eqs. (9.B35) and (9.B40) gives the required orthogonality relation

$$\sum_{(m)}N_{(m)}\chi_{(m)}^{(\lambda)}\chi_{(m)}^{(\xi)} = N!\,\delta^{(\lambda)(\xi)}$$

(9.B41)

The orthogonality relations alone are not quite enough to show that the $\chi_{(m)}^{(\lambda)}$ are simple characters. There is a possible sign ambiguity, since $\chi_{(m)}^{(\lambda)} \to -\chi_{(m)}^{(\lambda)}$ (for all λ) is consistent with orthogonality. But the graphical calculation in Example 9.55 in the next section shows that the sign is given correctly.

Anticipating this calculation, we have thus derived the *Frobenius generating function*

$$S_{(m)}(x_1,\ldots,x_N)A(x_1,\ldots,x_N) = \sum_{(\lambda)}\chi_{(m)}^{(\lambda)}A^{(\lambda)}(x_1,\ldots,x_N)$$

(9.B42)

for the simple characters $\chi_{(m)}^{(\lambda)}$ of $\mathcal{S}_N$. The graphical methods of the next section flow from this generating function.

B.2   Graphical Calculation of the Characters $\chi_{(m)}^{(\lambda)}$

Suppose $(m) = (m_1 m_2 \cdots m_t)$ and $(\lambda) = (\lambda_1 \cdots \lambda_s)$ are two partitions of N. The Frobenius generating function (Eq. (9.B42)) tells us that the simple character $\chi_{(m)}^{(\lambda)}$ of the class $K_{(m)}$ of $\mathcal{S}_N$ in the irreducible representation $\Gamma^{(\lambda)}$ associated with (λ) is given by the coefficient of the monomial

$$x_1^{\lambda_1+N-1} \, x_2^{\lambda_2+N-2} \cdots x_N^{\lambda_N} \tag{9.B43}$$

in the product

$$S_{(m)}(x_1, x_2, \ldots, x_N) A(x_1, x_2, \ldots, x_N) =$$

$$= \left[\sum_{i=1}^{N} (x_i)^{m_1} \right] \left[\sum_{i=1}^{N} (x_i)^{m_2} \right] \times \cdots \times \left[\sum_{i=1}^{N} (x_i)^{m_t} \right] \tag{9.B44}$$

$$\times \sum_{\mathbf{P}=(i_1 i_2 \cdots i_N)} \varepsilon_{\mathbf{P}} \, x_{i_1}^{N-1} x_{i_2}^{N-2} \cdots x_{i_{N-1}}$$

A graphical method to evaluate this coefficient is based on the observation that the product

$$\left(\prod_{j=1}^{t} s_{m_j} \right) x_{i_1}^{N-1} x_{i_2}^{N-2} \cdots x_{i_{N-1}}$$

can be depicted as the construction of the Young diagram $\mathcal{Y}^{(\lambda)}$ of the partition (λ) from an empty diagram by successive addition of $m_1, \ldots, m_t$ nodes. This construction must be subject to rules that exclude monomials in which two variables have the same exponent, since such monomials disappear from the sum over permutations due to the antisymmetry introduced by the $\varepsilon_{\mathbf{P}}$ factor.

Now suppose $(\mu) = (\mu_1 \mu_2 \cdots \mu_q)$ is a partition of m, and consider the product

$$\left[\sum_{i=1}^{N} (x_i)^{\ell} \right] A^{(\mu)}(x_1, \ldots, x_N) \equiv \sum_{(\nu)} c_{\ell}^{(\mu)(\nu)} A^{(\nu)}(x_1, \ldots, x_N) \tag{9.B45}$$

where the summation is over partitions (ν) of $n = \ell + m$. In a term of the form

$$\left[\sum_{i=1}^{N} (x_i)^{\ell} \right] x_1^{\mu_1+N-1} \, x_2^{\mu_2+N-2} \cdots x_N^{\mu_N}$$

there will be nominal contributions from partitions of the form

$$(\mu_1 + \ell \;\; \mu_2 \;\; \cdots \;\; \mu_N), (\mu_1 \;\; \mu_2 + \ell \;\; \cdots \;\; \mu_N), \; \ldots, \; (\mu_1 \;\; \mu_2 \;\; \cdots \;\; \mu_N + \ell)$$

However, one or more of these modified partitions may not satisfy the ordering condition $(\nu_k \geq \nu_{k+1})$. If a partition does not satisfy the condition, then either

(i) $\mu_k = \mu_{k+1} + \ell - 1$, in which case two variables have the same exponent in the new monomial, and the term disappears from the sum, or

(ii) $\mu_k < \mu_{k+1} + \ell - 1$, when the variables x_k and x_{k+1} must be transposed, providing a factor (-1), and the partition modified so that

$$\mu'_k = \mu_{k+1} + \ell - 1 \quad \text{and} \quad \mu'_{k+1} = \mu_k + 1 \tag{9.B46}$$

This continues until the partition is in standard order (with an overall sign factor ± 1), or the partition is dropped because condition (i) has been encountered at some stage.

In graphical terms, this analysis leads to the result that the coefficient $c_\ell^{(\mu)(\nu)}$ in Eq. (9.B45) vanishes unless the diagram $\mathcal{Y}^{(\nu)}$ can be constructed from the diagram $\mathcal{Y}^{(\mu)}$ by the regular application of ℓ nodes, as defined below. If the construction is possible, then $c_\ell^{(\mu)(\nu)} = \pm 1$ according to whether the application is even or odd.

Definition 9.29.   The addition of ℓ nodes to the Young diagram associated with a partition $(\mu) = (\mu_1, \ldots, \mu_q)$ of m is a *regular application* (of ℓ nodes) if nodes are added to one row, row r say, until the number of nodes in row r is equal to $\mu_{r-1} + 1$, after which nodes are added to row $r - 1$ until the number of nodes in this row is equal to $\mu_{r-2} + 1$. The process continues until *either* the nodes are exhausted at the diagram of a properly ordered partition *or* row 1 is reached, in which case the remaining nodes are added to row 1. The parity of the application is even $(+1)$ or odd (-1) according to whether nodes are added in an odd or even number of rows of the diagram. ∎

The parity rule is based on the fact that changing rows in a diagram during the application of nodes corresponds to a transposition of variables, with the associated sign factor.

❑ **Example 9.53.**   The product

$$\left[\sum_{i=1}^{N} (x_i)^3 \right] A^{(21)}(x) \tag{9.B47}$$

corresponds to adding three nodes to the partition (21) of 3. This leads to the four nonvanishing contributions shown (with appropriate signs) at the right. Algebraically, we look at a term of the form (with $N = 5$ variables)

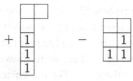

$$\left[\sum_{i=1}^{5} (x_i)^3 \right] x_1^6 x_2^4 x_3^2 x_4$$

which gives after expansion

$$x_1^9 x_2^4 x_3^2 x_4 + x_1^6 x_2^7 x_3^2 x_4 + x_1^6 x_2^4 x_3^5 x_4 + x_1^6 x_2^4 x_3^2 x_5^3$$

(a vanishing term $x_1^6 x_2^4 x_3^2 x_4^4$ has been dropped). The second and third terms each require a single transposition to order the exponents correctly, so they require a minus sign, while the last term requires two transpositions ($x_5 \rightarrow x_4 \rightarrow x_3$), so it has a positive sign. ∎

To compute the character $\chi_{(m)}^{(\lambda)}$, we then enumerate the possible constructions of $\mathcal{Y}^{(\lambda)}$ starting from the empty diagram and making successive regular applications of $m_1, \ldots, m_t$ nodes. The number of such constructions, counted with weight ± 1 equal to the product of the parities of the t regular applications, is equal to the character $\chi_{(m)}^{(\lambda)}$. The order of the t regular applications does not matter, so long as it is the same for each construction.

❑ **Example 9.54.** Evidently $\chi_{(m)}^{(N)} = 1$ for every partition (m) of N, so the partition (N) is identified with the symmetric representation. Also, $\chi_{(m)}^{(1^N)} = \varepsilon_{(m)}$, where $\varepsilon_{(m)}$ is the parity of the permutations in the class $K_{(m)}$, since the parity of a regular application of p nodes in a single column is the same as the parity of a p-cycle. Thus we identify the partition (1^N) with the antisymmetric representation.   ∎

❑ **Example 9.55.** The dimension $d[(\lambda)]$ of the irreducible representation $\Gamma^{(\lambda)}$ is just the character of the class $K_{(1^N)}$ of the identity, i.e., the trace of the unit matrix. By the rule just given, this is equal to the number of ways of constructing the diagram $\mathcal{Y}^{(\lambda)}$ by N successive regular applications of a single node. This in turn is the number of regular tableaux (defined in Section 9.3.2) associated with $\mathcal{Y}^{(\lambda)}$. Note that this fixes the sign of the character $\chi_{(1^N)}^{(\lambda)}$, removing the sign ambiguity in the orthogonality relations.   ∎

❑ **Example 9.56.** To calculate the dimension of the irreducible representation $\Gamma^{(321)}$ of $\mathcal{S}_6$, consider

$$(6)\;\begin{array}{|c|c|}\hline 1 & 2 \\\hline 3 & \\\hline \end{array} \;+\; (6)\;\begin{array}{|c|c|}\hline 1 & 3 \\\hline 2 & \\\hline \end{array} \;+\; (2)\;\begin{array}{|c|c|c|}\hline 1 & 2 & 3 \\\hline 4 & & \\\hline \end{array} \;+\; (2)\;\begin{array}{|c|c|}\hline 1 & 4 \\\hline 2 & \\\hline 3 & \\\hline \end{array} \;=\; 16$$

In the first two diagrams, the numbers $4, 5$, and 6 can be entered in any of $3! = 6$ ways. In the last two, the number $5, 6$ can be entered in either of two ways. Thus the dimension of the irreducible representation is 16.   ∎

❑ **Example 9.57.** The symmetric group $\mathcal{S}_4$ has five classes, and five inequivalent irreducible representations. Of the permutations, we know that 12 are even, 12 are odd, and the parities of the classes and the number h of elements of each class are shown in the table at the right. The dimensions of the representations have been computed in Section 9.3.2. From Problem 14, we have

$$\chi_{(m)}^{(\tilde{\lambda})} = \varepsilon_{(m)} \chi_{(m)}^{(\lambda)}$$

where $\varepsilon_{(m)}$ is the parity of the permutations in the class $K_{(m)}$. The remaining characters of the group are computed with the graphical methods, and produce the character table shown at the right.

	(1^4)	(21^2)	(2^2)	(31)	(4)
ε	$+$	$-$	$+$	$+$	$-$
h	1	6	3	8	6
4	1	1	1	1	1
31	3	1	-1	0	-1
2^2	2	0	2	-1	0
21^2	3	-1	-1	0	1
1^4	1	-1	1	1	-1

Character table for $\mathcal{S}_4$.

We have

$$\chi^{(31)}_{(21^2)} = \boxed{\begin{array}{|c|c|c|}1&2&2\\\hline\end{array}}\!\!\Big/\boxed{3} = 1 \qquad \chi^{(31)}_{(2^2)} = \boxed{\begin{array}{|c|c|c|}1&2&2\\\hline\end{array}}\!\!\Big/\boxed{1} = -1$$

$$\chi^{(31)}_{(31)} = \boxed{\begin{array}{|c|c|c|}\bullet&&\\\hline\end{array}} = 0 \qquad \chi^{(31)}_{(4)} = \boxed{\begin{array}{|c|c|c|}1&1&1\\\hline\end{array}}\!\!\Big/\boxed{1} = -1$$

where the "•" signifies that it is impossible to proceed with another regular application, so the character vanishes (we have used the freedom to order the regular applications at will). Also

$$\chi^{(2^2)}_{(2^2)} = \boxed{\begin{array}{|c|c|}1&1\\\hline 2&2\end{array}} + \boxed{\begin{array}{|c|c|}1&2\\\hline 1&2\end{array}} = 2 \qquad \chi^{(2^2)}_{(21^2)} = \boxed{\begin{array}{|c|c|}1&1\\\hline 2&3\end{array}} - \boxed{\begin{array}{|c|c|}1&2\\\hline 1&3\end{array}} = 0$$

The first character shows that the elements of the class $K_{(2^2)}$ are represented by the unit matrix. How is this possible? The second is an explicit example to show that the character of an odd class vanishes in a self-conjugate partition. ∎

The path traced out by the nodes added in a regular application of nodes can be described as a *hook*, or *skew hook*; it is a sequence of nodes that traces a path of single steps, each of which is upward or to the right. For example, the set of added nodes in each of the diagrams in Example 9.53 is a hook of length three. In general, a regular application of ℓ nodes to a Young diagram is equivalent to adding a hook of length ℓ.

With each node in a Young diagram is associated a hook consisting of the node together with all the nodes to the right of it in the same row, and all the nodes below it in the same ·column. The number of nodes in this hook is the *hook length* of the node. The *hook diagram* of a Young diagram is obtained by assigning to each node in the diagram the hook length of the node. The *hook product* $H[(\lambda)]$ of a partition (λ) is the product of the hook lengths assigned to the nodes of the corresponding Young diagram $\mathcal{Y}^{(\lambda)}$.

❏ **Example 9.58.** The hook diagrams associated with the partitions of 3 are

$$\boxed{\begin{array}{|c|c|c|}3&2&1\\\hline\end{array}} \qquad \boxed{\begin{array}{|c|c|}3&1\\\hline 1&\\\cline{1-1}\end{array}} \qquad \boxed{\begin{array}{|c|}3\\\hline 2\\\hline 1\end{array}}$$

with hook products 6, 3, and 6, respectively. ∎

❏ **Example 9.59.** The hook diagrams associated with the partitions of 4 are

$$\boxed{\begin{array}{|c|c|c|c|}4&3&2&1\\\hline\end{array}} \quad \boxed{\begin{array}{|c|c|c|}4&2&1\\\hline 1&&\\\cline{1-1}\end{array}} \quad \boxed{\begin{array}{|c|c|}3&2\\\hline 2&1\end{array}} \quad \boxed{\begin{array}{|c|c|}4&1\\\hline 2&\\\cline{1-1} 1&\\\cline{1-1}\end{array}} \quad \boxed{\begin{array}{|c|}4\\\hline 3\\\hline 2\\\hline 1\end{array}}$$

with hook products 24, 8, 12, 8, and 24, respectively. ∎

➜ **Exercise 9.B2.** Construct the hook diagrams and the corresponding hook products $H^{(\lambda)}$ for each of the partitions of $N = 5$ and $N = 6$. ☐

The dimension of the irreducible representation $\Gamma^{(\lambda)}$ can be calculated without counting diagrams by means of the *hook formula*

$$d[(\lambda)] = \frac{N!}{H[(\lambda)]} \tag{9.B48}$$

where $H[(\lambda)]$ is the hook product just introduced. Note that this works for the examples above, but it is left as an exercise for the reader to find a general proof of this formula.

B.3   Outer Products of Representations of $\mathcal{S}_m \otimes \mathcal{S}_n$

If $\Gamma^{(\mu)}$ and $\Gamma^{(\nu)}$ are irreducible representations of $\mathcal{S}_m$ and $\mathcal{S}_n$, respectively, then the *outer product* $\Gamma^{(\mu)} \circ \Gamma^{(\nu)}$ of $\Gamma^{(\mu)}$ and $\Gamma^{(\nu)}$ is the representation of $\mathcal{S}_{m+n}$ induced by the irreducible representation $\Gamma^{(\mu)} \otimes \Gamma^{(\nu)}$ of $\mathcal{S}_m \otimes \mathcal{S}_n$, as introduced above in Definition 9.28. The outer product is important both in the context of the symmetric group and in the context of the classical Lie groups, whose Kronecker products are directly related to the outer products of the symmetric group. $\Gamma^{(\mu)} \circ \Gamma^{(\nu)}$ is in general reducible; we have

$$\Gamma^{(\mu)} \circ \Gamma^{(\nu)} = \sum_{(\lambda)} K\{(\mu)(\nu)|(\lambda)\}\Gamma^{(\lambda)} \tag{9.B49}$$

where the summation is over partitions (λ) of $m + n$. The expansion coefficients $K\{(\mu)(\nu)|(\lambda)\}$ can be evaluated by graphical methods that we will explain without giving a full derivation.

First suppose $(\lambda) = (\lambda_1 \cdots \lambda_p)$ is a partition of some integer t into p parts. Define

$$F^{(\lambda)}(x_1, \ldots, x_N) \equiv \frac{A^{(\lambda)}(x_1, \ldots, x_N)}{A(x_1, \ldots, x_N)} \tag{9.B50}$$

where $A^{(\lambda)}(x_1, \ldots, x_N)$ has been defined in Eq. (9.B19) and $A(x_1, \ldots, x_N)$ is the alternant. If $(\xi) = (\xi_1 \ldots \xi_q)$ is another partition of t, then the Frobenius generating function (9.B42) is equivalent to

$$S_{(\xi)}(x_1, \ldots, x_N) = \sum_{(\lambda)} \chi_{(\xi)}^{(\lambda)} F^{(\lambda)}(x_1, \ldots, x_N) \tag{9.B51}$$

where $S_{(\xi)}$ is defined in Eq. (9.B23). Then orthogonality relation (9.B41) then gives

$$F^{(\lambda)}(x_1, \ldots, x_N) = \sum_{(\xi)} \frac{N_{(\xi)}}{N!} \chi_{(\xi)}^{(\lambda)} S_{(\xi)}(x_1, \ldots, x_N) \tag{9.B52}$$

Thus the functions $F_{(\xi)}$, also known as Schur functions, are actually homogeneous symmetric polynomials of degree N.

Now it is true that

$$F^{(\mu)}(x_1, \ldots, x_N) F^{(\nu)}(x_1, \ldots, x_N) = \sum_{(\lambda)} K\{(\mu)(\nu)|(\lambda)\} F^{(\lambda)}(x_1, \ldots, x_N) \tag{9.B53}$$

This is not obvious, and actually requires a careful derivation of its own, but we do not provide that here. However, we now derive a graphical expression of the product $F^{(\mu)}(x)F^{(\nu)}(x)$ that will allow us to calculate the reduction coefficients $K\{(\mu)(\nu)|(\lambda)\}$.

To this end, consider first the partition of m with a single part. Then we have

$$mA^{(m)}(x_1,\ldots,x_N) = \sum_{k=1}^{m} s_k(x_1,\ldots,x_N)A^{(m-k)}(x_1,\ldots,x_N) \tag{9.B54}$$

To show this, recall Eq. (9.B45) and the subsequent discussion, from which we have

$$s_1(x)A^{(m-1)}(x) = A^{(m)}(x) + A^{(m-1\,1)}(x)$$

$$s_2(x)A^{(m-2)}(x) = A^{(m)}(x) + A^{(m-2\,2)}(x) - A^{(m-2\,1^2)}(x)$$

$$\vdots \tag{9.B55}$$

$$s_{m-1}(x)A^{(1)}(x) = A^{(m)}(x) - A^{(m-2\,2)}(x) + A^{(m-3\,2\,1)}(x) + \cdots - (-1)^m A^{(1^m)}$$

$$s_m(x)A(x) = A^{(m)}(x) - A^{(m-1\,1)}(x) + A^{(m-2\,1^2)}(x) + \cdots + (-1)^m A^{(1^m)}$$

In each of the m products, $A^{(m)}(x)$ appears with coefficient $+1$. If (μ) is any other partition of m, then either

(i) the diagram $\mathcal{Y}^{(\mu)}$ has at most two rows containing more than one node, in which case $A^{(\mu)}(x)$ appears in exactly two products, once each with coefficient $+1$ and -1, or

(ii) the diagram $\mathcal{Y}^{(\mu)}$ has more than two rows with more than one node, in which case $A^{(\mu)}(x)$ appears in no product, since there is no way to construct the diagram $\mathcal{Y}^{(\mu)}$ by a regular application of k nodes to the diagram $\mathcal{Y}^{(m-k)}$.

Remark.   To show the point of (i), suppose for example that $m = 6$, and note that the partition (42) appears in the products $s_2 A^{(4)}$ and $s_5 A^{(1)}$, shown graphically as

while it seems clear without a picture that a regular application of nodes to a graph with only one row leads to a graph with at most a second row containing more than one node.   □

Remark.   An equivalent statement is that in the product $s_k A^{(m)}$, there can appear only graphs with at most two hooks, one of which has only one row $[A^{(m-k)}]$.   □

Remark.   Note that Eq. (9.B54) is equivalent to

$$mF^{(m)}(x_1,\ldots,x_N) = \sum_{k=1}^{m} s_k(x_1,\ldots,x_N)F^{(m-k)}(x_1,\ldots,x_N) \tag{9.B56}$$

after dividing both sides by the alternant $A(x_1,\ldots,x_N)$.   □

Now consider the product

$$F^{(m)}(x)F^{(\nu)}(x) = \sum_{(\lambda)} K\{(m)(\nu)|(\lambda)\}F^{(\lambda)}(x) \qquad (9.\text{B}57)$$

The coefficient $K\{(m)(\nu)|(\lambda)\}$ is equal to one if and only if the diagram $\mathcal{Y}^{(\lambda)}$ can be constructed from the diagram $\mathcal{Y}^{(\nu)}$ by the addition of m nodes, no two of which appear in the same column of $\mathcal{Y}^{(\lambda)}$. Otherwise, the coefficient is zero.

To show this, note that the result is obviously true for $m = 0, 1$. Suppose the result is true for $0, 1, \ldots, m - 1$. Then, with Eq. (9.B56) in mind, we have three following possibilities.

(i) $\mathcal{Y}^{(\lambda)}$ can be constructed from $\mathcal{Y}^{(\nu)}$ by the addition of m nodes, no two of which appear in the same column of $\mathcal{Y}^{(\lambda)}$. Then $F^{(\lambda)}(x)$ appears in the product $s_k(x)F^{(m-k)}(x)F^{(\nu)}(x)$ with coefficient equal to the number of different rows of $\mathcal{Y}^{(\lambda)}$ that contain at least k added nodes. Thus $F^{(\lambda)}(x)$ appears in the product

$$m\, F^{(m)}(x)\, F^{(\nu)}(x) = [s_1(x)F^{(m-1)(x)} + \cdots + s_m(x)F(x)]\, F^{(\nu)}(x)$$

with total coefficient m, or

(ii) $\mathcal{Y}^{(\lambda)}$ is constructed from $\mathcal{Y}^{(\nu)}$ by the addition of m nodes, of which at least two appear in the same column of $\mathcal{Y}^{(\lambda)}$. Then *either*

(a) $F^{(\lambda)}(x)$ appears in no product $s_k(x)F^{(m-k)}(x)F^{(\nu)}(x)$, *or*

(b) it appears in exactly two products, once with coefficient $+1$ and once with coefficient -1 (an illustration of this is given below), or

(iii) $\mathcal{Y}^{(\lambda)}$ cannot be constructed from $\mathcal{Y}^{(\nu)}$ by the addition of m nodes, in which case $F^{(\lambda)}(x)$ appears in no product $s_k(x)F^{(m-k)}(x)F^{(\nu)}(x)$.

Remark.   To illustrate the argument in (ii)(b), consider the product of (3) and (31^2). We have

since the second product corresponds to a term x_3^2 $(x_1^6 x_2^2 x_3)$ that acquires a minus sign when variables are reordered. Thus the product $(3) \times (31^2)$ does not contain the partition (42^2). $\square$

Thus we have a graphical rule to construct the outer product of the identity representation of $\mathcal{S}_m$ with an arbitrary representation of $\mathcal{S}_n$. Some examples were given in Section 9.5.2; here we offer a few more.

❑ **Example 9.60.**   We have

$$\square\square\square \circ \square\square\square = \square\square\square\boxed{1\,1\,1} \oplus \cdots \oplus \cdots \oplus \cdots$$

This is a special case of the rule given in Exercise 9.22.   ∎

Definition 9.30.   Suppose (ν) is a partition of n with Young diagram $\mathcal{Y}^{(\nu)}$. The construction of the diagram $\mathcal{Y}^{(\lambda)}$ of a partition (λ) of $m + n$ by the addition of m nodes to $\mathcal{Y}^{(\nu)}$, no two of that appear in the same column of $\mathcal{Y}^{(\lambda)}$, is a *normal application of m nodes* to $\mathcal{Y}^{(\nu)}$.   ∎

Remark. Thus what we have shown above is that the outer product $\Gamma^{(m)} \circ \Gamma^{(\nu)}$ contains the irreducible representations of $\mathcal{S}_{m+n}$ whose Young diagrams can be constructed from the diagram $\mathcal{Y}^{(\nu)}$ by the normal application of m nodes.   □

→ Exercise 9.B3. Explain with words and diagrams the difference between a *normal* and a *regular* application of m nodes to a Young diagram.   □

Now suppose $(\xi) = (\xi_1 \xi_2 \cdots \xi_q)$ is a partition of n. The principal representation $\Gamma_{(\xi)} \equiv \Gamma_{(\xi_1 \xi_2 \cdots \xi_q)}$ induced by the subgroup $\mathcal{G}(\xi) \equiv \mathcal{S}_{\xi_1} \otimes \mathcal{S}_{\xi_2} \otimes \cdots \otimes \mathcal{S}_{\xi_q}$ is reducible; let

$$\Gamma_{(\xi)} = \sum_{(\lambda)} H\{(\xi)|(\lambda)\}\Gamma^{(\lambda)} \tag{9.B58}$$

From the preceding discussion, it is clear that $H\{(\xi)|(\lambda)\}$ is the number of distinct constructions of the diagram $\mathcal{Y}^{(\lambda)}$ from the empty diagram by successive normal applications of $\xi_1, \ldots, \xi_q$ nodes. Note that $H\{(\xi)|(\lambda)\} = 0$ if (λ) comes after (ξ) in the dictionary ordering of the partitions, as seen in the next example.

❑ **Example 9.61.** The principal representation of $\mathcal{S}_6$ induced by the subgroup $\mathcal{S}_3 \otimes \mathcal{S}_2 \otimes \mathcal{S}_1$ is given by

This example was chosen to show that an irreducible representation can occur more than once if the partition (ξ) has more than two parts. Note that the partitions on the right-hand side have been given in order, and (321) is the last partition to occur.   ∎

Then we have also (see Eq. (9.B53))

$$F^{(\xi_1)}(x_1, \ldots, x_N) \cdots F^{(\xi_q)}(x_1, \ldots, x_N) = \sum_{(\lambda)} H\{(\xi)|(\lambda)\}F^{(\lambda)}(x_1, \ldots, x_N) \tag{9.B60}$$

This reduction allows us the compute the outer product $\Gamma^{(\mu)} \circ \Gamma^{(\nu)}$ by proceeding down the list of partitions of m. If $(\mu) = (\mu_1 \mu_2 \ldots \mu_q)$ is a partition of m, we can reduce the product

$$\Gamma^{(\mu_1)} \circ \Gamma^{(\mu_2)} \circ \cdots \circ \Gamma^{(\mu_1 q)} \circ \Gamma^{(\nu)}$$

by finding the number of constructions of each diagram $\mathcal{Y}^{(\lambda)}$ (corresponding to a partition of $m + n$) from $\mathcal{Y}^{(\nu)}$ by successive normal applications of $\mu_1, \mu_2, \ldots, \mu_q$ nodes. Subtracting the contributions of those partitions of m that precede (μ) – these having been previously computed – from this product leaves the reduction of $\Gamma^{(\mu)} \circ \Gamma^{(\nu)}$. This leads to the result given in the following theorem, but we leave out further details of the proof.

Theorem 9.5. Suppose $(\mu) = (\mu_1 \mu_2 \cdots \mu_q)$ is a partition of m and (ν) a partition of n. Then the coefficient $K\{(\mu)(\nu)|(\lambda)\}$ in the reduction (Eq. (9.B49))

$$\Gamma^{(\mu)} \circ \Gamma^{(\nu)} = \sum K\{(\mu)(\nu)|(\lambda)\}\Gamma^{(\lambda)}$$

of $\Gamma^{(\mu)} \circ \Gamma^{(\nu)}$ into a sum of irreducible representations of $\mathcal{S}_{m+n}$ is equal to the number of distinct constructions of the diagram $\mathcal{Y}^{(\lambda)}$ from the diagram $\mathcal{Y}^{(\nu)}$ by successive application of μ_1 nodes labeled $1, \ldots, \mu_q$ nodes labeled q such that

(N) no two nodes with the same label appear in the same column of $\mathcal{Y}^{(\lambda)}$ (the applications of the $\mu_1, \ldots, \mu_q$ nodes are normal applications), *and*

(R) if we scan $\mathcal{Y}^{(\lambda)}$ from left to right across the rows from top to bottom, then the ith node labeled $k + 1$ appears in a lower row of $\mathcal{Y}^{(\lambda)}$ than the ith node labeled k.

→ **Exercise 9.B4.** Show that

$$K\{(\widetilde{\mu})(\widetilde{\nu})|(\widetilde{\lambda})\} = K\{(\mu)(\nu)|(\lambda)\}$$

This symmetry relation reduces the number of explicit calculations that need to be done. ☐

❑ **Example 9.62.** For $\mathcal{S}_2 \otimes \mathcal{S}_3$, direct application of the graphical rule gives

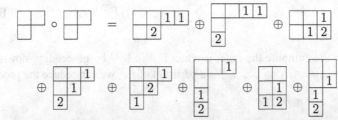

although this is more easily derived from the result for $\Gamma^{(2)} \circ \Gamma^{(\lambda)}$ given above using the symmetry relation in Exercise 9.B4. Note that diagrams such as

are eliminated by rule (R).   ∎

❑ **Example 9.63.** For $\mathcal{S}_3 \otimes \mathcal{S}_3$, we have

Note that the partition (321) appears twice, since there are the two distinct constructions shown of the diagram of (321) from that of (21) consistent with the rule (R).   ∎

→ **Exercise 9.B5.** Compute the remaining outer products for $\mathcal{S}_3 \otimes \mathcal{S}_3$. Use the relation given in Exercise 9.B4 to simplify the calculations.   ☐

Bibliography and Notes

An old classic that emphasizes many physical applications is

> Morton Hamermesh, *Group Theory and its Application to Physical Problems,*
> Addison-Wesley (1962), reprinted by Dover (1989).

Chapter 7 of this book has an especially detailed treatment of the symmetric group S_N, upon which much of our discussion here is based. There is also a solid discussion of the classical Lie groups, and the connection of the representations of Lie groups with those of S_N.

Another classic is

> Michael Tinkham, *Group Theory and Quantum Mechanics,* McGraw-Hill (1964),
> reprinted by Dover (2003).

This book has a thorough treatment of discrete symmetry groups that are relevant to molecules and crystal structure. In addition, there is a useful discussion of the full rotational symmetry as applied to atomic systems. The physical relevance of group theory to selection rules and to level structure in systems perturbed by external fields is clearly explained.

A brief introduction is

> H. F. Jones, *Groups, Representations, and Physics,* Adam Hilger (1990).

A longer introduction that covers a broad range of applications is

> W. Ludwig and C. Falter, *Symmetries in Physics: Group Theory Applied to Physical Problems* (2nd extended edition) , Springer (1996).

A new book that deals with the crystallographic groups is

> Richard L. Liboff, *Primer for Point and Space Groups,* Springer (2004).

There are many other books that combine the analysis of the symmetric group with the study of rotational symmetry and the angular momentum algebra to the spectra of atoms, molecules, and nuclei. A sampling of these is listed at the end of Chapter 10.

Problems

1. Show that if K is a class of the group $\mathcal{G}$, then the set $\overline{K}$ of elements g^{-1} with g in K is also a class of $\mathcal{G}$.

   Remark. $\overline{K}$ is the *inverse class* of K. If $\overline{K} = K$, the class K is *ambivalent.* □

2. Suppose the finite group $\mathcal{G}$ has classes $K_1 = \{e\}, K_2, \ldots, K_p$.

   (i) Show that the product $K_k K_\ell$ contains only complete classes of $\mathcal{G}$.

   Remark. Thus we can write

   $$K_k K_\ell = \sum_m c_{k\ell}^m K_m$$

   with integer coefficients $c_{k\ell}^m$ that are the *class multiplication coefficients* of $\mathcal{G}$. □

(ii) Show that if $K_{\bar{k}}$ denotes the inverse class to K_k, then the $c_{k\ell}^m$ satisfy

$$h_m c_{k\ell}^m = h_\ell c_{k\bar{m}}^{\bar{\ell}}$$

(iii) Find the class multiplication coefficients of S_3 and S_4.

3. $\mathcal{H}$ is an invariant subgroup of $\mathcal{G}$ if and only if every class of $\mathcal{H}$ is also a class of $\mathcal{G}$.

4. Find the number of distinct Abelian groups of order 720.

5. Consider the group $\mathcal{Q}$ with elements $1, -1, a, -a, b, -b, c, -c$ such that

$$a^2 = b^2 = c^2 = -1 \qquad abc = 1$$

(i) Complete the multiplication table of $\mathcal{Q}$.

(ii) Find the conjugacy classes of $\mathcal{Q}$.

(iii) Show that an irreducible two-dimensional representation of $\mathcal{Q}$ is given by

$$1 = 1 \qquad a = i\sigma_x \qquad b = i\sigma_y \qquad c = i\sigma_z$$

where $\sigma_x, \sigma_y, \sigma_z$ are the 2×2 Pauli matrices introduced in Eq. (2.81) of Chapter 2.

Remark. This group is the *quaternion group* discovered by Hamilton. The elements a, b, and c are three independent roots of $x^2 = -1$, but they do not commute under multiplication. The (noncommutative) algebra generated by linear combinations of $1, a, b, c$ with multiplication as given here is the *quaternion algebra*.   □

6. Describe the inequivalent groups of order 8 (there are five of them).

7. The benzene ring has a hexagonal structure with symmetry group D_6 generated by the elements

   a—rotation through angle $\pi/3$ about an axis normal to the plane of the hexagon passing through its center, and

   σ—reflection of an axis joining two opposite vertices of the hexagon.

   These elements satisfy the relations

$$a^6 = 1 \qquad \sigma^2 = 1 \qquad (\sigma a)^2 = 1$$

(i) Show that D_6 is of order 12.

(ii) Find the classes of D_6.

(iii) What are the dimensions of the irreducible representations of D_6?

(iv) Construct the character table for D_6.

(v) Express the matrices for the two-dimensional irreducible representation(s) of D_6 in terms of the 2×2 unit matrix and the Pauli matrices.

8. Consider the symmetries of the tetrahedron (see Section 9.2) as permutations of the four vertices of the tetrahedron.

(i) Show that the group T is isomorphic to the alternating group A_4.

(ii) Show that each of the six reflections corresponds to a transposition in S_4, while the combined reflection–rotations correspond to the 4-cycles.

(iii) Finally, show that the complete symmetry group T_d is isomorphic to the symmetric group S_4.

9. Consider the group of rotational symmetries of the cube in three dimensions. Draw a right-handed set of axes through the center of the cube, with each axis bisecting a pair of opposite faces. Let a, b, c denote rotations through angle $\pi/2$ about X, Y, Z axes, respectively.

(i) Show that

$$a^4 = b^4 = c^4 = e$$

$$a^2 b^2 c^2 = e = c^2 b^2 a^2$$

$$(ab)^3 = (bc)^3 = (ca)^3 = e = (ac)^3 = (cb)^3 = (ba)^3$$

(ii) Express each element of the group as a (nonunique) product of powers of a, b, c. What is the order of the group?

(iii) Express each element of the group as a permutation of the six faces of the cube.

(iv) Express each element of the group as a permutation of the eight vertices of the cube.

(v) Find the classes of the group.

(vi) Is the group isomorphic to S_4?

10. The cube is also symmetric under the operation of reflection through the origin, denoted by $\mathbf{P}$. Show that $\mathbf{P}$ commutes with all the rotations of the preceding problem, and thus that the complete symmetry group of the cube is the direct product $\mathcal{O} \otimes \mathbf{Z}_2 \equiv \mathcal{O}_d$ (here the cyclic group $\mathbf{Z}_2 = \{1, \mathbf{P}\}$ – note that $\mathbf{P}^2 = 1$).

Remark. The symmetry group of the cube is the same as the symmetry group of the octahedron whose vertices are the centers of the six faces of the cube. Hence $\mathcal{O}$ and $\mathcal{O}_d$ are also known as the *octahedral groups*. □

11. Show that the coefficients of composition of a group $\mathcal{G}$ defined by Eq. (9.98) satisfy

$$\sum_{c=1}^{p} \left[C_c^{ab} \right]^2 = \sum_{c=1}^{p} C_c^{a\bar{a}} \, C_{\bar{c}}^{b\bar{b}}$$

12. Show that the coefficient a_n of x^n in the formal power series expansion

$$\mathcal{E}(x) \equiv \prod_{k=1}^{\infty} \left(\frac{1}{1 - x^k} \right) = \sum_{n=1}^{\infty} a_n x^n$$

is equal to $\pi(n)$, the number of partitions of n. What is the radius of convergence of this power series?

Remark. Thus $\mathcal{E}(x)$ is a generating function for the $\pi(n)$.   □

13. Let $\mathbf{P} = (i_1 i_2 \cdots i_N)$ be a permutation of degree N. Associated with $\mathbf{P}$ is the $N \times N$ *permutation matrix* $\mathbf{A} = \mathbf{A}(\mathbf{P})$ with elements

$$A_{jk} = \delta_{ji_k}$$

(see Eq. (9.27)). Show that $\mathbf{P} \to \mathbf{A}(\mathbf{P})$ is a reducible representation of $\mathcal{S}_N$, and express it as a direct sum of irreducible representations.

14. If $(m) = (m_1 \cdots m_t)$ and (λ) are partitions of N, and $\chi^{(\lambda)}_{(m)}$ is the character of the class $K_{(m)}$ in the irreducible representation $\Gamma^{(\lambda)}$ of $\mathcal{S}_N$, then

$$\chi^{(\widetilde{\lambda})}_{(m)} = \varepsilon_{(m)} \chi^{(\lambda)}_{(m)}$$

where $(\widetilde{\lambda})$ is the partition conjugate to (λ), and $\varepsilon_{(m)}$ is the parity of the permutations in the class $K_{(m)}$.

Hint. Show that for every construction of the diagram $\mathcal{Y}^{(\lambda)}$ by regular applications of $m_1, \ldots, m_t$ modes to the empty graph, there is a similar construction of the diagram $\mathcal{Y}^{(\widetilde{\lambda})}$, not necessarily of the same parity.

Remark. Thus the character of the irreducible representation associated with the conjugate partition $(\widetilde{\lambda})$ is simply related to the character of (λ). Note that this result implies that the character of an odd permutation vanishes in an irreducible representation associated with a self-conjugate partition.   □

15. (i) Compute the characters for the representation of $\mathcal{S}_4$ induced by the antisymmetric representation of the $\mathcal{S}_2 \otimes \mathcal{S}_2$ subgroup.

(ii) Reduce this representation to a sum of irreducible representations of $\mathcal{S}_4$.

16. Find a set of matrices representing the transpositions in each of the irreducible representations of $\mathcal{S}_4$.

17. (i) Find the parity and the number of elements $N_{(m)}$ for each class $K_{(m)}$ of $\mathcal{S}_6$.

(ii) Compute the character table of $\mathcal{S}_6$.

18. Let $X_1 = \mathbf{Y}_a$ and $X_2 = (23)\mathbf{Y}_a$,   where

$$\mathbf{Y}_a = e + (12) - (13) - (123)$$

is the projection operator introduced in Example 9.43.

(i) Find the action of each of the transpositions in $\mathcal{S}_3$ on X_1 and X_2, and thus show that X_1 and X_2 form a basis for the two-dimensional representation of $\mathcal{S}_3$.

(ii) Transform this basis to that in which the transpositions are represented by orthogonal matrices.

19. We know from Example 9.48 that the Kronecker product of the mixed symmetry two-dimensional irreducible representation of $\mathcal{S}_3$ with itself is given by

Explicit matrices for the two-dimensional irreducible representation were given in Example 9.40. Now let ϕ_1, ϕ_2 and ψ_1, ψ_2 be orthonormal bases for copies of this representation on the vector spaces $\mathcal{V}^1$ and $\mathcal{V}^2$. Find basis vectors for each of the irreducible representations in the Kronecker product defined on $\mathcal{V}^1 \otimes \mathcal{V}^2$ in terms of the basis vectors $\phi_a \otimes \psi_b$.

20. Reduce the outer products of the irreducible representations of $\mathcal{S}_4 \otimes \mathcal{S}_4$ in $\mathcal{S}_8$.

21. Reduce the Kronecker products of the irreducible representations of $\mathcal{S}_5$ and $\mathcal{S}_6$

10 Lie Groups and Lie Algebras

It has long been understood that conservation laws of energy, momentum, and angular momentum are related to the invariance of physical laws of a closed system under translations and rotations of the system. These invariance principles are described by continuous groups (*Lie groups*), as already noted in Chapter 9, with examples given in Section 9.1. In quantum mechanics, the connection between conservation laws and the corresponding symmetries is made explicit, as the conserved quantities are represented by operators that are directly related to the generators of translations and rotations:

$$\mathbf{P} = i\hbar\nabla \quad \text{and} \quad \mathbf{L} = i\hbar\mathbf{r} \times \nabla$$

Here Planck's constant $\hbar$ sets a scale for quantum physics. Having introduced it, we follow standard usage and set $\hbar = 1$.

Symmetries other than spacetime symmetries have become increasingly important in modern physics. The special nature of two standard dynamical systems, the harmonic oscillator and the Kepler–Coulomb problem, is related to the existence of higher symmetries. The harmonic oscillator in n dimensions has a $U(n)$ symmetry associated with rotations in the $2n$-dimensional phase space of the oscillator. The Kepler–Coulomb problem of motion in an inverse square potential has an $SO(4)$ symmetry associated with the existence of a second conserved vector, the *Runge–Lenz vector*, in addition to the usual angular momentum.

Shell models of atomic and nuclear structure have approximate symmetries associated with states of several fermions in a single shell, that are useful in understanding atomic and nuclear spectra of atoms and nuclei. Rotational symmetry also leads to the existence of rotational bands in molecular spectra. Applications of group theory to atomic, molecular, and nuclear systems are described in the books cited in the bibliography.

The similarity of proton and neutron, apart from a small difference in mass and the absence of electric charge on the neutron, led Heisenberg in the 1930s to propose an approximate $SU(2)$ symmetry (*isotopic spin*, or simply *isospin*) in which proton and neutron form an elementary doublet. Discovery of exotic ("strange") baryons and mesons led to attempts to generalize this symmetry to $SU(3)$ and beyond. The approximate $SU(3)$ symmetry, known today as *flavor $SU(3)$*, is an important tool for analyzing the spectra of mesons and baryons, and stimulated conjectures in the 1960s about the existence of quarks that eventually led to the standard model of quarks and leptons as we know it today.

Higher symmetries are also of interest in condensed matter physics. The Hubbard model, in which electrons move freely on a lattice except for a strong repulsion when two electrons occupy the same site, is used as a starting point for the analysis of many solids, notably including high T_c superconductors as well as antiferromagnetic systems. This model has an

Introduction to Mathematical Physics. Michael T. Vaughn
Copyright © 2007 WILEY-VCH Verlag GmbH & Co. KGaA, Weinheim
ISBN: 978-3-527-40627-2

exact $SO(4)$ symmetry that was explicitly recognized only after the model had been studied for over twenty years. Further work has led to introduction of groups as large as $SO(8)$ to classify the spectrum of states in the Hubbard model.

Conservation of electric charge implies a continuity equation for the charge and current densities. In quantum mechanics, this is equivalent to invariance under arbitrary phase transformations of the wave function of a charged particle. These transformations form a group $U(1)$, and allowing phase transformations that vary (smoothly) as a function of spacetime point (*gauge transformations*) leads to the principle of *gauge invariance*, from which follows the long-range nature of the Coulomb force and the existence of a massless photon.

Yang and Mills extended the idea of local gauge invariance to non-Abelian groups such as $SU(2)$, later enlarged to larger groups such as $SU(3)$ and beyond. This led to the unified gauge theory of the weak and electromagnetic interactions based on a gauge group $SU(2) \otimes U(1)$, after the problem of how to generate masses for the weak gauge fields was solved by Higgs and others. Further work led to the development of quantum chromodynamics as a theory of the strong interactions based on an unbroken gauge group $SU(3)$.

Generalizations of these symmetries are at the center of contemporary attempts to construct a unified theory of all interactions, beginning with the grand unified theories of the 1970s and early 1980s and continuing with the various string theories that are at the front of the stage in the early part of the 21st century. Less obvious symmetries have also been discovered in certain nonlinear systems that are described in Chapter 8. The infinite hierarchy of symmetries discovered in integrable systems have led to a deeper insight into the structure of these systems, and are also relevant to string theory research.

These higher symmetries are examples of Lie groups, which have been studied since the middle of the 19th century. The *classical groups* include the groups $GL(n, C)$ [$GL(n, R)$] of nonsingular linear operators on an n-dimensional linear vector space with complex [real] scalars. Of special importance in physics are the subgroups $U(n)$ and $O(n)$ of unitary and orthogonal operators, and the corresponding subgroups $SU(n)$ and $SO(n)$ of operators whose matrices have determinant $+1$. Also of interest are the groups $Sp(2n)$ of linear operators that leave invariant the canonical 2-form on the phase space of a Hamiltonian system with n coordinates. There are also exceptional Lie groups that are mentioned only in passing, although they may turn out to be important to elementary particle theory. Beyond that, there are groups associated with the infinite hierarchies in the integrable systems just mentioned, as well as with possible string theories, but these are relegated to the problems.

Closely related to a Lie group is its *Lie algebra*, which is obtained from the structure of the group elements near the identity element. The Lie algebra is a linear vector space of operators whose commutators are also elements of the vector space. The example best-known in physics is the algebra of angular momentum operators $\mathbf{L} = (\mathbf{L}_x, \mathbf{L}_y, \mathbf{L}_z)$. These operators satisfy the commutation relations

$$\mathbf{L} \times \mathbf{L} = i\mathbf{L}$$

They also generate rotations of the coordinate axes in three dimensions.

One method to find the irreducible representations of a Lie group is to start from irreducible representations of its Lie algebra. To construct these, we start with a maximal set of commuting elements of the algebra (a *Cartan subalgebra*) and find simultaneous eigenvectors of this set. The commutation rules then determine the action of the other elements of the Lie algebra on these eigenvectors. In general, we can find elements that serve as *ladder operators*, transforming eigenvectors of the Cartan subalegbra into eigenvectors with shifted eigenvalues. In finite-dimensional representations, the chains of states connected by the ladder operators must terminate. This condition leads to a discrete set of finite-dimensional irreducible representations of the classical groups.

The simplest example of this method is the construction the irreducible representations of the algebra of angular momentum operators. We look for eigenstates $\mid m \rangle$ of one component, $\mathbf{L}_z$ say, with

$$\mathbf{L}_z \mid m \rangle = m \mid m \rangle$$

Then the operators $\mathbf{L}_\pm = \mathbf{L}_x \pm i\mathbf{L}_y$ have the property that they shift the eigenvalue of $\mathbf{L}_z$ by one unit of $\hbar$:

$$\mathbf{L}_z\mathbf{L}_\pm \mid m \rangle = (m \pm 1)\mathbf{L}_\pm \mid m \rangle$$

For a finite-dimensional unitary representation, m must have a maximum value, j, say, and a minimum value that must be $-j$ since the operators $\mathbf{L}$ must be Hermitian. The fact that the ladder operators shift the eigenvalues by integer steps then requires that the number $2j$ of steps from j to $-j$ be an integer.

We work out the application of this method to find some irreducible representations of the Lie algebra of the group $SU(3)$, and explain how this method can be extended to other representations and other Lie algebras. The flavor $SU(3)$ symmetry of baryons and mesons is used as an illustrative example.

The extension of symmetries to local gauge symmetries is important both in classical electromagnetic theory and in the $SU(3) \times SU(2) \times U(1)$ model (the *standard model*) of quarks and leptons, as well as further symmetries that are a focus of research in elementary particle physics. Here we offer a brief introduction to these symmetries.

An alternative approach to classifying irreducible representations of a Lie group is to reduce the group representations on tensor products of identical copies of the defining vector space. These tensor representations can be reduced according to their symmetry under permutations of the spaces in the product, leading to a correspondence between representations of the Lie group and irreducible representations of the symmetric group with their associated Young diagrams. The diagrammatic methods introduced for the symmetric groups can then be applied to the study of Lie group representations. For the general linear and unitary groups, this classification actually gives irreducible representations. For the orthogonal and symplectic groups, there are further reductions associated with the metric tensors left invariant by these groups. These ideas are developed in Appendix A.

An important problem is to reduce the tensor product, or Kronecker product, of irreducible representations of a Lie group into a sum of irreducible representations. This appears in quantum mechanics as the problem of finding the allowed states of total angular momentum of two or more particles each of which has a definite angular momentum. For the classical Lie groups, the irreducible representations are associated with those of the symmetric groups $\mathcal{S}_N$, and the tensor product corresponds to an outer product of irreducible representations of symmetric groups. We show how to use the graphical rules developed in Chapter 9 for the outer products of symmetric groups to reduce these tensor products.

Principles of relativity from Galileo to Einstein have been based on the idea that observers moving at a constant velocity relative to each other should observe the same physical laws. Application of this idea to Maxwell's equations led to the Lorentz transformation law relating the spacetime coordinates in two coordinate systems moving with a constant relative velocity, and invariance under Lorentz transformations is a key ingredient in constructing theories of elementary particles. At its deepest level, independence of physics from the coordinate system used to describe spacetime leads to Einstein's theory of gravitation (*general relativity*). In Appendix B, we discuss the Lorentz group and the Poincaré group (the Lorentz group augmented by translations of the spacetime coordinates) and their representations.

10.1   Lie Groups

In addition to the discrete groups discussed in Chapter 9, there are continuous groups that are of fundamental importance in physics. Space–time translations, rotations, and transformations between coordinate systems moving with constant relative velocity are all expected to be symmetries of closed systems; other continuous symmetries mentioned at the top of the chapter have become increasingly important in microscopic physics. These groups are characterized by the dependence of group elements on a set of parameters, or coordinates, that vary continuously as we move through the group. If we denote these parameters collectively by ξ, we can express the group multiplication law in the form

$$g(\xi) = g'(\xi')g''(\xi'') \tag{10.1}$$

with a relation

$$\xi = \phi(\xi', \xi'') \tag{10.2}$$

The group is *continuous* if we can choose coordinates so that the function $\phi(\xi', \xi'')$ is continuous in each of its variables. The group is a *Lie group* if we can choose coordinates so that $\phi(\xi', \xi'')$ has derivatives of all orders in its variables. Note that the coordinates are taken here to be real, but there are circumstances in which it is useful to extend the range of the coordinates into the complex plane.

❑ **Example 10.1.**   The translations of an n-dimensional vector space V^n form a group T^n, the *translation group* in n dimensions. If $\mathbf{T}(a)$ denotes translation of the origin by a, then the coordinates of a vector are transformed by

$$\mathbf{T}(a)\, x = x - a \tag{10.3}$$

Evidently

$$\mathbf{T}(a_2)\mathbf{T}(a_1) = \mathbf{T}(a_2 + a_1) \tag{10.4}$$

so that T^n is an n-parameter Abelian Lie group.   ∎

❏ **Example 10.2.**   The nonsingular linear operators on an n-dimensional linear vector space V^n form a group. This group is the *general linear group*, denoted by $GL(n, R)$ or $GL(n, C)$ depending on whether the scalars of V^n are real or complex. The groups $GL(n)$ are non-Abelian for $n > 1$, since matrix multiplication is noncommutative. The operators whose matrix has determinant equal to $+1$ form an invariant subgroup known as the *special linear group*, denoted by $SL(n, R)$ or $SL(n, C)$.   ∎

❏ **Example 10.3.**   Linear operators that preserve the vector space scalar product also form groups, as noted in Section 9.1. In a complex V^n, we have the *unitary group* $U(n)$ of unitary operators, and its subgroup $SU(n)$, the *special unitary group*, of unitary operators with determinant $+1$. On a real vector space, these groups are the *orthogonal group* $O(n)$ and the *special orthogonal group* $SO(n)$, since real unitary operators are orthogonal.   ∎

❏ **Example 10.4.**   There are spaces with a symplectic (antisymmetric) metric, notably the phase space of classical mechanics introduced in Chapter 3 (Section 3.5.3). These spaces are necessarily even-dimensional due to the antisymmetry of the metric. Linear operators that preserve the symplectic scalar product belong to the *symplectic group* $Sp(2n)$.   ∎

10.2   Lie Algebras

10.2.1   The Generators of a Lie Group

If $\mathbf{A}$ is a linear operator on the finite-dimensional vector space V^n that can be reached continuously from the identity, then $\mathbf{A}$ can be expressed as an exponential

$$\mathbf{A} = \exp\left(i\xi\mathbf{X}\right) \tag{10.5}$$

where $\mathbf{X}$ is another linear operator on V^n, and ξ is a real parameter.

The operators $\mathbf{A}(\xi)$ form a *one-parameter subgroup*, the *subgroup generated by* $\mathbf{X}$. $\mathbf{X}$ is the *generator* of the subgroup. The factor i in the exponential is conventional in physics, since it associates unitary operators $\mathbf{A}$ with self-adjoint generators $\mathbf{X}$ for a real parameter ξ. The factor is sometimes $-i$, as is the case for the rotation groups presented from the passive point of view in Section 2.2 (see Eqs. (2.2.107) and (2.2.113), and Problems 2.16 and 2.18, for example). It is often omitted in the mathematical literature, where unitary group elements are associated with anti-Hermitian generators.

The collection of operators $\mathbf{X}$ such that $\mathbf{A} = \exp(i\mathbf{X})$ is a group element forms the set of generators of the group. It is clear that a generator multiplied by a scalar is also a generator. Perhaps less obvious, but true, is that a linear combination of two generators is again a generator, so that the generators form a linear vector space. For example, a unitary matrix $\mathbf{U}$ on V^n can be expressed as

$$\mathbf{U} = \exp[i\mathbf{X}] \tag{10.6}$$

with $\mathbf{X}$ Hermitian. But the sum of two Hermitian operators $\mathbf{X}$ and $\mathbf{Y}$ is also Hermitian, so that $\exp[i(\mathbf{X} + \mathbf{Y})]$ is unitary.

Note however, that

$$e^{i(\mathbf{X}+\mathbf{Y})} \neq e^{i\mathbf{X}} e^{i\mathbf{Y}} \tag{10.7}$$

in general. The equality is true if and only if $\mathbf{X}$ and $\mathbf{Y}$ commute. Also, if

$$\mathbf{A} = \exp\left(i\xi\mathbf{X}\right) \qquad \mathbf{B} = \exp\left(i\eta\mathbf{Y}\right) \tag{10.8}$$

then we can expand the exponentials in formal power series to get

$$\mathbf{A}\mathbf{B}\mathbf{A}^{-1}\mathbf{B}^{-1} \cong 1 + 2\xi\eta\left[\mathbf{X}, \mathbf{Y}\right] + \cdots \tag{10.9}$$

where

$$\left[\mathbf{X}, \mathbf{Y}\right] \equiv \mathbf{X}\mathbf{Y} - \mathbf{Y}\mathbf{X} \tag{10.10}$$

is the *commutator* of $\mathbf{X}$ and $\mathbf{Y}$.

→ **Exercise 10.1.**   Show that the terms proportional to ξ^2 and η^2 in Eq. (10.9) vanish.   □

10.2.2   The Lie Algebra of a Lie Group

Since the left-hand side of Eq. (10.9) is a group element, the term proportional to the commutator on the right-hand side must be a group generator,

$$\left[\mathbf{X}, \mathbf{Y}\right] = i\mathbf{Z} \tag{10.11}$$

with $\mathbf{Z}$ a generator of the group. Thus the set of generators of the group is closed under commutation, and the generators form a *Lie algebra*, as defined in Section 9.1.3.

The commutation relations (10.11) can be given a concrete form if we introduce a basis $\mathbf{X}_1, \ldots, \mathbf{X}_N$ on the vector space defined by the group generators. Then we have

$$\left[\mathbf{X}_j, \mathbf{X}_k\right] = i\sum_\ell c_{jk}^\ell \mathbf{X}_\ell \tag{10.12}$$

which defines the *structure constants* $c_{jk}^\ell = -c_{kj}^\ell$ of the Lie algebra, also called the structure constants of the Lie group. The structure constants are real if the group coordinates are real.

❏ **Example 10.5.**   The operators $\mathbf{S}_k$ on $\mathcal{V}^2$ defined in terms of the matrices σ_k by

$$\mathbf{S}_k \equiv \tfrac{1}{2}\sigma_k \tag{10.13}$$

($k = 1, 2, 3$) satisfy the commutation relations

$$\left[\mathbf{S}_k, \mathbf{S}_\ell\right] = i\sum_m \varepsilon_{k\ell m}\mathbf{S}_m \tag{10.14}$$

(see Problem 2.18). These matrices generate the group $SU(2)$ of unitary 2×2 matrices with determinant equal to one (see Problem 2.17).   ∎

❏ **Example 10.6.** From Eq. (10.6), it follows that the group $U(n)$ of unitary $n \times n$ matrices is generated by the Hermitian $n \times n$ matrices. These matrices form a Lie algebra, since if $\mathbf{X}$ and $\mathbf{Y}$ are Hermitian, then the operator $\mathbf{Z}$ defined by Eq. (10.11) is also Hermitian. The unit matrix $\mathbf{1}$ commutes with every element of the algebra; it defines a (one-dimensional) invariant subalgebra as defined in Section 10.2.3. ∎

The exponentiation of a Lie algebra $\mathcal{A}$, as in Eq. (10.5), defines a Lie group $\mathcal{G}(\mathcal{A})$, but this may not be the only Lie group with $\mathcal{A}$ as its Lie algebra. For example, the 3×3 matrices $\mathbf{L}_k$ ($k = 1, 2, 3$) defined in Problem 2.16 form the Lie algebra of the group $SO(3)$ of rotations in three (real) dimensions, and satisfy the same commutation relations as the $\mathbf{S}_k$ in Example 10.5, so the Lie algebras of $SU(2)$ and $SO(3)$ are isomorphic. However, the groups $SU(2)$ and $SO(3)$ are not quite isomorphic. $SU(2)$ has an invariant subgroup $\mathbf{Z}_2 = \{1, -1\}$, the *center* of $SU(2)$. The group $SO(3)$ is isomorphic to the factor group $SU(2)/\mathbf{Z}_2$. There are two $SU(2)$ matrices $\pm\mathbf{U}(\mathbf{R})$ corresponding to each rotation $\mathbf{R}$ in $SO(3)$. Rotation about any axis through angle 2π, which is the identity in $SO(3)$, corresponds to the matrix $-\mathbf{1}$ in $SU(2)$.

Following Definition 9.20 of a group representation, we have for a Lie algebra:

Definition 10.1. A *representation* of a Lie algebra $\mathcal{A}$ is a one-to-one mapping between the Lie algebra and an algebra $\mathcal{R}$ of matrices that preserves the commutation relations (10.12). The representation is *reducible* if $\mathcal{R}$ can be expressed a direct sum $\mathcal{R}_1 \oplus \mathcal{R}_2$ of two commuting subalgebras, each of which is a representation of $\mathcal{A}$. Otherwise, $\mathcal{R}$ is *irreducible*. ∎

There may be groups other than $\mathcal{G}(\mathcal{A})$ that have the same Lie algebra, as seen in the example just discussed, but the problem of finding representations of a Lie group is essentially reduced to finding representations of its Lie algebra. For example, the $\mathbf{L}_k$ that generate $SO(3)$ also define a three-dimensional (irreducible) representation of the group $SU(2)$.

❏ **Example 10.7.** A 3×3 generalization of the Pauli matrices is the set $\{\lambda_A\}$ defined by

$$\lambda_0 = \sqrt{\frac{2}{3}}\, \mathbf{1} \qquad \lambda_A = \begin{pmatrix} \sigma_A & 0 \\ 0 & 0 \end{pmatrix} \qquad (A = 1, 2, 3)$$

$$\lambda_4 = \begin{pmatrix} 0 & 0 & 1 \\ 0 & 0 & 0 \\ 1 & 0 & 0 \end{pmatrix} \qquad \lambda_5 = \begin{pmatrix} 0 & 0 & -i \\ 0 & 0 & 0 \\ i & 0 & 0 \end{pmatrix}$$

$$\lambda_6 = \begin{pmatrix} 0 & 0 & 0 \\ 0 & 0 & 1 \\ 0 & 1 & 0 \end{pmatrix} \qquad \lambda_7 = \begin{pmatrix} 0 & 0 & 0 \\ 0 & 0 & -i \\ 0 & i & 0 \end{pmatrix} \qquad (10.15)$$

$$\lambda_8 = \sqrt{\frac{1}{3}} \begin{pmatrix} 1 & 0 & 0 \\ 0 & 1 & 0 \\ 0 & 0 & -2 \end{pmatrix}$$

These matrices are Hermitian, and normalized so that

$$\operatorname{tr} \lambda_A \lambda_B = 2\delta_{AB} \qquad (10.16)$$

consistent with the standard normalization of the Pauli matrices for $A = 1, 2, 3$. ∎

→ **Exercise 10.2.** (i) Show that any 3×3 matrix $\mathbf{M}$ can be expressed as

$$\mathbf{M} = \sum_{A=0}^{8} \alpha_A \lambda_A \quad \text{with} \quad \alpha_A = \tfrac{1}{2} \operatorname{tr} \lambda_A \mathbf{M} \quad (A = 0, 1, \ldots, 8) \tag{10.17}$$

(ii) Show that $\mathbf{M}$ is Hermitian if and only if all the α_A are real.

(iii) Show that any unitary 3×3 matrix $\mathbf{U}$ can be expressed as

$$\mathbf{U} = \exp\left(i \sum_{A=0}^{8} \alpha_A \lambda_A \right) \tag{10.18}$$

with the α_A real ($A = 0, 1, \ldots, 8$). Express $\det \mathbf{U}$ in terms of the α_A.   ☐

It follows from Exercise 10.2 that the $\lambda_0, \lambda_1, \ldots, \lambda_8$ generate the group $U(3)$ of unitary 3×3 matrices, and the $\lambda_1, \ldots, \lambda_8$ generate the subgroup $SU(3)$ of unitary matrices with determinant equal to one. The matrix λ_0 commutes with all matrices since it is a multiple of the identity; the commutators of the $\lambda_1, \ldots, \lambda_8$ have the form

$$[\lambda_A, \lambda_B] = 2i \sum_{C=1}^{8} f_{ABC} \lambda_C \tag{10.19}$$

which defines the structure constants f_{ABC} of $SU(3)$—the factor two on the right-hand side is to insure that the structure constants for the $SU(2)$ subgroup generated by $\lambda_1, \lambda_2, \lambda_3$ are the same as those introduced above ($f_{ABC} = \varepsilon_{ABC}$). We can introduce generators

$$\mathbf{F}_A \equiv \tfrac{1}{2} \lambda_A \tag{10.20}$$

satisfying commutation relations

$$[\mathbf{F}_A, \mathbf{F}_B] = i \sum_{C=1}^{8} f_{ABC} \mathbf{F}_C \tag{10.21}$$

by analogy to the $SU(2)$ generators $\mathbf{S}_k = \tfrac{1}{2} \sigma_k$.

→ **Exercise 10.3.** Show that the structure constants f_{ABC} defined in Eq. (10.19) are antisymmetric under any permutation of the indices A, B, C. Note that while antisymmetry under interchange of A, B follows from the antisymmetry of the commutator, the remaining antisymmetry does not. *Hint.* Use the Jacobi identity introduced in Problem 2.7.   ☐

→ **Exercise 10.4.** The anticommutators of the λ_A can be written as

$$\{\lambda_A, \lambda_B\} = 2\left(\delta_{AB} + \sum_{C=1}^{8} d_{ABC} \lambda_C \right) \tag{10.22}$$

($A, B, C = 0, 1, \ldots, 8$). Show that the d_{ABC} defined here are symmetric under any permutation of the indices A, B, C. *Hint.* Use Eq. (10.17)   ☐

Remark. Thus the product $\lambda_A \lambda_B$ is given by

$$\lambda_A \lambda_B = \delta_{AB} + \sum_{C=1}^{8} (d_{ABC} + i f_{ABC}) \lambda_C$$

This is the $SU(3)$ analog of the relation between Pauli matrices in Exercise 2.8. □

10.2.3 Classification of Lie Algebras

There are classifications of Lie algebras that are parallel to those of groups. For example,

Definition 10.2. The Lie algebra $\mathcal{A}$ is *Abelian* if

$$[\mathbf{X}, \mathbf{Y}] = 0 \tag{10.23}$$

for every $\mathbf{X}$, $\mathbf{Y}$ in $\mathcal{A}$. Evidently the Lie group generated by an Abelian Lie algebra is an Abelian group. A linear subspace $\mathcal{B}$ of the Lie algebra $\mathcal{A}$ is a *subalgebra* of $\mathcal{A}$ if the commutator $[\mathbf{X}, \mathbf{Y}]$ is in $\mathcal{B}$ for every pair $(\mathbf{X}, \mathbf{Y})$ in $\mathcal{B}$. It is an *invariant subalgebra* of $\mathcal{A}$ if $[\mathbf{X}, \mathbf{Y}]$ is in $\mathcal{B}$ for every $\mathbf{Y}$ in $\mathcal{B}$ and every $\mathbf{X}$ in $\mathcal{A}$. ∎

If $\mathcal{B}$ is a subalgebra of $\mathcal{A}$, then the Lie group $\mathcal{L}(\mathcal{B})$ generated by $\mathcal{B}$ is a subgroup of the Lie group $\mathcal{L}(\mathcal{A})$ generated by $\mathcal{A}$, and $\mathcal{L}(\mathcal{B})$ is an invariant subgroup of $\mathcal{L}(\mathcal{A})$ if $\mathcal{B}$ is an invariant subalgebra of $\mathcal{A}$.

❑ **Example 10.8.** The one-dimensional group with elements of the form $(e^{i\alpha})\mathbf{1}$, with α real, is an Abelian invariant subgroup of the group $U(n)$ of unitary $n \times n$ matrices. The corresponding generator, also proportional to the unit matrix, is an Abelian invariant subalgebra of the Lie algebra of $U(n)$, since it commutes with all the generators. ∎

Definition 10.3. The Lie algebra $\mathcal{A}$ is *simple* if it is non-Abelian, and contains no proper invariant subalgebra. It is *semisimple* if it contains no Abelian invariant subalgebra. A Lie group $\mathcal{L}(\mathcal{A})$ is simple (semisimple) if its Lie algebra $\mathcal{A}$ is simple (semisimple). ∎

The Lie algebra of $U(n)$ is not simple, or even semisimple, since the unit matrix T1 defines a (one-dimensional) Abelian invariant subalgebra. However, the Lie algebra of $SU(n)$ excludes the unit matrix, and is simple for any $n \geq 2$. The Lie algebra of $SO(n)$ is also simple, except in the case of $SO(4)$, whose Lie algebra can be expressed as a direct sum

$$\mathcal{A}[SO(4)] \sim \mathcal{A}[SO(3)] \oplus \mathcal{A}[SO(3)]$$

of two $SO(3)$ subalgebras (see Problem 7).

Note, however, that a Lie group can be simple even if it contains discrete Abelian invariant subgroups. For example, the group $SU(n)$ has an Abelian invariant subgroup isomorphic to the cyclic group $\mathbf{Z}_n$ with elements of the form $\mathbf{U} = \omega\mathbf{1}$, with $\omega^n = 1$ so that $\det \mathbf{U} = 1$. In fact, we have

$$U(n)/U(1) \sim SU(n)/\mathbf{Z}_n \tag{10.24}$$

Whether or not a Lie algebra $\mathcal{A}$ is semisimple can be determined by direct computation from the structure constants. Consider the matrix $\mathbf{g} = (g_{jk})$ with

$$g_{jk} = \tfrac{1}{2} \sum_{\ell,m} c_{j\ell}^m c_{mk}^\ell \tag{10.25}$$

$\mathcal{A}$ is semisimple if and only if $\mathbf{g}$ is nonsingular, though we do not prove that here. If $\mathbf{g}$ is nonsingular, it has an inverse

$$\bar{\mathbf{g}} = \mathbf{g}^{-1} = (g^{jk}) \tag{10.26}$$

and we can form the (quadratic) *Casimir operator*

$$\mathbf{C}_2 \equiv \sum_{jk} g^{jk} \mathbf{X}_j \mathbf{X}_k \tag{10.27}$$

→ **Exercise 10.5.**   Show that $\mathbf{C}_2$ commutes with every element of the Lie algebra, i.e., $[\mathbf{C}_2, \mathbf{X}] = 0$ for every $\mathbf{X}$ in $\mathcal{A}$. Thus $\mathbf{g}$ serves as a metric tensor on the Lie algebra.   □

→ **Exercise 10.6.**   Show that a Casimir operator for the Lie algebra of $SU(3)$ as defined by the operators $\mathbf{F}_A$ introduced in Eq. (10.20) is given by

$$\mathbf{C}_2 = \sum_{A=1}^{8} \mathbf{F}_A^2$$

Hint. Use the commutation relations (10.21) and the results of Exercise 10.3.   □

The matrix $\mathbf{g}$ is real and symmetric. If all its eigenvalues have the same sign (here positive), the algebra $\mathcal{A}$ is *compact*, since a constant value of $\mathbf{C}_2$ defines the surface of a generalized ellipsoid in $\mathcal{A}$. If one or more of the eigenvalues is negative, then $\mathcal{A}$ is *noncompact*, since a constant $\mathbf{C}_2$ then defines the surface of a generalized hyperboloid, which is unbounded.

❑ **Example 10.9.**   For the group $SU(2)$, with generators $\mathbf{S}_k = \tfrac{1}{2}\sigma_k$ and commutation relations given by Eq. (10.14), we have

$$g_{jk} = \tfrac{1}{2} \sum_{\ell,m} \varepsilon_{j\ell m} \varepsilon_{\ell m k} = \delta_{jk} \tag{10.28}$$

Thus the Casimir operator is given by

$$\mathbf{C}_2 = \vec{\mathbf{S}} \cdot \vec{\mathbf{S}} = \mathbf{S}_1^2 + \mathbf{S}_2^2 + \mathbf{S}_3^2 \tag{10.29}$$

Hence the Lie algebra is compact. To show that $\mathbf{C}_2$ commutes with all the $\mathbf{S}_k$, consider

$$[\mathbf{S}_k, \mathbf{C}_2] = [\mathbf{S}_k, \sum_\ell \mathbf{S}_\ell \mathbf{S}_\ell] = i \sum_{\ell,m} \varepsilon_{k\ell m} (\mathbf{S}_m \mathbf{S}_\ell + \mathbf{S}_\ell \mathbf{S}_m) = 0 \tag{10.30}$$

since $(\mathbf{S}_m \mathbf{S}_\ell + \mathbf{S}_\ell \mathbf{S}_m)$ is symmetric in ℓ and m. It is not enough to show that $\mathbf{C}_2$ is a multiple of $\mathbf{1}$ in the defining representation.   ∎

❑ **Example 10.10.** A basis of generators of the Lie group $SL(2, R)$ of real unimodular 2×2 matrices is

$$\mathbf{K}_1 = \tfrac{1}{2}i\sigma_3 \qquad \mathbf{K}_2 = \tfrac{1}{2}i\sigma_1 \qquad \mathbf{J}_3 = \tfrac{1}{2}\sigma_2 \tag{10.31}$$

Note the generators must be imaginary so that $\exp(i\xi\mathbf{X})$ will be real, and traceless so the determinant will be $+1$. These generators satisfy

$$[\mathbf{J}_3, \mathbf{K}_1] = i\mathbf{K}_2 \qquad [\mathbf{J}_3, \mathbf{K}_2] = -i\mathbf{K}_1 \qquad [\mathbf{K}_1, \mathbf{K}_2] = -i\mathbf{J}_3 \tag{10.32}$$

These commutation relations are similar to those for the algebra of $SU(2)$, but there is a critical sign difference in the commutator $[\mathbf{K}_1, \mathbf{K}_2]$. As a result of this sign difference, we have $g_{11} = g_{22} = -1$, $g_{33} = 1$, and the quadratic Casimir operator has the form

$$\mathbf{C}_2 = \mathbf{J}_3^2 - \mathbf{K}_1^2 - \mathbf{K}_2^2 \tag{10.33}$$

This Lie algebra is simple, but noncompact. It generates homogeneous Lorentz transformations in the plane (see Problem 25). ∎

❑ **Example 10.11.** Another three-dimensional Lie algebra is defined by elements $\mathbf{P}_1$, $\mathbf{P}_2$, and $\mathbf{J}_3$ that satisfy the commutation relations

$$[\mathbf{J}_3, \mathbf{P}_1] = i\mathbf{P}_2 \qquad [\mathbf{J}_3, \mathbf{P}_2] = -i\mathbf{P}_1 \qquad [\mathbf{P}_1, \mathbf{P}_2] = 0 \tag{10.34}$$

The matrix $\mathbf{g}$ is singular ($g_{33} = 1$, but all other matrix elements vanish), so the Lie algebra is not simple, or even semisimple. Here $\mathbf{P}_1$ and $\mathbf{P}_2$ generate a two-dimensional Abelian invariant subalgebra. ∎

→ **Exercise 10.7.** Show that the group of rotations and translations in two dimensions has a Lie algebra defined by generators $\mathbf{P}_1$, $\mathbf{P}_2$, and $\mathbf{J}_3$, with commutators given by Eq. (10.34). In particular, show that $\mathbf{J}_3$ generates rotations, while $\mathbf{P}_1$ and $\mathbf{P}_2$ generate translations in the plane. Then show that

$$\mathbf{P}^2 = \mathbf{P}_1^2 + \mathbf{P}_2^2 \tag{10.35}$$

commutes with all three generators. □

Remark. The group of translations and rotations in the plane is the *Euclidean group* $\mathbf{E}(2)$. Its generalization to n dimensions is denoted by $\mathbf{E}(n)$. □

Remark. $\mathbf{P}^2$ is a Casimir operator for the Lie algebra of $\mathbf{E}(2)$. In general, any function of the generators that commutes with every element of the Lie algebra is a Casimir operator. □

Definition 10.4. A Lie group is compact or noncompact according to whether or not its Lie algebra is compact or noncompact. ∎

Definition 10.5. The *rank* r of a semisimple Lie algebra $\mathcal{A}$ is the dimension of a maximal Abelian subalgebra of $\mathcal{A}$ (one that is not contained in a larger Abelian subalgebra). Such a maximal Abelian subalgebra is a *Cartan subalgebra* of $\mathcal{A}$. The rank of a semisimple Lie group is the same as the rank of its Lie algebra. ∎

The rank of a semisimple Lie algebra $\mathcal{A}$ is the maximum number of linearly independent commuting elements of the algebra. The rank of the Lie algebra $SU(2)$ is one; any of the $\mathbf{S}_k$ ($\mathbf{S}_3$, say) commutes with itself, but then $\mathbf{S}_1$ and $\mathbf{S}_2$ do not commute with $\mathbf{S}_3$. The rank of the Lie algebra $SU(3)$ is two, since λ_3 and λ_8 commute, but no independent element commutes with both of these. In general, the rank of the Lie algebra of $SU(n)$ is $n - 1$. There are n linearly independent diagonal $n \times n$ matrices. But since

$$\det{(\exp{i\mathbf{X}})} = \exp{(i\,\mathrm{tr}\,\mathbf{X})}$$

we must have $\mathrm{tr}\,\mathbf{X} = 0$ in order to have $\det{(\exp{i\mathbf{X}})} = 1$ This eliminates the unit matrix $\mathbf{1}$, and there are only $n - 1$ independent diagonal traceless $n \times n$ matrices.

Remark.   For a compact semisimple Lie algebra of rank r, there are actually r independent Casimir operators. For the Lie algebras of $GL(n)$ and $U(n)$ of rank n, this is equivalent to the statement that the trace $\mathrm{tr}(\mathbf{A}^k)$ of any power of a matrix $\mathbf{A}$ in invariant under unitary transformations, and the traces are independent for $k = 1, \ldots, n$.   □

It is true that every semisimple Lie algebra is a direct sum of simple Lie algebras, though we again omit the proof. The compact simple Lie algebras have been completely classified since the work of E. Cartan in the late 19th century. There are four infinite series:

A_n $(n = 1, 2, \ldots)$ containing the generators of the group $SU(n + 1)$,

B_n $(n = 1, 2, \ldots)$ containing the generators of the group $SO(2n + 1)$,

C_n $(n = 2, 3, \ldots)$ containing the generators of the group $Sp(2n)$, and

D_n $(n = 3, 4, \ldots)$ containing the generators of the group $SO(2n)$.

Here $Sp(2n)$ denotes the *symplectic group* in $2n$ dimensions, the group of matrices on $\mathcal{V}^{2n}$ that leave invariant an antisymmetric (*symplectic*) quadratic form

$$\langle x, y \rangle \equiv \sum_{k=1}^{n} \left(x_{2k-1}\,y_{2k} - x_{2k}\,y_{2k-1} \right) \tag{10.36}$$

(see also Section 3.5.3 and Problem 3.12). In addition to these infinite series of Lie algebras associated with classical groups, there are five *exceptional algebras* G_2, F_4, E_6, E_7, and E_8. These are of interest in elementary particle theory and string theory; the report by Slansky and books cited in the bibliography contain more details.

Remark.   The subscript on the label denotes the rank of the algebra. We have just shown that the rank of $SU(n)$ is $n - 1$. The rank of the orthogonal groups is examined in Problem 6.   □

Noncompact Lie algebras and groups are also important in physics. The most prominent noncompact group is the group of Lorentz transformations that appears in the special theory of relativity. Both the Lorentz group and its extension to include spacetime translations (the *Poincaré group*), are discussed in Appendix B. Other useful noncompact groups are described in the book by Wybourne cited in the bibliography.

10.3 Representations of Lie Algebras

Irreducible representations of Lie algebras can be constructed on vector spaces in which the basis vectors are eigenstates of the Cartan subalgebra. For simple or semisimple algebras, the other independent elements of the Lie algebra can be expressed in terms of *ladder operators* that transform these eigenstates into eigenstates with new eigenvalues shifted by characteristic values (the *roots* of the algebra) associated with the ladder operators. An example of the use of ladder operators was seen in Problem 7.6, for example. Here we use ladder operators to construct irreducible representations of $SU(2)$ and $SU(3)$.

It is also important to reduce the tensor product, or Kronecker product, of irreducible representations, as introduced in Section 9.4.3. One physical context where this reduction is needed is in the study of shell models of atoms or nuclei. We want to characterize the allowed states of multiparticle systems in which each particle has a definite angular momentum in terms of states of definite total angular momentum, and also in terms of the symmetry of the states under permutations of the particles. The ladder operators also play an important role in this reduction, as we show in examples again chosen from $SU(2)$ and $SU(3)$.

Irreducible representations of any of the simple Lie algebras can be built up from the defining representation using Kronecker products of this representation with itself. Once again we use $SU(2)$ and $SU(3)$ as examples, but in Appendix A we consider the analysis of irreducible representations of the classical Lie algebras in terms of tensors whose components have definite symmetry properties under permutations of the indices. This analysis relates representations of the classical algebras to those of the symmetric groups S_N, which allows us to make use of the graphical methods for the symmetric group introduced in Chapter 9 to study the simple Lie algebras.

10.3.1 Irreducible Representations of $SU(2)$

To see how this method works for the Lie algebra of $SU(2)$ [or $SO(3)$], consider a standard basis $\mathbf{J}_1, \mathbf{J}_2, \mathbf{J}_3$ with commutators

$$[\mathbf{J}_3, \mathbf{J}_1] = i\mathbf{J}_2 \qquad [\mathbf{J}_3, \mathbf{J}_2] = -i\mathbf{J}_1 \qquad [\mathbf{J}_1, \mathbf{J}_2] = i\mathbf{J}_3 \tag{10.37}$$

Introduce the operators

$$\mathbf{J}_\pm \equiv \mathbf{J}_1 \pm i\mathbf{J}_2 \tag{10.38}$$

These satisfy the commutation relations

$$[\mathbf{J}_3, \mathbf{J}_\pm] = \pm\mathbf{J}_\pm \qquad [\mathbf{J}_+, \mathbf{J}_-] = 2\mathbf{J}_3 \tag{10.39}$$

If $|m\rangle$ is an eigenvector of $\mathbf{J}_3$ with eigenvalue m, then the vectors $\mathbf{J}_\pm|m\rangle$ are, if nonzero, eigenvectors of $\mathbf{J}_3$ with eigenvalues $m \pm 1$. That is, the $\mathbf{J}_\pm$ act as ladder operators for $\mathbf{J}_3$, raising ($\mathbf{J}_+$) or lowering ($\mathbf{J}_-$) the eigenvalue of $\mathbf{J}_3$ by one.

The Casimir operator $\mathbf{C}_2$ for the Lie algebra can be expressed as

$$\mathbf{C}_2 = \mathbf{J} \cdot \mathbf{J} = \mathbf{J}^2 = \mathbf{J}_1^2 + \mathbf{J}_2^2 + \mathbf{J}_3^2 = \mathbf{J}_+\mathbf{J}_- + \mathbf{J}_3^2 - \mathbf{J}_3 = \mathbf{J}_-\mathbf{J}_+ + \mathbf{J}_3^2 + \mathbf{J}_3 \tag{10.40}$$

Since this operator commutes with the entire algebra, it will be a multiple of the identity on any irreducible representation, by Schur's lemma. For a unitary representation of the group $SU(2)$, the $\mathbf{J}_k$ must be represented by Hermitian matrices. Then

$$\mathbf{J}_- = \mathbf{J}_+^\dagger \tag{10.41}$$

so that $\mathbf{J}_+\mathbf{J}_-$ and $\mathbf{J}_-\mathbf{J}_+$ are nonnegative operators. Hence the eigenvalues of $\mathbf{J}_3$ must be bounded on an irreducible representation. Thus there must be eigenvectors $|\pm j\rangle$ of $\mathbf{J}_3$ with eigenvalues $\pm j$, such that

$$\mathbf{J}_+|j\rangle = 0 \qquad \mathbf{J}_-|-j\rangle = 0 \tag{10.42}$$

These vectors are also eigenvectors of the Casimir operator, with eigenvalue $j(j+1)$. Since we must be able to reach the vector $|j\rangle$ from the vector $|-j\rangle$ by repeated application of the operator $\mathbf{J}_+$, the difference $2j$ must be an integer. Hence j is restricted to the set

$$j = 0, \tfrac{1}{2}, 1, \tfrac{3}{2}, 2, \ldots \tag{10.43}$$

of integer or half-integer values, and the dimension of the representation is $2j+1$. Hence for any dimension $n = 2j+1$, there is an irreducible representation $\Gamma^{(j)}$ of $SU(2)$, with basis vectors $\{|j\,m\rangle, m = -j, -j+1, \ldots, j-1, j\}$ such that

$$\mathbf{J}^2|j\,m\rangle = j(j+1)|j\,m\rangle \qquad \mathbf{J}_3|j\,m\rangle = m|j\,m\rangle \tag{10.44}$$

Also,

$$\mathbf{J}_\pm|j\,m\rangle = C_\pm(j,m)|j\,m\pm 1\rangle \tag{10.45}$$

with coefficients $C_\pm(jm)$ determined using Eqs. (10.40) and (10.41). This gives

$$|C_\pm(j,m)|^2 = j(j+1) - m(m\pm 1) = (j\mp m)(j\pm m+1) \tag{10.46}$$

The standard phase convention is to choose the coefficients $C_\pm(jm)$ to be real and positive, so we have

$$C_\pm(j,m) = \sqrt{(j\mp m)(j\pm m+1)} \tag{10.47}$$

In the context of quantum mechanics, the operators $\mathbf{J}$ correspond to the components of the angular momentum of a system (in units of Planck's constant $\hbar$). For a system in a state $|jm\rangle$, j is called the *angular momentum*, or simply *spin*, of the system. The eigenvalue m of $\mathbf{J}_3$ is the Z-component of the angular momentum. It is sometimes called the *magnetic quantum number*, because the energy levels of a system in a magnetic field are split by the interaction of the magnetic moment of the system with the magnetic field, and the interaction energy is proportional to the component of the angular momentum along the magnetic field direction.

10.3.2 Addition of Angular Momenta

Consider two quantum-mechanical systems with angular momentum states belonging to irreducible representations $\Gamma^{(j_1)}$ and $\Gamma^{(j_2)}$ of $SU(2)$ defined on vector spaces $\mathcal{V}^{(1)}$ and $\mathcal{V}^{(2)}$ of dimension $2j_1 + 1$ and $2j_2 + 1$, respectively. Let $\mathbf{J}^{(1)}$ and $\mathbf{J}^{(2)}$ denote the angular momentum operators acting on $\mathcal{V}^{(1)}$ and $\mathcal{V}^{(2)}$. If we now consider these two systems as a single compound system defined on the product space $\mathcal{V}^{(1)} \otimes \mathcal{V}^{(2)}$, we want to find states that correspond to definite values of the total angular momentum

$$\mathbf{J} = \mathbf{J}^{(1)} + \mathbf{J}^{(2)} \tag{10.48}$$

These states are obtained by reducing the representation $\Gamma^{(j_1)} \otimes \Gamma^{(j_2)}$ in terms of irreducible representations of $SU(2)$.

❑ **Example 10.12.** An elementary example is the problem of combining states of the orbital angular momentum $\mathbf{L}$ of a particle with states of the intrinsic spin $\mathbf{S}$ (often $s = \frac{1}{2}$) to find states of total angular momentum $\mathbf{J} = \mathbf{L} + \mathbf{S}$. ∎

❑ **Example 10.13.** An example often encountered in shell models of atomic or nuclear structure is the problem of finding the allowed angular momentum states of two or more particles with angular momentum j in a single shell, subject to the constraint of the Pauli principle that the state be antisymmetric under exchange of any pair of particles. ∎

Suppose then that we start with two sets of states, $\{|j_1\ m_1\rangle\ (m_1 = j_1, j_1 - 1, \ldots, -j_1)\}$ (angular momentum j_1) and $\{|j_2\ m_2\rangle\ (m_2 = j_2, j_2 - 1, \ldots, -j_2)\}$ (angular momentum j_2), on which

$$\mathbf{J}^{(k)} \cdot \mathbf{J}^{(k)}|j_k\ m_k\rangle = j_k(j_k + 1)|j_k\ m_k\rangle \qquad J_3^{(k)}|j_k\ m_k\rangle = m_k|j_k\ m_k\rangle \tag{10.49}$$

$(k = 1, 2)$. In the tensor product space, define states

$$|j_1\ j_2\ ; m_1\ m_2\rangle \equiv |j_1\ m_1\rangle \otimes |j_2\ m_2\rangle \tag{10.50}$$

We are looking for states $|j_1\ j_2\ J; M\rangle$ that belong to irreducible representations of $SU(2)$, i.e., states of definite total angular momentum. These must satisfy

$$\mathbf{J} \cdot \mathbf{J}\,|j_1\ j_2\ J; M\rangle = J(J + 1)|j_1\ j_2\ J; M\rangle \tag{10.51}$$

$$J_3\,|j_1\ j_2\ J; M\rangle = M|j_1\ j_2\ J; M\rangle \tag{10.52}$$

The $|j_1\ j_2\ J; M\rangle$ can be expanded in terms of the $|j_1\ j_2\ ; m_1\ m_2\rangle$ as

$$|j_1\ j_2\ J; M\rangle = \sum_{m_1, m_2} |j_1\ j_2\ ; m_1\ m_2\rangle\langle j_1\ j_2\ ; m_1\ m_2|j_1\ j_2\ J; M\rangle \tag{10.53}$$

where j_1 and j_2 are often omitted when their values are clearly understood. The coefficients in this expansion, also denoted by

$$C(j_1\ j_2\ J; m_1\ m_2\ M) = \langle j_1\ j_2\ ; m_1\ m_2|j_1\ j_2\ J; M\rangle \tag{10.54}$$

are called *vector coupling coefficients*, or *Clebsch–Gordan coefficients*.

To determine the allowed values of total angular momentum J, note first that

$$M = m_1 + m_2 \tag{10.55}$$

in view of Eq. (10.48). The largest value of m_1 (m_2) is j_1 (j_2); hence the largest value of M is $j_1 + j_2$. Then the largest value of J is also given by

$$J_{\max} = j_1 + j_2 \tag{10.56}$$

and we have

$$|J_{\max} \, J_{\max}\rangle = |J = J_{\max} \, M = J_{\max}\rangle = |j_1 \, j_2\rangle \tag{10.57}$$

There are two independent states with $M = j_1 + j_2 - 1$, namely $|j_1 \, j_2 - 1\rangle$ and $|j_1 - 1 \, j_2\rangle$. One linear combination of these is the state $|J = J_{\max} \, M = J_{\max} - 1\rangle$, and it is given by

$$
\begin{aligned}
\mathbf{J}_- |J_{\max} \, J_{\max}\rangle &= C_-(J_{\max}, J_{\max})|J_{\max} \, J_{\max} - 1\rangle \\
&= C_-(j_2, j_2)|j_1 \, j_2 - 1\rangle + C_-(j_1, j_1)|j_1 - 1 \, j_2\rangle
\end{aligned}
\tag{10.58}
$$

where the coefficients $C_-(J, M)$ were introduced in Eq. (10.47). The state orthogonal to this must belong to $J = J_{\max} - 1$. It can be constructed either by orthogonality or by noting that

$$\mathbf{J}_+ |J_{\max} - 1 \, J_{\max} - 1\rangle = 0 \tag{10.59}$$

We can continue to look at states with smaller values of M, and we find that the number of states increases by one each time we lower M by one until we reach $M = |j_1 - j_2|$. Thus the smallest value of J is given by

$$J_{\min} = |j_1 - j_2| \tag{10.60}$$

and the allowed values of J are given by

$$J = j_1 + j_2, j_1 + j_2 - 1, \ldots, |j_1 - j_2| + 1, |j_1 - j_2| \tag{10.61}$$

Thus j_1, j_2 and J must satisfy the triangle inequality even while restricted to integer or half-integer values. As a dimension check, note that (and prove as an exercise).

$$\sum_{J=|j_1-j_2|}^{j_1+j_2} (2J + 1) = (2j_1 + 1)(2j_2 + 1) \tag{10.62}$$

→ **Exercise 10.8.** Show that

$$\mathbf{J}_- |j \, j\rangle = \sqrt{2j}|j \, j - 1\rangle \qquad \text{and} \qquad \mathbf{J}_+ |j \, j - 1\rangle = \sqrt{2j}|j \, j\rangle$$

This result is very useful, and it is used in the following examples.   □

❑ **Example 10.14.** For two spin-$\frac{1}{2}$ particles, the allowed values of total spin are $S = 0, 1$. The eigenstates $|S\ M\rangle$ of total spin can be expressed in terms of eigenstates $|m_1\ m_2\rangle$ of individual spin Z-components as

$$|1\ 1\rangle = |\tfrac{1}{2}\ \tfrac{1}{2}\rangle \qquad |1\ 0\rangle = \sqrt{\tfrac{1}{2}}\,(|\tfrac{1}{2}\ -\tfrac{1}{2}\rangle + |-\tfrac{1}{2}\ \tfrac{1}{2}\rangle) \qquad |1\ -1\rangle = |-\tfrac{1}{2}\ -\tfrac{1}{2}\rangle$$

$$|0\ 0\rangle = \sqrt{\tfrac{1}{2}}\,(|\tfrac{1}{2}\ -\tfrac{1}{2}\rangle - |-\tfrac{1}{2}\ \tfrac{1}{2}\rangle) \tag{10.63}$$

Note that the $S = 1$ states are symmetric, the $S = 0$ state antisymmetric.   ∎

❑ **Example 10.15.** Consider a spin-$\frac{1}{2}$ particle with orbital angular momentum $\ell = 1$ (an electron in an excited state of a hydrogen atom, for example). The total angular momentum j of the particle can be $\frac{1}{2}$ or $\frac{3}{2}$. To construct the states $|j\ m\rangle$ from the states $|m_\ell\ m_s\rangle$, we can start with

$$|\tfrac{3}{2}\ \tfrac{3}{2}\rangle = |1\ \tfrac{1}{2}\rangle \tag{10.64}$$

Then , using Eq. (10.47), we have

$$\mathbf{J}_-|\tfrac{3}{2}\ \tfrac{3}{2}\rangle = \sqrt{3}\,|\tfrac{3}{2}\ \tfrac{1}{2}\rangle = |1\ -\tfrac{1}{2}\rangle + \sqrt{2}\,|0\ \tfrac{1}{2}\rangle \tag{10.65}$$

As noted above, the state $|\tfrac{1}{2}\ \tfrac{1}{2}\rangle$ can then be constructed either by orthogonality or by the requirement that $\mathbf{J}_+|\tfrac{1}{2}\ \tfrac{1}{2}\rangle$ must be zero. The result is

$$|\tfrac{1}{2}\ \tfrac{1}{2}\rangle = \sqrt{\tfrac{2}{3}}\,|1\ -\tfrac{1}{2}\rangle - \sqrt{\tfrac{1}{3}}\,|0\ \tfrac{1}{2}\rangle) \tag{10.66}$$

Here we have used the convention (*Condon–Shortley convention*) that the Clebsch–Gordan coefficient $C(j_1\ j_2\ J; j_1\ J - j_1\ J)$ must be positive (provided that j_1, j_2 and J satisfy the triangle inequality, of course). The remaining states can then be constructed by applying the lowering operator $\mathbf{J}_-$ to the states already given. The construction of the states for $j = \ell \pm \frac{1}{2}$ for arbitrary ℓ is left to Problem 13.   ∎

❑ **Example 10.16.** The allowed values of total angular momentum from combining $j_1 = \frac{3}{2}$ and $j_2 = 1$ are $j = \frac{5}{2}, \frac{3}{2}, \frac{1}{2}$. To construct states $|j\ m\rangle$ in terms of the states $|m_1\ m_2\rangle$, we can follow the line outlined in the previous examples. However, if we only need the state $|\tfrac{1}{2}\ \tfrac{1}{2}\rangle$, for example, then a more direct route is to let

$$|\tfrac{1}{2}\ \tfrac{1}{2}\rangle = a|\tfrac{3}{2}\ -1\rangle + b|\tfrac{1}{2}\ 0\rangle + c|-\tfrac{1}{2}\ 1\rangle \tag{10.67}$$

Then $\mathbf{J}_+|\tfrac{1}{2}\ \tfrac{1}{2}\rangle = 0$ requires

$$(\sqrt{2}a + \sqrt{3}b)|\tfrac{3}{2}\ 0\rangle + (\sqrt{2}b + 2c)|\tfrac{1}{2}\ 1\rangle = 0 \tag{10.68}$$

Thus we need

$$b = \sqrt{2}c \qquad \text{and} \qquad a = \sqrt{3}c$$

so that after normalizing the state, we have

$$|\tfrac{1}{2}\ \tfrac{1}{2}\rangle = \sqrt{\tfrac{1}{2}}\,|\tfrac{3}{2}\ -1\rangle - \sqrt{\tfrac{1}{3}}\,|\tfrac{1}{2}\ 0\rangle + \sqrt{\tfrac{1}{6}}\,|-\tfrac{1}{2}\ 1\rangle \tag{10.69}$$

The remaining states are left to Problem 14.   ∎

10.3.3 $\mathcal{S}_N$ and the Irreducible Representations of $SU(2)$

The irreducible representations of $SU(2)$ can also be constructed from tensor products of the fundamental two-dimensional (spin-$\frac{1}{2}$) representation with itself. Example 10.14 shows that the product of two spin-$\frac{1}{2}$ representations leads to the $j = 1$ and $j = 0$ representations.

For three spin-$\frac{1}{2}$ particles, the total angular momentum $j = \frac{1}{2}, \frac{3}{2}$. It is clear that the $j = \frac{3}{2}$ representation is symmetric in the three particles. There are two independent $j = \frac{1}{2}$ representations, one in which the first two particles have total spin $j_{12} = 0$ and another in which they have $j_{12} = 1$. These two representations will transform under the (21) mixed symmetry representation of $\mathcal{S}_3$ under permutations of the three particles.

To see this explicitly, start with the states $|m_1\, m_2\, m_3\rangle$ defined by

$$|m_1\, m_2\, m_3\rangle = |\tfrac{1}{2}\, m_1\rangle \otimes |\tfrac{1}{2}\, m_2\rangle \otimes |\tfrac{1}{2}\, m_3\rangle \tag{10.70}$$

($j_1 = j_2 = j_3 = \frac{1}{2}$ is understood) with $m_1, m_2, m_3, = \pm\frac{1}{2}$, so there are eight states in total. The states with definite total angular momentum can be denoted by $|\frac{3}{2}\, m\rangle$ $(m = \pm\frac{1}{2}, \pm\frac{3}{2})$ and $|\frac{1}{2}\, (j_{12})\, m\rangle$ $(m = \pm\frac{1}{2})$, where we need the extra label $j_{12} = 0, 1$ to distinguish the independent $j = \frac{1}{2}$ states. We have

$$|\tfrac{3}{2}\, \tfrac{3}{2}\rangle = |\tfrac{1}{2}\, \tfrac{1}{2}\, \tfrac{1}{2}\rangle \qquad\qquad |\tfrac{3}{2}\, -\tfrac{3}{2}\rangle = |-\tfrac{1}{2}\, -\tfrac{1}{2}\, -\tfrac{1}{2}\rangle$$

$$|\tfrac{3}{2}\, \tfrac{1}{2}\rangle = \sqrt{\tfrac{1}{3}}\left(|\tfrac{1}{2}\, \tfrac{1}{2}\, -\tfrac{1}{2}\rangle + |\tfrac{1}{2}\, -\tfrac{1}{2}\, \tfrac{1}{2}\rangle + |\tfrac{1}{2}\, \tfrac{1}{2}\, -\tfrac{1}{2}\rangle\right) \tag{10.71}$$

$$|\tfrac{3}{2}\, -\tfrac{1}{2}\rangle = \sqrt{\tfrac{1}{3}}\left(|\tfrac{1}{2}\, -\tfrac{1}{2}\, -\tfrac{1}{2}\rangle + |-\tfrac{1}{2}\, \tfrac{1}{2}\, -\tfrac{1}{2}\rangle + |-\tfrac{1}{2}\, -\tfrac{1}{2}\, \tfrac{1}{2}\rangle\right)$$

Also,

$$|\tfrac{1}{2}(1)\, \tfrac{1}{2}\rangle = \sqrt{\tfrac{1}{6}}\left(2|\tfrac{1}{2}\, \tfrac{1}{2}\, -\tfrac{1}{2}\rangle - |\tfrac{1}{2}\, -\tfrac{1}{2}\, \tfrac{1}{2}\rangle - |\tfrac{1}{2}\, \tfrac{1}{2}\, -\tfrac{1}{2}\rangle\right) \tag{10.72}$$

$$|\tfrac{1}{2}(1)\, -\tfrac{1}{2}\rangle = \sqrt{\tfrac{1}{6}}\left(|\tfrac{1}{2}\, -\tfrac{1}{2}\, -\tfrac{1}{2}\rangle + |-\tfrac{1}{2}\, \tfrac{1}{2}\, -\tfrac{1}{2}\rangle - 2|-\tfrac{1}{2}\, -\tfrac{1}{2}\, \tfrac{1}{2}\rangle\right)$$

and finally,

$$|\tfrac{1}{2}(0)\, \pm\tfrac{1}{2}\rangle = \sqrt{\tfrac{1}{2}}\left(|\tfrac{1}{2}\, -\tfrac{1}{2}\, \pm\tfrac{1}{2}\rangle - |-\tfrac{1}{2}\, \tfrac{1}{2}\, \pm\tfrac{1}{2}\rangle\right) \tag{10.73}$$

→ **Exercise 10.9.** Find the matrices representing the permutations of the three particles in the two-dimensional space spanned by $|\frac{1}{2}(0)\, \frac{1}{2}\rangle$ and $|\frac{1}{2}(1)\, \frac{1}{2}\rangle$ ☐

In general, the allowed values of total angular momentum j for N spin-$\frac{1}{2}$ particles are

$$j = \tfrac{1}{2}\, N, \tfrac{1}{2}\, N - 1, \ldots, 0\left(\tfrac{1}{2}\right) \tag{10.74}$$

where $j = 0\left(\frac{1}{2}\right)$ is the smallest allowed value for N even (odd). Each value of j corresponds to a definite irreducible representation of $\mathcal{S}_N$ and thus to a Young diagram as introduced in Chapter 9. The distinct N-particle states associated with a given Young diagram can be enumerated by counting the number distinct ways of assigning a single-particle label to each node of the diagram, noting that the N-particle state must be symmetric (antisymmetric) with respect to permutations of labels on the same row (column) of the diagram.

In this counting, we can have n_+ states $|\frac{1}{2}\ \frac{1}{2}\rangle$ and n_- states $|\frac{1}{2}\ -\frac{1}{2}\rangle$ in the first row of the diagram and $n_2 \le n_+$ states $|\frac{1}{2}\ -\frac{1}{2}\rangle$ in the second row of the diagram. If we let "$\pm$" be shorthand labels for the single-particle states $|\frac{1}{2}\ \pm\frac{1}{2}\rangle$, then the allowed states correspond to assigning "+" to the first n_+ nodes in the first row, "$-$" to the remaining n_- nodes in the first row, and "$-$" to the n_2 nodes in the second row, which necessarily have a "+" assigned to the nodes above them in the first row. To illustrate for $N = 3$, we have

Young diagrams with more than two rows, (i.e., partitions into more than two parts) do not correspond to irreducible representations of $SU(2)$, as there are only two single-particle states with which to fill a column. Thus the partitions of N that correspond to irreducible representations of $SU(2)$ have the form $(m_1\ m_2)$ with $m_1 \ge m_2$ and $m_1 + m_2 = N$, so that $m_1 \ge \frac{1}{2}N$. Such a partition corresponds to an angular momentum $j = m_1 - m_2$, and the allowed values of j are precisely those given by Eq. (10.74).

Addition of angular momenta can also be expressed in terms of Young diagrams. Two irreducible representations with angular momenta j_1 and j_2 can be described by partitions of $n_1 = 2j_1$ and $n_2 = 2j_2$ with a single part, i.e., whose Young diagrams have only a single row. The tensor product of these representations is described by a set of partitions of $n_1 + n_2$ obtained by taking the outer product $\Gamma^{(n_1)} \circ \Gamma^{(n_2)}$ of the corresponding irreducible representations of the symmetric groups $\mathcal{S}_{n_1}$ and $\mathcal{S}_{n_2}$ and discarding partitions with more than two rows (in fact there are none of these). This gives exactly the result of Eq. (10.61).

❏ **Example 10.17.** The outer product corresponding to the addition of $j_1 = j_2 = 1$ was given in Exercise 9.22. The addition of $j_1 = 2$ and $j_2 = \frac{3}{2}$, for example, is shown graphically as

$$\square\square\square\square \times \square\square\square = \square\square\square\square\square\square\square + \begin{array}{c}\square\square\square\square\square\square\\\square\end{array}$$

$$+ \begin{array}{c}\square\square\square\square\square\\\square\square\end{array} + \begin{array}{c}\square\square\square\square\\\square\square\square\end{array}$$

What values of total angular momentum J are present on the right-hand side? ∎

The use of methods based on Young diagrams is somewhat superfluous for $SU(2)$, since the Lie algebra has rank one, and all the irreducible representations are simply derived from the spin-$\frac{1}{2}$ representation. However, the graphical methods are more essential when we deal to $SU(3)$ and even larger groups; understanding how they work in $SU(2)$ may be helpful then.

Beyond that, understanding the symmetry properties of tensor products is important because of the Pauli exclusion principle that forces the total state of a system of identical spin-$\frac{1}{2}$ particles to be antisymmetric under permutations of the particles. Since a typical state is a product of a space wave function and a spin function (and, in the case of nuclei, an isospin function), it is not necessary that each component be antisymmetric. Thus it is necessary to analyze the symmetry properties when identical angular momenta for two or more particles are combined. Here we leave this to Problems 16 and 17.

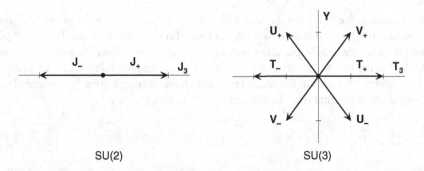

SU(2)   SU(3)

Figure 10.1: Roots and root vectors for the Lie algebras of $SU(2)$ and $SU(3)$. The root vectors for $SU(2)$ are defined in Eq. (10.38); those for $SU(3)$ in Eq. (10.76).

10.3.4 Irreducible Representations of $SU(3)$

The method used in Section 10.3.1 to generate irreducible representations of $SU(2)$ can be generalized to larger Lie algebras. For example, consider the Lie algebra of $SU(3)$ with standard generators $\mathbf{F}_A$ defined by Eq. (10.20) ($A = 1, \ldots, 8$). Here there are two commuting generators, and we can choose, for example,

$$\mathbf{T}_3 = \mathbf{F}_3 \quad \text{and} \quad \mathbf{Y} = \tfrac{2}{\sqrt{3}}\,\mathbf{F}_8 \tag{10.75}$$

Then we can use simultaneous eigenstates of $\mathbf{T}_3$ and $\mathbf{Y}$ as basis vectors in a representation; these can be denoted by $|\,T_3\,Y\,\rangle$. The eigenvalue pairs (T_3, Y) of states in a representation are the *weights* of the representation. In some irreducible representations there are multiple states with the same weight; then further labels are needed to identify the states.

The remaining generators can be rearranged to form ladder operators for the eigenvalues of $\mathbf{T}_3$ and $\mathbf{Y}$. Define

$$\mathbf{T}_\pm = \mathbf{F}_1 \pm i\mathbf{F}_2 \qquad \mathbf{U}_\pm = \mathbf{F}_6 \pm i\mathbf{F}_7 \qquad \mathbf{V}_\pm = \mathbf{F}_4 \pm i\mathbf{F}_5 \tag{10.76}$$

Then

$$[\mathbf{T}_3, \mathbf{T}_\pm] = \pm\mathbf{T}_\pm \qquad [\mathbf{T}_3, \mathbf{U}_\pm] = \mp\tfrac{1}{2}\mathbf{U}_\pm \qquad [\mathbf{T}_3, \mathbf{V}_\pm] = \pm\tfrac{1}{2}\mathbf{V}_\pm \tag{10.77}$$

and

$$[\mathbf{Y}, \mathbf{T}_\pm] = 0 \qquad [\mathbf{Y}, \mathbf{U}_\pm] = \pm\mathbf{U}_\pm \qquad [\mathbf{Y}, \mathbf{V}_\pm] = \pm\mathbf{V}_\pm \tag{10.78}$$

The generators $\mathbf{T}_\pm, \mathbf{T}_3$ define an $SU(2)$ subalgebra of $SU(3)$ that commutes with $\mathbf{Y}$; hence the $\mathbf{T}_\pm$ change the eigenvalue of $\mathbf{T}_3$ by ± 1 and leave the eigenvalue of $\mathbf{Y}$ unchanged. The generators $\mathbf{U}_\pm$ and $\mathbf{V}_\pm$ change the eigenvalue of $\mathbf{Y}$ by ± 1; $\mathbf{U}_-$ and $\mathbf{V}_+$ ($\mathbf{U}_+$ and $\mathbf{V}_-$) change the eigenvalue of $\mathbf{T}_3$ by $+\tfrac{1}{2}$ ($-\tfrac{1}{2}$). The changes in the eigenvalues induced by the ladder operators are the *roots* of the algebra. The ladder operators are also known as *root vectors*; there is one for each root. The roots for $SU(2)$ and $SU(3)$ are shown in Fig. 10.1.

Remark. The notation for the generators is based on the historical development of $SU(3)$ as an approximate symmetry of the strong interactions. The $SU(2)$ subalgebra generated by $\mathbf{T}_\pm, \mathbf{T}_3$ is the *isotopic spin* (or simply *isospin*) introduced by Heisenberg in the 1930s as an approximate symmetry of protons and neutrons in atomic nuclei. The later extension to $SU(3)$ was suggested by the discovery of "strange" baryons and mesons that fit into multiplets corresponding to irreducible representations of $SU(3)$ described here and in Exercises 10.13 and 10.16. The generator $\mathbf{Y}$ corresponds to what is now known as *hypercharge*. □

➜ **Exercise 10.10.** Define operators $\mathbf{U}_3, \mathbf{V}_3$ by

$$[\mathbf{U}_+, \mathbf{U}_-] = 2\mathbf{U}_3 \qquad [\mathbf{V}_+, \mathbf{V}_-] = 2\mathbf{V}_3 \tag{10.79}$$

$\mathbf{U}_\pm, \mathbf{U}_3$ ($\mathbf{V}_\pm, \mathbf{V}_3$) define an $SU(2)$ subalgebra of $SU(3)$ called U-spin (V-spin).

(i) Express $\mathbf{U}_3$ and $\mathbf{V}_3$ in terms of $\mathbf{T}_3$ and $\mathbf{Y}$.

(ii) There are "hypercharge" operators $\mathbf{Y}_U$ ($\mathbf{Y}_V$) that commute with all components of U-spin (V-spin). Express these operators in terms of $\mathbf{T}_3$ and $\mathbf{Y}$. □

To construct irreducible representations of $SU(3)$, we can start with the three-dimensional representation defining the λ_A in Example 10.7. This is the *fundamental representation* of $SU(3)$, often denoted simply by its dimension **3**. The basis vectors of this representation are eigenstates of $\mathbf{T}_3$ and $\mathbf{Y}$, and can be expressed as

$$u = |\tfrac{1}{2} \ \tfrac{1}{3}\rangle \qquad d = |-\tfrac{1}{2} \ \tfrac{1}{3}\rangle \qquad s = |0 \ -\tfrac{2}{3}\rangle$$

where u, d, s denote the quarks corresponding to these states in the flavor $SU(3)$ classification of quarks.

The complex conjugate of a group representation is obtained by changing all generators $\mathbf{F}_A$ to $-\mathbf{F}_A^*$. Thus we have the *conjugate fundamental representation* **3*** with basis vectors

$$\bar{s} = |0 \ \tfrac{2}{3}\rangle \qquad \bar{d} = |\tfrac{1}{2} \ -\tfrac{1}{3}\rangle \qquad \bar{u} = |-\tfrac{1}{2} \ -\tfrac{1}{3}\rangle$$

with $\bar{u}, \bar{d}, \bar{s}$ corresponding to antiquarks in the flavor $SU(3)$ scheme.

The basis vectors of a representation of a Lie algebra can be shown graphically as points on a *weight diagram*. The weight diagrams for the irreducible representations **3** and **3*** are shown in Fig. 10.2, together with the matrix elements of the root vectors $\mathbf{T}_\pm, \mathbf{U}_\pm,$ and $\mathbf{V}_\pm$.

Remark. We adopt the sign convention that the matrix elements of $\mathbf{T}_\pm$ and $\mathbf{U}_\pm$ must be positive. The matrix elements of $\mathbf{V}_\pm$ are then determined by the commutation relations derived in the following exercise. □

➜ **Exercise 10.11.** Show that the ladder operators satisfy the commutation relations

$$[\mathbf{T}_+, \mathbf{U}_+] = \mathbf{V}_+ \qquad [\mathbf{T}_+, \mathbf{V}_+] = 0 \qquad [\mathbf{U}_+, \mathbf{V}_+] = 0$$

$$[\mathbf{T}_+, \mathbf{U}_-] = 0 \qquad [\mathbf{T}_+, \mathbf{V}_-] = -\mathbf{U}_- \qquad [\mathbf{U}_+, \mathbf{V}_-] = \mathbf{T}_- \tag{10.80}$$

and then derive the signs for the matrix elements of $\mathbf{V}_\pm$ in the representations **3** and **3***. □

To proceed further, note that in any irreducible representation Γ of $SU(3)$, there must be an eigenvector $|w_*(\Gamma)\rangle$ of $\mathbf{T}_3$ and $\mathbf{Y}$ such that

$$\mathbf{T}_+|w_*(\Gamma)\rangle = 0 = \mathbf{U}_+|w_*(\Gamma)\rangle \tag{10.81}$$

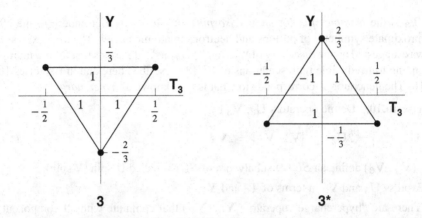

Figure 10.2: Weight diagrams for the irreducible representations **3** and **3*** of $SU(3)$. Also shown are the matrix elements of the root vectors $\mathbf{T}_\pm$, $\mathbf{U}_\pm$, and $\mathbf{V}_\pm$ in each representation.

The action of $\mathbf{V}_+$ then follows from the commutation rule in Eq. (10.81),

$$\mathbf{V}_+|w_*(\Gamma)\rangle = [\mathbf{T}_+, \mathbf{U}_+]|w_*(\Gamma)\rangle = 0 \tag{10.82}$$

$w_*(\Gamma)$ is the *highest weight* of Γ, and $|w_*(\Gamma)\rangle$ is the state of highest weight. The states of highest weight for the irreducible representations **3** and **3*** are

$$|w_*(\mathbf{3})\rangle = u = |\tfrac{1}{2}\,\tfrac{1}{3}\rangle \quad \text{and} \quad |w_*(\mathbf{3}^*)\rangle = \bar{s} = |0\,\tfrac{2}{3}\rangle \tag{10.83}$$

respectively, so that

$$w_*(\mathbf{3}) = (\tfrac{1}{2}\,\tfrac{1}{3}) \quad \text{and} \quad w_*(\mathbf{3}^*) = (0\,\tfrac{2}{3}) \tag{10.84}$$

Now consider the tensor product of two copies of the fundamental representation **3**. This nine-dimensional representation splits into a six-dimensional symmetric part and a three-dimensional antisymmetric part. Each part is in fact irreducible, and looking at the weight diagrams in Fig. 10.2 shows that the antisymmetric part is equivalent to the representation **3***, while the symmetric part defines a new irreducible representation **6** with highest weight

$$w_*(\mathbf{6}) = (1\,\tfrac{2}{3}) \tag{10.85}$$

This reduction is expressed simply as

$$\mathbf{3} \otimes \mathbf{3} = \mathbf{6} \oplus \mathbf{3}^* \tag{10.86}$$

→ **Exercise 10.12.** Find the eigenstates $|T_3\,Y\rangle$ of $\mathbf{T}_3$ and Y in the representations **6** and **3*** contained in the product $\mathbf{3} \otimes \mathbf{3}$ in terms of "quark states" $|q_a\,q_b\rangle$. □

Remark. If states in representation **3** are identified as "quark states," and the states in **3*** as "antiquark states," then the states in **6** are identified as "diquark states." □

→ **Exercise 10.13.**   Draw the weight diagrams for the irreducible representations **6** and **6*** of $SU(3)$. Indicate graphically the effects of the operators $\mathbf{T}_\pm$, $\mathbf{U}_\pm$, and $\mathbf{V}_\pm$ on the eigenstates of T_3 and Y in each of these representations, including numerical values of the relevant matrix elements as in Fig. 10.2.   □

In general, the tensor product of N copies of the fundamental representation splits into irreducible representations $\Gamma(m_1\ m_2\ m_3)$ of $SU(3)$ corresponding to partitions $(m_1\ m_2\ m_3)$ of N into not more than three parts, as explained in greater detail in Appendix A. In fact, since the antisymmetric product of three fundamental representations is equivalent to the identity representation, $\Gamma(m_1, m_2, m_3)$ is equivalent to representation $\Gamma(m_1 - m_3, m_2 - m_3)$ associated with a partition into at most two parts. This representation is contained in the product of $\mu_1 = m_1 - m_2$ copies of **3** and $\mu_2 = m_2 - m_3$ copies of **3***, and its highest weight $w(\mu_1, \mu_2)$ is given by

$$w(\mu_1, \mu_2) = \mu_1 w_*(\mathbf{3}) + \mu_2 w_*(\mathbf{3}^*) \tag{10.87}$$

The dimension of the irreducible representation $\Gamma(m_1, m_2)$ is equal to the number of ways of filling the diagram $\mathcal{Y}^{(m_1\ m_2)}$ with the numbers 1, 2, 3 such that the numbers are (i) nondecreasing across each row (since other orderings of the indices will be included after symmetrization of the row indices), and (ii) strictly increasing down each column (since antisymmetrization will eliminate indices that are the same in two nodes of the column).

❑ **Example 10.18.**   Consider the irreducible representation of $SU(3)$ associated with the partition (21). Using the quark labels (with $u \sim 1$, $d \sim 2$, and $s \sim 3$), we have

$$\begin{array}{c}\square\square\\\square\end{array} \sim \begin{array}{|c|c|}\hline u & u \\\hline d \\\hline\end{array} + \begin{array}{|c|c|}\hline u & u \\\hline s \\\hline\end{array} + \begin{array}{|c|c|}\hline u & d \\\hline d \\\hline\end{array} + \begin{array}{|c|c|}\hline u & d \\\hline s \\\hline\end{array} + \begin{array}{|c|c|}\hline u & s \\\hline d \\\hline\end{array} + \begin{array}{|c|c|}\hline u & s \\\hline s \\\hline\end{array} + \begin{array}{|c|c|}\hline d & d \\\hline s \\\hline\end{array} + \begin{array}{|c|c|}\hline d & s \\\hline s \\\hline\end{array}$$

so the corresponding irreducible representation $\Gamma(2,1)$ is eight-dimensional (hence it is known as the *octet* representation **8**). The eight tableaux correspond to the baryon octet— can you identify the proton and neutron? the remaining members of the octet?   ∎

The weight diagram for the octet representation is shown in Fig. 10.3, together with the matrix elements of the root vectors, where again we adopt the sign convention that matrix elements of $\mathbf{T}_\pm$ and $\mathbf{U}_\pm$ must be positive. Note that there are two states at the center of the diagram, corresponding to the weight (0 0). These can be distinguished by the total isospin quantum number T; one state is the $T_3 = 0$ member of a $T = 1$ triplet and the other has $T = 0$. We denote these states by $|(1)0\ 0\rangle$ and $|(0)0\ 0\rangle$, respectively.

→ **Exercise 10.14.**   Derive the matrix elements in Fig. 10.3.   □

→ **Exercise 10.15.**   The two states with (0 0) can also be distinguished as eigenstates of U-spin with $U = 0, 1$ or V-spin with $V = 0, 1$ (see Exercise 10.10). Express these eigenstates in terms of the isospin eigenstates.   □

Remark.   The octet representation has a special role in $SU(3)$ corresponding to the role of the $j = 1$ representation of $SU(2)$. In each case, the dimension of the representation is equal to the dimension of the Lie algebra. This is not coincidental; the structure constants of the

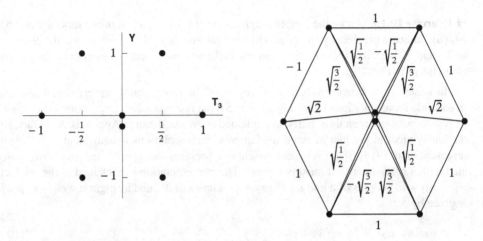

Figure 10.3: Left: Weight diagram for the irreducible representation **8** (the *octet*) of $SU(3)$. Right: Matrix elements of the root vectors $\mathbf{T}_\pm$, $\mathbf{U}_\pm$ and $\mathbf{V}_\pm$ in the octet representation.

algebra define a matrix representation on the algebra itself, as explained more fully in Appendix A, and the generators themselves form a basis for the representation. This representation is known as the *adjoint representation*, and it is a necessary ingredient of gauge theories—the vector gauge fields in these theories necessarily belong to the adjoint representation, and there is a one-to-one correspondence between generators and gauge fields.

The relation between the generators and the basis vectors of the adjoint representation can be seen by comparing Figs. 10.3 and 10.1. The weights correspond to the roots except for the weight (0 0) at the center, which corresponds to the two commuting generators $\mathbf{T}_3$ and $\mathbf{Y}$. The state $|(1)0\,0\rangle$ corresponds to $\mathbf{T}_3$; the state $|(0)0\,0\rangle$ to $\mathbf{Y}$.   □

❑ **Example 10.19.** The irreducible representation of $SU(n)$ associated with the partition (3) is symmetric in the three indices. The number of independent components of such a tensor is given in general by $\frac{1}{6}n(n+1)(n+2)$, which is ten for $SU(3)$; hence the irreducible representation $\Gamma(3,0)$ of $SU(3)$ is known as the *decuplet*, denoted by **10**.   ∎

Tensor products in $SU(3)$ can be reduced using the rules for constructing outer products of representations of symmetric groups. For example, the product $\mathbf{3} \otimes \mathbf{3}^*$ is given by

$$\square \;\times\; \begin{array}{c}\square\\\square\end{array} \;=\; \begin{array}{cc}\square&\square\\\square&\end{array} \;+\; \begin{array}{c}\square\\\square\\\square\end{array} \qquad \rightarrow \quad \mathbf{3} \otimes \mathbf{3}^* = \mathbf{8} \oplus \mathbf{1}$$

while the product $\mathbf{6} \otimes \mathbf{3}$ is given by

$$\square\square \;\times\; \square \;=\; \square\square\square \;+\; \begin{array}{cc}\square&\square\\\square&\end{array} \qquad \rightarrow \quad \mathbf{6} \otimes \mathbf{3} = \mathbf{10} \oplus \mathbf{8}$$

Thus in the product $\mathbf{3} \otimes \mathbf{3} \otimes \mathbf{3}$ of three fundamental representations, we have a symmetric **10**, an antisymmetric **1** and two octets that transform as the mixed symmetry representation under permutations of the three fundamentals:

$$\mathbf{3} \otimes \mathbf{3} \otimes \mathbf{3} = \mathbf{10}_S \oplus (2 * \mathbf{8})_m \oplus \mathbf{1}_A \tag{10.88}$$

Remark.   These states can be identified as flavor states for three quarks. The total wave function of a three-quark system can be expressed as a product of a space wave function and a spin-flavor function for the three spin-$\frac{1}{2}$ quarks. For the lowest state of such a system, the space wave function is symmetric, so one might expect to find a spin-$\frac{1}{2}$ octet and a spin-$\frac{3}{2}$ flavor singlet in order to be consistent with the antisymmetry of the three-quark wave function required the exclusion principle. Instead, the spin-$\frac{3}{2}$ baryons form a flavor decuplet, and the spin-$\frac{1}{2}$ octet has a symmetric product of spin and flavor functions (see also Problem 20). This puzzle was resolved by the introduction of the $SU(3)$ "color group" for quarks. An antisymmetric singlet color wave function for the three quarks in the observed baryons restored the consistency with the Pauli exclusion principle.   □

As a final example, we reduce the product of two octet representations. The outer product of Young diagrams leads to

(see Exercise 9.23) where we have omitted the partitions (31^3) and (2^21^2) since they have more than three rows and thus do not correspond to irreducible representations of $SU(3)$. Also, the partitions (41^2), (321), and (2^3) have three rows, and for $SU(3)$ these partitions are equivalent to the partitions (3), (21), and (0), respectively. Finally, the partition (3^2) is dual to the partition (3) in $SU(3)$; hence the irreducible representation (3^2) is equivalent to the complex conjugate of (3). Thus we end up with the reduction

$$\mathbf{8} \otimes \mathbf{8} = (\mathbf{27} \oplus \mathbf{8} \oplus \mathbf{1})_S \oplus (\mathbf{10} \oplus \mathbf{10}^* \oplus \mathbf{8})_A \tag{10.89}$$

where the subscripts S and A refer to the symmetric and antisymmetric parts of the product. The split into symmetric and antisymmetric parts is fairly straightforward. The **27** must be symmetric, since it is the only representation that can contain the product of two copies of the highest weight state of **8**, and the remaining assignments follow directly from counting (the singlet is clearly symmetric, and of the two octets, one must be symmetric, the other antisymmetric; then the only possibility for **10** and **10*** is antisymmetric).

→ **Exercise 10.16.**   Draw the weight diagrams for the irreducible representations **10** and **10*** of $SU(3)$. and indicate graphically the effects of the operators $\mathbf{T}_\pm$, $\mathbf{U}_\pm$, and $\mathbf{V}_\pm$ on the eigenstates of $\mathbf{T}_3$ and $\mathbf{Y}$ in each of these representations, including numerical values of the relevant matrix elements as in Fig. 10.2.   □

→ **Exercise 10.17.**   Reduce the $SU(3)$ tensor product representation **10** $\otimes$ **8** using Young diagrams. Compute the dimension of any new irreducible representation(s) that appear.   □

A   Tensor Representations of the Classical Lie Groups

A.1   The Classical Lie Groups

The *classical groups* are the linear groups $GL(n)$ and $SL(n)$, the unitary groups $U(n)$ and $SU(n)$ defined on a complex n-dimensional linear vector space V^n, the orthogonal groups $O(n)$ and $SO(n)$ defined on a real n-dimensional space, and the symplectic groups $Sp(2n)$ defined on a $2n$-dimensional space. These groups were introduced briefly in Section 10.1. Here we construct representations of these groups using tensor methods whose origins can be traced to the works of Herman Weyl. Another approach is to study representations of the Lie algebras, generalizing the standard treatment of angular momentum in quantum mechanics. This has been worked out in detail for $SU(2)$ and $SU(3)$ in Section 10.3.

The *fundamental representation* of each of these groups is the defining representation, in which each element $\mathbf{A}$ of the group is represented by itself in the vector space V^n on which it is defined ($\mathbf{R}^n$ for the orthogonal groups and the real linear groups, $\mathbf{C}^n$ for the others). There are other representations on the same vector space V^n, namely,

$$\mathbf{A} \to \mathbf{A}^* \qquad \mathbf{A} \to \widetilde{\mathbf{A}}^{-1} \qquad \mathbf{A} \to (\mathbf{A}^\dagger)^{-1} \tag{10.A1}$$

where $\mathbf{A}^*$ denotes the complex conjugate of $\mathbf{A}$ (this definition depends on the basis in V^n, but all such representations are evidently equivalent), and $\widetilde{\mathbf{A}}$ denotes the transpose of $\mathbf{A}$.

For the group $SL(2, C)$, the representation $\mathbf{A} \to \widetilde{\mathbf{A}}^{-1}$ is equivalent to the defining representation (see Problem 2). We note that the group $SL(2, C)$ is closely related to the group of homogeneous Lorentz transformations. This relations is similar to the relation between $SU(2)$ and the three-dimensional rotations group $SO(3)$, as explained in detail in Appendix B. However, for $n > 2$, Eq. (10.A1) defines four inequivalent irreducible representations of the groups $GL(n, C)$, and $SL(n, C)$. For the groups $GL(n, R)$ and $SL(n, R)$, the representation matrices are real ($\mathbf{A} = \mathbf{A}^*$), but $\mathbf{A} \to \widetilde{\mathbf{A}}^{-1}$ provides an inequivalent representation except in the case of $SL(2, R)$.

For the unitary groups $U(n)$ and $SU(n)$, unitarity of the matrices implies

$$\mathbf{A} = (\mathbf{A}^\dagger)^{-1} \qquad \mathbf{A}^* = \widetilde{\mathbf{A}}^{-1} \tag{10.A2}$$

so that of the representations in Eq. (10.A1), only the map $\mathbf{A} \to \mathbf{A}^*$ gives a new inequivalent irreducible representation, the *conjugate fundamental representation*, that is not equivalent to the fundamental representation, again apart from the special case of $SU(2)$ (see Problem 1). For the orthogonal groups $SO(n)$, the four representations are identical, since

$$\mathbf{A} = \mathbf{A}^* = \widetilde{\mathbf{A}}^{-1} = (\mathbf{A}^\dagger)^{-1} \tag{10.A3}$$

for real orthogonal matrices. For the symplectic groups $Sp(2n)$, the four representations are again equivalent, though we will not prove that here.

An irreducible representation of the general group $GL(n, C)$ $[SL(n, C)]$ is also irreducible for the subgroup $U(n)$ $[SU(n)]$ of unitary matrices, since the Lie algebra of $GL(n, C)$ is obtained from the real Lie algebra of $U(n)$ by allowing the parameters of the algebra to be complex. Thus we consider here only the irreducible representations of $U(n)$ and $SU(n)$. The reduction of these irreducible representations under restriction to the subgroups $SO(n)$ or $Sp(2n)$ will also be described.

A.2 Tensor Representations of $U(n)$ and $SU(n)$

To construct irreducible representations of $U(n)$ and its subgroup $SU(n)$, we start with a basis for the complex linear vector space V^n on which these groups are defined. Then a linear operator $\mathbf{A}$ has a matrix representation $\mathbf{A} = (a^j{}_k)$ so that when acting on a vector x with components $\xi^1, \ldots, \xi^n$, it gives the vector $\mathbf{A}x$ with components

$$[\mathbf{A}x]^j \equiv \xi'^j = a^j{}_k \xi^k \qquad (10.\text{A4})$$

Remark. We recall the (Einstein) *summation convention* from Chapter 3, which requires us to sum over a repeated pair of indices (one subscript and one superscript) such as the index k in the last equation, unless explicitly instructed not to by writing

$$S = a_k b^k \quad \text{(no sum)}$$

Since we deal here only with linear manifolds, we do not need the full geometric apparatus of Chapter 3, but we use superscripts to denote the components of a vector in V^n, and subscripts to denote components of vectors in the dual space to V^n. In Chapter 3, elements of the dual space are identified with the space differential forms on a manifold; here they are linear functionals on V^n as introduced in Chapter 2. For a complex vector space, the dual space can be envisioned as the "complex conjugate" space, though that identification depends on an explicit basis. In a real unitary vector space, forms and vectors are equivalent, and we then sum over *any* pair of repeated indices. □

If we now take the tensor product of V^n with itself N times,

$$\otimes_N V^n \equiv \underbrace{V^n \otimes \cdots \otimes V^n}_{N \text{ times}} \qquad (10.\text{A5})$$

we have a representation of $U(n)$ on the space $\otimes_N V^n$. The elements X of this space are *tensors* of *rank* N[1] whose components $\xi^{j_1 \cdots j_N}$ are labeled by a set of N indices. When acted on by a linear operator $\mathbf{A}$ from $U(n)$, these components are transformed according to the rule

$$(\mathbf{A}X)^{j_1 \cdots j_N} \equiv \xi'^{j_1 \cdots j_N} = a^{j_1}{}_{k_1} \cdots a^{j_N}{}_{k_N} \, \xi^{k_1 \cdots k_N} \qquad (10.\text{A6})$$

This representation is reducible, since the transformation coefficients

$$A^{j_1 \cdots j_N}{}_{k_1 \cdots k_N} \equiv a^{j_1}{}_{k_1} \cdots a^{j_N}{}_{k_N} \qquad (10.\text{A7})$$

are unchanged if we apply any permutation of $\{1, \ldots, N\}$ to both sets of indices $j_1 \cdots j_N$ and $k_1 \cdots k_N$. That is, permutations of the indices commute with the tensor transformation law.

Thus we can collect the tensor components into sets that transform among themselves under permutations of the indices. For the unitary groups $U(n)$ and $SU(n)$, these sets define *irreducible tensor representations*, though we do not prove irreducibility here. An irreducible tensor of rank N is associated with an irreducible representation of the symmetric group S_N, with a partition of N and its Young diagram (see Section 9.5) that define the symmetry type of the tensor.

---

[1] *Caution.* The rank of a tensor is *not* the same as the rank of a Lie algebra defined in Definition 10.5.

Not every partition of N corresponds to an irreducible representation of $U(n)$, however. Since the tensor corresponding to a Young diagram is antisymmetric when we interchange elements in the same column of the diagram, the partition of N for a $U(n)$ tensor of rank N can have no more than n rows. Furthermore, an antisymmetric tensor of rank n is necessarily proportional to the *Levi-Civita symbol*

$$\varepsilon^{i_1\cdots i_n} = \begin{cases} +1 & \text{if } i_1\cdots i_n \text{ is an even permutation of } \{1\ldots n\} \\ -1 & \text{if } i_1\cdots i_n \text{ is an odd permutation of } \{1\ldots n\} \\ 0 & \text{if any two indices are equal} \end{cases} \tag{10.A8}$$

which is antisymmetric in its n indices. Under transformations of $U(n)$, we have

$$\varepsilon^{i_1\cdots i_n} \rightarrow A^{i_1\cdots i_n}{}_{k_1\cdots k_n}\, \varepsilon^{k_1\cdots k_n} = (\det \mathbf{A})\, \varepsilon^{i_1\cdots i_n} \tag{10.A9}$$

so the Levi-Civita symbol is transformed into a multiple of itself by an element of $U(n)$. It is a *numerical tensor*, the *ε-tensor*, which provides a one-dimensional representation of $U(n)$. Under $SU(n)$, whose elements have $\det \mathbf{A} = 1$, the Levi-Civita tensor is transformed exactly into itself; it is an *invariant* of $SU(n)$.

Now suppose $(m) = (m_1 m_2 \cdots m_p)$ is a partition of N into p parts. Then

- for $p > n$, there are no irreducible representations of $U(n)$,

- for $p \leq n$, there are two irreducible representations of $U(n)$ on the tensor product space, corresponding to the direct tensor product representation and the conjugate representation introduced in Eq. (10.A1).

In the case $p \leq n$, we can use the Young diagram $\mathcal{Y}^{(m)}$ of (m) to compute the dimension of the corresponding irreducible representation of $U(n)$. Recall that indices are symmetric (antisymmetric) with respect to permuting the elements of a row (column) of the diagram. Then the dimension of the irreducible representation of $U(n)$ corresponding to a partition (m) is equal to the number of ways to fill the boxes of $\mathcal{Y}^{(m)}$ with numbers in the range $\{1, \ldots, n\}$, not necessarily distinct, such that

    (i) the indices are nondecreasing across each row, and

    (ii) the indices are strictly increasing down each column.

This procedure counts the number of independent components of a tensor of the symmetry type (m); the remaining components are determined by permuting indices of the tensor.

❑ **Example 10.20.** The irreducible representations of $U(n)$ associated with the partitions (2) and (1^2) are the symmetric and antisymmetric tensors of rank two, with $\frac{1}{2}n(n+1)$ and $\frac{1}{2}n(n-1)$ independent components, respectively. The diagrammatic rule gives exactly this result for the dimension of the corresponding irreducible representation. ∎

❑ **Example 10.21.** In Example 10.18, the dimension of the irreducible representation of $SU(3)$ associated with the partition (21) and (3) was found to be 8, using the procedure just described. ∎

→ **Exercise 10.A1.**   Show that the dimension of the irreducible representation $\Gamma(m_1 \, m_2)$ of $SU(3)$ is

$$d(m_1, m_2) = \tfrac{1}{2}(m_1 + 1)(m_2 + 1)(m_1 - m_2 + 1)$$

by actually counting the number of ways to fill the boxes of of $\mathcal{Y}^{(m_1 \, m_2)}$ with numbers $1, 2, 3$ consistent with the rules given above. Then use this result to compute the dimension of the irreducible representation $\Gamma(4 \, 2)$.   □

Remark.   The irreducible representations of $U(n)$ can alternatively be defined in terms of an n-tuple $[m] = [m_1 \, m_2 \cdots m_n]$ of integers with

$$m_1 \geq m_2 \geq \cdots \geq m_n \geq 0 \tag{10.A10}$$

The difference between this label and the label by a partition is that the n-tuple may have zeros at the end, while the zeros are omitted from a partition); we can think of this n-tuple as an *extended partition*. These representations can be denoted by $\Gamma[m]$ and $\Gamma^*[m]$, In this notation, the irreducible representations of $U(3)$ associated with the partitions (21) and (321) are denoted by $\Gamma[210]$ and $\Gamma[321]$, while the irreducible representations of $U(4)$ associated with the same partitions are denoted by $\Gamma[2100]$ and $\Gamma[3210]$, respectively.   □

   If we restrict $U(n)$ to the subgroup $SU(n)$, then a column with n rows can be omitted from the Young diagram for the symmetry type. Such a column is antisymmetric in n indices, and hence must be correspond to a factor proportional to the ε-tensor, which is invariant under $SU(n)$. The column can then be eliminated from the diagram. Thus only diagrams with at most $n - 1$ rows give distinct irreducible representations of $SU(n)$. These irreducible representations can then be labeled by an $(n-1)$-tuple $(m]) = (m_1 m_2 \cdots m_{n-1})$ of integers, and denoted by $\Gamma(m_1 \, m_2 \cdots m_{n-1}])$.

❑ **Example 10.22.**   In $SU(3)$, the partition $[321]$ reduces to (21), shown graphically as

$$[321] \; = \; \begin{array}{l}\square\square\square\\\square\square\\\square\end{array} \; \rightarrow \; \begin{array}{l}\square\square\\\square\end{array} \; = \; [21]$$

   Thus $[321] \sim [21]$ for $SU(3)$ and the representation is denoted simply by $\Gamma(2 \, 1)$.   ∎

   Now let

$$\bar{\mathbf{A}} \equiv \mathbf{A}^{-1} = \left(\bar{a}^j{}_k\right) \tag{10.A11}$$

and introduce *dual vectors* y, with components $\eta_1, \ldots, \eta_n$ that transform under $\mathbf{A}$ by the rule

$$y \rightarrow y\bar{\mathbf{A}} \; : \; [y\bar{\mathbf{A}}]_k = \eta'_k = \bar{a}^j{}_k \eta_j \tag{10.A12}$$

Thus the representation $\mathbf{A} \rightarrow \widetilde{\mathbf{A}}^{-1} (= \mathbf{A}^*$ for unitary operators), the *dual*, or *conjugate*, *fundamental* representation, acts in a natural way on the space $\mathcal{V}^{n*}$ dual to $\mathcal{V}^n$. *Dual tensors* of rank N can be introduced on the space $\otimes_N \mathcal{V}^{n*}$, with components $\eta_{j_1 \cdots j_N}$ labeled with N superscripts. Dual tensors can be split into irreducible representations associated with an irreducible representation of $\mathcal{S}_N$ in the same way as ordinary tensors.

Tensor indices that are antisymmetric can be raised and lowered using the ε-tensor. If $\mathbf{T}$ is an antisymmetric $SU(n)$ tensor of rank p with components $T^{i_1\cdots i_p}$, then there is a corresponding antisymmetric dual tensor $^*\mathbf{T}$ of rank $(n-p)$ with components given by

$$^*T_{i_{p+1}\cdots i_n} = \bar{\varepsilon}_{i_1\cdots i_n}\, T^{i_1\cdots i_p} \tag{10.A13}$$

where $\bar{\varepsilon}_{i_1\cdots i_n} = \varepsilon^{i_1\cdots i_n}$ is the completely antisymmetric dual tensor. This dual $^*\mathbf{T}$ is the same as the Hodge dual introduced in Chapter 3. Conversely, the antisymmetric $SU(n)$ dual tensor $^*\mathbf{T}$ of rank $(n-p)$ corresponds to an antisymmetric tensor $\mathbf{T}$ of rank p, with components related by

$$T^{i_1\cdots i_p} = \varepsilon^{i_1\cdots i_n}\, {}^*T_{i_{p+1}\cdots i_n} \tag{10.A14}$$

The construction of dual tensors extends to tensors of general symmetry. If $\mathbf{T}$ is a $SU(n)$ tensor of rank N, with symmetry associated with the partition $(m) = (m_1 \cdots m_q)$ of N, then the dual tensor has the symmetry of a partition $(m)^*$ of $nm_1 - N$ constructed by assigning to $(m)^*$ those boxes of a rectangle with n rows and m_1 columns that are not in the partition (m), and rotating the diagram to the standard position. Note that $(m)^*$ depends both on n and (m). The dual representation is equivalent to the (complex) conjugate representation of $SU(n)$. Tensors whose dual is equivalent to the original tensor are *self-dual*; the corresponding representation of $SU(n)$ is then real.

❑ **Example 10.23.**   The fundamental representation of $SU(3)$ is $\mathbf{3} = (10)$. The six-dimensional irreducible representation $\mathbf{6} = (20)$ is a symmetric tensor of rank two; the corresponding dual tensor is $\mathbf{6}^* = (22)$. The antisymmetric rank two tensor is associated with the partition (11); it is equivalent to the conjugate fundamental $\mathbf{3}^*$ for $SU(3)$.   ∎

Irreducible representations can be denoted simply by their dimension $(\mathbf{3}, \mathbf{3}^*, \ldots)$ when the context is clear. We have shown in Example 10.18 that the irreducible representation (21) of $SU(3)$ is eight-dimensional; it can be denoted by $\mathbf{8}$. It is also self-dual, and hence real.

❑ **Example 10.24.**   Consider a rank three tensor of symmetry type (21). For an $SU(3)$ tensor, we can construct the diagram

so we conclude that the $SU(3)$ tensor dual to (21) also has symmetry type (21) and is thus self-dual. For an $SU(4)$ tensor, on the other hand, we have

Hence the $SU(4)$ tensor dual to (21) has symmetry type $(2^2 1)$, and is *not* self-dual.   ∎

→ **Exercise 10.A2.**   Find the symmetry types of the $SU(3)$ tensors dual to tensors of types (21^2), (2^2), (321), and (42). Repeat for $SU(4)$ tensors of the same symmetry types.   □

Invariants can be formed in the usual way by combining vectors with dual vectors. For example, the scalar product

$$(y, x) = \eta_k \xi^k \tag{10.A15}$$

is invariant under any $U(n)$ transformation $\mathbf{A}$, since under the transformation

$$\eta_k \xi^k \to \eta_j \,\bar{\mathbf{A}}^j_{\;k} \mathbf{A}^k_{\;\ell} \,\xi^\ell = \eta_j \,(\mathbf{A}^{-1}\mathbf{A})^j_{\;\ell} \,\xi^\ell = \eta_\ell \xi^\ell \tag{10.A16}$$

In fact, any summation over a pair of indices, one vector (superscript) and one dual vector (subscript), reduces the rank of a mixed tensor by two. This will be significant when we discuss the reduction of Kronecker products of representations.

Remark.   The invariance of the scalar product is equivalent to the existence of a second numerical tensor, the *Kronecker delta*, with components $\delta_j^{\;k} = \delta^k_{\;j}$, since

$$\mathbf{A}^j_{\;\ell} \delta^\ell_{\;m} \bar{\mathbf{A}}^m_{\;k} = (\mathbf{A}\mathbf{A}^{-1})^j_{\;k} = \delta^j_{\;k} \tag{10.A17}$$

is equivalent to Eq. (10.A16).   □

A.3   Irreducible Representations of $SO(n)$

The orthogonal groups $SO(2n)$ and $SO(2n+1)$ each have rank n. To see this, note that any rotation matrix in $2n$ dimensions can be brought to the standard form

$$\mathbf{R} = \begin{pmatrix} \cos\theta_1 & -\sin\theta_1 & 0 & 0 & \cdots & 0 & 0 \\ \sin\theta_1 & \cos\theta_1 & 0 & 0 & \cdots & 0 & 0 \\ 0 & 0 & \cos\theta_2 & -\sin\theta_2 & \cdots & 0 & 0 \\ 0 & 0 & \sin\theta_2 & \cos\theta_2 & \cdots & 0 & 0 \\ \vdots & \vdots & \vdots & \vdots & \ddots & \vdots & \vdots \\ 0 & 0 & 0 & 0 & \cdots & \cos\theta_n & -\sin\theta_n \\ 0 & 0 & 0 & 0 & \cdots & \sin\theta_n & \cos\theta_n \end{pmatrix} \tag{10.A18}$$

as noted in Problem 2.20, while the standard form in $2n+1$ dimensions has an added row $2n+1$ and added column $2n+1$ in which all the entries are zero except for the diagonal element in the lower right corner. The matrices

$$\mathbf{H}_k = i \frac{\partial}{\partial\theta_k} \mathbf{R}\Big|_{\theta_k=0} \tag{10.A19}$$

($k = 1, \ldots, n$) define a set of n commuting generators of $SO(2n)$ or $SO(2n+1)$, and it is clear that there are no other generators that commute with all of these. As with the unitary groups, we can now construct tensors of rank N, with components $\xi^{j_1 \cdots j_N}$ labeled by a set of N indices, and sort these into sets of components with definite symmetry type under permutations of the indices. These sets are associated with a partition of N and its Young diagram. However, there are two distinctions compared with the unitary groups:

1. There is no distinction between upper and lower indices, so that the invariant δ_{jk} can be used to contract pairs of indices (recall that this contraction corresponds to taking the trace of a matrix). Thus, in addition to the original Young diagram, we also have irreducible parts corresponding to Young diagrams constructed from the original by removing pairs of boxes from the diagram, subject only to the condition that the two boxes are not removed from the same column (recall that the tensor is antisymmetric with respect to indices in the same column, so that contracting with respect to those indices gives zero).

❑ **Example 10.25.** We provide some simple reductions to illustrate the method:

- For $n \geq 4$, an $SU(n)$ tensor corresponding to partition (4) is reduced to irreducible $SO(n)$ tensors by

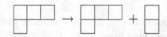

For $n = 4$, this corresponds to

$$35 \;\to\; 25 + 9 + 1$$

- Again for $n \geq 4$, the reduction of the $SU(n)$ tensor corresponding to partition (31) is given by

which for $n = 4$ corresponds to

$$45 \;\to\; 39 + 6$$

where we note that the partition (1^2) is irreducible both for $SU(4)$ and $SO(4)$ since no contractions are possible.   ∎

2. The ε-tensor transforms an antisymmetric tensor of rank p into a antisymmetric tensor of rank $2n - p$. Hence we only need to consider tensors that are antisymmetric in at most n indices. That is, we only need to consider partitions into at most n parts, or Young diagrams that have at most n rows.

B   Lorentz Group; Poincaré Group

B.1   Lorentz Transformations

The principle of relativity, that the laws of physics should appear the same to observers moving with a constant velocity relative to each other, is traced to Galileo. In its early versions, this principle took for granted that the relation between the spacetime coordinates $(t, \vec{x})$ of an event seen by an observer $\mathcal{O}$ and the coordinates $(t', \vec{x}')$ of the same event seen by an observer $\mathcal{O}'$ moving with constant velocity $\vec{v}$ relative to $\mathcal{O}$ is

$$t' = t \qquad \vec{x}' = \vec{x} - \vec{v}t \tag{10.B20}$$

However, the success of Maxwell's equations for the electromagnetic field cast doubt on this transformation law. Maxwell's equations are not invariant under transformation (10.B20), and the Michelson–Morley experiment failed to detect an absolute frame in which the equations were valid.

As we now know, extending the principle of relativity to Maxwell's equations requires a modification of this transformation law: The relation between the spacetime coordinates must be given by the *Lorentz transformation law*

$$\vec{x}'_\perp = \vec{x}_\perp \qquad x'_\parallel = \gamma \left(x_\parallel - vt \right) \qquad t' = \gamma \left(t - vx_\parallel / c^2 \right) \tag{10.B21}$$

where

$$\gamma = 1/\sqrt{1 - v^2/c^2} \tag{10.B22}$$

(c = speed of light). Here $x_\parallel$ denotes the component of $\vec{x}$ along the velocity $\vec{v}$, while $\vec{x}_\perp$ denotes the components perpendicular to $\vec{v}$. The transformation to a moving coordinate system, without rotation of the coordinate axes, is a *pure Lorentz transformation* (or *boost*), with velocity $\vec{v}$, to be denoted by $\mathbf{B}(\vec{v})$. See Problem 24 for the transformation properties of the electromagnetic field needed to make Maxwell's equations invariant under the boost.

The physical interpretation and consequences of the transformation law (10.B21) are discussed in many places. Here we are concerned with the group theoretic structure of the Lorentz transformations, including as well translations of the spacetime origin, and the information that can be extracted from the principle of invariance under these transformations (*Lorentz invariance*). In particular, we construct some irreducible representations of various groups that include Lorentz transformations.

Since the Lorentz transformation mixes space and time coordinates, we can imagine these transformations acting on a four-dimensional spacetime (*Minkowski space*) in which points $\mathbf{x}$ (*events*) are labeled by four coordinates,

$$\mathbf{x} = (x^0 = ct, \vec{x}) = \{x^\mu\} \tag{10.B23}$$

that form the components of a *four-vector*, with x^0 the time component and $\vec{x} = (x^1, x^2, x^3)$ the space components, The index μ ranges over the values $\mu = 0, 1, 2, 3$; a popular convention is to use Greek letters $\mu, \nu, \lambda, \ldots$ to denote four-vector indices ranging over $0, 1, 2, 3$, and Roman letters $j, k, \ell, \ldots$ for the space components with range $1, 2, 3$. Repeated indices are to be summed over the appropriate range, following the Einstein summation convention. The use of superscripts rather than subscripts is conventional, but see Chapter 3 for further discussion of the relation between superscripts and subscripts as labels for components.

Linear transformations of the spacetime coordinates will be denoted by Λ, with explicit transformation law

$$\mathbf{x} \to \mathbf{x}' = \Lambda \mathbf{x} \; : \; x'^{\mu} = \Lambda^{\mu}{}_{\nu} \, x^{\nu} \tag{10.B24}$$

with the summation convention in effect. Note that when a matrix has one subscript and one superscript index, the row index should be placed clearly to the left of the column index.

The difference between four-vectors and vectors in a vector space $\mathcal{V}^4$ of Chapter 2 is that here we define the scalar product of two four-vectors

$$\mathbf{a} = (a^0, \vec{a}\,) \qquad \mathbf{b} = (b^0, \vec{b}\,) \tag{10.B25}$$

as

$$(\mathbf{a}, \mathbf{b}) = \mathbf{a} \cdot \mathbf{b} = a^{\mu} g_{\mu\nu} b^{\nu} = a^0 b^0 - \vec{a} \cdot \vec{b} \tag{10.B26}$$

The matrix

$$\mathbf{g} = (g_{\mu\nu}) = \mathrm{diag}(1, -1, -1, -1) \tag{10.B27}$$

is a *metric tensor*, the *Minkowski metric*, and $\mathbf{a} \cdot \mathbf{b}$ is a *Lorentz scalar*. The metric $\mathbf{g}$ is not positive definite, and a four-vector $\mathbf{a}$ is

   (i) *timelike* if $\mathbf{a} \cdot \mathbf{a} > 0$,

   (ii) *spacelike* if $\mathbf{a} \cdot \mathbf{a} < 0$, or

   (iii) *lightlike* if $\mathbf{a} \cdot \mathbf{a} = 0$ ($\mathbf{a} \neq 0$).

❑ **Example 10.26.** An important four-vector is the four-momentum $\mathbf{p}$ of a particle, which has components

$$\mathbf{p} = (E/c, \vec{p}\,) \tag{10.B28}$$

where E is the energy of the particle. We have the Lorentz scalar

$$\mathbf{p} \cdot \mathbf{p} = m^2 c^2 \tag{10.B29}$$

where m is the mass of the particle. ∎

Remark. In the older literature m was often called the "rest mass" of the particle, and E/c^2 the "relativistic mass." The modern view refers to the invariant m simply as the "mass." The energy

$$E = \sqrt{(pc)^2 + (mc^2)^2} \tag{10.B30}$$

includes a contribution mc^2 from the *rest energy* of the particle. ☐

The rotations of the spatial coordinate axes, as described in Section 2.2.4, define the rotation group. Including the transformations to moving coordinate systems leads to a larger group, the *homogeneous Lorentz group*, which contains those linear transformations Λ of four-vectors such that

$$(\Lambda \mathbf{a}, \Lambda \mathbf{b}) = (\mathbf{a}, \mathbf{b}) \tag{10.B31}$$

for all four-vectors $\mathbf{a}$ and $\mathbf{b}$. In terms of the matrix elements of Λ, this requires

$$\Lambda^\mu_{\ \alpha}\, g_{\mu\nu}\, \Lambda^\nu_{\ \beta} = g_{\alpha\beta} \tag{10.B32}$$

Any such transformation is a *Lorentz transformation*. As already noted, the transformation to a coordinate system moving with constant velocity is a boost, or pure Lorentz transformation.

→ **Exercise 10.B3.**   (i) Show that the matrix elements of the inverse transformation $\overline{\Lambda}$ satisfy

$$g_{\mu\alpha}\, \overline{\Lambda}^\alpha_{\ \nu} = \Lambda^\beta_{\ \mu}\, g_{\beta\nu}$$

(ii) Show that the matrix elements $\Lambda^\mu_{\ \nu}$ satisfy

$$\left|\Lambda^0_{\ 0}\right|^2 - \Lambda^0_{\ j}\Lambda^j_{\ 0} = 1$$

Remark.   Thus $\left|\Lambda^0_{\ 0}\right| \geq 1$. If $\left|\Lambda^0_{\ 0}\right| = 1$, then Λ is just a rotation of the coordinate axes.   □
(iii) Show that

$$\det \Lambda = \pm 1$$

for any Lorentz transformation.   □

The preceding exercise shows that $\det \Lambda = \pm 1$ and $\left|\Lambda^0_{\ 0}\right| \geq 1$, but only *restricted* Lorentz transformations, those with $\det \Lambda = +1$ and $\Lambda^0_{\ 0} \geq 1$, can be reached continuously from the identity. Other Lorentz transformations are obtained as the product of a restricted Lorentz transformation and one or both of the discrete transformations

> *space reflection* $\Sigma : \vec{x} \rightarrow -\vec{x},\ t \rightarrow t$
>
> *time inversion* $\Theta : \vec{x} \rightarrow \vec{x},\ t \rightarrow -t.$

Remark.   Note that *time reversal* in quantum mechanics is not simply time inversion. Reversing the role of initial and final states in a scalar product

$$\langle \phi | \psi \rangle \rightarrow \langle \psi | \phi \rangle = \langle \phi | \psi \rangle^* \tag{10.B33}$$

involves complex conjugation as well.   □

Thus the full Lorentz group $\mathcal{L}$ has four connected components:

$$\mathcal{L}^\uparrow_\pm \ :\ \det \Lambda = \pm 1 \qquad \Lambda^0_{\ 0} \geq 1 \qquad \text{and} \qquad \mathcal{L}^\downarrow_\pm \ :\ \det \Lambda = \mp 1 \quad \Lambda^0_{\ 0} \leq -1$$

and the subgroups $\mathcal{L}^\uparrow_+$, the *restricted* Lorentz group, $\mathcal{L}_+ \equiv \mathcal{L}^\uparrow_+ \cup \mathcal{L}^\downarrow_+$, the *proper* Lorentz group, $\mathcal{L}^\uparrow \equiv \mathcal{L}^\uparrow_+ \cup \mathcal{L}^\uparrow_-$, the *orthochronous* Lorentz group, and $\mathcal{L}_0 \equiv \mathcal{L}^\uparrow_+ \cup \mathcal{L}^\downarrow_+$.

The restricted Lorentz group is also characterized as $SO(3, 1)$ in a notation where $SO(p, q)$ denotes the group that leaves invariant a metric of the form

$$\mathbf{g} = \text{diag}(\underbrace{1, \ldots, 1}_{p \text{ times}}, \underbrace{-1, \ldots, -1}_{q \text{ times}}) \tag{10.B34}$$

The matrix Λ corresponding to a rotation of the coordinate axes has the form

$$\Lambda = \begin{pmatrix} 1 & 0 \\ 0 & \mathbf{R}(\phi, \theta, \psi) \end{pmatrix} \tag{10.B35}$$

where $\mathbf{R}(\phi, \theta, \psi)$ is the rotation matrix given in Eq. (2.113), and any Lorentz transformation with $\Lambda^0_0 = 1$ is a rotation (proper or improper, as the case may be). For a boost in the direction $\mathbf{n}$ with velocity

$$\vec{v} = (c \tanh \chi) \, \mathbf{n} \tag{10.B36}$$

the matrix elements of $\Lambda \equiv \mathbf{B_n}(\chi)$ are given by

$$\Lambda^0_0 = \cosh \chi \qquad \Lambda^j_0 = -(\sinh \chi) \, \mathbf{n}_j = \Lambda^0_j$$

$$\Lambda^j_k = \delta^j_k + (\cosh \chi - 1) \, \mathbf{n}_j \mathbf{n}_k \tag{10.B37}$$

Remark.   The matrix for a boost is symmetric. The converse is almost true: A symmetric Lorentz transformation is a pure boost, possibly accompanied by a rotation through a multiple of π about the boost axis. The proof of this statement is left to Problem 26.   □

The generator $\mathbf{L_n}$ for rotations about the axis $\mathbf{n}$ are obtained from the rotation matrix $\mathbf{R_n}(\Phi)$ by (see Problem 2.16)

$$\mathbf{L_n} = i\mathbf{R}'_\mathbf{n}(\Phi)\Big|_{\Phi=0} \tag{10.B38}$$

Similarly, the generator $\mathbf{K_n}$ for boosts in the direction $\mathbf{n}$ is given by

$$\mathbf{K_n} = i\mathbf{B}'_\mathbf{n}(\chi)\Big|_{\chi=0} \tag{10.B39}$$

The generators satisfy the commutation rules

$$[\mathbf{L}_\alpha, \mathbf{L}_\beta] = i \sum_\gamma \varepsilon_{\alpha\beta\gamma} \mathbf{L}_\gamma \qquad [\mathbf{L}_\alpha, \mathbf{K}_\beta] = i \sum_\gamma \varepsilon_{\alpha\beta\gamma} \mathbf{K}_\gamma$$

$$[\mathbf{K}_\alpha, \mathbf{K}_\beta] = -i \sum_\gamma \varepsilon_{\alpha\beta\gamma} \mathbf{L}_\gamma \tag{10.B40}$$

where $\varepsilon_{\alpha\beta\gamma}$ is the usual antisymmetric symbol on three indices. The minus sign on the right-hand side of the $[\mathbf{K}_\alpha, \mathbf{K}_\beta]$ commutator distinguishes the Lorentz algebra from that of $SO(4)$ (see Problem 7).

→ **Exercise 10.B4.**   (i) Find explicit matrices for the $\mathbf{L}_\alpha$, $\mathbf{K}_\alpha$. Then verify the commutators (10.B40)

(ii) The $\mathbf{L}_\alpha$, $\mathbf{K}_\alpha$ span the Lie algebra of the group $\mathcal{L}_+^\uparrow$. Find quadratic Casimir operators for the Lie algebra using the results of Section 10.2.3. Is $\mathcal{L}_+^\uparrow$ compact?   □

Now introduce the operators

$$\vec{\mathbf{M}} \equiv \tfrac{1}{2}(\vec{\mathbf{L}} + i\,\vec{\mathbf{K}}) \qquad \vec{\mathbf{N}} \equiv \tfrac{1}{2}(\vec{\mathbf{L}} - i\,\vec{\mathbf{K}}) \tag{10.B41}$$

The operators $\vec{\mathbf{M}}$ and $\vec{\mathbf{N}}$ each satisfy standard angular momentum commutation rules

$$[\mathbf{M}_\alpha, \mathbf{M}_\beta] = i \sum_\gamma \varepsilon_{\alpha\beta\gamma} \mathbf{M}_\gamma \qquad [\mathbf{N}_\alpha, \mathbf{N}_\beta] = i \sum_\gamma \varepsilon_{\alpha\beta\gamma} \mathbf{N}_\gamma \tag{10.B42}$$

while commuting with each other ($[\mathbf{M}_\alpha, \mathbf{N}_\beta] = 0$).

Thus finite-dimensional representations of the homogeneous Lorentz algebra can be constructed from the finite-dimensional representations of the angular momentum algebra. There are irreducible representations $\Gamma(m, n)$ of the homogeneous Lorentz algebra with basis vectors $\Phi_{\mu,\nu}^{m,n}$ such that

$$\vec{\mathbf{M}} \cdot \vec{\mathbf{M}} \, \Phi_{\mu,\nu}^{m,n} = m(m+1)\Phi_{\mu,\nu}^{m,n} \qquad \vec{\mathbf{N}} \cdot \vec{\mathbf{N}} \, \Phi_{\mu,\nu}^{m,n} = n(n+1)\Phi_{\mu,\nu}^{m,n} \tag{10.B43}$$

$$\mathbf{M}_3 \, \Phi_{\mu,\nu}^{m,n} = \mu \, \Phi_{\mu,\nu}^{m,n} \qquad \mathbf{N}_3 \, \Phi_{\mu,\nu}^{m,n} = \nu \, \Phi_{\mu,\nu}^{m,n} \tag{10.B44}$$

$$(\mathbf{M}_1 \pm \mathbf{M}_2) \, \Phi_{\mu,\nu}^{m,n} = \sqrt{(m \mp \mu)(m \pm \mu + 1)} \, \Phi_{\mu\pm1,\nu}^{m,n} \tag{10.B45}$$

$$(\mathbf{N}_1 \pm \mathbf{N}_2) \, \Phi_{\mu,\nu}^{m,n} = \sqrt{(n \mp \nu)(n \pm \nu + 1)} \, \Phi_{\mu,\nu\pm1}^{m,n} \tag{10.B46}$$

Here $m, n = 0, \tfrac{1}{2}, 1, \tfrac{3}{2}, \ldots$; $\mu = -m, -m+1 \cdots, m-1, m$; $\nu = -n, -n+1, \ldots, n-1, n$ as worked out in Section 10.3.1.

The representations $\Gamma(m, n)$ do *not* generate unitary representations of $\mathcal{L}_+^\uparrow$, since the representations of the $\mathbf{K}_\alpha$ are not Hermitian. This is consistent with the general rule that a nontrivial unitary representation of a noncompact group is necessarily infinite-dimensional (a finite-dimensional unitary representation would be compact). However, these representations are still useful in constructing representations of the Poincaré group.

B.2 $SL(2, C)$ and the Homogeneous Lorentz Group

The fundamental representation of $SL(2, C)$ is defined on a two-dimensional vector space whose elements are two-component *spinors*

   $u \equiv (u_a)$ ($a = 1, 2$ is a *spinor index*). Under an $SL(2, C)$ transformation $\mathbf{A}$,

$$u \to u' = \mathbf{A}u \; : \; u'_a = A_a{}^b \, u_b \tag{10.B47}$$

(summation convention here on spinor indices). There are also conjugate spinors $v \equiv (v_{\bar{a}})$ ($\bar{a} = 1, 2$ is a *conjugate spinor index*, sometimes called a *dotted index* and expressed as $\dot{a}$). Conjugate spinors transform under the conjugate representation

$$v \to v' = \mathbf{A}^* v \; : \; v'_{\bar{a}} = A^{*}_{\bar{a}}{}^{\bar{b}} \, u_{\bar{b}} \tag{10.B48}$$

These representations can be ntified with the irreducible representations $\Gamma(\frac{1}{2}, 0)$ and $\Gamma(0, \frac{1}{2})$ of the $\mathcal{L}_+^\uparrow$ once we have derived the equivalence between $\mathcal{L}_+^\uparrow$ and $SL(2, C)$ transformations. Under restriction the subgroup $SU(2)$, the representations are both equivalent to the $j = \frac{1}{2}$ representation (see Exercise 1).

The analysis of rotations, and more generally Lorentz transformations, is simplified if we introduce a 2×2 matrix representation for four-vectors. Then Lorentz transformations can also be represented as complex 2×2 matrices, which are easier to manipulate than 3×3 rotation matrices or 4×4 Lorentz transformation matrices. The representation of rotations by 2×2 matrices was introduced in Problem 2.18. We can extend the Pauli matrices to form a basis

$$\sigma = \{\sigma_\mu\} \equiv (\sigma_0 = 1, \vec{\sigma}) \equiv \bar{\sigma}^\mu \tag{10.B49}$$

of four-vectors, and a basis $\bar{\sigma}$ of dual four-vectors, with elements

$$\bar{\sigma}_\mu = \zeta \sigma_\mu^* \zeta^{-1} = (\bar{\sigma}_0 = 1, -\vec{\sigma}) \equiv \sigma^\mu \tag{10.B50}$$

Then to each real four-vector $\mathbf{x} = (x^0, \vec{x})$ corresponds a Hermitian 2×2 matrix

$$\mathbf{X}(\mathbf{x}) \equiv \sigma \cdot \mathbf{x} = \sigma_\mu x^\mu = \sigma_0 \, x^0 + \vec{\sigma} \cdot \vec{x} = \begin{pmatrix} x^0 + x^3 & x^1 - ix^2 \\ x^1 + ix^2 & x^0 - x^3 \end{pmatrix} \tag{10.B51}$$

and a dual matrix

$$\overline{\mathbf{X}}(\mathbf{x}) \equiv \bar{\sigma} \cdot \mathbf{x} = \sigma_0 \, x^0 - \vec{\sigma} \cdot \vec{x} \tag{10.B52}$$

This correspondence can be inverted: To each Hermitian 2×2 matrix $\mathbf{X}$ corresponds a four-vectors $\mathbf{x}(\mathbf{X})$ and $\bar{\mathbf{x}}(\mathbf{X})$ with components given by

$$x^\mu = \tfrac{1}{2} \operatorname{tr}(\bar{\sigma}^\mu \mathbf{X}) \qquad \bar{x}^\mu = \tfrac{1}{2} \operatorname{tr}(\sigma^\mu \mathbf{X}) \tag{10.B53}$$

→ **Exercise 10.B5.** Show that

$$\overline{\mathbf{X}}(\mathbf{x})\mathbf{X}(\mathbf{x}) = (\mathbf{x} \cdot \mathbf{x})\mathbf{1} = \mathbf{X}(\mathbf{x})\overline{\mathbf{X}}(\mathbf{x})$$

$$\det \mathbf{X}(\mathbf{x}) = \mathbf{x} \cdot \mathbf{x} = \det \overline{\mathbf{X}}(\mathbf{x})$$

for any four-vector $\mathbf{x}$.   □

Now suppose $\mathbf{A}$ is an element of $SL(2, C)$, that is, a 2×2 matrix with $\det \mathbf{A} = 1$. Then the transformation

$$\mathbf{X} \to \mathbf{X_A} \equiv \mathbf{A} \mathbf{X} \mathbf{A}^\dagger \tag{10.B54}$$

of the 2×2 matrix $\mathbf{X}$ corresponds to a linear transformation

$$\mathbf{x} = \mathbf{x}(\mathbf{X}) \to \mathbf{x_A} \equiv \mathbf{x}(\mathbf{A} \mathbf{X} \mathbf{A}^\dagger) \tag{10.B55}$$

of four-vectors, with $\det \mathbf{X} = \det \mathbf{X_A}$ since $\det \mathbf{A} = 1$, and hence

$$\mathbf{x} \cdot \mathbf{x} = \mathbf{x_A} \cdot \mathbf{x_A} \tag{10.B56}$$

Thus $\mathbf{x} \rightarrow \mathbf{x_A}$ is a Lorentz transformation, to be denoted by $\Lambda(\mathbf{A})$. It follows from Eq. (10.B53) that

$$x_A^\mu = \tfrac{1}{2} \operatorname{tr}(\bar{\sigma}^\mu \mathbf{A} \sigma_\nu \mathbf{A}^\dagger) \, x^\nu \tag{10.B57}$$

so that the Lorentz transformation matrix is given by

$$\Lambda^\mu{}_\nu(\mathbf{A}) = \tfrac{1}{2} \operatorname{tr}(\bar{\sigma}^\mu \mathbf{A} \sigma_\nu \mathbf{A}^\dagger) \tag{10.B58}$$

Remark. In terms of components, the four-vector transformation law (10.B54) is

$$X_{a\bar{c}} \rightarrow A_a{}^b \, X_{b\bar{d}} \, A_{\bar{c}}^{*\,\bar{d}} \tag{10.B59}$$

with one spinor and one conjugate spinor index. This is exactly the transformation law for the irreducible representation $\Gamma(\tfrac{1}{2}, \tfrac{1}{2})$ of $SL(2,C)$. □

Now

$$\Lambda^0{}_0(\mathbf{A}) = \tfrac{1}{2} \operatorname{tr}(\mathbf{A}\mathbf{A}^\dagger) \geq 1 \tag{10.B60}$$

(see Problem 2.19), so that $\Lambda(\mathbf{A})$ is orthochronous, and $\det \Lambda(\mathbf{A}) = 1$ since the determinant is a continuous function of $\mathbf{A}$. Thus we have a mapping from $SL(2,C)$ onto the restricted Lorentz group $\mathcal{L}_+^\uparrow$. The mapping is two-to-one, since the matrices $\pm\mathbf{A}$ correspond to the same Lorentz transformation; we have

$$\mathcal{L}_+^\uparrow \simeq SL(2,C)/\mathbf{Z}_2 \tag{10.B61}$$

Note that $\Lambda(\mathbf{A})$ is a rotation ($\Lambda^0{}_0 = 1$) if and only if $\mathbf{A}\mathbf{A}^\dagger = \mathbf{1}$, that is, if and only $\mathbf{A}$ is unitary (Problem 2.19 again). We have already noted the relation between $SU(2)$ and $SO(3)$ in Section 10.2.2. Also, if $\mathbf{A}$ is Hermitian, then $\Lambda(\mathbf{A})$ is symmetric, and thus represents a boost in some direction, perhaps with rotation through a multiple of π about the boost axis (see Problem 26).

→ **Exercise 10.B6.** Show that the matrix

$$\mathbf{A_n}(\chi) \equiv \exp\left(-\tfrac{1}{2}\vec{\sigma} \cdot \mathbf{n}\chi\right)$$

corresponds to a boost in the direction $\mathbf{n}$ with velocity $\vec{v} = (c \tanh \chi)\, \mathbf{n}$. □

Remark. Thus in the fundamental representation, the generators are represented by

$$\vec{\mathbf{L}} = \tfrac{1}{2}\vec{\sigma} \qquad \vec{\mathbf{K}} = -\tfrac{1}{2}i\vec{\sigma} \tag{10.B62}$$

and hence

$$\vec{\mathbf{M}} = \tfrac{1}{2}\vec{\sigma} \qquad \vec{\mathbf{N}} = 0 \tag{10.B63}$$

consistent with the identification of the fundamental representation as $\Gamma(\tfrac{1}{2}, 0)$. □

B.3   Inhomogeneous Lorentz Transformations; Poincaré Group

Translation of the origin in Minkowski space by a four-vector **a** changes coordinates of a spacetime point **x** by

$$x^\mu \rightarrow x'^\mu = x^\mu - a^\mu \tag{10.B64}$$

Since this translation has the same effect on all spacetime points, the relative spacetime co-ordinate (**x**–**y**) of two points **x** and **y** is invariant under the translation. Since interactions between particles (both classical and quantum) generally depend only on this relative coor-dinates, translation invariance is a general principle of physics. This principle is so deeply ingrained that the freedom to translate a system is often implicitly removed from considera-tion, by working in the center-of-mass system at the start, for example. Only when we deal with cosmological scales does translation invariance become problematic.

The translations form an Abelian group, as already noted. They can be combined with the Lorentz group to form a larger group, the *inhomogeneous Lorentz group*, or *Poincaré group* $\mathcal{P}$. The general Poincaré transformation acts on the coordinates of a spacetime point according to

$$x^\mu \rightarrow x'^\mu = \Lambda^\mu_{\ \nu}\, x^\nu - a^\mu \tag{10.B65}$$

combining a translation **a** with a homogeneous Lorentz transformation Λ. The transformation (10.B65) is denoted by $(\mathbf{a}, \Lambda)$; we have the rules

$$(\mathbf{a}, \Lambda) \ = \ (\mathbf{a}, 1)(0, \Lambda) \tag{10.B66}$$

$$(\mathbf{a_2}, \Lambda_2)(\mathbf{a_1}, \Lambda_1) \ = \ (\mathbf{a_2} + \Lambda_2\, \mathbf{a_1}, \Lambda_2\Lambda_1) \tag{10.B67}$$

➜ **Exercise 10.B7.**   Show that the inverse transformation to $(\mathbf{a}, \Lambda)$ is

$$(\mathbf{a}, \Lambda)^{-1} = (-\Lambda^{-1}\mathbf{a}, \Lambda^{-1})$$

➜ **Exercise 10.B8.**   Show that the translations form an invariant subgroup of $\mathcal{P}$, i.e., if g is a Poincaré transformation and $(\mathbf{a}, 1)$ is a translation, then $g(\mathbf{a}, 1)g^{-1}$ is also a translation.   □

The Poincaré group has subgroups $\mathcal{P}^\uparrow_+$, the *restricted* Poincaré group, $\mathcal{P}_+ \equiv \mathcal{P}^\uparrow_+ \cup \mathcal{P}^\downarrow_+$, the *proper* Poincaré group, $\mathcal{P}^\uparrow \equiv \mathcal{P}^\uparrow_+ \cup \mathcal{P}^\uparrow_-$, the *orthochronous* Poincaré group, and $\mathcal{P}_0 \equiv \mathcal{P}^\uparrow_+ \cup \mathcal{P}^\downarrow_+$. These subgroups consist of Poincaré transformations $(\mathbf{a}, \Lambda)$ with Λ from the corre-sponding subgroup of the homogeneous Lorentz group.

The translations can be expressed as

$$(\mathbf{a}, 1) = \exp\left(-iP_\mu a^\mu\right) \tag{10.B68}$$

where the generators $\mathbf{P} = \{P_\mu\} = (P_0, \overrightarrow{P})$ are identified with the four-momentum of a system. These generators satisfy the commutation rules

$$[\mathbf{L}_\alpha, P_0] = 0 \qquad [\mathbf{L}_\alpha, P_\beta] = i\varepsilon_{\alpha\beta\gamma}\, P_\gamma \tag{10.B69}$$

$$[\mathbf{K}_\alpha, P_0] = P_\alpha \qquad [\mathbf{K}_\alpha, P_\beta] = \delta_{\alpha\beta}\, P_0 \tag{10.B70}$$

and

$$[P_\mu, P_\nu] = 0 \tag{10.B71}$$

(verify these as an exercise).

The quadratic Casimir operators for the homogeneous Lorentz group do not commute with the P_μ. However, the operator

$$\mathbf{P}^2 \equiv \mathbf{P} \cdot \mathbf{P} = P_0^2 - \vec{P} \cdot \vec{P} \tag{10.B72}$$

commutes with $\vec{\mathbf{L}}$ and $\vec{\mathbf{K}}$, since it is by construction a scalar, invariant under homogeneous Lorentz transformations. It also commutes with the P_μ, so it provides a Casimir operator for the Poincaré group. For a single particle of mass m, this invariant is proportional to m^2; in general it is the square of the total energy of a system in a frame where the total momentum $\vec{P} = 0$.

To find a second invariant, introduce the four-vector

$$\mathbf{W} \equiv \left(W_0 = \vec{\mathbf{L}} \cdot \vec{P}, \vec{W} = \vec{\mathbf{L}} P_0 + \vec{\mathbf{K}} \times \vec{P} \right) \tag{10.B73}$$

The components $\{W_\mu\}$ satisfy the standard four-vector transformation rules (10.B69) and (10.B70)

$$[\mathbf{L}_\alpha, W_0] = 0 \qquad [\mathbf{L}_\alpha, W_\beta] = i\varepsilon_{\alpha\beta\gamma} W_\gamma \tag{10.B74}$$

$$[\mathbf{K}_\alpha, W_0] = W_\alpha \qquad [\mathbf{K}_\alpha, W_\beta] = \delta_{\alpha\beta} W_0 \tag{10.B75}$$

They also commute with the P_μ. Hence the scalar

$$\mathbf{W}^2 \equiv \mathbf{W} \cdot \mathbf{W} = W_0^2 - \vec{W} \cdot \vec{W} \tag{10.B76}$$

is a second quadratic Casimir operator for the Poincaré group. In the rest frame of a particle, we have $W_0 = 0$ and

$$\vec{W} = m\vec{L} \tag{10.B77}$$

where $\vec{L}$ is the angular momentum. Thus $\vec{W}$ is proportional to the instrinsic spin of a particle or the total angular momentum of a composite system.

Bibliography and Notes

Three classic books on the angular momentum algebra and the coupling schemes for addition of angular momenta in atomic and nuclear physics are

A. R. Edmonds, *Angular Momentum in Quantum Mechanics* (2nd edition), Princeton University Press (1960, reissued 1996),

Morris E. Rose, *Elementary Theory of Angular Momentum*, Wiley (1960), reprinted by Dover (1995),

D. M. Brink and G. R. Satchler, *Angular Momentum* (3rd edition), Oxford, Clarendon Press (1994).

These have more detail than the book by Tinkham cited in Chapter 9.

An unsurpassed pedagogical introduction to Lie algebras and the use of ladder operators in constructing their representations is

Harry J. Lipkin, *Lie Groups for Pedestrians* (2nd edition), North-Holland (1966), reprinted by Dover (2002).

Two useful books that cover the classical Lie groups and algebras in physics contexts are

Brian G. Wybourne, *Symmetry Principles and Atomic Spectroscopy*, Wiley (1970),

Brian G. Wybourne, *Classical Groups for Physicists*, Wiley (1974).

The first of these has a long description of the properties of the symmetric group and the rotational and unitary groups, together with a detailed discussion of the use of these groups in the classification of atomic states. The second book has a detailed description of both the classical Lie groups and the exceptional groups, at a somewhat more advanced mathematical level, with many applications to physics as of the early 1970s, including atomic and nuclear physics, but with less emphasis on particle physics. There are also some useful examples of noncompact algebras, including the group theory underlying the full spectrum of the hydrogen atom.

A fairly elementary introduction that is focused on applications to particle physics is

Fl. Stancu, "Group Theory in Subnuclear Physics," Oxford University Press (1997).

An excellent review that goes into the more advanced details of working with arbitrary Lie algebras in a particle theory context is

Richard Slansky, *Group Theory for Unified Model Building,* Physics Reports **79** (1981) 1–128.

A recent introduction to group theoretical methods in physics applications is

J. F. Cornwell, *Group Theory in Physics: An Introduction*, Academic Press (1997)

This is a short and updated version of the more extensive three-volume treatise

J. F. Cornwell, *Group Theory in Physics* (3 volumes), Academic Press (1984).

The third volume of this series contains an extensive treatment of supersymmetry, which is important in modern particle theory, as well as having some applications in other fields—nuclear theory, for example, as well as some integrable models of interest in condensed matter physics. It also has an extended discussion of infinite-dimensional algebras, which are important both in string theory and in the study of integrable systems.

Problems

1. To show the equivalence of the complex conjugate representations of $SU(2)$, we need a matrix ζ such that

$$\zeta \mathbf{U} \zeta^{-1} = \mathbf{U}^* \qquad\qquad (*)$$

for every $\mathbf{U}$ in $SU(2)$. In fact, such a matrix is given by

$$\zeta = -i\sigma_2 = \begin{pmatrix} 0 & 1 \\ -1 & 0 \end{pmatrix}$$

   (i) First, show that $\zeta \mathbf{M} \zeta^{-1} = -\mathbf{M}^*$ for every hermitian 2×2 matrix $\mathbf{M}$.

   (ii) Then show that $(*)$ is satisfied for every unitary 2×2 matrix $\mathbf{U}$ with determinant $+1$.

   (iii) Why does the equivalence require $\det \mathbf{U} = +1$?

2. Show that for the group $SL(2, C)$, the representation $\mathbf{A} \to \tilde{\mathbf{A}}^{-1}$ is equivalent to the defining representation. That is, find a matrix $\mathbf{U}$ such that

$$\mathbf{U} \mathbf{A} \mathbf{U}^{-1} = \tilde{\mathbf{A}}^{-1}$$

for every $\mathbf{A}$ in $SL(2, C)$. *Hint.* See Problem 1.

3. Using the Jacobi identity (see Problem 2.7)

$$[[\mathbf{X}, \mathbf{Y}], \mathbf{Z}] + [[\mathbf{Y}, \mathbf{Z}], \mathbf{X}] + [[\mathbf{Z}, \mathbf{X}], \mathbf{Y}] = 0$$

show that the structure constants defined in Eq. (10.12)) satisfy

$$\sum_{\ell} \left(c_{jk}^{\ell} c_{\ell m}^{n} + c_{km}^{\ell} c_{\ell j}^{n} + c_{mj}^{\ell} c_{\ell k}^{n} \right) = 0$$

4. Consider the coefficients $e_{jk\ell}$ defined by

$$e_{jk\ell} \equiv \frac{1}{2} \sum_{m,p,q} c_{jk}^{m} c_{\ell p}^{q} c_{qm}^{p} = \sum_{m} c_{jk}^{m} g_{m\ell}$$

where the $g_{m\ell}$ are elements of the metric introduced in Eq. (10.25). Use the relation between the structure constants derived from the Jacobi identity in the preceding problem to show that the $e_{jk\ell}$ are completely antisymmetric in the indices j, k, ℓ.

Remark. Antisymmetry in j, k follows from the antisymmetry of the commutator, of course, but the remaining antisymmetry does not. □

5. To construct a standard basis for the Lie algebras of $U(n)$ and $SU(n)$, let $\mathbf{T}^{ab}$ be the $n \times n$ matrix with 1 in position (a, b) and zeros elsewhere, so that

$$(\mathbf{T}^{ab})_{jk} = \delta_j^a \delta_k^b$$

and introduce the Hermitian matrices ($a \neq b$)

$$\mathbf{X}^{ab} \equiv \mathbf{T}^{ab} + \mathbf{T}^{ba} = \mathbf{X}^{ba} \qquad \mathbf{Y}^{ab} \equiv -i\left(\mathbf{T}^{ab} - \mathbf{T}^{ba}\right) = -\mathbf{Y}^{ba} \qquad \mathbf{Z}^a \equiv \mathbf{T}^{aa}$$

(i) Express commutators $[\mathbf{X}^{ab}, \mathbf{X}^{cd}]$, $[\mathbf{X}^{ab}, \mathbf{Y}^{cd}]$, $[\mathbf{Y}^{ab}, \mathbf{Y}^{cd}]$, $[\mathbf{X}^{ab}, \mathbf{Z}^c]$, $[\mathbf{Y}^{ab}, \mathbf{Z}^c]$ and $[\mathbf{Z}^a, \mathbf{Z}^b]$ in terms of the $\mathbf{X}^{ab}$, $\mathbf{Y}^{ab}$ and $\mathbf{Z}^a$, and thus show that these matrices define a Lie algebra.

Remark. Since there are $\frac{1}{2}n(n-1)$ independent $\mathbf{X}^{ab}$, the same number of $\mathbf{Y}^{ab}$, and n distinct $\mathbf{Z}^a$, there are n^2 matrices in all, which form a basis for the Hermitian $n \times n$ matrices. Exponentiating these matrices generates the unitary $n \times n$ matrices, so the Lie algebra here is that of the group $U(n)$.   □

(ii) The matrix

$$\mathbf{Z} = \sum_{a=1}^{n} \mathbf{Z}^a = 1$$

generates the Abelian $U(1)$ invariant subgroup of $U(n)$ whose elements are of the form $\mathbf{U} = \exp^{i\alpha} 1$, and commutes with all the generators $\mathbf{X}^{ab}$, $\mathbf{Y}^{ab}$, and $\mathbf{H}^n$. In order to find a set of generators of the group $SU(n)$ of unitary $n \times n$ matrices with determinant equal to one, we need to find a set $\{\mathbf{H}^1, \mathbf{H}^2, \dots, \mathbf{H}^{n-1}\}$ of traceless diagonal matrices in addition to the $\mathbf{X}^{ab}$, $\mathbf{Y}^{ab}$. Show that one such set is given by the matrices

$$\mathbf{H}^k = \sqrt{\frac{2}{k(k+1)}} \left\{ \sum_{a=1}^{k} \mathbf{Z}^a - k\mathbf{Z}^{k+1} \right\}$$

($k = 1, \dots, n-1$), and that the $\mathbf{H}^k$ satisfy

$$\operatorname{tr} \mathbf{H}^k = 0 \qquad \operatorname{tr} \mathbf{H}^k \mathbf{H}^\ell = 2\delta^{k\ell}$$

The $\mathbf{X}^{ab}$, $\mathbf{Y}^{ab}$ and $\mathbf{H}^k$ form a standard $n \times n$ generalization of the Pauli matrices.

6. (i) Show that the $\frac{1}{2}n(n-1)$ independent $\mathbf{Y}^{ab}$ introduced in the preceding problem form by themselves a Lie algebra.

Remark. Since the Hermitian $\mathbf{Y}^{ab}$ are imaginary, exponentiation (with the factor $\pm i$) leads to real unitary (i.e., orthogonal) matrices. Thus the $\mathbf{Y}^{ab}$ define the Lie algebra of the group $SO(n)$ of orthogonal $n \times n$ matrices with determinant equal to one.   □

(ii) How many independent commuting matrices are there among the $\mathbf{Y}^{ab}$? Find a maximal set of commuting matrices.

7. Show that the Lie algebra of $SO(4)$ can be expressed as a direct sum of two commuting copies of the Lie algebra of $SO(3)$ [or $SU(2)$].

Hint. With generators of $SO(4)$ denoted by $\mathbf{Y}^{ab}$ ($a \neq b$) as in the preceding problems, consider the generators

$$J_k^{\pm} \equiv \tfrac{1}{2}\left(\mathbf{Y}^{\ell m} \pm \mathbf{Y}^{k4}\right)$$

where $k = 1, 2, 3$ and $(k\ell m)$ is a cyclic permutation of (123).

8. (i) How many independent generators are there for the group $SU(4)$? How many for the group $SO(6)$?

(ii) Use the results of Problems 5 and 6 to find explicit bases of generators for each of these groups.

(iii) Find an isomorphism between these two sets of generators.

Remark. This isomorphism between the Lie algebras of $SU(4)$ and $SO(6)$ is similar to the isomorphism between those of $SU(2)$ and $SO(3)$, but in each case the groups are not exactly isomorphic because they have different discrete invariant subgroups. □

(iv) Does this pattern of isomorphism persist? Is there an isomorphism between the Lie algebras of $SU(6)$ and $SO(9)$?

9. Consider a particle of mass m moving in a potential

$$V(r) = -\frac{\kappa}{r}$$

The *Runge–Lenz* vector for the particle is defined by

$$\mathbf{A} = \mathbf{p} \times \mathbf{L} - m\kappa \, \frac{\mathbf{r}}{r}$$

where $\mathbf{L} = \mathbf{r} \times \mathbf{p}$ is the angular momentum. In quantum mechanics, $\mathbf{r}$ and $\mathbf{p}$ are represented by operators that satisfy the *canonical commutation relations*

$$[r_j, p_k] = i\delta_{jk}$$

(i) Show that $\mathbf{L}$ and $\mathbf{A}$ satisfy the commutation relations

$$[L_j, A_k] = i\varepsilon_{jk\ell} A_\ell$$

Remark. This commutation relation with $\mathbf{L}$ is actually satisfied by any vector $\mathbf{A}$; it is one definition of a vector. □

(ii) With Hamiltonian given by

$$\mathbf{H} = \frac{p^2}{2m} + V(r)$$

show that

$$[\mathbf{H}, \mathbf{A}] = 0$$

Remark. Thus $\mathbf{A}$ is a constant of motion for this Hamiltonian. □

(iii) Show that

$$[A_j, A_k] = -2im\mathbf{H}\varepsilon_{jk\ell} L_\ell$$

Remark.   From this result it follows that on an eigenmanifold of $\mathbf{H}$ with energy E, the operators

$$\mathbf{D} \equiv \mathbf{A}/\sqrt{2m\,|E|}$$

together with the $\mathbf{L}$ define a Lie algebra.   □

(iv) Use the results of Problem 7 to show that for $E < 0$, the $\mathbf{L}$ and $\mathbf{D}$ together form a Lie algebra of $SO(4)$.

Remark.

Thus the energy levels of the hydrogen atom correspond to representations (irreducIble, in fact) of $SO(4)$. For extra credit, describe these representations.   □

(v) What is the Lie algebra defined by the $\mathbf{L}$ and $\mathbf{D}$ for $E > 0$?

10. Let $a_\sigma^\dagger$, a_σ be the creation and annihilation operators for a spin-$\frac{1}{2}$ fermion in spin state σ ($\sigma = \pm\frac{1}{2}$), as introduced in Problem 7.8. The spin operators for the fermion are

$$\vec{\mathbf{S}} \equiv \tfrac{1}{2} \sum_{\alpha,\beta} a_\alpha^\dagger \vec{\sigma}_{\alpha\beta} a_\beta$$

(i) Show that the $\mathbf{S}_k$ are generators of $SU(2)$ [see Eq. (10.14)].

(ii) Show that the fermion number operator

$$\mathbf{N} = \sum_\sigma a_\sigma^\dagger a_\sigma$$

commutes with the $\mathbf{S}_k$.

11. Introduce fermion pairing operators

$$\mathbf{B}^\dagger \equiv \tfrac{1}{2} \sum_{\alpha,\beta} a_\alpha^\dagger \zeta_{\alpha\beta} a_\beta^\dagger = a_\uparrow^\dagger a_\downarrow^\dagger$$

$$\mathbf{B} \equiv \tfrac{1}{2} \sum_{\alpha,\beta} a_\beta \zeta_{\alpha\beta} a_\alpha = a_\downarrow^\dagger a_\uparrow^\dagger$$

where $\zeta = i\sigma_2$ has been introduced in Problem 1 (in some contexts, ζ serves as a spinor metric). $\mathbf{B}^\dagger$ creates a pair of fermions, one with spin up and one with spin down, in the same mode, while $\mathbf{B}$ annihilates a pair.

(i) Show that the fermion number operator

$$\mathbf{N} = \sum_\sigma a_\sigma^\dagger a_\sigma$$

satisfies the commutation rules

$$[\mathbf{N}, \mathbf{B}^\dagger] = 2\mathbf{B}^\dagger \qquad [\mathbf{N}, \mathbf{B}] = -2\mathbf{B}$$

(ii) Show that

$$[\mathbf{B}^\dagger, \mathbf{B}] = \mathbf{N} - 1 \equiv 2\mathbf{B}_3$$

so that the operators $\mathbf{B}^\dagger$, $\mathbf{B}$ and $\mathbf{B}_3$ generate an $SU(2)$ group (the *pairing group*).

(iii) Show that this $SU(2)$ pairing algebra commutes with the $SU(2)$ spin algebra just introduced in Problem 10, so the two algebras can be combined to form an $SO(4)$ Lie algebra.

12. We want to consider the group of $2N \times 2N$ matrices that define linear transformations of the coordinates $q^1, \ldots, q^N$ and momenta $p_1, \ldots, p_N$ such that the canonical 2-form

$$\omega \equiv \sum_{k=1}^{N} dp_k \wedge dq^k$$

is invariant. Such matrices are *symplectic matrices*, and the corresponding group is the *symplectic group* $Sp(2N)$.

(i) For $N = 1$, use your detailed knowledge of the properties of 2×2 matrices to give a complete characterization of $Sp(2)$. Find the subgroups of $Sp(2)$ for which the matrices are (a) orthogonal and (b) unitary.

(ii) For $N = 2$, try to construct a set of generators of the group $Sp(4)$ (such a set would correspond to the angular momenta that generate the group of rotation matrices). Note that symplectic matrices $\mathbf{A}$ must satisfy

$$\mathbf{A}^T \zeta \mathbf{A} = \zeta$$

where $\mathbf{A}^T$ is the transpose of $\mathbf{A}$ and ζ is the *symplectic metric* defined by

$$\zeta \equiv \begin{pmatrix} 0 & 1 & 0 & 0 \\ -1 & 0 & 0 & 0 \\ 0 & 0 & 0 & 1 \\ 0 & 0 & -1 & 0 \end{pmatrix}$$

Find also the generators of the subgroups of $Sp(4)$ for which the matrices are (a) orthogonal and (b) unitary.

Remark.

Having done the construction for $N = 2$, you might want to extend it to arbitrary N, but that is not needed here. □

13. Consider a spin-$\frac{1}{2}$ particle with orbital angular momentum ℓ (the case $\ell = 1$ was discussed in Example 10.15). The total angular momentum j of the particle can be $\ell \pm \frac{1}{2}$. Derive general formulas for the states $|j\ m\rangle$ in terms of the states $|m_\ell\ m_s\rangle$.

14. Consider the addition of angular momenta $j_1 = \frac{3}{2}$ and $j_2 = 1$. Construct the eigenstates $|j\ m\rangle$ of $\mathbf{J} \cdot \mathbf{J}$ and $\mathbf{J}_3$ in terms of the eigenstates $|m_1\ m_2\rangle$ of $\mathbf{J}_3^{(1)}$ and $\mathbf{J}_3^{(2)}$.

Note. In Example 10.16, we gave a direct method to compute the state $|\frac{1}{2}\ \frac{1}{2}\rangle$. Now we want all the eigenstates $|j\ m\rangle$.

15. The interaction between an electron and a nucleus has a relatively small term that can be expressed as

$$\mathbf{H}_{int} = a\mathbf{L} \cdot \mathbf{S}$$

where $\mathbf{L}$ is the orbital angular momentum operator, $\mathbf{S}$ the spin, and a is a constant that can be evaluated with some precision. Show that states of definite total angular momentum j (such as those constructed in Problem 13 are also eigenstates of $\mathbf{H}_{int}$, and find the eigenvalues of $\mathbf{H}_{int}$ in terms of l, s, j (here we do not require $s = \frac{1}{2}$).

Remark. $\mathbf{H}_{int}$ is the *fine structure* Hamiltonian, responsible for splitting atomic energy levels that are degenerate when only the classical Coulomb interaction is considered. □

16. Consider the addition of angular momentum for two particles each with $j = 1$.

    (i) What are the allowed values of the total angular momentum J of the two particles?

    (ii) Which of these values correspond to symmetric states? antisymmetric states?

    Now consider three $j = 1$ particles.

    (iii) What are the allowed values of the total angular momentum J of the three particles, and how many independent states are there for each J?

    (iv) Classify the states into irreducible representations of $\mathcal{S}_3$.

17. Consider the addition of angular momentum of $j = \frac{3}{2}$ particles.

    (i) What are the allowed values of the total angular momentum J of two particles? Which of these values correspond to symmetric (antisymmetric) states?

    (ii) What are the allowed values of the total angular momentum J of three particles? Classify these states into irreducible representations of $\mathcal{S}_3$. What value(s) of J correspond to antisymmetric states?

    (iii) What are the allowed values of the total angular momentum J of four particles? Classify these states into irreducible representations of $\mathcal{S}_4$. What value(s) of J correspond to antisymmetric states?

18. Consider the linear operators $\mathbf{A}_u, \mathbf{A}_d, \mathbf{A}_s, \bar{\mathbf{A}}_u, \bar{\mathbf{A}}_d, \bar{\mathbf{A}}_s$ such that

$$[\mathbf{A}_a, \bar{\mathbf{A}}_b] = \delta_{ab}$$

$(a, b = u, d, s)$ and introduce the operators

$$\mathbf{T}_+ \equiv \bar{\mathbf{A}}_u \mathbf{A}_d \,, \ \mathbf{U}_+ \equiv \bar{\mathbf{A}}_d \mathbf{A}_s \,, \ \mathbf{V}_+ \equiv \bar{\mathbf{A}}_u \mathbf{A}_s$$

$$\mathbf{T}_- \equiv \bar{\mathbf{A}}_d \mathbf{A}_u \,, \ \mathbf{U}_- \equiv \bar{\mathbf{A}}_s \mathbf{A}_d \,, \ \mathbf{V}_- \equiv \bar{\mathbf{A}}_s \mathbf{A}_u$$

as well as

$$\mathbf{T}_3 \equiv \tfrac{1}{2} \left(\bar{\mathbf{A}}_u \mathbf{A}_u - \bar{\mathbf{A}}_d \mathbf{A}_d \right)$$

$$\mathbf{Y} \equiv \tfrac{1}{3} \left(\bar{\mathbf{A}}_u \mathbf{A}_u + \bar{\mathbf{A}}_d \mathbf{A}_d - 2\bar{\mathbf{A}}_s \mathbf{A}_s \right)$$

(i) Show that these eight operators define a Lie algebra, and that $\mathbf{T}_3$ and $\mathbf{Y}$ can be simultaneously diagonalized.

(ii) Indicate graphically the effects of the operators $\mathbf{T}_\pm$, $\mathbf{U}_\pm$ and $\mathbf{V}$ on the eigenvalues of $\mathbf{T}_3$ and $\mathbf{Y}$.

(iii) Show that each of the eight operators commutes with the operator

$$\mathbf{N} \equiv \bar{\mathbf{A}}_u \mathbf{A}_u + \bar{\mathbf{A}}_d \mathbf{A}_d + \bar{\mathbf{A}}_s \mathbf{A}_s$$

so that subspaces corresponding to definite eigenvalues N of $\mathbf{N} = \mathbf{N}_u + \mathbf{N}_d + \mathbf{N}_s$ are invariant under the Lie algebra. What further subspaces are invariant? (For definiteness, discuss the cases of $N = 2$ and $N = 3$.)

(iv) Express the operators introduced in part (i) in terms of the standard generators of $SU(3)$ introduced in Eq. (10.20).

Remark. This explicit formulation of the $SU(3)$ Lie algebra works just as well if the operators $\mathbf{A}_a$ and $\bar{\mathbf{A}}_a$ satisfy anticommutation rules

$$\{\mathbf{A}_a, \bar{\mathbf{A}}_b\} = \mathbf{A}_a \bar{\mathbf{A}}_b + \bar{\mathbf{A}}_b \mathbf{A}_a = \delta_{ab}$$

(show this). These operators can be visualized as creation and annihilation operators for quarks, and the picture can be extended to describe $SU(n)$ for any value of n. □

19. There are eight spin-$\frac{1}{2}$ baryons, including the proton and neutron, that form an octet under the old flavor $SU(3)$ approximate symmetry of the strong interactions. Draw the weight diagram of the octet (see Fig. 10.3) and label the weights with the symbols for the baryons.

20. Consider an octet state of three spin-$\frac{1}{2}$ quarks u, d, s. Both spin and flavor states have mixed symmetry under permutations of the three quarks, so the combined spin-flavor states can be symmetric, antisymmetric or mixed symmetry (see Example 9.48). These states can be denoted by $|s\, j_3; T_3\, Y\rangle$, where $\alpha = S, A, (j_{12})$ characterizes the symmetry type of the product function (j_{12} identifies a basis vector in the mixed representation).

(i) Construct both symmetric and antisymmetric product states for $j_3 = \frac{1}{2}$ in terms of the basic quark states (see Problem 9.19).

(ii) Suppose the magnetic moment operator $\vec{\mu}$ for quarks is

$$\vec{\mu} = \left\{ \left(\tfrac{2}{3} + \mathbf{Y}\right) \left[\left(\tfrac{1}{2} + \mathbf{T}_3\right)\mu_u + \left(\tfrac{1}{2} - \mathbf{T}_3\right)\mu_d \right] + \left(\tfrac{1}{3} - \mathbf{Y}\right)\mu_s \right\} \vec{\sigma}$$

where $\mu_{u,d,s}$ denote the quark magnetic moments. Compute the magnetic moment matrix elements

$$\langle |\alpha\, \tfrac{1}{2}; (T)\, T_3\, Y|\, \mu_3\, |\alpha\, \tfrac{1}{2}; (T)\, T_3\, Y\rangle$$

between the octet states for possibilities $\alpha = S, A$.

Remark. The actual baryon magnetic moments can be explained very well if the product wave function is symmetric. By comparison, the magnetic moments of the nuclei ^{3}H and ^{3}He are very close to the values obtained for the antisymmetric wave function with μ_u and μ_d replaced by measured proton and neutron magnetic moments. The observed baryon moments provided another piece of evidence for the existence of an $SU(3)$ color degree of freedom; as already noted, this allows an antisymmetric singlet color wave function to provide the antisymmetry required by the Pauli principle.   □

21. Compute the dimensions of irreducible representations of $SU(4)$ associated with the partitions (21^2) and (2^2).

22. (i) Find the conjugate representations to the irreducible representations of $SU(6)$ corresponding to each of the partitions (2), (1^2), (3), (21), (1^3), and (21^4). Which representations are self-conjugate?

    (ii) Find the dimension of each of these irreducible representations.

23. Show that the irreducible representation of $SL(n, C)$ dual to $\Gamma(m_1\, m_2 \cdots m_{n-1})$ is $\Gamma(m_1\, m_1 - m_{n-1} \,\cdots\, m_1 - m_2)$.

24. Consider the behavior of Maxwell's equations (3.141), (3.142), (3.145), (3.148) under the Lorentz transformation (10.B21). Find the transformation laws for the electric and magnetic fields needed for the equations to be invariant under the transformation.

    Remark. One hint is to look at the four-vector potential $\{A_\mu\} = (\phi, \overrightarrow{A})$ as a one-form, and the electromagnetic field as a two-form $F = dA$ or, with explicit components,

    $$F_{\mu\nu} = \frac{\partial A_\nu}{\partial x^\mu} - \frac{\partial A_\mu}{\partial x^\nu}$$

    The transformation laws for the two-form follow directly from the Lorentz transformation (10.B21).   □

25. Show that the generators of $SL(2, R)$ introduced in Eq. (10.31), with commutators given by Eq. (10.32), are suitable generators for the homogeneous Lorentz transformations in two space dimensions (+time). In particular, J_3 generates rotations in the plane, while K_1 and K_2 generate transformations to coordinate frames moving with constant velocity in the plane. Show further that the quadratic Casimir operator

    $$C_2 = J_3^2 - K_1^2 - K_2^2$$

    actually commutes with all three generators.

26. (i) Show that a boost commutes with a rotation about the boost axis.

    (ii) Show that a symmetric Lorentz transformation is a boost, possibly accompanied by a rotation through a multiple of π about the boost axis (see also Problem 2.15).

27. Use the fact that the operators $\overrightarrow{M}$ and $\overrightarrow{N}$ defined in Appendix B are angular momentum operators to construct two quadratic Casimir operators for the homogeneous Lorentz group in terms of the generators $\overrightarrow{L}$ and $\overrightarrow{K}$.

Index

Introduction to Mathematical Physics. Michael T. Vaughn
Copyright © 2007 WILEY-VCH Verlag GmbH & Co. KGaA, Weinheim
ISBN: 978-3-527-40627-2

Related Titles

Trigg, G. L. (ed.)
Mathematical Tools for Physicists
686 pages with 98 figures and 29 tables
2005
Hardcover
ISBN: 978-3-527-40548-0

Tinker, M., Lambourne, R.
Further Mathematics for the Physical Sciences
744 pages
2000
Softcover
ISBN: 978-0-471-86723-4

Lambourne, R., Tinker, M.
Mathematics for the Physical Sciences
688 pages
2000
Softcover
ISBN: 978-0-471-85207-0

Kusse, B., Westwig, E. A.
Mathematical Physics
Applied Mathematics for Scientists and Engineers
680 pages
1998
Hardcover
ISBN: 978-0-471-15431-0

Courant, R., Hilbert, D.
Methods of Mathematical Physics
Volume 1
575 pages with 27 figures
1989
Softcover
ISBN: 978-0-471-50447-4

Courant, R., Hilbert, D.
Methods of Mathematical Physics
Volume 2
575 pages with 27 figures
1989
Softcover
ISBN: 978-0-471-50439-9